Dairy Science and Technology Handbook

2 Product Manufacturing

Dairy Science and Technology Handbook

2 Product Manufacturing

Y. H. Hui
EDITOR

Dr. Y. H. Hui
3006 "S" Street
Eureka, California 95501
U.S.A.

Library of Congress Cataloging-in-Publication Data
Dairy science and technology handbook / editor, Y. H. Hui.
p. cm.
Includes bibliographical references and index.
ISBN 1-56081-078-5
1. Dairy processing. 2. Dairy products. I. Hui, Y. H. (Yiu H.)
SF250.5.D35 1992
637—dc20 92-30191

Printed in the United States of America

ISBN 1-56081-078-5 VCH Publishers
ISBN 3-527-28162-2 VCH Verlagsgesellschaft

Printing History:
10 9 8 7 6 5 4 3 2 1

Published jointly by:

VCH Publishers, Inc.
220 East 23rd Street
New York, New York 10010

VCH Verlagsgesellschaft mbH
P.O. Box 1-11 16
D-6940 Weinheim
Federal Republic of Germany

VCH Publishers (UK) Ltd.
8 Wellington Court
Cambridge CB1 1HZ
United Kingdom

Contents

PREFACE

Although there are many professional reference books on the science and technology of processing dairy products, this 3-volume set is unique in its coverage (topics selected, emphasis, and latest development) and its authors (experts with diversified background and experience).

Volume I discusses four important properties and applications of milk and dairy ingredients: chemistry and physics, analyses, sensory evaluation, and protein. Each chapter is not a comprehensive treatment of the subject, since more than one reference book has been written on each of the four disciplines. Rather, each chapter discusses the basic information in reasonable details that are supplemented by new research data and advances. This assures that each chapter contributes new information not available in many reference books already published.

Volume II discusses the manufacture technology for yogurt, ice cream, cheese, and dry and concentrated dairy products. The direction of each chapter is carefully designed to provide two types of information. Each chapter details the currently accepted procedures of manufacturing the product and then explores new advances in technology and their potential impact on the processing of such products in the future. The fifth chapter in this volume discusses microbiology and associated health hazards for dairy products. The goal of this chapter is obvious, since there are so much new information on this topic in the last few years. The authors have done an excellent job in reviewing available data on this highly visible field.

Volume III is unique because it covers five topics not commonly found in professional reference books for dairy manufacture: quality assurance, biotechnology, computer application, equipment and supplies, and processing plant designs. The length of each chapter is limited by the size of the book. As a result, I assume full responsibility for any missing details since I assigned a fixed length to each chapter.

The appendix to Volume I alphabetically lists products and services in the dairy industry. Under each product or service, the appendix describes the names of companies that provide those products and services. In Volume III, the appendix provides information for each company listed in Volume I. This includes contact data and the types of products and services for each company. The appendixes for Volumes I and III are not repeated in Volume II in order to assure a reasonable price for the books.

As for the expertise of the authors, you are the best judge since most of them are known among scientists, technologists, and engineers in the dairy discipline.

This three-volume set is a reference book and will benefit dairy professionals in government, industry, and academia. The information is useful to individuals engaged in research, manufacturing, and teaching. In general, the texts form an excellent background source for professionals who just enter the field. For expert dairy professionals, these books serve as a subject review as well as a summary of what is new. Any chapter in the three volumes can be used as a supplement material for a class teaching a specific topic in or an overview of the science and technology of processing dairy products.

Y. H. Hui
October 1992

Contributors

Marijana Carić, Faculty of Technology, University of Novi Sad, 2100 Novi Sad, Bulevar, Yugoslavia

Ramesh C. Chandan, James Ford Bell Technical Center, General Mills, Inc., 9000 Plymouth Avenue North, Minneapolis, MN 55427, U.S.A.

Maribeth A. Cousin, Department of Food Science, Purdue University, Lafayette, IN 47906, U.S.A.

Rafael Jiménez-Flores, Agricultural Bioprocessing Laboratory, University of Illinois, Urbana, IL 61801-4726, U.S.A.

Norman J. Klipfel, Baskin-Robbins International Company, Glendale, CA, U.S.A.

K. Rajinder Nath, Kraft General Foods, 801 Waukegan Road, Glenview, IL 60025, U.S.A.

Khem Shahani, Department of Food Science and Technology, Food Industry Complex, University of Nebraska, Lincoln, NE 68583-0919, U.S.A.

Joseph Tobias, Agricultural Bioprocessing Laboratory University of Illinois, Urbana, IL 61801-4726, U.S.A.

P.C. Vasavada, Department of Animal and Food Science, University of Wisconsin, River Falls, WI 54022

CHAPTER

1

Yogurt

Ramesh C. Chandan and Khem M. Shahani

1.1 Introduction

Yogurt has emerged as a significant dairy product of modern times. Historically, fermented milks have constituted a vital component of the human diet in many regions of the world. The main objective of fermenting milk has been to preserve the precious fluid milk which otherwise would deteriorate rapidly under the high

Table 1.1 CONSUMPTION OF YOGURT AND OTHER FERMENTED MILKS IN CERTAIN COUNTRIES IN 1988

	Annual Per Capita Consumption (kg)		Annual Total Consumption (1000 Tons)	
Country	All Fermented Milks	Yogurt	All Fermented Milks	Yogurt
Australia	3.6	3.6	60.8	60.8
Austria	9.8	7.2	73.6	54.2
Belgium	8.4	6.9	83.6	68.3
Bulgaria	42.2	42.2	379.0	379.0
Canada	3.3	3.3	86.6	86.6
Czechoslovakia	6.6	3.2	102.8	49.3
Denmark	14.8	7.8	75.7	39.8
Finland	39.0	11.4	192.8	56.3
France	15.2	—	846.6	—
Germany (West)	11.2	10.8	690.0	638.0
Hungary	3.0	1.5	31.7	15.9
Iceland	23.0	8.6	5.7	2.1
India	4.3	4.3	3410.0	3410.0
Ireland	3.3	3.3	11.6	11.6
Israel	22.1	9.4	98.0	41.8
Italy	3.7	2.4	210.2	135.0
Japan	8.0	3.8	520.0	465.0
Luxembourg	6.8	—	2.5	—
Netherlands	18.9	18.9	278.5	278.5
Norway	15.3	4.3	64.3	18.0
Poland	1.8	—	70.0	—
South Africa	3.6	1.6	105.9	47.2
Spain	7.9	7.9	297.7	297.7
Sweden	29.1	6.8	245.5	188.4
Switzerland	16.9	16.9	114.0	114.0
United Kingdom	3.9	3.9	220.0	220.0
USA	—	2.1	—	517.9
USSR	7.9	—	2250	—

Source: International Dairy Federation (1990).[5]

ambient temperatures of the Middle East, where it is likely to have originated. Conversion of milk to yogurt with a distinctive thicker consistency, smooth texture, and unmistakable flavor has added safety, portability, and novelty to the nutrition of milk for the consumer.

The objective of this chapter is to furnish basic information, including recent trends, on various aspects of the yogurt industry. It is not intended to serve as a treatise on yogurt science and technology. For detailed information, the reader is referred to various books and chapters on the subject.[1–3] Vedamuthu,[4] in a series of articles, has reviewed various technological aspects of yogurt manufacture.

Yogurt and other fermented milks have been particularly popular in countries located in the Mediterranean region; in central, southern, and southwestern Asia; and in central and eastern Europe. Table 1.1 shows the per capita consumption of

Table 1.2 ANNUAL TOTAL AND PER CAPITA SALES OF REFRIGERATED YOGURT IN THE UNITED STATES

Year	Total Sales (Millions of pounds)	Per Capita Sales (Pounds)
1972	281	1.3
1977	533	2.4
1982	613	2.6
1987	1094	4.5
1988	1142	4.6
1989	1030	4.2

Source: Milk Industry Foundation (1990).[6]

yogurt and yogurtlike products. In many parts of the world yogurt is still made at home by traditional kitchen recipes involving milk of various mammals, mainly cows, water buffaloes, goats, sheep, mare, or camel. The milk is boiled, cooled, and inoculated with yogurt left over from the previous day and incubated at ambient temperature for 4 to 6 h until it acquires a thick consistency. It is then utilized for consumption in the fresh state as a snack, as an accompaniment as a salad containing fresh vegetables (carrots, cucumber, boiled potatoes, etc.), as a sweet or savory drink, or as a dessert containing sugar and fresh sliced banana and other seasonal fruits.

In the United States the past two decades have witnessed a dramatic rise in the annual yogurt consumption from nearly 1 lb to 4.2 lb per capita. The increase in yogurt consumption may be attributed to its perceived natural and healthy image, providing to the consumer convenience, taste, and wholesomeness attributes. Table 1.2 summarizes recent trends in consumption of refrigerated yogurt in the United States.

The popularity of yogurt consumption is also related to sophisticated marketing techniques in response to consumer demand. Figure 1.1 illustrates the point. Diversification of the yogurt category has created niches to fill the needs of various consumer segments (Table 1.3). The total yogurt market (refrigerated and frozen) in the United States has grown sixfold during the last 20 years. Total sales for refrigerated yogurt alone are over 1 billion dollars. Frozen yogurt sales are estimated to reach 2 billion dollars in 1991. According to the USDA,[7] sales of frozen yogurt in 1989 reached almost 83 million gallons. In 1990, the sales increased 45%. The frozen yogurt market comprises soft-serve yogurt and hard-pack yogurt. All the major segments of the yogurt market are expected to grow moderately in the future.

The success of yogurt in the market place can be attributed to various factors, including[4]:

- Scientific evidence is mounting to corroborate consumer perception of yogurt's good-for-you image. Indeed, clinical studies have established that yogurt is well tolerated by lactose-intolerant individuals who generally have distressing symptoms of flatulence and diarrhea associated with the maldigestion of milk sugar

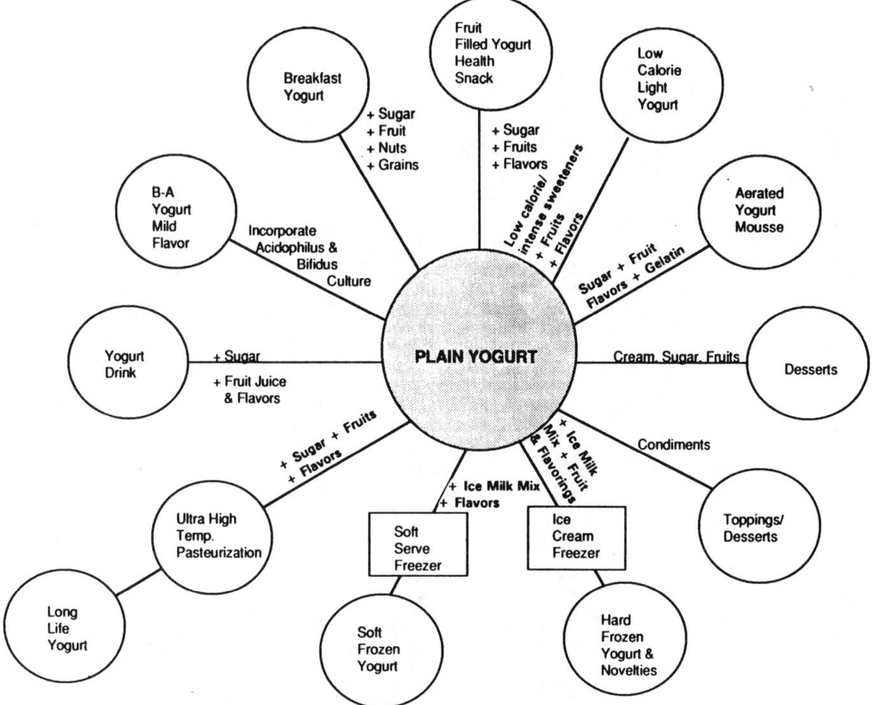

Figure 1.1 Segmentation of yogurt market.

(lactose) present in most dairy products. This effectively provides an opportunity for all consumers to benefit from the protein, calcium, B vitamins, and other significant nutrients available in milk and milk products through the consumption of yogurt. Also, recent data in the literature have suggested that yogurt containing live and active cultures may provide immunostimulatory effects. Furthermore, studies are indicating that yogurt bacteria may provide protection from pathogenic and undesirable bacteria introduced via food intake into the gastrointestinal tract.

- Use of sweeteners such as sugar and high-fructose corn syrups in yogurt manufacture adds a very desirable dimension to yogurt taste and tends to moderate harsh acidic flavor. Furthermore, intense sweeteners such as aspartame impart the desirable attribute without incurring caloric buildup in the product.
- Addition of fruit preparations, fruit flavors, and fruit purees further enhances versatility of taste, color, and texture. Fruits generally are perceived as healthy by the consumer. Their association with yogurt endorses the healthy image of yogurt even more.
- Incorporation of nuts and grains gives yogurt multiple textures and flavors, thus providing a packaged convenient and wholesome breakfast food.
- Development and availability of nonfat, low-fat, and reduced fat yogurts has encouraged consumers to benefit from the health-driven trends currently in vogue.

Table 1.3 TRENDS IN YOGURT STYLE AND PACKAGE SIZE IN THE UNITED STATES (PERCENT OF TOTAL PRODUCTION)

	Package Size			Fat Content			Style					
Year	8 oz	5.1–6.0 oz	Other	Full Fat	Low Fat	Nonfat	Fruit-on-Bottom	Swiss	French	Plain	Breakfast	Other
1984	—	—	—	30	66	4	41	28	16	10	3	2
1987	59.8	17.3	22.9	17	73	10	28	50	5	13	0.2	5

Source: Milk Industry Foundation (1990).[5]

Table 1.4 TYPICAL CHEMICAL COMPOSITION AND NUTRIENT PROFILE OF YOGURT

Constituent (per 100g)	Skim Milk	Yogurt					
		Plain			Fruit-Flavored		
		Full Fat	Low Fat	Nonfat	Full Fat	Low Fat	Nonfat
Protein (g)	3.50	3.88	3.55	4.35	3.90	3.60	3.80
Fat (g)	0.10	3.50	1.60	0.1	2.62	1.33	0.11
Lactose (g)	5.00	3.9	4.10	4.20	3.08	3.11	2.98
Galactose (g)	0.00	1.50	1.50	1.50	1.20	1.20	1.20
Total carbohydrate (g)	5.00	5.42	5.60	5.70	15.50	13.51	12.83
Lactic acid (g)	0.00	1.00	1.00	1.00	1.00	1.00	1.00
Citric acid (g)	0.20	0.30	0.30	0.30	—	—	—
Sodium (g)	0.05	0.07	0.07	0.07	0.05	0.05	0.06
Potassium (g)	0.15	0.20	0.20	0.20	0.16	0.16	0.18
Calcium (g)	0.12	0.18	0.18	0.17	0.13	0.15	0.17
Phosphorus (g)	0.10	0.14	0.14	0.12	0.10	0.10	0.10
Chloride (g)	0.10	0.12	0.12	0.12	0.10	0.10	0.10
Energy value (KJ)	150	307	221	165	432	343	289
(calories)	38	73	53	39	103	82	69
Bacterial mass (g)	0	0.15	0.15	0.15	0.15	0.15	0.15

Source: Sellars (1989),[8] Souci et al. (1990).[9]

- Marketing and merchandising practices have accelerated consumer acceptance and desirability of the product.
- Proliferation of sister yogurt products such as hard-frozen and soft-serve yogurt have provided alternatives perceived healthier than their counterpart ice cream product.

1.2 Definition of Yogurt

Yogurt is a semisolid fermented product made from a standardized milk mix by the activity of a symbiotic blend of *Streptococcus salavarius* subsp. *thermophilus* and *Lactobacillus delbruechii* subsp. *bulgaricus* cultures. For the sake of brevity we shall term the yogurt culture organisms as ST and LB.

Milk of various mammals is used for making yogurt in various parts of the world. However, most of the industrialized production of yogurt uses cow's milk. It is common to boost the solids-not-fat fraction of the milk to about 12% with added nonfat dry milk or condensed skim milk. The increased protein content in the mix results in a custardlike consistency following the fermentation period.

The typical composition and nutrient profile of yogurt are shown in Table 1.4. In general, yogurt contains more protein, calcium, and other nutrients than milk, reflecting extra solids-not-fat content. Bacterial mass content and products of lactic fermentation further distinguish yogurt from milk. Fat content is standardized commensurate with consumer demand of lowfat to fat-free foods.

1.2.1 Standard of Identity and Regulatory Aspects of Yogurt

Grandstrand[10] discussed the current U.S. Food and Drug Administration standards of identity for refrigerated yogurt promulgated in September 1982, effective July 1, 1985. A summary of the requirements excerpted from the Code of Federal Regulations, April 1991[11] is presented below.

1.2.1.1 Yogurt

Description

Yogurt is the food produced by culturing one or more of the optional dairy ingredients specified below with a characterizing bacterial culture that contains the lactic acid-producing bactera, *Lactobacillus bulgaricus* and *Streptococcus thermophilus*. One or more of the other optional ingredients described below may also be added. All ingredients used are safe and suitable. Yogurt, before the addition of bulky flavors, contains not less than 3.25% milkfat and not less than 8.25% milk-solids-not-fat, and has a titratable acidity of not less than 0.9%, expressed as lactic acid. In a subsequent action, the FDA stayed the titratable acidity requirement. The food may be homogenized and shall be pasteurized or ultrapasteurized prior to the addition of the bacterial culture. Flavoring ingredients may be added after pasteurization or ultrapasteurization. To extend the shelf life of the food, yogurt may be heat-treated after culturing is completed, to destroy viable microorganisms.

Optional Ingredients

Vitamins. (1) If added, Vitamin A shall be present in such quantity that each 946 ml (quart) of the food contains not less than 2000 International Units thereof, within limits of current good manufacturing practice. (2) If added, Vitamin D shall be present in such quantity that each 946 ml (quart) of the food contains 400 International Units thereof, within limits of current good manufacturing practice.

Dairy Ingredients. Cream, milk, partially skimmed milk, or skim milk, used alone or in combination.

Other Optional Ingredients. (1) Concentrated skim milk, nonfat dry milk, buttermilk, whey, lactose, lactalbumins, lactoglobulins, or whey modified by partial or complete removal of lactose and/or minerals, to increase the nonfat solids content of the food, *provided that* the ratio of protein to total nonfat solids of the food and the protein efficiency ratio of all protein present shall not be decreased as a result of adding such ingredients. (2) Nutritive carbohydrate sweeteners. Sugar (sucrose), beet or cane; invert sugar (in paste or syrup form); brown sugar, refiner's syrup; molasses (other than blackstrap); high-fructose corn syrup; fructose; fructose syrup; maltose; maltose syrup, dried maltose syrup; malt extract, dried malt extract; malt syrup, dried malt syrup; honey; maple sugar, except table syrup. (3) Flavoring ingredients. (4) Color additives. (5) Stabilizers.

Methods of Analysis

The following referenced methods of analysis are from *Official Methods of Analysis of the Association of Official Analytical Chemists*, 13th edit. (1980), which is incorporated by reference. Copies are available from the Association of Official Analytical Chemists, 2200 Wilson Blvd., Suite 400, Arlington, VA 22201-3301, or available for inspection at the Office of the Federal Register, 1100 L St. NW, Washington, D.C. 20408. (1) Milkfat content—as determined by the method prescribed in Section 16.059 ''Roese-Gottlieb Method (Reference Method) (11)-Official Final Action,'' under the heading ''Fat.'' (2) Milk solids-not-fat content—calculated by subtracting the milkfat content from the total solids content as determined by the method prescribed in Section 16.032, ''Method I—Official Final Action,'' under the heading ''Total Solids.'' (3) Titratable acidity—as determined by the method prescribed in Section 16.023, ''Acidity (2)—Official Final Action,'' or by an equivalent potentiometric method.

Nomenclature

The name of the food is ''yogurt.'' The name of the food shall be accompanied by a declaration indicating the presence of any characterizing flavoring. (1) The following terms shall accompany the name of the food wherever it appears on the principal display panel or panels of the label in letters not less than one-half of the height of the letters used in such name: (a) The word ''sweetened'' if nutritive carbohydrate sweetener is added without the addition of characterizing flavor. (b) The parenthetical phrase ''(heat-treated after culturing)'' shall follow the name of the food if the dairy ingredients have been heat-treated after culturing. (c) The phrase ''Vitamin A'' or ''Vitamin A added,'' or ''Vitamin D'' or ''vitamin D added,'' or ''Vitamins A and D added,'' as appropriate. The word ''vitamin'' may be abbreviated ''vit.'' (2) The term ''homogenized'' may appear on the label if the dairy ingredients used are homogenized.

Label Declaration

Each of the ingredients used in the food shall be declared on the label as required by the applicable sections of Part 101.

1.2.1.2 Low-Fat Yogurt

Low-fat yogurt is the food produced according to the description given in the previous section for yogurt, except the milkfat content before the addition of bulky flavors shall be not less than 0.5% and not more than 2%. Percent milkfat shall be declared on the principal display panel in ½% increments closest to the actual fat content of the food. All other provisions for yogurt apply for the nomenclature Low-Fat Yogurt.

1.2.1.3 Nonfat Yogurt

Nonfat yogurt is the food produced as per the previous description for yogurt, except the milkfat content before the addition of bulky flavors shall be <0.5%. All other provisions for yogurt apply for the nomenclature Nonfat Yogurt.

Until further action by the FDA, the following four provisions were stayed in 1985.

1. Exclusion of the use of reconstituted dairy ingredients as the basic ingredients in yogurt.
2. The requirement for a minimum titratable acidity of 0.9%, expressed as lactic acid.
3. The exclusion of preservatives as functional ingredients in yogurt.
4. The presence of 3.25% milkfat prior to the addition of bulky flavors for full-fat yogurt.

Besides compliance with the FDA standard of identity and labeling requirements under the Fair Packaging and Labeling Act, yogurt manufacture is regulated by two other agencies in the United States. First, the state regulatory agencies of the State Department of Agriculture (Dairy Division) or the Department of Health require plant inspection and conformation to each state's standards pertaining to dairy plants in general, and yogurt in particular. Second, National Conference on Interstate Milk Shippers (NCIMS), in compliance with Pasteurized Milk Ordinance (PMO), is involved in developing methods for sanitation ratings of milk supplies, sanitation requirements for Grade A condensed/dry milk, and condensed/dry whey, in fabrication of single-service containers and closures for milk and milk products, and in the evaluation of milk laboratories. The main function of NCIMS is to make unrestricted and uniform milk supply available in interstate shipment.

The PMO outlines requirements for Grade A milk production at the farm, dairy processing facility and equipment product standards, sanitation aspects, and product handling (45°F or below) to ensure compliance to Grade A regulations. Also, requirements are defined for coliforms (not to exceed 10/ml), passing the phosphatase test, and antibiotics in milk supply (e.g., no zone greater than or equal to 16 mm with the *Bacillus stearothermophilus* disc assay procedure).

Mareschi and Cueff[12] outlined regulatory requirements in 21 countries for defining yogurt. Three essential criteria for yogurt label include (1) the type of milk and quantity of its constituents; (2) type, amount, and live-active nature of yogurt culture in the product; and (3) the technological process involving extent of fermentation involved.

1.2.2 National Yogurt Association Criteria for Live and Active Culture Yogurt

The integrity of yogurt must be maintained in the product to fulfill consumer expectations. Accordingly, the National Yogurt Association has defined yogurt as follows:

Live and active culture yogurt (refrigerated cup and frozen yogurt) is the food produced by culturing permitted dairy ingredients with a characterizing bacterial culture in accordance with the FDA standards of identity for yogurt (21 C.F.R. S 131.200), lowfat yogurt (21 C.F.R. S 131.203), and nonfat yogurt (21 C.F.R. S 131.206). In addition to the use of the bacterial cultures required by the referenced federal standards of identity and by these National Yogurt Association criteria, live and active culture yogurt may contain other safe and suitable food grade bacterial cultures. Declaration of the presence of cultures on the label of live and active culture yogurt is optional.

Heat treatment of live and active yogurt is inconsistent with the maintenance of live active cultures in the product; accordingly, heat treatment that is intended to kill the live and active organisms shall not be undertaken after fermentation. Likewise, manufacturers of live and active culture yogurt should undertake their best efforts to ensure that distribution practices, code dates, and handling instructions are conducive to the maintenance of living and active cultures.

In order to meet these criteria, live and active culture yogurt must satisfy each of these requirements:

1. The product must be fermented with both *L. delbruechii* subsp. *bulgaricus* and *S. thermophilus.*
2. The cultures must be active at the end of the stated shelf life as determined by the activity test described in item 3. Compliance with this requirement shall be determined by conducting an activity test on a representative sample of yogurt that has been stored at temperatures between 32 and 45°F for refrigerated cup yogurt and at temperatures of 0°F, or colder for frozen yogurt for the entire stated shelf life of the product.
3. The activity test is carried out by pasteurizing 12% solids nonfat dry milk (NFDMS) at 92°C (198°F) for 7 min, cooling to 110°F, adding 3% inoculum of the material under test, and fermenting at 110°F for 4 h. The total organisms are to be enumerated in the test material both before and after fermentation by IDF methodology.[14] The activity test is met if there is an increase of 1 log or more during fermentation.
4. a. In the case of refrigerated cup yogurt, the total population of organisms in live and active culture yogurt must be at least 10^8 per gram at the time of manufacture. b. In the case of frozen yogurt, the total population of organisms in live and active culture yogurt must be at least 10^7 at the time of manufacture. (It is anticipated that if proper distribution practices and handling instructions are followed, the total organisms in both refrigerated cup and frozen live and active culture yogurt at the time of consumption will be at least 10^7.
5. The product shall have a total titratable acidity expressed as lactic acid at least 0.3% at all times. At least 0.15% of total acidity must be obtained by fermentation. This is confirmed by demonstrating the presence of both D-(−) and L-(+) forms of lactic acid.

1.2.3 Frozen Yogurt

The Food and Drug Administration standards for frozen yogurt are under development. In the advanced notice of proposed rule-making FDA[15] published proposed standards which are summarized below:

1.2.3.1 Frozen Yogurt

Description

(1) Frozen yogurt is the food produced by freezing, while stirring, a mix containing safe and suitable ingredients including, but not limited to, dairy ingredients. The mix may be homogenized, and all of the dairy ingredients shall be pasteurized or ultra-pasteurized. All or a portion of the dairy ingredients shall be cultured with a characterizing live bacterial culture that shall contain the lactic acid-producing bacteria *Lactobacillus bulgaricus* and *Streptococcus thermophilus* and may contain other lactic acid-producing bacteria. After culturing, the unflavored frozen yogurt mix shall have a titratable acidity of not less than 0.3%, calculated as lactic acid. Where the titratable acidity of the frozen yogurt mix is <0.3%, the manufacturer may establish compliance with this section by disclosing to the Federal Food and Drug Administration (FDA) quality control records that demonstrate that as a result of bacterial culture fermentation, there has been at least a 0.15% increase in the titratable acidity, calculated as lactic acid, of the product above the apparent titratable acidity of the uncultured dairy ingredients in the frozen yogurt mix. The direct addition of food grade acids or other acidogens for the purpose of raising the titratable acidity of the frozen yogurt mix to comply with the prescribed minimum is not permitted, and no chemical preservation treatment or other preservation process, other than refrigeration, may be utilized that results in reduction of the live culture bacteria. Sweeteners, flavorings, color additives, and other characterizing food ingredients, unless otherwise provided in the regulations of the FDA, may be added to the mix before or after pasteurization or ultrapasteurization, provided that any ingredient addition after pasteurization or ultrapasteurization is done in accordance with current good manufacturing practice. Any dairy ingredients added after pasteurization or ultrapasteurization shall have been pasteurized. (2) Frozen yogurt may be sweetened with any sweetener that has been affirmed as generally regarded as safe (GRAS) or approved as a food additive for this use by FDA and may or may not be characterized by the addition of flavoring ingredients. (3) Frozen yogurt, before the addition of bulky characterizing ingredients or sweeteners, shall contain not less than 3.25% milkfat and 8.25% milk-solids-not-fat. Frozen yogurt shall contain not less than 1.3 lb of total solids/gal and shall weigh not less than 4.0 lb/gal.

Nomenclature

The name of the food is ''frozen yogurt.'' The name of the food shall be accompanied by a declaration indicating the presence of any characterizing flavoring.

Label Declaration

(1) Each of the ingredients used in the food shall be declared on the label as required by the applicable sections of part 101 of the CFR chapter. (2) If the food purports to be or is represented for special dietary use, it shall be labeled in accordance with the requirements of part 105 of the CFR chapter.

1.2.3.2 Frozen Low-Fat Yogurt

Description

Frozen low-fat yogurt is the food that is prepared from the same ingredients and in the same manner prescribed for frozen yogurt, and complies with all of the provisions of Frozen Yogurt, except that the milkfat level is not less than 0.5% nor more than 2.0%.

Nomenclature

The name of the food is ''frozen low-fat yogurt'' or, alternatively, ''low-fat frozen yogurt.''

1.2.3.3 Frozen Nonfat Yogurt

Description

Frozen nonfat yogurt is the food that is prepared from the same ingredients and in the same manner prescribed for Frozen Yogurt, except that the milkfat level is <0.5%.

Nomenclature

The name of the food is ''frozen nonfat yogurt'' or, alternatively, ''nonfat frozen yogurt.''

1.3 Yogurt Starters

Milk is a normal habitat of a number of lactic acid bacteria which cause spontaneous souring of milk held at bacterial growth temperatures for an appropriate length of time. Depending on the type of lactic acid bacteria gaining entry from the environmental sources (air, utensils, milking equipment, milkers, cows, feed, etc.), the sour milk attains uncontrollable flavor and texture characteristics. Modern industrial processes utilize defined lactic acid bacteria as a starter for yogurt production. Details of the starters are discussed elsewhere.[14]

A starter consists of food-grade microorganism(s) that on culturing in milk produce predictable attributes characterizing yogurt. The composition of yogurt starter is shown in Table 1.5. Also shown are some additional organisms found in yogurt or yogurtlike products marketed in various parts of the world. Most of the yogurt in the United States is fermented with ST and LB. In Europe and Japan, optional bacteria, especially those of intestinal origin, are incorporated in the starter or the product. ST and LB are fairly compatible as well as symbiotic for growth in milk medium. However, the optional organisms do not necessarily exhibit compatibility with LB and ST. Judicious selection of strains of LB, ST, and the optional organisms is necessary to ensure survival and growth of all the component organisms of the

Table 1.5 REQUIRED AND OPTIONAL COMPOSITION OF YOGURT BACTERIA

Required by FDA Standard Identity for Yogurt	Optional Additional Bacteria Used or Suggested
Streptococcus salivarius ssp. *thermophilus* (ST)	*Lactobacillus acidophilus*
Lactobacillus delbruéchii ssp. *bulgaricus* (LB)	*Lactobacillus casei*
	Lactobacillus helveticus
	Lactobacillus jugurti
	Lactobacillus lactis
	Bifidobacterium longum
	Bifidobacterium bifidum
	Bifidobacterium infantis

Source: Ming et al. (1989),[16] Vedamuthu (1991).[4]

starter. Nevertheless, product characteristics, especially flavor, are significantly altered from traditional yogurt flavor when yogurt culture is supplemented with optional bacteria.

Commercial production of yogurt relies heavily on fermentation ability and characteristics imparted by the starter. Sellars[8] stated the criteria essential for commercial success for a starter are:

- Strain selection
- Maintenance of desirable ST and LB ratios
- Survival and viability during manufacture of starter, preservation, storage, and distribution.

The starter performance factors are:

- Rapid acid development
- Typical yogurt flavor, body, and texture
- Exopolysaccharide secreting strains to enhance viscosity of yogurt
- Scaleup possibilities in various production conditions, including compatibility to a variety and levels of ingredients used
- Fermentation times and temperatures
- Survival of culture viability during shelf life of yogurt
- Possess probiotic properties and exhibit survival in the human gastrointestinal tract for certain health attributes
- Minimum acid production during distribution and storage at 40 to 50°F until yogurt is consumed.

The activity of a starter culture is determined by direct microscopic counts of a methylene blue-stained culture slide. This exercise also indicates physiological state of the culture cells. Cells of ST grown fresh in milk or broth display pairs or long chains of spherical coccal shape. Under stressed condition of nutrition and age (old cells, cells exposed to excessive acid, solid media colonies, inhibitor containing milk), the cells appear oblong in straight chains, somewhat resembling rods. The acid producing ability is measured by pH drop and titratable acidity rise in 12% reconstituted nonfat dry milk medium (sterilized at 116°C/18 min) incubated at 40°C for 8 h. A ratio of 3 parts of ST and 1 part of LB gives a pH of 4.20 and % TA of 1.05[8] under these conditions.

Microbiological specifications of commercial cultures are also outlined by Sellars.[8] In general, counts of mesophilic lactics, yeasts and molds, coliforms, anaerobic sporeformers, and salt-tolerant micrococci should not exceed 10 colony-forming units (cfu)/g. *E. coli, S. faecium*, and coagulase-positive staphylococci should be <1 cfu/g. The culture must be free of salmonella, listeria, and other pathogenic contaminants.

1.3.1 Taxonomy of Yogurt Bacteria

Data relative to various taxonomic factors characterizing yogurt bacteria are presented in Table 1.6. The effect of temperature of incubation on the growth of yogurt bacteria is shown in Table 1.7. Acid production is normally used as a means of growth of yogurt culture. However, growth of the organisms is not necessarily synonymous with their acid producing ability. Differences in acid liberated per unit cell mass have been recorded, which are both environmental and genetic in origin.

1.3.1.1 Streptococcus salivarius Subsp. *thermophilus*

Originally described by Orla-Jensen,[19] *Streptococcus salivarius* subsp. *thermophilus* is the new name of *Streptococcus thermophilus* (ST). ST is characterized through its typical attributes which distinguish it from lactococci (or lactic streptococci). ST originates exclusively from the dairy environment, from which it can be easily isolated.

Methylene blue-stained ST cells grown on solid media or aged cells could exhibit a rodlike shape under microscope. However, display of abnormal shapes of cells obtained from liquid media are indicators of stress conditions on the organism such as bacteriophage attack or presence of inhibitors (sanitizers, antibiotics, cleaning compounds, etc.) in the growth medium.

Recent work on DNA–DNA homology has questioned its classification as a subspecies of *Streptococcus salivarius.*[20]

1.3.1.2 Lactobacillus delbruechii Ssp. *bulgaricus*

This organism was originally described by Orla Jensen[19] as *Thermobacterium bulgaricum.* Based on DNA homology studies, four subspecies of *L. delbruechii* are classified as *bulgaricus*, *leichmannii*, *lactis*, and *delbruechii* in Bergey's Manual.[17]

Younger LB cells do not show metachromatic granules under microscopic examination. Nutritional stress leads to copious granules in the rods, which under the microscope could be confused with cocci.

Table 1.6 TAXONOMIC DATA ON YOGURT BACTERIA

Characteristic	*Streptococcus salivarius* ssp. *thermophilus*	*Lactobacillus delbruechii* ssp. *bulgaricus*
Shape	Ovid–spherical 0.7–0.9 μm diameter Pairs/long chains. Long chains in acidic medium and at higher temperature of growth	Slender rods, 0.8–1.0 μm wide, 4–6 μm long. Single/chains. Aged culture shows granules and long chains.
Gram reaction	+	+
Catalase	−	−
Fermentation	Homolactic	Homolactic
Granules	−	+
Growth at:		
Below 20°C	−	−
Above 45°C	+	+
Growth in:		
2% NaCl	+ (2.5)%	−
4% NaCl	−	−
Urease	+	−
Arginine	−	−
Milk % TA	0.7–1.0	1.8
Survives 60°C (140°F) for 30 min.	+	−
Acid from:		
Glucose	+	+
Galactose	+	−
Lactose	+	+
Sucrose	+	−
Maltose	±	−
Lactic acid isomer	L-(+)	D-(−)
Mucopolysaccharide	+	+
Mol. % G + C of the DNA	40	49–51

Source: Bergey's Manual of Determinative Bacteriology, 9th ed. (1986),[17] Reinbold (1989.[18]

Yogurt fermentation constitutes the most important step in its manufacture. To optimize parameters for yogurt production, an understanding of factors involved in the growth of yogurt bacteria is important to manage uniformity of product quality and cost effectiveness of the manufacturing operation.

1.3.1.3 Collaborative Growth of ST and LB

Yogurt starter organisms display obligate symbiotic relationships during their growth in milk medium. Although they can grow independently, they utilize each other's

Table 1.7 GROWTH TEMPERATURE PROFILE OF YOGURT BACTERIA

	°C/F	
Growth Temperature	ST	LB
Minimum	20 (68)	>15 (59)
Maximum	50 (122)	50–52 (122–125)
Optimum	39–46 (102–115)	40–47 (104–117)

Source: Reinbold (1989).[18]

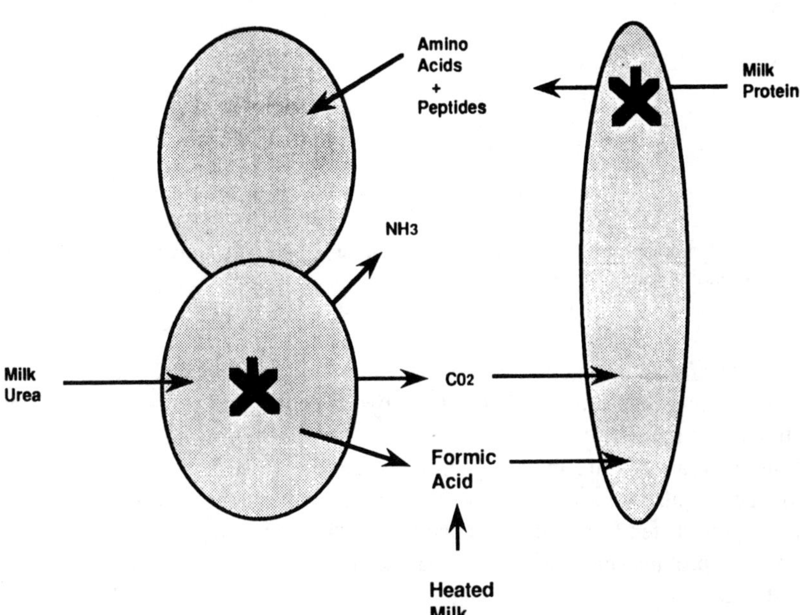

Figure 1.2 Symbiosis of yogurt bacteria. (From ref. 21.)

metabolites to effect remarkable efficiency in acid production. Figure 1.2 illustrates this. In general, LB has significantly more cell-bound proteolytic enzyme activity, producing stimulatory peptides and amino acids for ST. The relatively high aminopeptidase and cell-free and cell-bound dipeptidase activity of ST is complementary to strong proteinase and a low peptidase activity of LB. Urease activity of ST produces CO_2 which stimulates LB growth. Concomitant with CO_2 production, urease liberates ammonia which acts as a weak buffer. Consequently, milk cultured by ST alone exhibits considerably low TA or high pH of coagulated mass. Formic acid formed by ST as well as by heat treatment of milk accelerates LB growth.

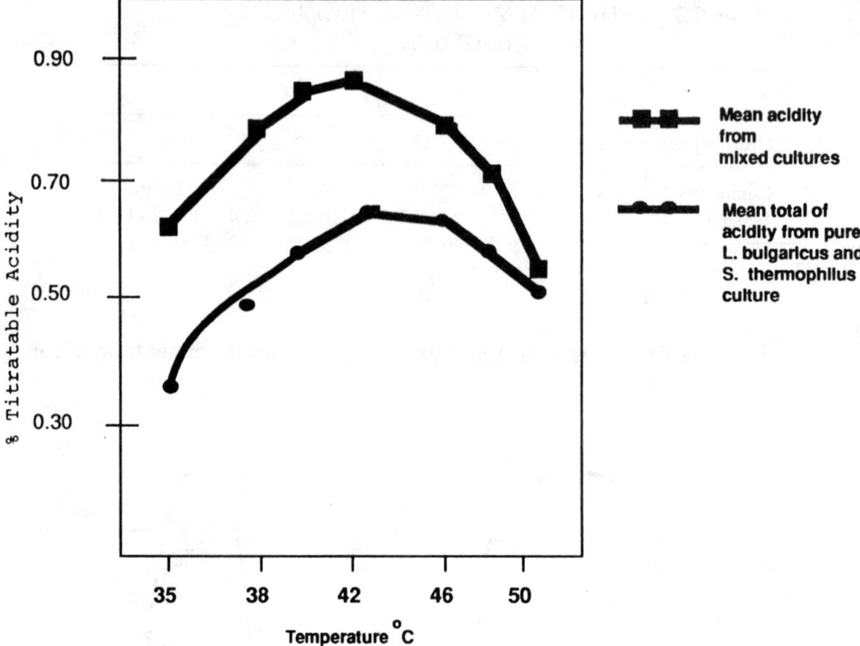

Figure 1.3 Comparison of acid production by mixed *S. thermophilus* and *L. bulgaricus* by the corresponding pure cultures.

The rate of acid production by yogurt starter containing both ST and LB is considerably higher than that by either of the two organisms grown separately. This is illustrated in Figure 1.3

Yogurt organisms are microaerophilic in nature. Heat treatment of milk drives out oxygen. It also wipes out competitive flora. Furthermore, heat-produced sulfhydryl compounds tend to generate reducing conditions in the medium. Accordingly, rate of acid production in high-heat-treated milk is considerably higher than in raw or pasteurized milk.

1.3.1.4 Inhibiting Factors

Inherent Inhibitors

Proper selection of ST and LB strains is necessary to avoid possible antagonism between the two organisms. Also, certain abnormal milks (mastitic cows, hydrolytic rancidity in milk) are inhibitory to their growth. Seasonal variations in milk composition resulting in lower micronutrients (trace elements, nonprotein nitrogenous compounds) may affect starter performance. Natural inhibitors secreted in milk (lactoperoxidase thiocyanate system, agglutinins, lysozyme) are generally destroyed by proper heat treatment.

Table 1.8 SENSITIVITY OF YOGURT BACTERIA TO VARIOUS INHIBITORS IN MILK

	Inhibitory Level		
Inhibitor	ST	LB	Mixed Culture
I. Antibiotics (per ml)			
Penicillin	0.004–0.010 IU	0.02–0.100 IU	0.01 IU
Streptomycin	0.380 IU	0.380 IU	1.00 IU
Streptamycin	12.5–21.0 μg	6.6 μg	
Tetracycline	0.130–0.500 μg	0.34–2.000 μg	1.00 IU
Chlortetracycline	0.060–1.000 μg	0.060–1.000 μg	0.10 IU
Oxytetracycline	0.400 IU	0.700 IU	0.40 IU
Bacitracin	0.040–0.120 IU	0.040–0.100 IU	0.04 IU
Erythromycin	0.300–1.300 mg	0.070–1.300 mg	0.10 IU
Chloramphenicol	0.800–13.000 mg	0.800–13.000 mg	0.50 IU
II. Disinfectant/detergent (mg/L)			
Chlorine compounds	100	100	50–2500
QAC	100–500	50–100	>250
Ampholyte			>1000
Idophor	60	60	>2000
Alkaline detergent			>500–1000
III. Insecticides (PPM)			
Malathion			200
N-Methylcarbamates			20
IV. Miscellaneous (PPM)			
Fatty acids	1000		
Ethylenedichloride	10–100		
Methylsulfone	10–100		
Acetonitrile	10		
Chloroform	10		
Ether	10		

Source: Tamime and Robinson, 1985.[3]

Antibiotics

Antibiotic residues in milk and entry of sanitation chemicals (quaternary compounds, iodophors, hypochlorites, hydrogen peroxide) have a profound inhibitory impact on the growth of yogurt starter.

Table 1.8 summarizes the degree of sensitivity of yogurt bacteria to residual quantities of various inhibitors.

Sweeteners

Yogurt mixes designed for manufacture of refrigerated or frozen yogurt may contain appreciable quantities of sucrose, high fructose corn syrup, dextrose, and various dextrose equivalent (DE) corn syrups. The sweeteners exert osmotic pressure in the system, leading to progressive inhibition and decline in the rate of acid production by the culture. Being a colligative property, the osmotic based inhibitory effect would be directly proportional to concentration of the sweetener and inversely related to the molecular weight of the solute. In this regard, solutes inherently present in milk-solids-not-fat part of yogurt mix accruing from starting milk and added milk solids and whey products would also contribute toward the total potential inhibitory effect on yogurt culture growth.

Acid producing ability of yogurt culture has been reported in mixes containing 4.0% sucrose.[4] Commercial strains that are relatively osmotolerant may allow higher usage levels without interruption in acid production during yogurt manufacture.

Bacteriophages

Phage infections and accompanying loss in rate of acid production by lactic cultures results in flavor and texture defects as well as major product losses in fermented dairy products. Serious economic losses have been attributed to phage attack in the cheese industry. So far, thermophilic starters have not been threatened as much as mesophilic starters used largely in cheese production. However, production volumes for mozzarella cheese, Swiss cheese, and yogurt have more recently escalated in response to consumer demand with a concomitant appearance of a number of reports of phage inhibition in recent literature.[22] It is known that specific phages affect ST and LB, and that ST is relatively more susceptible than LB.

Yogurt fermentation process is relatively fast (3 to 4 h). It is improbable that both ST and LB would be simultaneously attacked by phages specific for the two organisms. In the likelihood of a phage attack on ST, acid production may be carried on by LB, causing little or no interruption in production schedule. In fact, lytic phage may lyse ST cells, spilling cellular contents in the medium, which could conceivably supply stimulants for LB growth. This rationale may explain partially why the yogurt industry has experienced a low incidence of phage problems. Nonetheless, most commercial strains of yogurt cultures have been phage typed. Specific phage sensitivity has been determined to facilitate starter rotation procedures as a practical way to avoid phage threats in yogurt plants. Reinbold[18] reported that ST phage is destroyed by heat treatment of 74°C for 23 s. This phage proliferates much faster at pH 6.0 than at 6.5 or 7.0. Methods used for phage detection include plaque assay, inhibition of acid production (litmus color change), enzyme immunoassay, ATP assay by bioluminescence, and changes in impedance and conductance measurement.

Phage problem in yogurt plants cannot be ignored. Accordingly, adherence to strict sanitation procedures would ensure prevention of phage attack.

1.3.2 Production of Yogurt Starters

Frozen culture concentrates available from commercial culture suppliers have received wide acceptance in the industry. Reasons for their use include convenience and ease of handling, reliable quality and activity, and economy. The concentrates are shipped frozen in dry ice and stored at the plant in special freezers at −40°C or below for a limited period of time specified by the culture supplier.

Table 1.9 APPROXIMATE COMPOSITION AND FOOD VALUES OF NONFAT DRY MILK

Constituents	Amount
Protein (N × 6.38) %	36.0
Lactose (milk sugar) %	51.0
Fat %	0.7
Moisture %	3.0
Minerals (ash) %	8.2
Calcium %	1.3
Phosphorus %	1.0
Vitamin A (IU/lb)	165.0
Riboflavin (mg/lb)	9.2
Thiamine (mg/lb)	1.6
Niacin (mg/lb)	4.2
Niacin equivalents[a] (mg/lb)	42.2
Pantothenic acid (mg/lb)	15.0
Pyridoxine (mg/lb)	2.0
Biotin (mg/lb)	0.2
Choline (mg/lb)	500.0
Energy (calories/lb)	1630.0

Source: American Dairy Products Institute (1990).[23]

[a] Includes contribution of tryptophan.

The starter is the most crucial component in the production of yogurt of high quality and uniformity of consumer attributes. Culture preparation room should be separate from the rest of plant activities. An effective sanitation program including filtered air and positive pressure in the culture and fermentation area should significantly control airborne contamination. The result would be controlled fermentation time and consistently high-quality product.

The medium for bulk starter production in most yogurt plants is antibiotic-free, nonfat dry milk reconstituted in water at 10 to 12% solids level. Pretesting for the absence of inhibitory principles (antibiotics, sanitizers) is advisable to ensure desirable growth of the starter in the medium. Other quality attributes associated with the nonfat dry milk are low heat powder with not less than 6.0 mg of whey protein nitrogen/g of powder. Typical composition of nonfat dry milk is shown in Table 1.9. The standards for Extra Grade spray-dried nonfat dry milk are given in Table 1.10.

The starter medium is not generally fortified with growth activators such as yeast extract, beef extract, and protein hydrolysates because they tend to impart undesirable flavor to the starter and eventually yogurt. Following reconstitution of nonfat dry milk in water, the medium is heated to 90 to 95°C and held for 30 to 60 min. Then the medium is cooled to 110°F in the vat. During cooling, the air drawn into the vat should be free of airborne contaminants (phages, bacteria, yeast, and mold spores). Accordingly, use of proper filters (e.g., High Efficiency Particulate Air) on the tanks to filter-sterilize incoming air is desirable.

Table 1.10 STANDARDS FOR EXTRA GRADE SPRAY-DRIED NONFAT DRY MILK

	Not Greater Than
Milkfat	1.25%
Moisture	4.0%
Titratable acidity	0.15%
Solubility index	1.25 ml
Bacterial estimate	50,000 per g
Scorched particles	Disc B (15.0 mg)

Source: American Dairy Products Institute (1990).[23]

Extra Grade nonfat dry milk shall be entirely free from lumps, except those that break up readily under slight pressure. The reliquefied product shall have a sweet and desirable flavor, but may possess the following flavors to a slight degree: chalky, cooked, feed, and flat.

The next step is inoculation of frozen bulk culture. Instruction for handling the frozen culture as prescribed by the supplier should be followed carefully. The frozen can is thawed by placing the can in cold or lukewarm water containing a low level of sanitizer until the contents are partially thawed. The culture cans are emptied into the starter vat as aseptically as possible and bulk starter medium is pumped over the partially thawed culture to facilitate mixing and achieving uniformity of dispersion.

The incubation period for yogurt bulk starter ranges from 4 to 6 h and the temperature of 43°C is maintained by holding hot water in the jacket of the tank. The fermentation must be quiescent (lack of agitation and vibrations) to avoid phase separation in the starter following incubation. The progress of fermentation is monitored by titratable acidity measurements at regular intervals. When the TA is 0.85 to 0.90%, the fermentation is terminated by turning the agitators on and replacing warm water in the jacket with ice water. Circulating ice water drops the temperature of starter to 4 to 5°C. The starter is now ready to use following a satisfactory microscopic examination of methylene blue-stained slide of the starter. Morphological view helps to ensure healthy cells in the starter and maintenance of desirable ST/LB ratio. In the earlier literature, a ratio of 1:1 was considered desirable, but a more recent trend is in favor of ST predomination (60 to 80%). An organoleptic examination is also helpful to detect unwanted flavors in the starter. Figure 1.4 shows the steps involved in bulk starter production.

1.4 General Principles of Manufacture

1.4.1 Ingredients and Equipment

Yogurt and other cultured dairy products are produced in various parts of the world from the milk of several species of mammals. The animals include cow (*Bos taurus*),

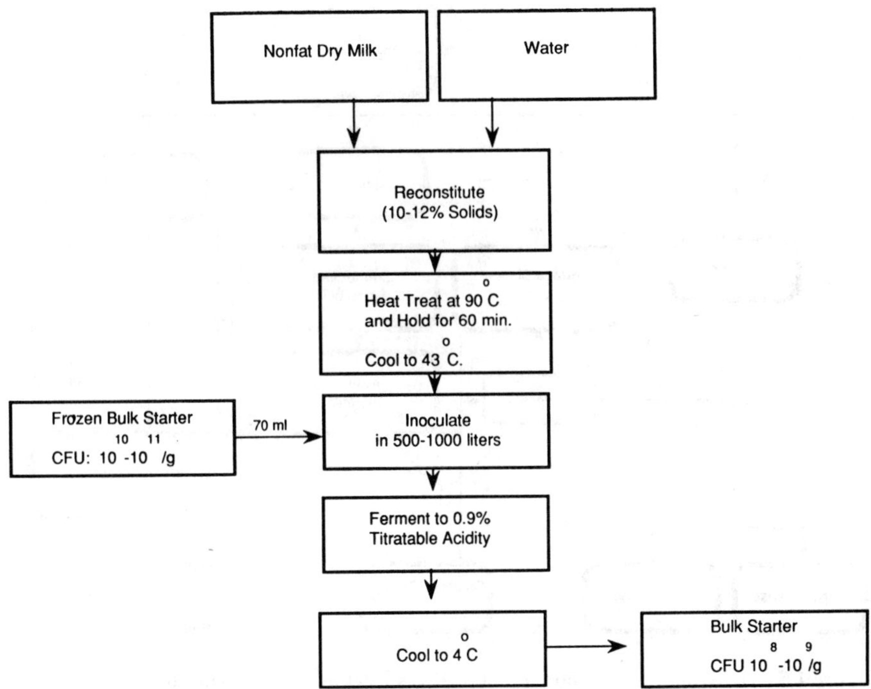

Figure 1.4 Preparation of bulk starter for yogurt manufacture.

Table 1.11 COMPOSITION OF MILKS USED IN THE PREPARATION OF CULTURED DAIRY FOODS IN VARIOUS PARTS OF THE WORLD

Mammal	Fat (%)	Caseins (%)	Whey Proteins (%)	Lactose (%)	Ash (%)	Total Solids (%)
Cow	3.7	2.8	0.6	4.8	0.7	12.7
Water buffalo	7.4	3.2	0.6	4.8	0.8	17.2
Goat	4.5	2.5	0.4	4.1	0.8	13.2
Sheep	7.4	4.6	0.9	4.8	1.0	19.3
Mare	1.9	1.3	1.2	6.2	0.5	11.2
Sow	6.8	2.8	2.0	5.5	—	18.8

Source: Chandan (1982).[2]

water buffalo (*Bubalus bubalis*), goat (*Capra hircus*), sheep (*Ocis aries*), mare (*Equus cabalus*), and sow (*Sus scrofa*). The composition of these milks is summarized in Table 1.11. Because the total solids in milk of various species range from 11.2 to 19.3%, the cultured products derived from them vary in consistency from a fluid to a custardlike gel. The range in casein content also contributes to the gel

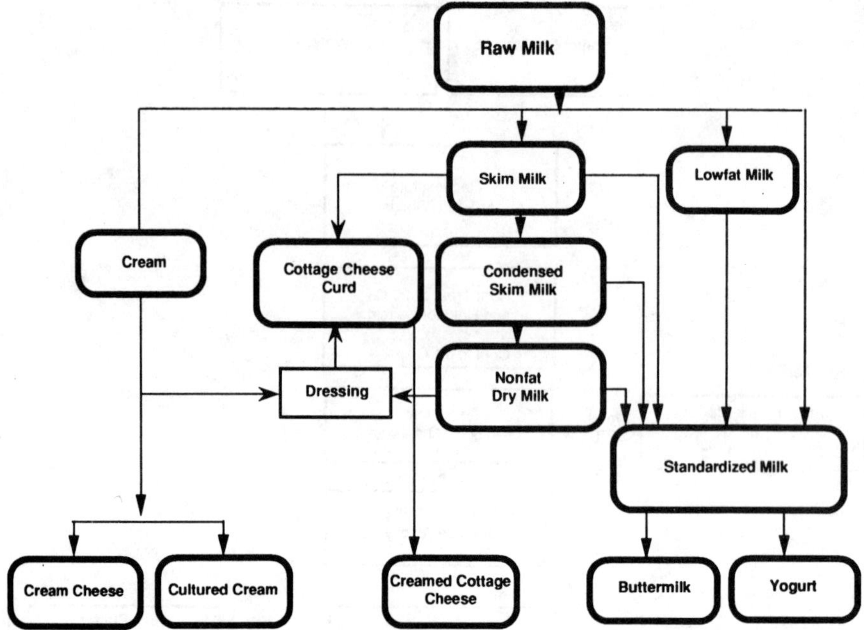

Figure 1.5 Dairy ingredients and their derivatives used in cultured dairy foods.

formation because on souring this class of proteins coagulates at its isoelectric point of pH 4.6. The whey proteins are considerably denatured and insolubilized by heat treatments prior to culturing. The denatured whey proteins are also precipitated along with caseins to exert an effect on the water binding capacity of the gel.

In the United States, bovine milk is practically the only milk employed in the industrial manufacture of cultured dairy products. Figure 1.5 shows the relationship among various forms of milk raw materials used in yogurt and other cultured dairy foods. For optimum culture growth, the raw materials must be free from culture inhibitors such as antibiotics, sanitizing chemicals, mastitis milk, colostrum, and rancid milk. Microbiological quality should be excellent for developing the delicate and clean flavor associated with top quality yogurt. The raw materials generally include whole milk, skim milk, condensed skim milk, nonfat dry milk, and cream. In addition, other food materials such as sweeteners, stabilizers, flavors, fruit preparations, etc. are required as components of yogurt mix. These materials are blended together in proportions to obtain a standardized mix conforming to the particular product to be manufactured.

A yogurt plant requires a special design to minimize contamination of the products with phage and spoilage organisms. Filtered air is useful in this regard. The plant is generally equipped with a receiving room to receive, meter or weigh, and store milk and other raw materials. In addition, a culture propagation room along with a control laboratory, a dry storage area, a refrigerated storage area, a mix processing room, a fermentation room, and a packaging room form the backbone of the plant. The mix processing room contains equipment for standardizing and separating milk, pasteurizing and heating, and homogenizing along with the necessary pipelines, fittings, pumps, valves, and controls. The fermentation room housing fermentation tanks is isolated from the rest of the plant. Filtered air under positive pressure is supplied to the room to generate clean room conditions. A control laboratory is generally set aside where culture preparation, process control, product composition, and shelf life tests may be carried out to ensure adherence to regulatory and company standards. Also, a quality control program is established by laboratory personnel. A utility room is required for maintenance and engineering services needed by the plant. The refrigerated storage area is used for holding fruit, finished products, and other heat-labile materials. A dry storage area at ambient temperature is primarily utilized for temperature-stable raw materials and packaging supplies.

The sequence of stages of processing in a yogurt plant is given in Table 1.12.

1.4.2 Mix Preparation

Milk is commonly stored in silos which are large vertical tanks with a capacity up to 100,000 l. A silo consists of an inner tank made of stainless steel containing 18% chromium, 8% nickel, and <0.07% carbon. Acid and salt resistance in the steel is attained by incorporating 3% molybdenum. To minimize corrosion, this construction material is used for the storage of acidic products. The stainless steel tank is usually covered with 50 to 100 mm of insulation material which in turn is surrounded by an outer shell of stainless or painted mild steel or aluminum. The silo tanks generally have an agitation system (60 to 80 rpm), spray balls mounted in the center for cleaning in place (CIP), an air vent, and a manhole. The air vent must be kept open during cleaning with hot cleaning solutions. This precaution is necessary to prevent a sudden development of vacuum in the tank and consequent collapse of the inner tank upon rinsing with cold water.

For reconstitution of dry powders, such as nonfat dry milk, sweeteners, and stabilizers, the use of a powder funnel and recirculation loop, or a special blender is convenient.

1.4.3 Heat Treatment

The common pasteurization equipment consists of vat, plate, triple-tube, scraped, or swept surface heat exchanger. In case of milk, vat pasteurization is conducted at 63°C with a minimum holding time of 30 min. This temperature is raised to 66°C in the presence of sweeteners in the mix. For a high temperature–short time (HTST) system, the equivalent temperature–time combination is 73°C for 15 s, or 75°C for 15 s in the presence of sweeteners. An ultra-high temperature (UHT) system employs temperatures >90°C and as high as 148°C for 2 s. Alternatively, culinary steam may be used directly by injection or infusion to raise the temperature to 77 to 94°C, but allowance must be made for an increase in water content of the mix due to steam condensation in this process. In some plants, steam volatiles are continuously re-

Table 1.12 SEQUENCE OF PROCESSING STAGES IN THE MANUFACTURING OF YOGURT

Step	Salient Feature
1. Milk procurement	Sanitary production of Grade A milk from healthy cows is necessary. For microbiological control, refrigerated bulk milk tanks should cool to 10°C in 1 h and <5°C in 2 h. Avoid unnecessary agitation to prevent lipolytic deterioration of milk flavor. Milk pickup is in insulated tanks at 48-h intervals.
2. Milk reception and storage in manufacturing plant	Temperature of raw milk at this stage should not exceed 10°C. Insulated or refrigerated storage up to 72 h helps in raw material and process flow management. Quality of milk is checked and controlled.
3. Centrifugal clarification and separation	Leucocytes and sediment are removed. Milk is separated into cream and skim milk or standardized to desired fat level at 5°C or 32°C.
4. Mix preparation	Various ingredients to secure desired formulation are blended together at 50°C in a mix tank equipped with powder funnel and an agitation system.
5. Heat treatment	Using plate heat exchangers with regeneration systems, milk is heated to temperatures of 85–95°C for 10–40 min, well above pasteurization treatment. Heating of milk kills contaminating and competitive microorganism, produces growth factors by breakdown of milk proteins, generates microaerophilic conditions for growth of lactic organisms, and creates desirable body and texture in the cultured dairy products.
6. Homogenization	Mix is passed through extremely small orifice at pressure of 2000–2500 psi, causing extensive physicochemical changes in the colloidal characteristics of milk. Consequently, creaming during incubation and storage of yogurt is prevented. The stabilizers and other components of a mix are thoroughly dispersed for optimum textural effects.
7. Inoculation and incubation	The homogenized mix is cooled to an optimum growth temperature. Inoculation is generally at the rate of 0.5–5% and the optimum temperature is maintained throughout incubation period to achieve a desired titratable acidity. Quiescent incubation is necessary for product texture and body development.
8. Cooling, fruit incorporation, and packaging	The coagulated product is cooled down to 5–22°C, depending on the product. Using fruit feeder or flavor tank, the desired level of fruit and flavor is incorporated. The blended product is then packaged.
9. Storage and distribution	Storage at 5°C for 24–48 h imparts in several yogurt products desirable body and texture. Low temperatures ensure desirable shelf life by slowing down physical, chemical, and microbiological degradation.

moved by vacuum evaporation to remove certain undesirable odors (feed, onion, garlic) associated with milk.

In yogurt processing, the mix is subjected to much more severe heat treatment than conventional pasteurization temperature–time combinations. Heat treatment at 85°C for 30 min or 95°C for 10 min is an important step in manufacture. The heat treatment (1) produces a relatively sterile medium for the exclusive growth of the starter; (2) removes air from the medium to produce a more conducive medium for microaerophilic lactic cultures to grow; (3) effects thermal breakdown of milk constituents, especially proteins, releasing peptones and sulfhydryl groups which provide nutrition and anaerobic conditions for the starter; and (4) denatures and coagulates milk albumins and globulins which enhance the viscosity and produce custardlike consistency in the product.

1.4.4 Homogenization

The homogenizer is a high-pressure pump forcing the mix through extremely small orifices. It includes a bypass for safety of operation. The process is usually conducted by applying pressure in two stages. The first stage pressure, of the order of 2000 psi, reduces the average milkfat globule diameter size from approx. 4 μm (range 0.1 to 16 μm) to <1 μm. The second stage uses 500 psi and is designed to break the clusters of fat globules apart with the objective of inhibiting creaming in milk. Homogenization aids in texture development and additionally it alleviates the surface creaming and wheying off problems. Ionic salt balance in milk is also involved in the wheying off problem.

1.4.5 Fermentation

Fermentation tanks for the production of cultured dairy products are generally designed with a cone bottom to facilitate draining of relatively viscous fluids after incubation.

For temperature maintenance during the incubation period, the fermentation vat is provided with a jacket for circulating hot or cold water or steam located adjacent to the inner vat containing the mix. This jacket is usually insulated and covered with an outermost surface made of stainless steel. The vat is equipped with a heavy-duty, multispeed agitation system, a manhole containing a sight glass, and appropriate spray balls for CIP. The agitator is often of swept surface type for optimum agitation of relatively viscous cultured dairy products. For efficient cooling after culturing, plate or triple-tube heat exchangers are used.

The fermentation vat is designed only for temperature maintenance. Therefore, efficient use of energy requires that the mix not be heat treated in the culturing vat.

1.4.6 Packaging

Most plants attempt to synchronize the packaging lines with the termination of the incubation period. Generally, textural defects in yogurt products are caused by excessive shear during pumping or agitation. Therefore, positive drive pumps are preferred over centrifugal pumps for moving the product after culturing or ripening. For incorporation of fruit, it is advantageous to use a fruit feeder system adapted from the frozen dessert industry.[24] Various packaging machines of suitable speeds (up to 400 cups per minute) are available to package various kinds and sizes of yogurt products.

1.5 Yogurt Production

The manufacture of yogurt has recently been reviewed by Chandan,[1] IDF,[13] Rasic and Kurmann,[2] and Tamime and Robinson.[3]

1.5.1 Yogurt Ingredients and Flavor, Texture, and Rheological Aspects

1.5.1.1 Dairy Ingredients

Yogurt is generally made from a mix standardized from whole, partially defatted milk, condensed skim milk, cream, and nonfat dry milk. In rare practice, milk may be partly concentrated by removal of 15 to 20% water in a vacuum pan. Supplementation of milk-solids-not-fat with nonfat dry milk is the preferred industrial procedure. All dairy raw materials should be selected for high bacteriological quality. Ingredients containing mastitis milk and rancid milk should be avoided. Also, milk partially fermented by contaminating organisms and milk containing antibiotic and sanitizing chemical residues cannot be used for yogurt production. The procurement of all ingredients should be based on specifications and standards that are checked and maintained with a systematic sampling and testing program by the quality control laboratory. Because yogurt is a manufactured product, it is likely to have variations according to the quality standards established by marketing considerations. Nonetheless, it is extremely important to standardize and control the day-to-day product in order to meet consumer expectations and regulatory obligations associated with a certain brand or label.

1.5.1.2 Sweeteners

Nutritive carbohydrates used in yogurt manufacture are similar to the sweeteners used in ice cream and other frozen desserts described by Arbuckle.[24] Sucrose is the major sweetener used in yogurt production. Sometimes corn sweeteners may also be used, especially in frozen yogurt mixes. The level of sucrose in yogurt mix appears to affect the production of lactic acid and flavor by yogurt culture. A decrease in characteristic flavor compound (acetaldehyde) production has been reported at 8% or higher concentration of sucrose.[1] Sucrose may be added in a dry, granulated, free-flowing, crystalline form or as a liquid sugar containing 67% sucrose. Liquid sugar is preferred for its handling convenience in large operations. However, storage capability in sugar tanks along with heaters, pumps, strainers, and meters is required. The corn sweeteners, primarily glucose, usually enter yogurt via the processed fruit flavor in which they are extensively used for their flavor enhancing characteristics. Up to 6% corn syrup solids are used in frozen yogurt. High-intensity sweeteners (e.g., aspartame) have been used to produce a ''light'' product containing about 60% of the calories of normal sweetened yogurt.

Commercial yogurts have an average of 4.06% lactose, 1.85% galactose, 0.05% glucose, and pH of 4.40.

1.5.1.3 Stabilizers

The primary purpose of using a stabilizer in yogurt is to produce smoothness in body and texture, impart gel structure, and reduce wheying off or syneresis. The stabilizer increases shelf life and provides a reasonable degree of uniformity of the product. Stabilizers function through their ability to form gel structures in water, thereby leaving less free water for syneresis. In addition, some stabilizers complex with casein. A good yogurt stabilizer should not impart any flavor, should be effective at low pH values, and should be easily dispersed in the normal working temperatures in a dairy plant. The stabilizers generally used in yogurt are gelatin; vegetable gums such as carboxymethyl cellulose, locust bean, and Guar; and seaweed gums such as alginates and carrageenans.

Gelatin is derived by irreversible hydrolysis of the proteins collagen and ossein. It is used at a level of 0.3 to 0.5% to get a smooth shiny appearance in refrigerated yogurt. Gelatin is a good stabilizer for frozen yogurt. The term Bloom refers to the gel strength as determined by a Bloom gelometer under standard conditions. Gelatin of a Bloom strength of 225 or 250 is commonly used. The gelatin level should be geared to the consistency standards for yogurt. Amounts above 0.35% tend to give yogurt of relatively high milk solids a curdy appearance on stirring. At temperatures below 10°C, the yogurt acquires a puddinglike consistency. Gelatin tends to degrade during processing at ultrahigh temperatures and its activity is temperature dependent. The yogurt gel is considerably weakened by a rise in temperature.

The seaweed gums impart a desirable viscosity as well as gel structure to yogurt. Algin and sodium alginate are derived from giant sea kelp. Carrageenan is made from Irish moss and compares with 250 Bloom gelatin in stabilizing value. These stabilizers are heat stable and promote stabilization of the yogurt gel by complex formation with Ca^{2+} and casein.

Among the seed gums, locust beam gum or carob gum is derived from the seeds of a leguminous tree. Carob gum is quite effective at low pH levels. Guar gum is also obtained from seeds and is a good stabilizer for yogurt. Guar gum is readily soluble in cold water and is not affected by high temperatures used in the pasteurization of yogurt mix. Carboxymethyl cellulose is a cellulose product and is effective at high processing temperatures.

The stabilizer system used in yogurt mix preparations is generally a combination of various vegetable stabilizers to which gelatin may or may not be added. Their ratios as well as the final concentration (generally 0.5 to 0.7%) in the product are carefully controlled to get desirable effects. More recently, whey protein concentrate is being used as a stabilizer, exploiting the water binding property of denatured whey proteins.

For detailed descriptions of various industrial gums, the reader is referred to Tamime and Robinson.[3]

1.5.1.4 Fruit Preparations for Flavoring Yogurt

The fruit preparations for blending in yogurt are specially designed to meet the marketing requirements for different types of yogurt. They are generally present at levels of 10 to 20% in the final product. A majority of the fruits contain natural flavors.

Flavors and certified colors are usually added to the fruit-for-yogurt preparations for improved eye appeal and better flavor profile. The fruit base should meet the following requirements. It should (1) exhibit true color and flavor of the fruit when blended with yogurt, and (2) be easily dispersible in yogurt without causing texture defects, phase separation, or syneresis. The pH of the fruit base should be compatible with yogurt pH. The fruit should have zero yeast and mold population in order to prevent spoilage and to extend shelf life. Fruit preserves do not necessarily meet all these requirements, especially of flavor, sugar level, consistency, and pH. Accordingly, special fruit bases of the following composition are designed for use in stirred yogurt.

	%
Fruit flavor, artificial	0.1
or natural	1.25
Color	0.01 or to specification
Potassium sorbate	0.1
Citric acid to pH 3.8 to 4.2	—

$CaCl_2$ and certain food-grade phosphates are also used in several fruit preparations. The soluble solids range from 60 to 65% and viscosity is standardized to 5 ± 1.5 Bostwick units (cm), 30 s reading at 24°C. Standard plate counts on the fruit bases are generally <500/g. Coliform count, yeast, and mold counts of nonaseptic fruit preparations are <10/g. The fruit flavors vary in popularity in different parts of the country and during different times of the year. In general, more popular fruits are strawberry, raspberry, blueberry, peach, cherry, orange, lemons, purple plum, boysenberry, spiced apple, apricot, and pineapple. Blends of these fruits are also popular. Fruits used in yogurt base manufacture may be frozen, canned, dried, or combinations thereof. Among the frozen fruits are strawberry, raspberry, blueberry, apple peach, orange, lemon, cherry, purple plum, blackberry, and cranberry. Canned fruits are pineapple, peach, mandarin orange, lemon, purple plum, and maraschino cherry. The dried fruit category included apricot, apple, and prune. Fruit juices and syrups are also incorporated in the bases. Sugar in the fruit base functions in protecting fruit flavor against loss by volatilization and oxidation. It also balances the fruit and the yogurt flavor. The pH control of the base is important for fruit color retention. The color of yogurt should represent the fruit color in intensity, hue, and shade. The base should be stored under refrigeration to obtain optimum flavor and extend shelf life. The current trend is to use aseptically packaged sterilized fruit preparations.

The following types of yogurts are marketed in the United States.

1. *Fruit-on-the-bottom style yogurt.* In this type, typically, 59 ml (2 oz) of fruit preserves or special fruit preparations are layered at the bottom followed by 177 ml (6 oz) of inoculated yogurt mix on the top. The top layer may consist of yogurt mix containing stabilizers, sweeteners, and the flavor and color indicative of the fruit on the bottom. After lids are placed on the cups, incubation and setting of the yogurt takes place in the cups. When a desirable pH of 4.2 to 4.4 is attained, the cups are placed in refrigerated rooms for rapid cooling. For consumption, the fruit and yogurt layers are mixed by the consumer. If used, fruit preserves have a standard of identity. A fruit preserve consists of 55% sugar and a minimum of 45% fruit which is cooked until the final soluble solids content is 68% or higher (65% in the case of certain fruits). Frozen fruits and juices are the usual raw materials. Commercial pectin, 150 grade, is normally utilized at a level of 0.5% in preserves and the pH is adjusted to 3.0 to 3.5 with a food-grade acid such as citric during manufacturing of the preserves.
2. *Stirred style yogurt.* Also known as Continental, French, and Swiss yogurt, the fruit preparaton is thoroughly blended in yogurt after culturing. Stabilizers are commonly used in this form of yogurt unless milk-solids-not-fat levels are relatively high (14 to 16%). In this style, cups are filled with a blended mixture of yogurt and fruit. On refrigerated storage for 48 h, the clot is reformed to exhibit a fine body and texture. Overstabilized yogurt possesses a solidlike consistency and lacks a refreshing character. Spoonable yogurt should not have the consistency of a drink. It should melt in the mouth without chewing.

1.5.2 Yogurt Starter and Its Contribution to Texture and Flavor

The starter is a critical ingredient in yogurt manufacture. The rate of acid production by yogurt culture should be synchronized with plant production schedules. Using frozen culture concentrates, incubation periods of 5 hr at 45°C, 11 h at 32°C, or 14 to 16 h at 29 to 30°C are required for yogurt acid development. Using bulk starters at 4% inoculum level, the period is 2.5 to 3.0 h at 45°C, 8 to 10 h at 32°C, or 14 to 16 h at 20 to 30°C.

The production of flavor by yogurt cultures is a function of time as well as the sugar content of yogurt mix. Acetaldehyde production in yogurt takes place predominantly in the first 1 to 2 h of incubation. Eventually, 23 to 55 ppm of acetaldehyde are found in yogurt. The acetaldehyde level declines in later stages of incubation. Yogurt flavor is typically ascribed to the formation of lactic acid, acetaldehyde, acetic acid, and diacetyl.

The milk coagulum during yogurt production results from the drop in pH due to the activity of the yogurt culture. The streptococci are responsible for lowering the pH of a yogurt mix to 5.0 to 5.5 and the lactobacilli are primarily responsible for further lowering of the pH to 3.8 to 4.4 Attempts have been made to improve the viscosity and to prevent synerisis of yogurt by including a slime-producing strain. The texture of yogurt tends to be coarse or grainy if it is allowed to develop firmness prior to stirring or if it is disturbed at pH values higher than 4.6. Incomplete blending of mix ingredients is an additional cause of a coarse smooth texture. Homogenization treatment and high fat content tend to favor smooth texture. Gassiness in yogurt may be attributed to defects in starters or contamination with sporeforming *Bacillus* species, coliform, or yeast, producing excessive CO_2 and hydrogen. In comparison with plate heat exchangers, cooling with tube type heat exchangers causes less damage to yogurt structure. Further, loss of viscosity of yogurt may be minimized by well-designed booster pumps, metering units, and valves involved in yogurt packaging.

The pH of yogurt during refrigerated storage continues to drop. Higher temperature of storage accelerates the drop in pH.

1.5.3 Manufacturing Procedures

1.5.3.1 Plain Yogurt

Plain yogurt is an integral component of the manufacture of frozen yogurt. The steps involved in the manufacturing of set-type and stirred-type plain yogurts are shown in Figure 1.6. Plain yogurt normally contains no added sugar or flavors in order to offer the consumer natural yogurt flavor for consumption as such or an option of flavoring with other food materials of the consumer's choice. In addition, it may be used for cooking or for salad preparation with fresh fruits or grated vegetables. In most recipes, plain yogurt is a substitute for sour cream, providing lower calories and fat alternative. The fat content may be standardized to the levels preferred by the market. Also, the size of the package may be geared to the market demand. Plastic cups and lids are the chief packaging materials used in the industry.

1.5.3.2 Fruit-Flavored Yogurt

A general manufacturing outline for both set style and stirred style yogurts is presented in Figure 1.7. Several variations of this procedure exist in the industry. Fruit incorporation is conveniently effected by the use of a fruit feeder at a 10 to 20% level. Prior to packaging, the stirred-yogurt texture can be made smoother by pumping it through a valve or a stainless steel screen.

The incubation times and temperatures are coordinated with the plant schedules. Incubation temperatures lower than 40°C in general tend to impart a slimy or sticky appearance to yogurt.

1.5.3.3 Postculturing Heat Treatment

The shelf life of yogurt may be extended by heating yogurt after culturing to inactivate the culture and the constituent enzymes. Heating to 60 to 65°C stabilizes the product so the yogurt shelf life will be 8 to 12 weeks at 12°C. However, this treatment

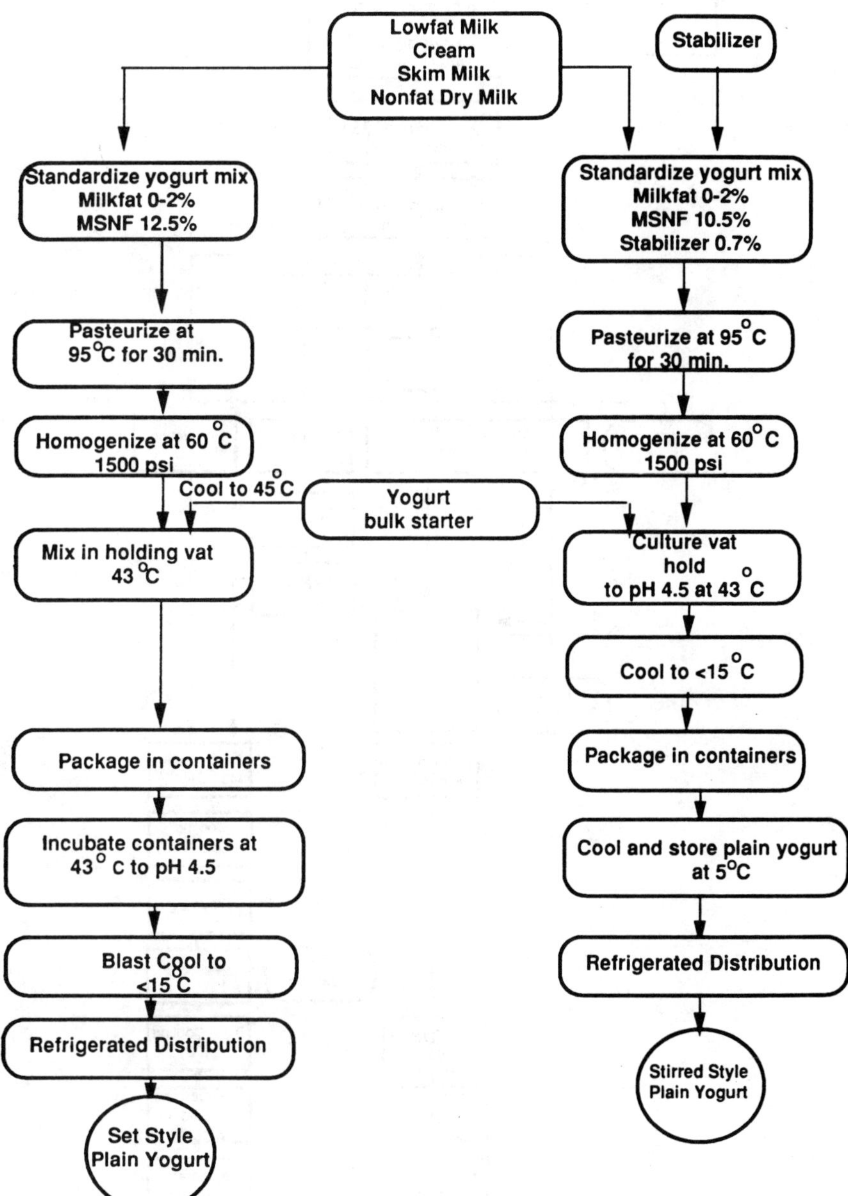

Figure 1.6 A flow sheet outline for the manufacture of plain yogurt.

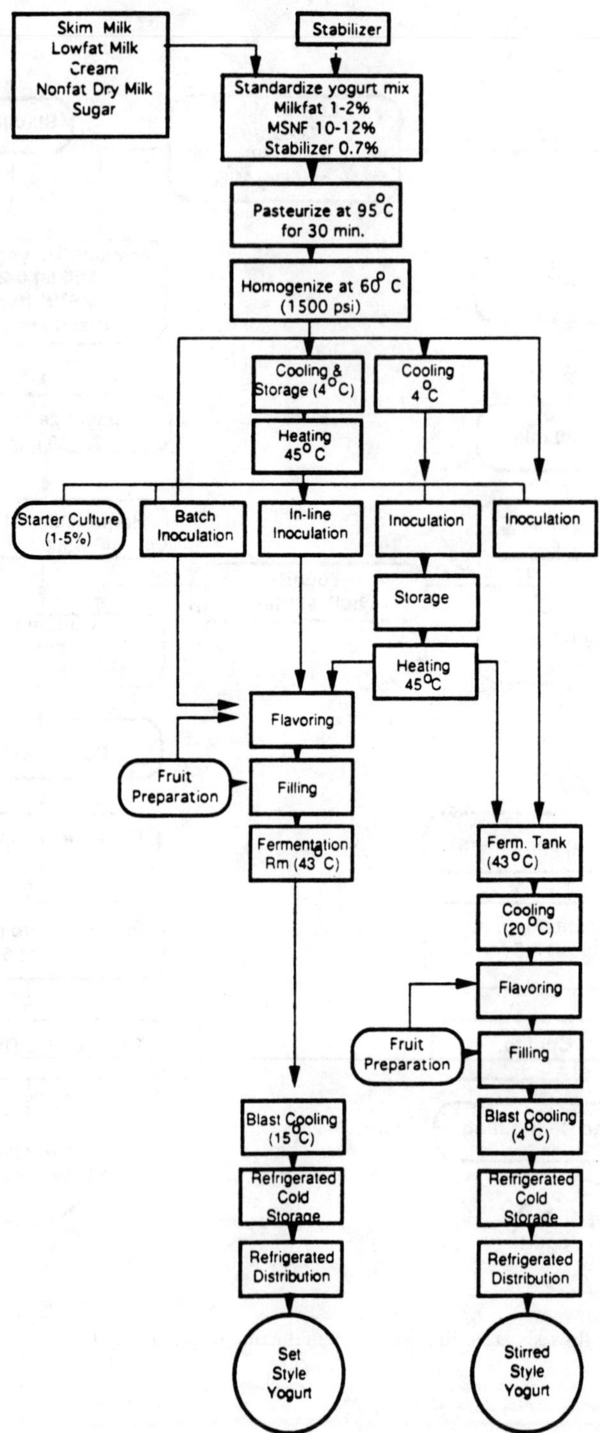

Figure 1.7 A flow sheet outline for the manufacture of fruit-flavored yogurt. [Adapted from Chandan (1982),[1] Larsen (1988).[25]]

destroys the "live" nature of yogurt, which may be a desirable consumer attribute to retain. Federal Standards of Identity for refrigerated yogurt permit the thermal destruction of viable organisms with the objective of shelf life extension, but the parenthetical phrase "heat treated after culturing" must show on the package following the yogurt labeling. The postripening heat treatment may be designed to (1) ensure destruction of starter bacteria, contaminating organisms, and enzymes; and (2) redevelop the texture and body of the yogurt by appropriate stabilizer and homogenization processes.

1.5.3.4 Frozen Yogurt

Both soft-serve and hard-frozen yogurts have gained immense popularity in recent years. Market value in frozen yogurt has exceeded that of refrigerated yogurt. Consumer popularity for frozen yogurt has been propelled by its low-fat and nonfat attribute. The recently developed frozen yogurt is a very low acid product resembling ice cream or ice milk in flavor and texture. A significant shift in reduced acidity in the product has been observed in relation to the products available 10 years before. Essentially, the industry standards require minimum titratable acidity of 0.3%, with a minimum contribution of 0.15% as a consequence of fermentation by yogurt bacteria.

The frozen yogurt base mix may be manufactured in a cultured dairy plant and shipped to a soft-serve operator or an ice cream plant. Alternatively, the mix may be prepared and frozen in an ice cream plant.

Technology for production of frozen yogurt involves limited fermentation in a single mix and arresting further acid development by rapid cooling, or a standardization of titratable acidity to a desirable level by blending plain yogurt with ice milk

Table 1.13 TYPICAL COMPOSITION OF NONFAT SOFT-SERVE AND HARD-PACK FROZEN YOGURT

	Soft-Serve (%)			Hard-Pack (%)		
Component	Stream 1 (20%)	Stream 2 (80%)	Blended Final Mix	Stream 1 (20%)	Stream 2 (80%)	Blended Final Mix
Milkfat	0	0	0	0	0	0
Milk-solids-not-fat	11	11	11	13	13	13
Sucrose	0	16.25	12	0	16.25	13
Corn syrup solids, 36 DE	0	7.50	6	0	7.5	6
Maltodextrin, 10 DE	0	2.5	2	0	2.5	2
Stabilizer	0	1.5	1.2	0	1.5	1.2
Total solids	11	38.75	32.2	13	40.75	35.20
Titratablc acidity	1.15	0.15	0.35	1.15	0.16	0.35
pH	4.4	6.7	5.5	4.4	6.7	5.5

Source: Germantown Manufacturing Co., Product Bulletin G-813.[26]

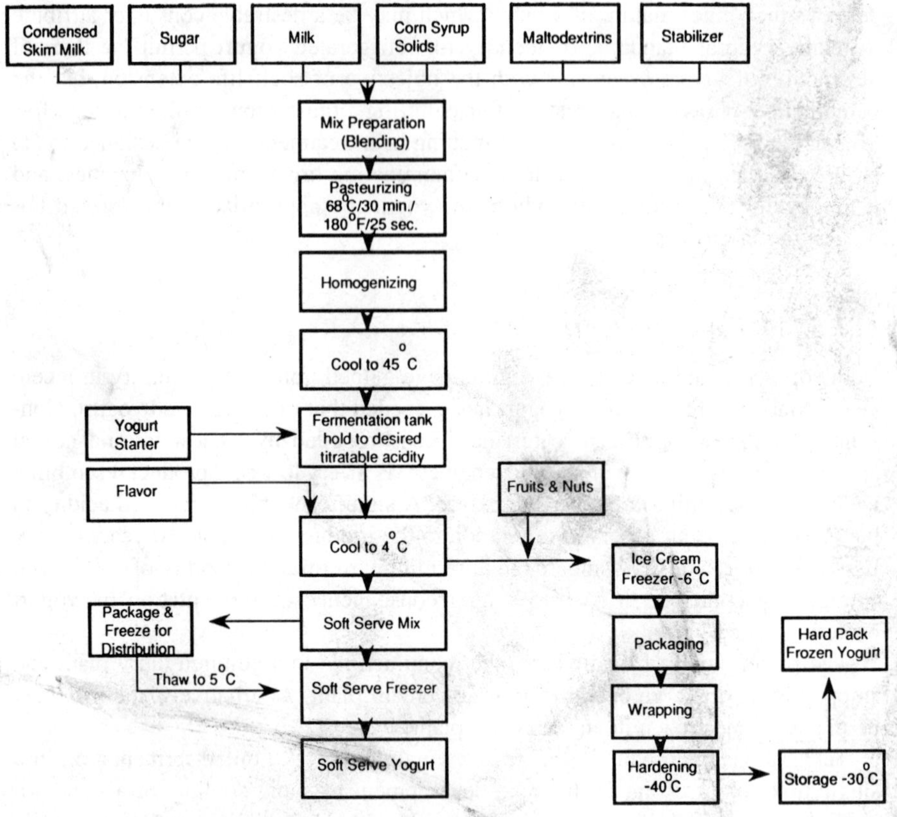

Figure 1.8 Flow chart for frozen yogurt (single-stream process).

mix (Table 1.13). In certain instances, the blend is pasteurized to ensure destruction of newly emerging pathogens, including listeria and campylobacter in the resulting low-acid food. To provide live and active yogurt culture in the finished product, frozen culture concentrate is blended with the pasteurized product. Alternatively, some processes are boosting the yogurt culture count by adding frozen culture concentrates to the fermented base. Figures 1.8 and 1.9 illustrates process suggested by Germantown Manufacturing Co.[26] for making frozen yogurt. Details of manufacture of soft frozen and hard pack mixes and frozen desserts are given by Arbuckle.[24]

1.6 Yogurt Quality Control

1.6.1 Refrigerated Yogurt

A well planned quality control program must be executed in the plant to maximize keeping quality of product. To deliver to the consumer yogurt with most desirable

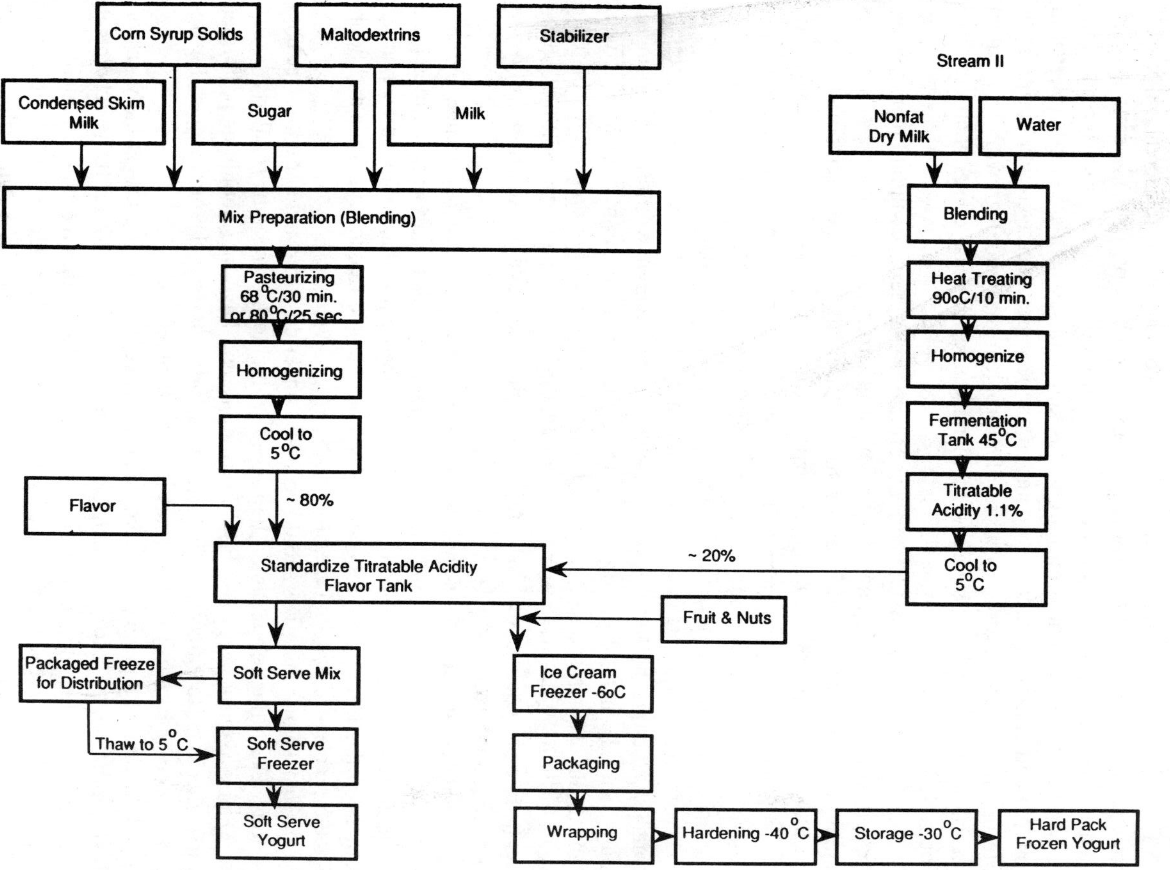

Figure 1.9 Flow chart for frozen yogurt (two-stream blending process).

attributes of flavor and texture, it is imperative to enforce a strict sanitation program along with good manufacturing practices. Shelf-life expectations from commercial yogurt vary but generally approximate a month from the date of manufacture, provided temperature during distribution and retail marketing channels does not exceed 45°F. Lactic acid and some other metabolites produced by fermentation process protect yogurt from most Gram-negative psychotrophic organisms. In general, most quality issues in a yogurt plant are not related to proliferation of spoilage bacteria.

Most spoilage flora in yogurt are yeasts and molds, which are highly tolerant to low pH and can grow under refrigeration temperatures. Yeast growth during shelf life of the product constitutes more of a problem than mold growth. The fungal growth manifests within 2 weeks of manufacture, if yeast contamination is not controlled.

The control of yeast contamination is effected by aggressive sanitation procedures related to equipment, ingredients, and plant environment. CIP chemical solutions should be used with special attention to their strength and proper temperature. Hypochlorites and iodophors are effective sanitizing compounds for fungal control on the contact surfaces and in combating the environmental contamination. Hypochlorites at high concentrations are corrosive. Iodophors are preferred for their noncorrosive property as they are effective at relatively low concentrations.

Yeast and mold contamination may also arise from starter, packaging materials, fruit preparations, and packaging equipment. Organoleptic examination of yogurt starter may be helpful in eliminating the fungal contamination therefrom. If warranted, direct microscopic view of the starter may reveal the presence of budding yeast cells or mold mycelium filaments. Plating of the starter on acidified potato dextrose agar would confirm the results. Avoiding contaminated starter for yogurt production is necessary.

Efficiency of equipment and environmental sanitation can be verified by enumeration techniques involving exposure of poured plates to atmosphere in the plant or making a smear of the contact surfaces of the equipment, followed by plating. Filters on the air circulation system should be changed frequently. Walls and floors should be cleaned and sanitized frequently and regularly.

The packaging materials should be stored under dust-free and humidity-free conditions. The filling room should be fogged with chlorine or iodine regularly.

Quality control checks on fruit preparations and flavorings should be performed (spot checking) to minimize yeast and mold entry into fruit-flavored yogurt. Refrigerated storage of the fruit flavorings is recommended.

Quality control programs for yogurt include control of product viscosity, flavor, body and texture, color, fermentation process, and composition. Daily chemical, physical, microbiological, and organoleptic tests constitute the core of quality assurance. The flavor defects are generally described as too intense (acid), too weak (fruit flavor), or unnatural. The sweetness level may be excessive, weak, or may exhibit corn syrup flavor. The ingredients used may impart undesirable flavors such as stale, metallic, old ingredients, oxidized, rancid, or unclean. Lack of control in processing procedures may cause overcooked, caramelized, or excessively sour flavor notes in the product. Proper control of processing parameters and ingredient quality ensure good flavor. Product standards of fats, solids, viscosity, pH (or titratable acidity), and organoleptic characteristics should be strictly adhered to. Wheying off or appearance of watery layer on the surface of yogurt is undesirable and can be controlled by judicious selection of effective stabilizers and by following proper processing conditions.

1.6.2 Frozen Yogurt

In hard-pack frozen yogurt, a coarse and icy texture may be caused by formation of ice crystals due to fluctuations in storage temperatures. Sandiness may be due to lactose crystals resulting from too high levels of milk solids. A soggy or gummy defect is caused by too high a milk-solids-not-fat level or too high sugar content. A weak body results from too high overrun and insufficient total solids.

Color defects may be caused by the lack of intensity or authenticity of hue and shade. Proper blending of fruit purees and yogurt mix is necessary for uniformity of color. The compositional control tests are fat, moisture, pH, and overrun, and microscopic examination of yogurt culture to ensure desirable ratio in LB and ST. Good microbiological quality of all ingredients is necessary.

1.7 Physicochemical, Nutritional, and Health Properties of Yogurt

Conversion of milk base to yogurt is accompanied by intense metabolic activity of the fermenting organisms ST and LB. Yogurt is a unique product in that it supplies the consumer vital nutrients of milk as well as metabolic products of fermentation along with abundant quantities of live and active yogurt cultures. As a result of culture growth, transformation of chemical, physical, microbiological, sensory, nutritional, and physiological attributes in basic milk medium is noted.

To appreciate the nutritional and health properties of yogurt, an understanding of the transformation of milk into yogurt is necessary. We shall first look at major changes brought about during the yogurt process including those by the bacteria, followed by specific health benefits documented in the scientific literature.

Loones[21] summarized changes in the milk constituents during yogurt manufacture. The changes are related to various steps in the yogurt process. Figure 1.10 should facilitate grasping the changes at various stages of transformation of yogurt mix to yogurt: prefermentation, fermentation, and postfermentation.

1.7.1 Prefermentation Changes

1.7.1.1 Mix Preparation

Standardization of milk for fat content and milk-solids-not-fat in yogurt industry results in fat reduction and an increase of 30 to 35% lactose, protein, mineral, and vitamin content. Nutrient density of yogurt mix is concentrated and, thereby, con-

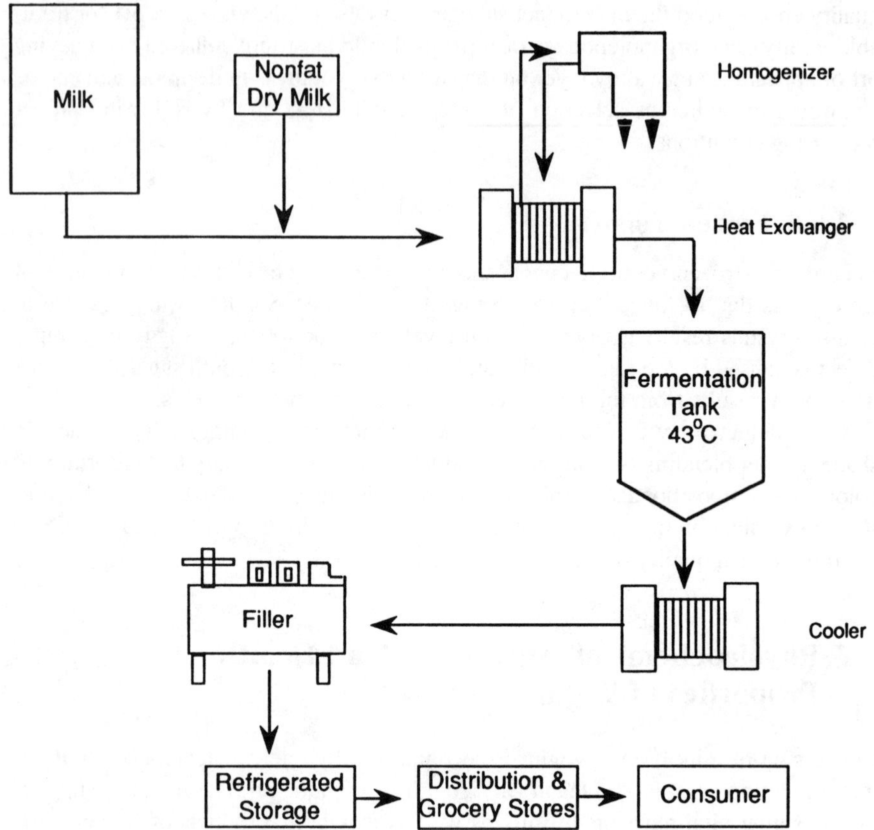

Figure 1.10 Key steps in yogurt processing related to major transformation of milk components. (From ref. 21.)

siderably higher than that of milk. Specific gravity changes from 1.03 to 1.04 g/ml at 20°C. Addition of stabilizers (gelatin, starch, pectin, agar, alginates, gums, and carrageenans) and sweeteners further impacts physical properties.

1.7.1.2 Heat Treatment

Yogurt processing requires intense heat treatment which destroys all the pathogenic flora and most vegetative cells of all microorganisms contained therein. In addition, milk enzymes inherently present are inactivated. Consequently, shelf life of yogurt is assured. From the microbiological standpoint, destruction of competitive organisms produces conditions conducive to the growth of desirable yogurt bacteria. Furthermore, expulsion of oxygen, creation of reducing conditions (sulfhydryl generation), and production of protein-cleaved nitrogenous compounds as a result of heat processing enhance the nutritional status of the medium for growth of the yogurt culture.

Physical changes in the proteins as a result of heat treatment have a profound effect on the viscosity of yogurt. Evidently whey protein denaturation, of the order of 70 to 95%, enhances water absorption capacity, thereby creating smooth consistency, high viscosity, and stability from whey separation in yogurt.

Nutritional changes include ease of digestion of denatured whey proteins in the gastrointestinal tract, soft curd in the stomach, and rapid gastric emptying rate attributed to viscous nature of yogurt.

1.7.1.3 Homogenization

Homogenization treatment reduces the fat globules to an average of <1 μm in diameter. Consequently, no distinct creamy layer (crust) is observed on the surface of yogurt produced from homogenized mix. In general, homogenized milk produces soft coagulum in the stomach, which may enhance digestibility.

1.7.2 Changes During Fermentation

1.7.2.1 Carbohydrates

Lactose content of yogurt mix is generally around 6%. During fermentation lactose is the primary carbon source, resulting in approximately 30% reduction. However, a significant level of lactose (4.2%) survives in yogurt. One mole of lactose gives rise to 1 mole of galactose, 2 moles of lactic acid, and energy for bacterial growth by the Embden–Meyerhof–Parnas pathway (Fig. 1.11).

Some strains of ST exhibit both β-galactosidase and phospho-β-D-galactosidase activity. Therefore, these strains also use a phosphoenolypyruvate-phosphotransferase system. Lactose is converted to lactose phosphate which is hydrolyzed by phospho-β-D-galactosidase to galactose-6-phosphate and glucose which on glycolysis gives lactic acid.

Although lactose is in large excess in the fermentation medium, lactic acid build up beyond 1.5% acts progressively as an inhibitor for further growth of yogurt bacteria. Normally, the fermentation period is terminated by a temperature drop to 4°C. At this temperature, the culture is live but its activity is drastically limited to allow fairly controlled flavor in marketing channels.

Lactic acid produced by ST is the L-(+) isomer which physiologically is more digestible than the D-(−) isomer produced by LB. Yogurt contains both isomers. The L-(+) isomer is normally 50 to 70% of the total lactic acid. Normal consumption levels of yogurt do not pose a hazard from D-(−) lactic acid, relatively large doses of which have been implicated in toxicity problems in small infants.

Lactic acid production results in coagulation of milk beginning at pH below 5.0 and completing at 4.6. Texture, body, and acid flavor of yogurt owe their origin to lactic acid produced during fermentation.

Small quantities of organoleptic moieties are generated through carbohydrate catabolism, via volatile fatty acids, ethanol, acetoin, acetic acid, butanone, diacetyl, and acetaldehyde. Homolactic fermentation in yogurt yields lactic acid as 95% of the fermentation output. Lactic acid acts as a preservative.

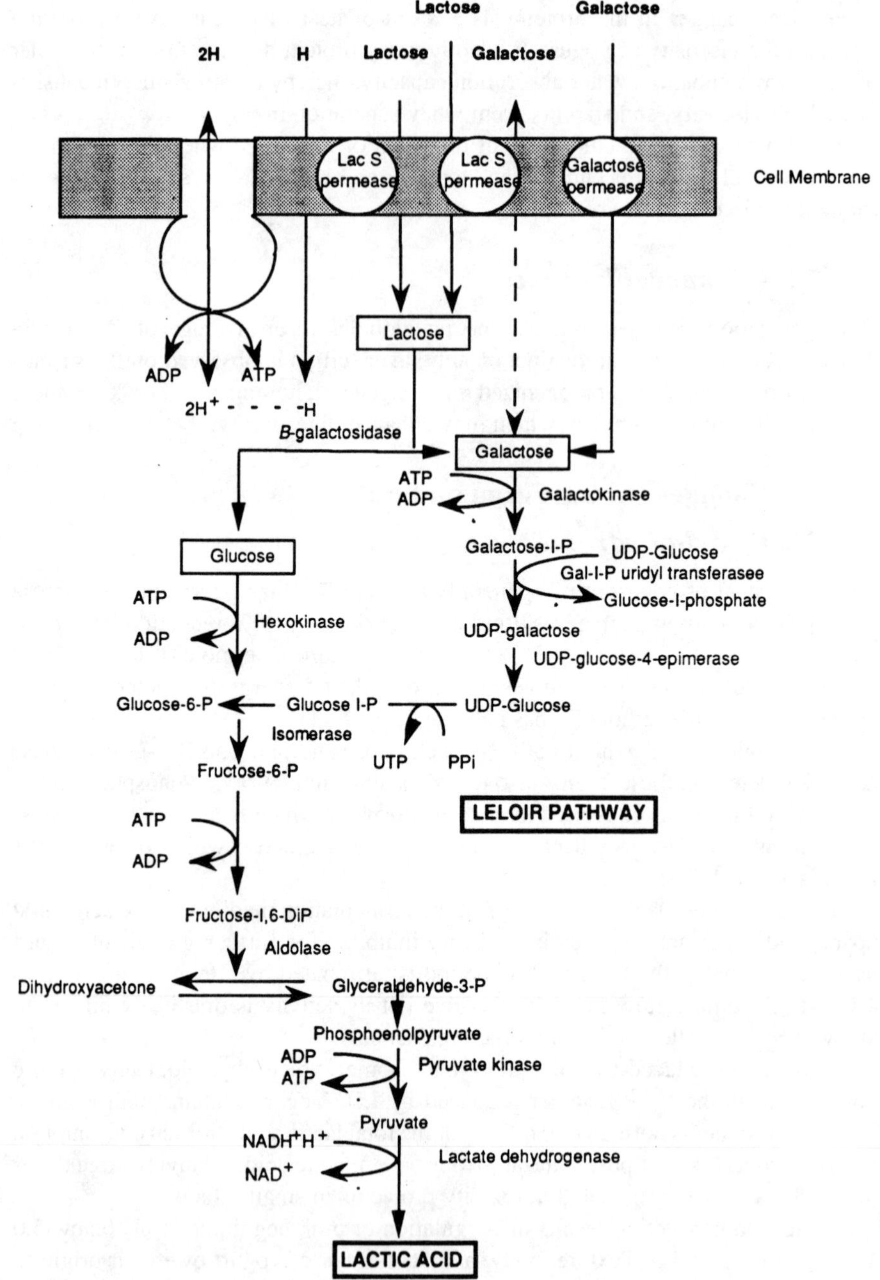

Figure 1.11 Embden–Meyerhof–Parnas Pathway for lactic acid production in yogurt. (From ref. 20.)

1.7.2.2 Proteins

Hydrolysis of milk proteins is easily measured by liberation of $-NH_2$ groups during fermentation. In his review, Loones[21] reported that free amino groups double in yogurt after 24 h. The proteolysis continues during the shelf life of yogurt, with the free amino group doubling again in 21 days of storage at 7°C. The major amino acids liberated are proline and glycine. The essential amino acids liberated increase 3.8- to 3.9-fold during storage of yogurt, indicating that various proteolytic enzymes and peptidases remain active throughout the shelf life of yogurt. The proteolytic activity of the two yogurt bacteria is moderate but is quite significant in relation to symbiotic growth of the culture and production of flavor compounds.

1.7.2.3 Lipids

A weak lipase activity results in the liberation of minor amounts of free fatty acids, particularly stearic and oleic acids. Individual esterases and lipases of yogurt bacteria appear to be more active toward short-chain fatty acid glycerides than toward long-chain substrates. As nonfat and lowfat yogurts comprise the majority of yogurt marketed in the United States, lipid hydrolysis contributes little to the product attributes.

1.7.2.4 Formation of Yogurt Flavor Compounds

Lactic acid, acetaldehyde, acetone, diacetyl, and other carbonyl compounds produced by fermentation constitute key flavor compounds of yogurt. Acetaldehyde content varies from 4 to 60 ppm in yogurt. Diacetyl varies from 0.1 to 0.3 ppm and acetic acid varies from 50 to 200 ppm. These key compounds are produced by yogurt bacteria. Certain amino acids (threonine, methionine) are known precursors of acetaldehyde. For example, threonine in the presence of threonine aldolase yields glycine and acetaldehyde. Acetaldehyde can arise from glucose, via acetyl-CoA or from nucleic acids, via thymidine of DNA. Diacetyl and acetoin are metabolic products of carbohydrate metabolism in ST. Acetone and butane-2-one may develop in milk during prefermentation processing.

1.7.2.5 Synthesis of Oligosaccharides and Polysaccharides

Both ST and LB are documented in the literature to elaborate different oligosaccharides in yogurt mix medium. As much as 0.2% (by weight) of mucopolysaccharides have been observed in a 10-days storage period. In stirred yogurt, drinking yogurt, and reduced-fat yogurt, potential contributions of exopolysaccharides to impart smooth texture, higher viscosity, lower synerisis, and better mechanical handling are possible. Excessive shear during pumping destroys much of the textural advantage because the viscosity functionality property of the mucopolysaccharides is not too shear resistant. Most of the polysaccharides elaborated in yogurt contain glucose and galactose along with minor quantities of fructose, mannose, arabinose, rhamnose, xylose, or *N*-acetylgalactosamine, individually or in combination. The molecular weight is of the order of 0.5 to 1 million. Intrinsic viscosity range of 1.5 to 4.7 dl g^{-1} has been reported for exopolysaccharides of ST and LB.[20] The polysaccharides form a network of filaments visible under the scanning electron microscope. The bacterial cells are covered by part of the polysaccharide and the filaments bind the cells and milk proteins. On shear treatment, the filaments rupture off from the cells, but maintain links with casein micelles. Ropy strains of ST and LB are commercially available. They are especially appropriate for stirred yogurt production.

It is conceivable that some of the exopolysaccharides exert a physiological role in human nutrition because of their chemical structure resembling fiber of grains and vegetables.

1.7.2.6 Other Metabolites

Bacteriocins and several antimicrobial compounds are generated by yogurt organisms. Benzoic acid (15 to 30 ppm) in yogurt has been detected and associated with metabolic activity of the culture. These metabolites tend to exert a preservative effect by controlling the growth of contaminating spoilage and pathogenic organisms gaining entry postfermentation. As a result, the product attains extension of shelf life and reasonable degree of safety from foodborne illness.

1.7.2.7 Cell Mass

As a consequence of fermentation, yogurt organisms multiply to a count of 10^8 to 10^{10} cfu/g. Yogurt bacteria occupy some 1% of volume or mass of yogurt. These cells contain cell walls, enzymes, nucleic acids, cellular proteins, lipids, and carbohydrates. Lactase or β-galactosidase has been shown to contribute a major health-related property to yogurt. Clinical studies have concluded that live and active culture containing yogurt can be consumed by several millions of lactose-deficient individuals in the United States without developing gastrointestinal distress or diarrhea.

1.7.2.8 Minerals

Yogurt is an excellent dietary source of calcium phosphorus, magnesium, and zinc in human nutrition. Research has shown that bioavilability of the minerals from yogurt is essentially equal to that from milk. Because yogurt is a low pH product compared to milk, most of calcium and magnesium occurs in ionic form.

The complete conversion from colloidal form in milk to ionic from in yogurt may have some bearing on the physiological efficiency of utilization of the minerals.

1.7.2.9 Vitamins

Yogurt bacteria during and after fermentation affect the B-vitamin content of yogurt. The processing parameters and subsequent storage conditions affect the vitamin content at the time of consumption of the products. Incubation temperature and fermentation time exert significant balance between vitamin synthesis and utilization by the culture. In general, there is a decrease of Vitamin B_{12}, biotin, and pantothenic acid and an increase of folic acid during yogurt production. Nevertheless, yogurt is still an excellent source of vitamins inherent to milk.

1.7.3 Postfermentation Changes

These changes refer to the shelf life period of yogurt following manufacture.

1.7.3.1 Refrigerated Yogurt

The chain comprised of distribution, marketing, and retail leading to eventual consumption of product by the consumer may require 4 to 6 weeks of shelf life. Nutritional quality is reasonably preserved by temperatures of 4 to 6°C in this chain. Maintenance of product integrity by appropriate packaging is achieved. However, a slight increase in acidity (of the order of 0.2%) is noticeable during this period. Viability of the yogurt culture is also slightly reduced by one log cycle. These changes are relatively minor compared to the changes observed during fermentation.

1.7.3.2 Soft-Serve Mix and Soft-Serve Yogurt

Soft-serve mix may be marketed refrigerated or frozen until dispensed as soft-serve frozen yogurt by the operator. If marketed refrigerated, changes similar to those in refrigerated yogurt are projected in the mix until extrusion through the soft-serve freezer. If marketed frozen, the mix has to be thawed prior to extrusion. A loss of ½ to 1 log cycle in viable cell counts of yogurt culture may be noticed by the freeze–thaw cycle. Further destruction of cell viability is possible during the freezing process through the soft-serve freezer. Other than viable cell counts, no significant changes are known.

1.7.3.3 Hard-Pack Frozen Yogurt

Shelf-life requirements of 6 to 12 months are possible in this type of yogurt. A loss of ½ to 1 log cycle in viable counts may be attributed to the freezing process of the mix. During shelf-life storage conditions, especially fluctuation in temperatures could have deleterious effect on the viability and activity of yogurt cultures. The formation of crystals during frozen state conceivably may rupture bacterial cells, reducing live cell counts progressively.

1.7.4 Prophylactic and Therapeutic Properties

Yogurt dietetically is perceived as a health food of modern times. Historically, culturally rooted legends and anecdotes have characterized the health attributes of yogurt such as exceptional digestibility, curative use for pediatric diarrhea, protection and maintenance of healthy gut ecology, and even longevity. Metchnikoff[27] postulated that LB possesses therapeutic value exercised by suppressing toxin production

Table 1.14 POSSIBLE THERAPEUTIC VALUE OF FERMENTED MILKS, INCLUDING YOGURT

Disease	Comments
A. Human Alimentary Tract Diseases	
Colitis	Spastic inflammation of colon
Constipation	Geriatric use: plain, prune, or bifidus yogurt; acidophilus milk
Deficient microflora	Caused by antibiotic therapy, e.g., infantile diarrhea from *Escherichia coli*; radiotherapy side effects; bifidobacteria in the intestine are enriched (selected for) if the infants diet includes yogurt
Diarrhea	Infantile type induced by antibiotics and microbes; *Lactobacillus* cell preparation for traveller's type
Fistulence	
Gastric acidity	Hypochlorohydria and hyperchlorhydria
Gastroenteritis	
Indigestion	Fermentation improves digestibility of milk
Intoxication	Bacterial toxins
Starvation	Refeeding; sourness and blandness attractive; better digestibility
Stomatitis, gingivitis	Topical use of *Lactobacillus* cells includes herpes etiologic types
B. Other Human Diseases	
Diabetes	Reduced hyperglycemia and hyperglycosuria incidences
Hypercholesteremia	Prophylactic effect of milk ferment
Kidney and bladder disorders	
Lactose intolerance	Yogurt is well tolerated with no symptoms
Liver and bile disorders	Use of yogurt, bifidus milk, bifidogenic factor (lactulose)
Miscellaneous disorders	Catarrh, rheumatism, malaise, migraine, nervous fatigue
Obesity	Dietetic weight loss
Skin disorders	Topical therapeutic or cosmetic use for freckles, wrinkles, sunburn, ulcers and canker; infected cancers
Tuberculosis	Koumiss used in TB sanatoria; yogurt used for liver disorders secondary to extrapulmonary TB
Vaginitis and urinary tract infections	Consumption of yogurt containing acidophilus drastically reduces incidence

Source: Bourlioux and Pochart,[32] Driesen and DeBoer,[33] and Hitchins and McDonough.[34]

of putrefactive bacteria in the human intestine. This conclusion was based on his study of inhabitants of the Balkans, who consumed a rather large quantity of Bulgarian buttermilk in their diets and displayed extraordinary vigor and longevity.

This section summarizes state-of-the-art information related to health attributes of yogurt. Prophylaxis signifies protection or prevention against diseases whereas therapeutic aspects relate to cure following illness. Detailed information on health related effects of yogurt is available in recent literature.[1,28–31]

Table 1.14 lists possible therapeutic usage of fermented milk including yogurt. The list includes suggested applications which are not necessarily supported by scientific data. The possible mode of nutritional and health benefits of yogurt consumption are outlined in Figure 1.12.

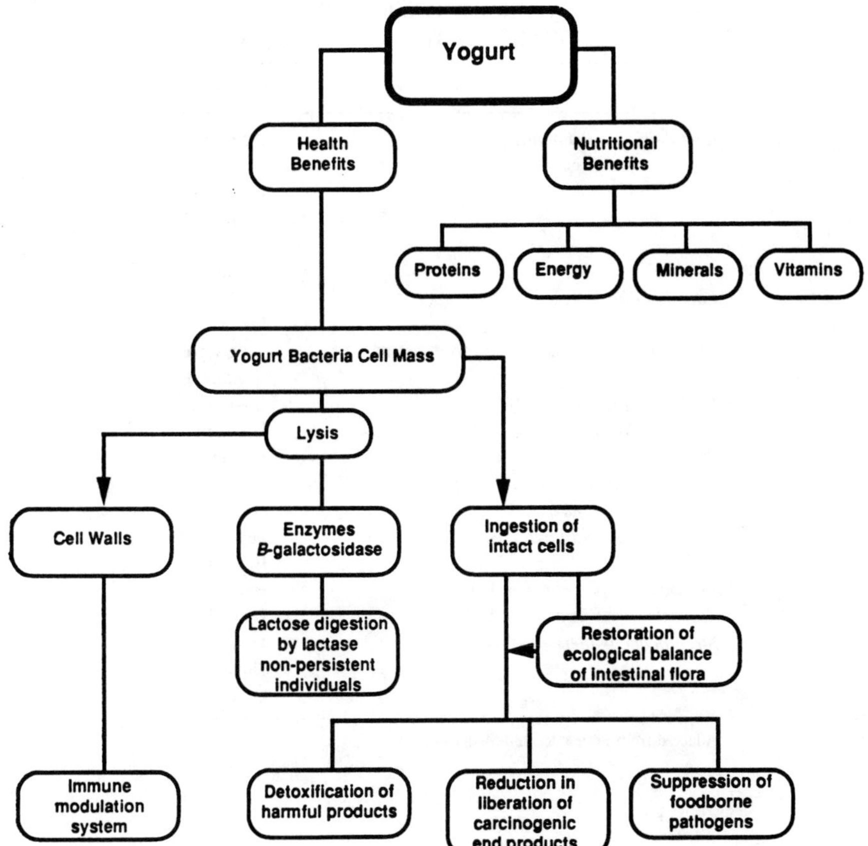

Figure 1.12 Possible mode of nutritional and health benefits via yogurt intake.

1.7.4.1 Antibiosis

The primary prophylactic and therapeutic properties of yogurt seem to be related to the antibiosis of yogurt attributed to fermentation products and bacterial enzymes. The antibiosis due to fermentation products include organic acids, oxidation–reduction (OR) potential, bacteriocins, and antibiotic substances (Table 1.15). The antibiosis due to bacterial enzymes includes bacterial deconjugation of bile salts. More than one factor may be responsible for antibiosis.

Antibiosis due to organic (e.g., acetic, lactic, and propionic) acids is possibly the most important. During growth, as organic acids are produced the acidity increases and pH decreases. The pH is a function of the acid dissociation constant. Organic acids dissociate weakly and the undissociated acidic species is detrimental to foodborne pathogens. The undissociated acidic species can penetrate into the bacterial cell. *S. enteritidis* and *E. coli* are reportedly inhibited by undissociated lactic acid. Also, acidophilus yogurt and traditional yogurt are bacteriocidal to *Yersinia*

Table 1.15 NATURAL ANTIMICROBIAL SUBSTANCES PRODUCED BY LACTIC ACID BACTERIA

Species	Compound
Lactobacillus acidophilus	Acidolin
	Acidophilin
	Lactocidin
	Lactacin B
	Protein
	Unnamed
Lactobacillus brevis	Lactobacillin
	Lactobrevin
Lactobacillus delbruechii ssp. *bulgaricus*	Bulgarican
	Unnamed
Lactobacillus fermenti 466	Bacteriocin
Lactobacillus helveticus LP27	Lactacin 27
481	Helvetican J
Lactobacillus plantarum	Lactolin
Lactococcus lactis	Nisin
Pediococcus acidilactici H	Pediocin AcH
PAC10	Bacteriocin
Pediococcus pentosaceus FBB61	Bacteriocin
L7230	Bacteriocin
Streptococcus thermophilus	Unnamed
	Unnamed
	Unnamed

Adapted from Fernandes and Shahani (1989).[35]

enterocolitca, and the effect is attributed to undissociated lactic acid. Antibiosis due to undissociated acid is efficacious in yogurt in vitro, but may be extremely weak in the gastrointestinal tract, as the high pH would neutralize the acid to its salt form.

Fermentation products produced during growth of yogurt culture lower the oxidation–reduction potential (E_h). A positive E_h favors aerobes whereas a negative E_h favors anaerobes. Some lactic acid bacteria produce hydrogen peroxide in small quantity. Because the gastrointestinal tract is anaerobic, it is doubtful if hydrogen peroxide per se would significantly lower the OR potential. However, hydrogen peroxide may be effective through the lactoperoxidase–thiocyanate system. The hydrogen peroxide oxidizes the thiocyanate to toxic oxidation products that are detrimental to foodborne pathogens.

The antibiosis due to bacteriocins and antibiotic substances may be greater in the gastrointestinal tract than in food systems. Lactobacilli produce bacteriocins with significant bacteriocidal effect toward foodborne pathogens. Although the effect has been elucidated in the in vitro food system its prophylactic role in the gastrointestinal tract has to be proven conclusively. Skepticism has been expressed regarding the antibiosis effect of bacteriocins in the gastrointestinal tract, which functionally abounds with proteolytic enzymes.

Antibiosis may be observed in the gastrointestinal tract due to deconjugation of bile salts, which are more detrimental to the growth of bacteria than conjugated bile salts. Viable lactic acid bacteria (LAB) can deconjugate bile salts in the intestine and thus suppress the foodborne pathogens.

1.7.4.2 Antibiosis and Diarrhea

There is a scientific consensus that LAB are antagonistic toward foodborne pathogens in vitro. The foodborne pathogens produce toxins in food resulting in food intoxication, or may multiply in food to cause infection. LAB may hinder the proliferation of some foodborne pathogens in the food system. Thus establishment of LAB in the gastrointestinal tract may provide prophylactic and therapeutic benefits against intestinal infections. Prophylaxis may have some beneficial role in circumventing travelers' diarrhea.[35]

One cause of gastrointestinal disturbance is the alteration in the intestinal microbiota following invasion or infection by foodborne pathogens. The observed decrease in the coliform count in yogurt has been attributed to the low pH produced by the lactic acid. Some pathogens must establish or colonize the gastrointestinal tract before the onset of the disease. LAB may hinder the colonization and subsequent proliferation of the foodborne pathogens, thereby preventing the disease state. This rationale has been used for treating some gastrointestinal diseases with yogurt. Further, milk products reduce the number or eliminate the foodborne pathogens that have a potential to produce toxin in the food system and the gastrointestinal tract by elaborating antimicrobial substances. The production of antimicrobial substances is dependent on the genera, species, strain, incubation medium, and other conditions.

Cultured milk foods containing viable lactobacilli have been used by humans primarily as a prophylactic aid and their use has been extended to intestinal infections. Dietary lactobacilli also have been used for the treatment of infantile diarrhea. Some scientific evidence exists to suggest that viable lactobacilli contained in fermented milk may be more efficacious to treat gastrointestinal disorders than the administration of antibiotics. More work involving controlled clinical studies using double-blind treatment with viable cultures of host-specific LAB is essential to clarify the therapeutic benefits. Further, diarrheal diseases may exhibit similar symptoms but there may be marked differences in their etiologies. The LAB may have more promise in prophylaxis than in therapy.

1.7.4.3 Cholesterol Reduction

There has been an increasing awareness that serum cholesterol and health are correlated. High serum cholesterol has been linked to an increase in the number of deaths from atherosclerotic heart diseases. Dietary practices have been modified to reduce serum cholesterol. Epidemiological evidence linking large daily intakes of fermented cow's milk (8 L/d) to low serum cholesterol level in Massai warriors exists in the literature.

Clinical results related to cholesterol reduction from human studies have been controversial. McNamara et al.[36] found no changes in serum cholesterol levels in young normolipidemic male subjects consuming low-fat yogurt as a part of an American Heart Association diet (low fat, low cholesterol). The controversial data in the literature may be related to factors such as strains of cultures, lipidemic status of the subjects, exercise–diet relationships, etc.

1.7.4.4 Anticarcinogenic Property

Epidemiological studies suggest that fermented milks suppress the onset of carcinogenesis. The alteration in intestinal microbiota is apparently responsible for the anticarcinogenic attribute. Animal models deployed to delineate the anticarcinogenic role may be broadly divided into prevention of cancer initiation and suppression of initiated tumor.

Prevention of Cancer Initiation

Consumption of yogurt containing viable LAB may reduce the possible initiation of colon cancer. The favorable change in intestinal microbiota can directly and indirectly reduce the conversion of procarcinogens to carcinogens.

Direct Reduction of Procarcinogens

Nitrites used in food processing can be converted into nitrosamines in the gastrointestinal tract. The conversion could possibly be reduced if LAB deplete nitrite through cellular uptake in the gastrointestinal tract. Secondly, bile salts and their derivatives may initiate colon carcinogenesis. *Clostridium, Bacteroides,* and *Eubacterium* are some of the genera that biotransform bile salt from the primary to secondary form. Studies have shown that *L. acidophilus* decreased the rate of conversion of the primary bile acid, chenodeoxycholic acid, to its secondary derivative in vitro. Based on this observation it was extrapolated that high numbers of viable *L. acidophilus* in the gastrointestinal tract may reduce the potential for cancer initiation.

Indirect Reduction of Procarcinogens

Bacterial procarcinogenic enzymes in feces (such as azoreductase, β-glucuronidase, and nitroreductase) are used to monitor mucosal carcinogenesis, as they convert the procarcinogens to carcinogens. The potential for initiation of carcinogenesis increases when the enzyme level is high. Ingestion of some fermented dairy products reduces the level of the enzymes in feces.

Suppression of Initiated Cancer

Feeding or injection of yogurt suppresses the growth of implanted tumors in experimental animals. Tumor growth was suppressed significantly in mice in a short-term study of 1 to 2 weeks but survival rate for rodents was not significantly increased in a long-term study.

Mechanism of Suppression of Tumors

The tumor suppression mechanism has been hypothesized partially in short-term studies. Whole viable and dead cells as well as cell wall fragments suppressed growth initiated tumor, whereas non-cell-wall solids did not show any effect. The antitumor effect is likely to be mediated through the immune response of the host.

1.7.4.5 Lactose Intolerance

The disaccharide lactose of milk is hydrolyzed by lactase and subsequently absorbed in the small intestine. The lactase is a constitutive, membrane-bound enzyme present in the brush borders of the small intestinal epithelial cells. When lactose enters the colon, the colonic flora ferment it generating organic acids, carbon dioxide, and hydrogen. The fermentation products, together with the osmotically driven excessive water drawn into the colon, are chiefly responsible for abdominal pain, bloating, cramps, loss of appetite, diarrhea, and flatulence. These symptoms are associated with lactose intolerance when lactose is not digested in the small intestine.

Lactose Digestion Status

The terms lactase deficiency, lactose intolerance, lactose malabsorption, and milk intolerance have often been used interchangeably. Low lactase activity can be broadly categorized into three main types:

1. Congenital lactase deficiency (or alactasia): In this extremely rare occurrence, lactase is missing throughout life, although the histology of the intestinal mucosa is normal.
2. Primary adult lactase deficiency (or hypolactasia): This condition refers to the normal development of age-related decrease in lactose digestion capacity. In a majority of the world's adult population, intestinal lactase activity is low, which is considered to be normal. Therefore, it has been recommended that primary adult lactase deficiency be renamed as "lactase non-persistence" and that lactase persistence be used to describe the individuals who retain abundant intestinal lactase due to an autosomal dominant trait.
3. Lactase deficiency: This is a transient state of low lactase in previously lactase-persistent individuals following injury to the small intestinal mucosa as a result of disease such as celiac sprue, infectious gastroenteritis, or protein malnutrition.

Lactose malabsorption implies the incomplete digestion of lactose that results in a flat or low rise in blood sugar following a lactose intolerance test. It reflects the outcome, but it is not the primary cause of the condition. Lactose intolerance is defined as the occurrence of clinical signs (diarrhea, bloating, flatulence) or subjective symptoms (abdominal pain, gaseousness) following intake of lactose in a standard lactose tolerance test in a person with proven lactose malabsorption.

Lactose Intolerance and Yogurt

Lactase nonpersistent individuals can ingest a large quantity of yogurt without exhibiting the symptoms associated with lactose intolerance. The increased lactose tolerance may be attributed primariy to the increased lactose digestibility in the gastrointestinal tract by the lactase of ST and LB.

Clinical data have shown that the amount of hydrogen (a measure of lactose intolerance) produced was significantly lower for yogurt than for milk. Diarrhea or flatulence experienced by the individuals ingesting milk is virtually eliminated in the individuals consuming yogurt. Lactase activity in the duodenal area is negligible just after ingestion but increases appreciably for an hour or so following ingestion of yogurt. Accordingly, yogurt is tolerated better by lactase nonpersistent individuals than milk.

Mechanism of Lactose Tolerance

Unheated yogurt containing live and active flora is tolerated better by lactase nonpersistent individuals than pasteurized yogurt.[37] Both yogurt and pasteurized yogurt contain equal concentrations of lactose. Pasteurization of yogurt reduces lactase activity significantly. Thus, viable yogurt organisms and their intact lactase are essential to increase lactose tolerance of the lactase nonpersistent individuals.

Differences in the lactase activity among yogurt cultures are known. The lactase activity also increases in the presence of bile salts which disrupt the bacterial cell and release lactase. To harness the increase in lactose tolerance attribute of yogurt to the full extent, it is recommended to screen the cultures for lactase activity.

Passage of LAB Through the Gastrointestinal Tract

The yogurt culture and the enzyme lactase must survive in the gastrointestinal tract to provide the beneficial properties. The health benefits will be sustained only if yogurt culture and its constituents are not killed or denatured by the acidic environment of stomach, nor lactase is hydrolyzed by the stomach proteases. Indeed, evidence is available to demonstrate that when viable culture is consumed, the dairy constituents offer excellent buffering capacity. Further, because the culture cells in yogurt are conditioned to a low pH environment, their survivability may be higher during their transit through the stomach's acidic environment.

The passage of yogurt culture through the gastrointestinal tract has also been studied in vivo. The survivability of *L. delbruechii* Subsp. *bulgaricus* and *S. thermophilus* may be higher when consumed through yogurt, as yogurt has a higher buffering capacity. The buffering capacity of yogurt and milk is principally due to proteins contained in higher milk solids content of yogurt. Pochart *et al.*[38] also studied the passage of yogurt's bacteria and lactase through the gastrointestinal tract of lactase nonpersistent individuals. The subjects were intubated with a simple lumen tube. Under radioscopic control, the tip of the tube was placed in the third portion of the duodenum. Subjects ingested live and active yogurt or pasteurized yogurt, along with polyethylene glycol and spores of *Bacillus stearothermophilus*. Both *B.*

stearothermophilus and yogurt bacteria were detected in the duodenal tract following ingestion. The *B. stearothermophilus* count was reduced by 1 log cycle, probably due to a dilution effect, whereas the LAB count was reduced by 5 log cycles due to dilution or cellular death. The lactase activity was also detected in the duodenal tract, but the activity was significantly lower after 90 min. This experiment provides the direct evidence for the presence of yogurt organisms and lactase in the gastrointestinal tract.

In a recent article, Marteau et al.[39] reported results of an in vivo study in lactase nonpersistent humans. They fed the subjects yogurt, heated yogurt, and milk to determine lactose absorption patterns. They confirmed that 18 g of lactose fed in yogurt was better absorbed from yogurt (1.7 g of 18 g or 10% of lactose unabsorbed) than from heated yogurt (2.8 g of 18 g or 15% of lactose unabsorbed) or from milk. Using an intestinal perfusion technique, they found a significantly lower amount of intact lactose in the ileum when yogurt was in the diet as compared to heated yogurt or milk. They indicated that >90% of the lactose in yogurt is digested in the small intestine of lactase nonpersistent subjects. They suggested that lactase activity contained in the yogurt culture and a slow oro–cecal transit time associated with yogurt ingestion are both involved in the excellent absorption of the lactose from yogurt.

1.7.4.6 Immune Modulation

Limited evidence has been presented in the reviews mentioned previously to suggest that yogurt may have potential role in augmenting the immune system of the host. Furthermore, the observed enhancement in immunocompetence markers is lost if yogurt is heat sterilized. If corroborated by future studies with humans, this observation could provide a scientific rationale for the suggested use of yogurt in prevention and treatment of gastrointestinal disturbances (diarrhea, enteritis, colitis). Feeding yogurt has been shown to boost serum immunoglobulins (Ig_2a) in mice in comparison with milk feeding. The immune response was not observed with heat-treated yogurt in which viability of yogurt culture was destroyed. In most of the human volunteers, yogurt consumption has been shown to enhance significantly the serum γ-interferon production and a boost in natural killer cell numbers in the peripheral blood. It is postulated that NK cells may further stimulate production of several cytokines helpful in maintaining activity and functional potency of the immune system of the host. This appears to be a plausible mechanism how the defense system of the host may modulate in yogurt mediated control of chronic infections and cancer (tumor) incidence. De Simone et al.[40] further suggested that yogurt bacteria may reduce intestinal monocytes and lymphocytes to elaborate Interluken 1 and 2, which in turn may activate resting NK cells to produce γ-interferon, proliferate NK cell production, and exert cytotoxicity (lymphokine activated killing). Certain strains of lactobacilli appear to exert a positive effect on activation of macrophage functions and antibody production.

Halpern et al.[41] reported the results obtained in a human study of long-term yogurt consumption in young adults. The subjects consumed 16 oz. of yogurt daily for four months. A remarkable increase in γ-interferon production by isolated T cells in subjects consuming yogurt containing live and active cultures was observed. This effect was not noticed in subjects consuming yogurt containing heat-inactivated culture. No negative side-effects were found in the haemotological and blood chemistry values as a result of high levels of consumption of yogurt. In contrast, the researchers found that yogurt consumption resulted in potentially beneficial increases in serum ionized calcium levels.

1.8 References

1. Chandan, R. C. (1982). Chapter 5. Other fermented dairy products. *In* G. Reed (ed.), *Prescott and Dunn's Industrial Microbiology,* 4th edit., pp. 113–184. AVI, Westport, CT.
2. Rasic, J. L. and J. A. Kurmann, (1978). *Yogurt: Scientific Grounds, Technology Manufacture and Preparations*. Technical Dairy Publishing House, Copenhagen, Denmark.
3. Tamime, A. Y., and R. K. Robinson, (1985). *Yogurt Science and Technology*. Pergamon Press, New York.
4. Vedamuthu, E. R. (1991). The yogurt story—past, present and future. *Dairy Food Environ. Sanit.* **11**:202–203, 265–276, 371–374, 513–514.
5. International Dairy Federation (1990). *Consumption Statistics for Milk and Dairy Products (1988)*. Bulletin No. 246/1990. Brussels, Belgium.
6. Milk Industry Foundation. (1990). *Milk Facts*. Washington, D.C.
7. United States Department of Agriculture. (1991). *Dairy Situation and Outlook Yearbook*. USDA Economic Research Service, D5-431. August, 1991, p. 7.
8. Sellars, R. L. (1989). Health Properties of Yogurt *In* R. C. Chandan (ed.), *Yogurt: Nutritional and Health Properties,* pp. 115–144. National Yogurt Association, McLean, VA.
9. Souci, S. W., W. Fachmann, and H. Kraut. (1989–90). *Food Composition and Nutrition Tables,* pp. 51–59. Wissenschaftliche Verlagsgesellschaft MbH, Stuttgart.
10. Grandstrand, D. T. (1989). Yogurt and the regulatory challenge in the U.S.—Current and future. *In* R. C. Chandan (ed.), *Yogurt: Nutritional and Health Properties,* pp. 1–10. National Yogurt Association, McLean, VA.
11. FDA. (1991) Code of Federal Regulation, Title 21, Sections 131.200, 131.203, 131.206. p. 177–181. U.S. Government Printing Office, Washington, D.C.
12. Mareschi, J.-P, and A. Cueff. (1989). Essential characteristics of yogurt and its regulations around the world. *In* R. C. Chandan (ed.), *Yogurt: Nutritional and Health Properties,* pp. 11–28. National Yogurt Association, McLean, VA.
13. International Dairy Federation (1988a). *Fermented Milks: Science and Technology*. Bulletin No. 277/1988. Brussels, Belgium.
14. International Dairy Federation. (1988b). *Yogurt: Enumeration of Characteristic Organisms: Colony Count Technique at 37°C*. IDF Standard No. 117A: 1988. Brussels, Belgium.
15. FDA. (1991). 21 CFR Parts 131 and 135. *Yogurt Products; Frozen Yogurt, Frozen Lowfat Yogurt, and Frozen Nonfat Yogurt; Petitions to Establish Standards of Identity and to Amend the Existing Standards*. Federal Register, Vol. 56, No. 105, May 31, 1991.
16. Ming, X., J. W. Ayres, and W. E. Sandine. (1989). Effect of yogurt bacteria on enteric pathogens. *In* R. C. Chandan (ed.), *Yogurt: Nutritional and Health Properties,* pp. 161–178. National Yogurt Association, McLean, VA.

17. *Bergey's Manual of Determinative Bacteriology,* 9th edit. (1986). William & Wilkins, Baltimore.

18. Reinbold, G. W. (1989). *Spare the Rod (or Coccus) and Spoil and Cheese. Dialogue 4(1).* Chr. Hansen's Laboratory, Inc., Milwaukee, WI.

19. Orla-Jensen, S. (1919). *The Lactic Acid Bacteria.* A. F. Host & Sons, Copenhagen.

20. Zourari, A., J.-P. Accolas, and M. J. Desmazeaud. (1992). Metabolism and biochemical characteristics of yogurt bacteria—a review. *Le Lait* 72(1):1–34.

21. Loones, A. (1989). Transformation of milk components during yogurt fermentation. *In* R. C. Chandan (ed.), *Yogurt: Nutritional and Health Properties,* pp. 95–114. National Yogurt Association, McLean, VA.

22. Sanders, M. E. (1989). Bacteriophage resistance and its applications to yogurt flora. *In* R. C. Chandan (ed.), *Yogurt: Nutritional and Health Properties,* pp. 57–67. National Yogurt Association, McLean, VA.

23. American Dairy Products Institute. (1990). *Standards for Grades of Dry Milk.* Bulletin 916, Chicago, IL.

24. Arbuckle, W. S. (1986). *Ice Cream,* 4th ed. AVI, Westport, CT.

25. Larsen, N. E. (1988). *Production of Yogurt.* APV Pasilac AS, Aarhus C, Denmark.

26. Germantown Manufacturing Co. (1991). *Pioneer Stabilizer/Emulsifer in No-Fat Frozen Yogurt.* G-813. Broomall, PA.

27. Metchnikoff, E. (1908). *The Prolongation of Life.* G. P. Putnam and Sons, The Knickerbocker Press, NY.

28. International Dairy Federation (1991). *Cultured Dairy Products in Human Nutrition.* Bulletin No. 255/1991. Brussels, Belgium.

29. Netherlands Institute for Dairy Research. (1989). *Fermented Milks and Health.* Nizo, Arnhem, The Netherlands.

30. Robinson, R. K. (edit.) (1991). *Therapeutic Properties of Fermented Milks.* Elsevier Applied Science, New York.

31. Syndifrais. (1989). *Fermented Milks: Current Research.* John Libbey Eurotext, Paris, France.

32. Bourlioux, P., and P. Pochart. (1988). Nutritional and Health Properties of Yogurt. *World Rev. Nutr. Diet.* **56**:217–258.

33. Driessen, F. M. and R. DeBoer. (1989). Fermented milks with selected intestinal bacteria. A healthy trend in new products. *Netherlands Milk Dairy J.* **43**:367–387.

34. Hitchins, A. D., and F. E. McDonough. (1989). Prophylactic and therapeutic aspects of fermented milk. *Am. J. Clin. Nutr.* **49**:675–684.

35. Fernandes, C. F., and K. M. Shahani. (1989). Modulation of antibiosis by lactobacilli and yogurt and its healthful and beneficial significance. *In* R. C. Chandan (ed.), *Yogurt: Nutritional and Health Properties,* pp. 145–159. National Yogurt Association, McLean, VA.

36. McNamara, D. J., A. E. Lowell, and J. E Sabb. (1989). Effect of yogurt intake on plasma lipid and lipoprotein levels in normal lipidemic males. *Atherosclerosis* **79**:167–171.

37. Savaiano, D. J. (1989). Lactose intolerance: dietary management with yogurt. *In* R. C. Chandan (ed.), *Yogurt: Nutritional and Health Properties,* pp. 215–223. National Yogurt Association, McLean, VA.

38. Pochart, P., O. Dewit, J. F. Desjeax, and P. Bourlioux. (1989). Viable starter culture, β-galactosidase activity and lactose in duodenum after yogurt ingestion in lactase-deficient humans. *Am. J. Clin. Nutr.* **49**:828–831.

39. Marteau, P., B. Flourie, P. Pochart, C. Chastang, J. F. Desjeax, and J.-C. Rambaud. (1990). Effect of microbial lactase (EC 3.2.1.23) activity in yogurt on the intestinal absorption of lactose: an *in vivo* study in lactase-deficient humans. *Br. J. Nutr.* **64**:71–79.

40. DeSimone, C., B. B. Salvadori, E. Jirillo, L. Baldinelli, F. Bitonti, and R. Vesely. (1989). Modulation of immune activities in humans and animals by dietary lactic acid bacteria. *In* R. C. Chandan (ed.), *Yogurt: Nutritional and Health Properties,* pp. 201–213. National Yogurt Association, McLean, VA.

41. Halpern, G. M., K. G. Vruwink, J. Van de Water, C. L. Keen, and M. E. Gershwin. (1991). Influence of long-term yoghurt consumption in young adults. International J. Immunotherapy 7(4):205–210.

CHAPTER

2

Ice Cream and Frozen Desserts

Rafael Jiménez-Flores, Norman J. Klipfel, and Joseph Tobias

2.1 Introduction

The historical aspects of the development of ice cream and the ice cream manufacturing industry will not be discussed here, as that information is available from other sources, particularly earlier books on the subject of ice cream.[1–13] Suffice it to say that the manufacture of ice cream has progressed from a homemaker's art to a sophisticated factory operation; from a largely manual to a more or less automated process; and from a product of variable composition to one whose composition is carefully selected and precisely monitored. From a limited number of options, the ice cream industry has engendered a whole family of products distinguished by a variety of shapes, flavors and flavor combinations, composition, packages, and consistency at serving time.

2.1.1 Steps in the Manufacture of Ice Cream

The basic steps in the manufacture of ice cream are generally as follows:

1. Selection of mix ingredients
   a. Dairy products
   b. Sweetening agents
   c. Stabilizers and emulsifiers
   d. Others including artificial color, flavorings that are incorporated into the mix such as cocoa and chocolate liquor, and in a few instances other optional generally regarded as safe (GRAS) additives
2. Weighing and assembly (mixing) of the ingredients
3. Pasteurization of the ice cream mix
4. Homogenization and cooling of the mix
5. Aging of the mix. If the mix is intended for sale to freezer operators, a mix packaging step is required.

The mix manufacturing process is now completed and the mix is either sold or frozen into ice cream, in-house. Following are the steps in the production of hard ice cream:

1. Selection of flavoring materials and any preliminary preparation of the flavoring for use (e.g., thawing of frozen fruit)
2. Adding flavoring to the mix
3. Freezing
4. When required, adding flavoring to the ice cream as it emerges from the freezer (e.g., incorporation of fruit, nut, candy, or syrup)
5. Packaging
6. Hardening.

When a soft-serve product is frozen in a restaurant or drive-in, the hardening step is omitted and the flavoring is generally restricted to a single flavor per freezer. The freezer is not of the same type as used in the manufacture of hard ice cream (see Vol. III, Chapters 3 and 4). Conventional ice cream requires freezers with rapid agitation during the freezing process and with a mechanism for controlled air incorporation. Certain types of products, such as ices on a stick, are frozen without agitation or air incorporation. Although these products are members of the frozen dessert family, they obviously have a different consistency and mouthfeel.

2.1.2 Ice Cream as a "Generic" Name

In conversation, the term ''ice cream'' may be used ''generically'' to include all products that resemble each other, that is, they are frozen desserts that have been frozen under similar conditions and are similar in appearance and consistency. They may differ in composition and source of food solids, but they are frozen under agitation and have varying amounts of air incorporated. Labeling requirements, however, clearly differentiate between these products. Without exception each must comply with any applicable local, state, and country labeling requirements. Because labeling laws are subject to change, it is important to be aware of all current applicable regulations.

2.1.3 Government Regulations

Regulations pertaining to ice cream and related products may originate at all levels of government, from local to national. This also applies to definitions of products and their composition. Among the regulations requiring compliance are labeling laws, pure food laws, public health regulations, and OSHA and EPA requirements. To determine whether plants comply with legal standards, regulatory agencies inspect plants and examine products for composition and bacterial content. Historically, the dairy industry has been highly regulated and closely monitored; ice cream manufacture is no exception.

The ice cream manufacturer not only must comply with all regulatory requirements, but must also be on guard against infringing on any existing patents and registered trade marks.

2.1.4 Types of Frozen Desserts

Frozen desserts defined under standards of identity (21CFR, see Appendix) include ice cream, ice milk, frozen custard (or French ice cream), fruit sherbet, nonfruit sherbet, water ices (fruit and nonfruit), and mellorine. Under consideration are proposed standards for frozen yogurt (including low-fat and nonfat) and reduced-fat, low-fat, and nonfat ice cream. The reduced-fat ice cream would replace and have essentially the same definition as the present ice milk. The suggested dividing line between reduced and low-fat is 2% fat. The fat content of the nonfat product could not exceed 0.5%. In response to the favorable acceptance of foods with a reduced fat content, many companies are offering a line of products under the designation of "light." It remains to be seen whether both terms, reduced-fat and light, will be acceptable in the event that a reduced-fat ice cream is legally defined. Because additional products may be defined in some states and localities, there is need to consult the appropriate agencies and comply with any applicable regulations before a product is marketed.

Frozen desserts with names that differ from those defined in the federal standards may be encountered. Some may comply with a particular state standard (e.g., Bisque Tortoni in Pennsylvania),[8] some may be nonstandardized products, and some may comply with the federal definition but use a name that conveys a special characteristic of the product. Gelato, for example, may comply with the federal definition but is perceived as being richly flavored, quite sweet, and somewhat softer than ice cream. Other examples are sorbet, a richly fruit flavored water ice, and Italian ice, which is expected to be somewhat coarser in texture. A mousse is perceived as a rich, creamy, and smooth-textured frozen dessert.

Items described as frozen novelties may also belong to one of the defined categories (federal or state) or be nonstandardized. They include coated and uncoated bars with or without sticks, and individual serving or "snack" size frozen desserts in a variety of shapes, combinations (flavoring, syrup, nutmeats, cones, wafers, etc.), and wrappers or packages.

The variety of the types of frozen desserts is illustrated by two other examples. Direct-draw milk shakes, or just plain "shakes," are defined by states to contain a specified fat and total milk solids content. The inclusion of the word "milk" in the name of the product generally triggers a fat content requirement similar to that of fluid milk. The definition of a "shake" may not include a fat content specification but the actual requirements should be ascertained by checking with local authorities. Parevine is a frozen dessert formulated to satisfy certain religious dietary requirements. It is an all-vegetarian product containing no milk solids. In this case, consultation with applicable religious authorities may be useful even if a state definition exists.

2.2 Selection of Ingredients

With some exceptions, ice cream and related products require ingredients that provide fat, milk-solids-not-fat (MSNF), sweetening agents, stabilizers, and emulsifiers.

Exceptions include products such as water ices and sorbets that contain no dairy ingredients, nonfat frozen desserts, and some "natural" products that contain no stabilizer or emulsifier. All ingredients must satisfy the safety and purity requirements of the Food and Drug Administration (FDA), but there is a wide choice of appropriate products from which to choose. The major reasons for selecting specific ingredients may be summarized as follows:

Cost
Availability
Quality
Desired product characteristics (flavor, body and texture, color etc.)
Consumer preference
Protection against heat shock
Desired freezing point
Desire to label as "natural"
Type of finished product (e.g., nonfat, soft-serve, sherbet, etc.)
Economy grade vs. premium product
Handling capability (e.g., availability of liquid storage vats)
Quality of service and technical support
Personal preference of managers

A product that physically resembles ice cream can be made with any number of ingredients, with or without dairy products. However, milkfat, milk-solids-not-fat, sugar with or without other sweeteners, and stabilizers (the traditional ingredients or components of ice cream) impart certain intrinsic properties to the taste and mouthfeel that cannot be easily duplicated by substitution with other ingredients. Thus, the success of new products that depart from the traditional composition depends on either how closely they approach the familiar ice cream properties, or how readily the consumer is willing to accept the new ones.

All of the ice cream components impart certain properties to ice cream. A summary is presented in Table 2.1. In-depth impact of flavoring components was not included in Table 2.1 because this extremely important subject will be addressed separately, later.

2.2.1 Sources of Dairy Products

Dairy products supply milkfat and milk-solids-not-fat (MSNF), sometimes also referred to as serum solids (SS). Because of a difference in composition, a distinction must be drawn between MSNF and whey solids (WS) although both provide nonfat milk solids. The quantitative composition of MSNF obtained from different sources of milk may not be exactly the same, but the same constituents are present, only at times in somewhat different concentrations. The composition of whey solids may also vary in protein, lactose, and mineral content, particularly when one of the approved modified whey products is used. The compositions of MSNF and WS are given in Table 2.2.

Ideally, the choice of ingredients should be based on considerations that address their sensory and bacteriological quality. Certain ingredients may be selected because they provide the functional properties needed to achieve some desired product characteristics. This is the case even to a greater extent with the nondairy ingredients. In practice, economic factors as well as product availability may affect the selection process, but one should never completely lose sight of quality when choosing a particular ingredient.

Sensory and bacteriological quality can be assessed only by actual test. As a general guideline, fresh dairy ingredients should provide the most desirable flavor characteristics, but they can do so reliably only if they actually are of high quality as determined by appropriate tests. As fresh dairy products are quite perishable, one must guard against quality deterioration. Although non-spore-forming bacteria, even if present in high numbers, are usually destroyed by pasteurization, bacterial enzymes released from the dead cells may remain active. Ingredients with low bacterial count therefore are favored because bacteria may cause deterioriation prior to pasteurization and bacterial enzymes may continue to be a problem after pasteurization.

Because ice cream is a precisely formulated product, the exact composition of the ingredients must be known. This information can only be obtained from an analysis, usually for percent fat and total solids (TS). (% TS minus % fat = % MSNF.) In some circumstances a quantitative analysis of casein, total protein, lactose, and ash also may be required. The composition of a number of ingredients is given in Table 2.3. However, due to the variability of the milk supply, these figures provide only a rough guide, something that must be kept in mind when formulating ice cream mixes.

2.2.2 Nonconcentrated Milk Products

Included in this group are milk and skim milk. When a plant is designed to receive raw milk from producers or has some arrangement to obtain fresh milk or skim milk of highest quality possibly from a sister milk plant in an adjoining or nearby location, advantages may be derived from the use of as much of these ingredients as a particular ice cream formulation permits. If freshness is a desired criterion, these ingredients are potentially the freshest, providing they are used without undue delay. It should be noted that the use of fresh milk or skim milk in formulating an ice cream mix is usually desirable but is not an absolute requirement, as other ingredients also can provide good quality milk solids. When working with raw ingredients, one must take precautions against the development of rancid flavor due to lipase (fat splitting enzyme) action. Prolonged exposure and the simultaneous presence of homogenized milkfat are major contributing factors. The subject is addressed further later in this chapter (Sections 2.9.8 and 2.9.9).

Another possible nonconcentrated dairy ingredient, sweet cream buttermilk, requires a source in the immediate vicinity of the ice cream plant and therefore is not commonly used. Even when readily available, it must be handled immediately and carefully to avoid quality problems, particularly those affecting flavor (oxidation).

Table 2.1 SOME PROPERTIES OF ICE CREAM COMPONENTS[a,b]

Component	Functions	Precautions
Milkfat	Imparts a pleasing body and mouthfeel; sensation of richness; pleasing flavor; flavor carrier and contributor to total flavor blend; its emulsion stabilizing membrane undergoes complex rearrangements as a result of homogenization and aging, and through interaction with other mix constituents.	Greasy, churned butter sensation due to excessive shear or faulty homogenization; source of rancid flavor due to lipolysis; source of oxidized, tallowy, cardboardy flavor
Milk-solids-not-fat (MSNF) (skim milk solids)	Level should be chosen to ''balance'' the formula to complement the flavor and yield the desired body; proteins are involved in interactions and complex formations with other mix constituents; ''structure'' buildup of ice cream; stabilization of fat emulsion; whipping properties; water binding; gel formation; meltdown properties. Lactose provides minimal sweet taste and is a reducing sugar. Minerals significantly affect the colloidal casein structure. Lactose and minerals are soluble solids that contribute to freezing point depression.	Excessive lactose may crystallize and produce a sandy texture; source of ''condensed milk,'' cooked and caramel flavor; proteins are believed to be involved in the ''shrinkage problem''; excessive levels may impart a salty taste; an imbalance of specific ions may affect protein behavior and consequent product properties.
Whey solids (WS)	Used as a partial replacement for MSNF; most commonly used for economic reasons. Modified wheys (e.g., products obtained by membrane filtration) have a higher protein content and potentially new functionality. Lactose and minerals are soluble solids that contribute to freezing point lowering.	Unless modified, have lower protein and higher lactose content than MSNF; there are legal limitations on permitted use level; excessive levels may impart ''whey flavor,'' salty taste, and encourage sandiness. Since the physical, chemical, and functional properties of whey proteins differ from those of casein, their optimum level should be determined for specific product applications.
Sucrose	Provides soluble solids and sweet taste; is pure sweet—has no other tastes or odors; complements flavors and contributes to desirable flavor blends; contributes to freezing point depression and body characteristics.	When used in excess, product becomes too sweet and possibly too soft; the opposite is true when sugar level is too low.
Dextrose (glucose or corn sugar)	Provides soluble solids and sweet taste; not as sweet as sucrose; lowers freezing point to a greater extent than sucrose.	Ice cream tends to be softer when drawn, and softer and faster melting after hardening.

(Continued)

Table 2.1 *(Continued)*

Component	Functions	Precautions
Corn syrup solids	Table 2.6 provides additional details on a number of products in the corn syrup group; provide soluble solids and sweetness; the group offers a choice of sweetening power levels, varying degrees of freezing point lowering, and water binding properties; convenient means for increasing total solids and improving heat shock resistance; provide one of the means by which the body of the ice cream may be ''tailored'' to possess desired characteristics.	Some may lack the desired bland flavor and impart a ''syrupy,'' caramel, or cereal flavor; there may be instances of a subdued flavor release in the ice cream.
Stabilizers	Affect tactile properties, thus aid in attaining the desired type of body; produce high viscosity in the unfrozen serum which should assist in maintaining a smooth texture; discouragement of the growth of lactose crystals by gums has been advanced as a possible reason for the reduced incidence of the sandy defect. Table 2.7 contains details on a number of thickening agents.	Excess may cause a gummy body, poor melting properties, and possibly flavor masking; some may cause whey separation in the mix and melted ice cream.
Emulsifiers	By ''controlled'' deemulsification, they promote a drier appearing product at the freezer outlet and a sensation of ''richness.'' Smaller air cells and improved whipping properties can also be observed. Additional details are presented in Table 2.7.	Excessive levels may cause churning; may cause excessive whipping in a batch freezer; fat separation may be problem particularly in soft-serve and high-fat-containing mixes; may impart off-flavors, especially if they contain oleic or other unsaturated fatty acids.
Egg yolk solids	Provide characteristic flavor to frozen custard (or French ice cream); provide ''natural'' emulsification.	Egg flavor may not be desired in ice cream; standards of identity limit to $<1.4\%$ in ice cream and $\geqslant 1.4\%$ in frozen custard; contributed emulsifier function may become excessive if used in addition to other emulsifiers.
Total solids content	Provides way to affect body, texture, and heat shock properties; high total solids (especially high fat) and low overrun are typical of premium ice cream	Low total solids and high overrun produce a weak body; too high total solids produce a heavy body.

(Continued)

Table 2.1 *(Continued)*

Component	Functions	Precautions
Bulking agents (maltodextrins, polydextrose, sorbitol, etc.)	Used in special applications to replace conventional solids (e.g., artificially sweetened products)	Level of usage must be carefully established; effect on mix viscosity, flavor, freezing point, hardness, and body characteristics should be evaluated; all specifically applicable regulations must be complied with.
Fat sparing agents (microcrystalline cellulose and proprietary products®)	Used to emulate mouthfeel properties of fat in products containing little or no fat.	Effect on mouthfeel, body, and flavor should be evaluated.
Artificial sweeteners	Used to provide sweetness in place of the conventional sugar(s).	Must use only products approved for the specific application in frozen desserts and comply with all required legal limitations and labeling.
Stabilizing salts (largely food grade complex phosphates)	Generally used only when difficulties attributable to protein stability or salt balance are encountered; by reacting with calcium they promote the disintegration of casein micelles into subunits (see Chapters 1 and 2, Vol. I).	Overcompensation can create new problems; must comply with legally imposed limits on quantities used.
Air (overrun)	Promotes typical body in ice cream; creamy, whipped-cream-like mouthfeel; blunts coldness; ingredient cost decreases as overrun increases.	Too high overrun produces a weak body, lacking resistance; low overrun may produce a too heavy body; standards of identity provide a limit to permissible overrun by requiring a minimum weight per gallon and a minimum weight of total solids per gallon; desired overrun varies in different products, e.g., lower in sherbet, ices, and soft-serve than in hard ice cream.
Flavoring	Consumer acceptance is a major consideration; quality and economics enter into the choice between natural and imitation flavors.	Each flavor has specific requirements that must be carefully monitored; standards of identity address compositional requirements for bulky flavors; body and texture is adversely affected when overrun of the mix portion is very high to compensate for weight of added flavoring substances.

[a] Properties of constituents are also susceptible to the consequences of quality variation and lack of uniformity.

[b] Ice cream properties are also affected by variables in mix processing, freezing, packaging, hardening, and storage.

Table 2.2 APPROXIMATE COMPOSITION OF MILK-SOLIDS-NOT-FAT (MSNF) AND SWEET WHEY SOLIDS (WS)[a,b]

Component	MSNF (%)	WS (%)
Protein[c]	37.6	13.5
Casein[d]	(30.1)	
Whey proteins	(7.5)	(13.5)
Lactose	54.1	77.8
Ash	8.2	8.7
Calcium	(1.31)	(0.83)
Potassium	(1.87)	(2.17)
Phosphorus	(1.01)	(0.97)
Sodium	(0.56)	(1.12)
Magnesium	(0.11)	(0.18)

[a] Values were calculated from *Nutritional Data, Agricultural Handbook 8-1*, 1976, U.S. Department of Agriculture. All calculations are on a moisture- and fat-free basis.

[b] Both MSNF and WS contain varying amounts of additional ash constituents and of the water-soluble vitamins ascorbic acid, thiamin, riboflavin, niacin, pantothenic acid, B_6, folacin, and B_{12}.

[c] Protein determined as N × 6.38 and includes nonprotein nitrogen.

[d] 80% of total protein is assumed to be casein, 20% whey protein.

2.2.3 Concentrated Milk Products

Ingredients in this category may be divided into highly perishable and relatively nonperishable products. The perishable products may have an expected refrigerated (<40°F) shelf life of 7 to 10 days, but the actual shelf life depends on such factors as postpasteurization bacterial contamination and storage temperature. The clear implication is that the quality must be monitored during storage. Although bacterial spoilage is the usual form of deterioration, chemical off-flavors such as stale and oxidized may also develop during storage. Nonperishable dairy ingredients may have a shelf life measured in months but they are subject to chemical deterioration which may affect their flavor and color. Such factors as moisture content of dry products, effectiveness of the air and moisture barrier provided by packaging, storage temperature and humidity, and mold growth on products such as sweetened condensed milk have an effect on the effective shelf life of a given ingredient. Due to their high fat content, frozen ingredients, such as frozen cream and butter, must be processed in such a way as to provide resistance to oxidation, that is, high pasteurization temperature, effective air barrier, and freedom from contamination with heavy metals, particularly copper and iron.

2.2.4 Perishable Concentrated Milk Products

Cream is the most common source of milkfat in ice cream. It usually contains from 30 to 40% fat, but cream of any fat content may be used. Its quality is determined

Table 2.3 APPROXIMATE COMPOSITION OF SOME INGREDIENTS[a,e]

Ingredient	Fat (%)	MSNF (%)	Sweetener (%)	Water (%)	Total Solids (%)	Density[b] (lb/gal)
Milk	3.25	8.35		88.4	11.6	8.59
	3.5	8.5		88.0	12.0	8.59
	3.75	8.65		87.6	12.4	8.6
	4.0	8.8		87.2	12.8	8.61
Skim milk	0.1	8.8		91.1	8.9	8.63
	0.07	9.23		90.7	9.3	8.64
	0.1	8.6		91.3	8.7	8.62
Cream	20.0	7.05		72.95	27.05	8.44
	30.0	6.2		63.8	36.2	8.35
	35.0	5.7		59.3	40.7	8.3
	40.0	5.3		54.7	45.3	8.26
	50.0	4.4		45.6	54.4	8.17
	80.0	1.75		18.25	81.75	7.92
Condensed skim milk	0.3	29.7		70.0	30.0	9.39
	0.3	31.7		68.0	32.0	9.47
	0.4	33.6		66.0	34.0	9.54
	0.4	35.6		64.0	36.0	9.62
Sweetened condensed skim milk	0.3	29.7	42.0	28.0	72.0	11.4
Sweetened condensed milk	8.5	21.5	42.0	28.0	72.0	10.86
Evaporated milk	7.5	18.0		74.5	25.5	8.9
Concentrated milk	10.0	25.0		65.0	35.0	9.13
Butter	80.5	0.75		18.75	81.25	
Anhydrous butteroil	99.9			0.1	99.9	
Dry whole milk	26.0	72.0		2.0	98.0	
Nonfat dry milk	1.0	96.0		3.0	97.0	
Dry whey (sweet)	1.0	96.0		3.0	97.0	
Dry buttermilk	5.0	92.0		3.0	97.0	
Liquid sugar (67.5 Brix)			67.5	32.5	67.5	11.10
Invert syrup (40.77 Be)			76.5	23.5	76.5	11.58
36 DE corn syrup[a,c]			79.6	20.1	79.9	11.81
Dried egg yolk[d]	62.5			6.0	94.0	
Chocolate liquor	51.0			5.0	95.0	
Cocoa	8–30			5.0	95.0	
Water						8.34

[a] Additional sweeteners are listed in Table 2.6.

[b] All values are temperature dependent. Density of milk products was calculated by the following equation:

$$\text{Specific gravity at } 60^{\circ}\text{F } (15.6^{\circ}\text{C}) = \frac{100}{\frac{\%\text{ fat}}{0.93} + \frac{\%\text{ all other solids}}{1.601} + \frac{\%\text{ water}}{1}}$$

$$\text{Density (lbs/gal)} = \text{specific gravity} \times 8.34$$

[c] Density at 100°F.

[d] Approximate composition of egg yolk is 51% H_2O, 33% fat, 15% protein, and 1% ash. Frozen egg yolk is commonly supplied in sweetened form (e.g., 10% sugar). Egg whites contain approximately 85% H_2O, 12% protein, and small quantities of fat, sugar, and minerals.

[e] For definitions and ingredient listing see CFR in the Appendix.

by the quality of the milk from which it was separated and by the level of adherence to good manufacturing practices during its processing and storage.

Condensed skim milk, when available and made from high quality skim milk, is an excellent source of MSNF. Properly separated skim milk yields a condensed product with a negligible amount of fat and 30 to 35% MSNF. The product is made by concentrating skim milk under vacuum until the desired concentration of solids is reached. The limit of concentration is about 35% if the danger of lactose (milk sugar) crystallization is to be avoided. The increased concentration of solids does not render the product less perishable than nonconcentrated skim milk. Therefore, it must be refrigerated and continuously checked for any signs of deterioration.

A related product is superheated condensed skim milk, which is equally perishable but has enhanced water binding properties that may be beneficial for the body and texture of the ice cream. An increased mix viscosity is immediately apparent when this ingredient is used. Superheating is accomplished by heating the condensed skim milk to a temperature of 180 to 190°F and holding until the desired "livery" body develops. Cooling must follow immediately and very rapidly to prevent a complete breakdown into protein and whey. The preheating temperature prior to concentration of the skim milk affects the rapidity of the subsequent superheating effort. Preheating temperatures above 150°F may slow or inhibit the viscosity buildup. The flavor imparted by this ingredient may be expected to be somewhat cooked or "custardy," but that does not necessarily presage a problem in acceptance. Although cooked flavor is technically considered a flavor defect in ice cream, it is the "burnt," "scorched," or "caramelized" variety of the off-flavor that is much more offensive. Many years ago, superheated condensed skim milk was in common use by ice cream makers, but its popularity has declined over the years. Lack of ready availability is one of the problems, but history frequently repeats itself and this product may be "rediscovered."

Condensed products can also be made with partially skimmed milk in which case they contribute some fat in addition to the MSNF. From the point of view of the technologist, there are no special concerns over the presence of fat providing there is no difference in quality. However, if the ingredient is to be used in the manufacture of a nonfat frozen dessert, only skim milk products can satisfy the formulation requirements.

2.2.5 Dehydrated Concentrated Milk Products

By far the most common ingredients in this category are nonfat dry milk (NDM), which is dried skim milk, and dried whey. Ice cream frequently contains both of them, although the content of whey solids is limited by the federal standards of identity to 25% of the MSNF content. Just as in the case of fluid ingredients, quality of powdered products cannot be assumed to be adequate without actual test. United States Department of Agriculture quality grades may be specified in purchasing but prudence dictates that in-house quality tests also be routinely conducted. Some manufacturers may demand even more rigorous quality standards than those specified for U.S. Extra Grade. The U.S. Grade classification for NDM and dried whey is

Table 2.4 U.S. GRADE CLASSIFICATION OF NONFAT DRY MILK[a,b,e]

Basis	U.S. Extra Grade	U.S. Standard Grade
Off-flavors (reliquified)[c]		
Bitter		Slight
Chalky	Slight	Definite
Cooked (spray and instant)	Slight	Definite
Feed	Slight	Definite
Flat	Slight	Definite
Oxidized		Slight
Scorched		
Roller	Slight	Definite
Spray and instant		Slight
Stale		Slight
Storage		Slight
Utensil		Slight
Physical appearance defects[c]		
Lumpy	Very slight	Slight
Unnatural color		Slight
Visible dark particles		
Spray	Very slight	Slight
Roller	Slight	Definite
Instant		
Reliquified		
Grainy		
Spray		Slight
Roller	Slight	Slight
Instant		

(Continued)

summarized in Tables 2.4 and 2.5, respectively. It can be seen that the U.S. Grades refer to both spray- and roller-dried products. In general, the powder made by the roller drying process may be expected to have a more intense cooked or scorched flavor, be somewhat less soluble, and contain more scorched particles. The spray-process product is the ingredient of choice. Flavor is obviously a key consideration but the importance of solubility should not be overlooked, particularly when the high temperature–short time system of mix pasteurization is employed. In this process the powder is dispersed in the cold and undissolved particles could pose a problem by adhering to heating surfaces or damaging the homogenizer valves.

As seen in the footnote to Table 2.4, the U.S. Grade classification also recognizes three classes based on heat treatment, namely U.S. high heat, U.S. low heat, and U.S. medium heat. A chemical test for the amount of undenatured whey proteins (whey proteins are progressively denatured when exposed to heat) provides an objective measure on which the classification is based.[14] High heat powder may be slightly less soluble, as shown by the higher solubility index, but it is the product preferred by bakers because of its favorable effect on loaf volume. The low heat

Table 2.4 *(Continued)*

Basis	U.S. Extra Grade	U.S. Standard Grade
Laboratory tests (or parameters)[d]		
Bacterial estimate, standard plate count per gram		
Spray and roller	50,000	100,000
Instant	30,000	
Milkfat content, %	1.25	1.5
Moisture content, %		
Spray and roller	4.0	5.0
Scorched particle content, mg		
Spray and instant	15.0	22.5
Roller	22.5	32.5
Solubility index, ml		
Spray	1.2	2.0
U.S. high heat	2.0	2.5
Roller	15.0	15.0
Instant	1.0	
Titratable acidity, %	0.15	0.17
Coliform count per gram		
Instant	10	
Dispersibility, %		
Instant	85	

[a] Only one grade, U.S. Extra, is recognized for instant NDM.
[b] Heat classification is as follows:

U.S. high heat ≤1.5 mg undenatured whey protein nitrogen per gram dry product
U.S. low heat ≥6.0 mg undenatured whey protein nitrogen per gram dry product
U.S. medium heat 1.51 to 5.99 mg undenatured whey protein nitrogen per gram dry product

[c] In general, the flavor shall be sweet, pleasing, and desirable; the dry product shall be white or light cream in color; and the intensity of indicated defects or characteristics shall not be exceeded.
[d] All numbers represent permissible maxima except that for dispersibility which is a minimum value.
[e] For more detailed information consult Title 7, Part and section 2858.2601, Subpart 0, Code of Federal Regulations.

product is preferred by the cottage cheese maker. Any of the products may be used in ice cream providing their quality is acceptable and the sensory characteristics that they impart to the ice cream conform to the "design" intended for it. High quality medium heat NDM is usually a good choice.

Instant NDM is easily dispersed in the cold and has more rigorous U.S. Grade requirements than noninstantized powders. Although there are no technological reasons why this product could not be used in ice cream, its increased cost generally prevents its consideration.

As seen in Table 2.2, the major constituents of MSNF are lactose, protein, and milk salts. Because NDM contains moisture and a small amount of fat, it is not quite a pure concentrated source of MSNF. In addition to any differences due to the moisture and fat content, the composition of the milk from which the skim milk was obtained provides another variable. A 36% protein content in NDM is typical but indications are that the range may be from 32 to 38% protein. This would obviously

Table 2.5 REQUIREMENTS FOR U.S. EXTRA GRADE DRY WHEY[a]

Basis	Requirement
Flavor	Shall have a normal whey flavor free from undesirable flavors, but may possess the following flavors to a slight degree: bitter, fermented, storage, and utensil; and the following to a definite degree: feed and weedy.
Physical appearance	Has a uniform color and is free-flowing, free from lumps that do not break up under slight pressure, and is practically free from visible dark particles.
Bacterial estimate	Not more than 50,000 per gram standard plate count
Coliform	Nor more than 10 per gram
Milkfat content	Not more than 1.5%
Moisture content	Not more than 5%
Optional tests:	
Protein content (N × 6.38)	Not less than 11%
Alkalinity of ash (sweet type whey only)	Not more than 225 ml of 0.1 *N* HCl per 100 g
Scorched particle content	Not more than 15 mg

[a] For more detailed information consult Title 7, Part and section 2858.2601, Subpart 0, Code of Federal Regulations.

cause some realignment of the other constituents. The differences in composition may be due to a preponderance of certain breeds of cows and seasonal or regional influences that affect the composition of milk. Aside from their contribution to nutritional requirements, proteins exert an effect on the whipping characteristics and other physical and sensory properties of ice cream. They bind water; interact with stabilizers, other proteins, and carbohydrates; stabilize the fat emulsion after homogenization; and, in general, contribute to the structure of the ice cream and to its mouthfeel characteristics. They are also a source of thiol groups which, when activated by heat, act as antioxidants and as precursors of a significant component of cooked flavor. (Details of protein composition are given in Vol. I.) For this reason, knowledge of the actual composition, source, and processing history of the NDM may provide useful data when, for instance, an explanation is sought for some unexpected changes in the properties of ice cream or unanticipated behavior during freezing and storage (e.g., shrinkage). One should be aware of the variables that can affect the product but that cannot always be controlled.

The question of how long NDM may be kept in storage encompasses another set of variables on which the answer is dependent. Of the compositonal factors, high fat and moisture contents have a limiting effect on stability. The age of the product at time of delivery and initial quality are important criteria. If the powder is marginally acceptable at time of delivery, its useful storage life may be quite short. Other obvious factors include quality of packaging and freedom from damage; clean, dry, and cool storage facilities; and freedom from such pests as insects, rodents, birds, and other animals.

When procuring NDM, the following are important factors to consider: composition; color (very close to white and no brown pigmentation); free-flowing and free of lumps and dark particles; flavor, appearance, and laboratory tests should comply with or exceed requirements for U.S. Extra Grade; freedom from pathogenic bacteria including but not limited to *Salmonella* and *Listeria,* and functional properties when incorporated into the ice cream. Shortages in the supply of NDM in recent years have caused some anxiety, but usually NDM of excellent quality is available. Even Grade A milk powder may be purchased when desired or local regulations require it.

2.2.6 Dry Whey

This ingredient is a byproduct of cheese manufacture. The actual composition of different types of cheese varies but with few exceptions cheese is made up of casein (the main protein of milk) and fat, unless it is a cheese made of skim milk. The residue, called whey, retains the whey proteins (α-lactalbumin, β-lactoglobulin, immunoglobulins, and others), lactose, water-soluble minerals and vitamins, and any residual fat that did not get incorporated into the cheese. The enzyme rennin, or a related enzyme with similar properties, is an essential coagulating agent in most cheeses. Because the enzyme causes the casein to coagulate as a calcium complex, some of the milk calcium will be missing in whey.

The flavor of whey and dry whey is affected by the quality of the milk originally used in cheesemaking and the care exercised in handling and processing of the whey after removal from the cheese vat. The dry whey is also subject to the usual factors involved in the quality deterioration of dry milk products. On storage, the color may progressively darken and the flavor may become stale or cereallike. When whey imparts an off-flavor to ice cream, usually the simple designation ''whey flavor'' is used to describe it. The requirements for U.S. Extra Grade Dry Whey are given in Table 2.5. Ice cream manufacturers may choose to specify more rigorous minimum quality criteria in procuring this ingredient. The absence of pathogenic bacteria, use of protective packaging, and other factors should also be part of the specifications. Spray-dried whey is the most common form of the product. Also available are several modified whey products, some of which have a protein content as high or higher than NDM (see CFR in the Appendix). The functional properties of these products should be evaluated for possible advantages in specific applications.

2.2.7 Dried Buttermilk

Dried sweet cream buttermilk is an acceptable ingredient for ice cream providing it is free of off-flavors. In its original form it is the liquid remaining after churning butter with a composition similar to skim milk. There are, however, some subtle but important differences in fat concentration and composition. When butter is churned, the fat is stripped of some of its phospholipid-rich membrane which is then lost to the buttermilk. The consequences of the higher concentration of phospholipids in the buttermilk are both desirable and undesirable. Their ability to act as an emulsifier is an advantage but their susceptibility to oxidation can present a problem. The flavor of dry buttermilk must therefore be carefully monitored to ensure that oxidation has not made the use of the product inadvisable. In an effort to control the problem, some have tried to set limits on the quantity of buttermilk powder that can be safely used in a given formulation. This proposal is too general because when the quality of the powder is poor, it is best not to use it at all. (This principle is actually applicable to all ingredients.) There is a U.S. Grade classification for dry buttermilk which may be found in the Code of Federal Regulations, Title 7, Part 58.

2.2.8 Other Dry Ingredients

Any dried dairy product such as dry whole milk, low-fat dry milk, and dry cream could theoretically be used in ice cream, but in practice this is seldom if ever done. The main reason is that dry dairy products containing fat readily develop stale and oxidized flavors even when stored under ideal conditions. Thus, they provide no advantage, but considerable risk, over NDM.

Edible forms of casein salts, such as sodium or calcium caseinate, are common ingredients in a number of food products, but their use in ice cream is limited by provisions of the federal standards of identity (see Appendix). They are classified as optional ingredients that do not satisfy the prescribed total milk solids requirements but may be added to a mix, within permissible limits, if used above and beyond these requirements.

According to Turnbow et al.,[13] sodium caseinate has a definite effect on mix whipping. They stated that home recipes made with this product and unhomogenized cream could be easily whipped to 100% overrun with no more elaborate equipment than that found in the household kitchen. They hypothesized that the effect was due to the creation of more elastic air cell walls which could be more resistant to rupture by large or clumped fat globules.

2.2.9 Preserved Fluid Concentrated Milk Products

The two products in this category, sweetened condensed milk and evaporated milk, are no longer common ingredients in ice cream, but may still be encountered. Sweetened condensed milk or skim milk is preserved by the addition of sufficient sugar to prevent bacterial growth, but problems with certain osmophilic bacteria, yeasts, and molds may at times be encountered. The undesirable development of large lactose crystals may be prevented by correct manufacturing steps at the condensery. Depending on the length of storage, the product may also brown and thicken, particularly at storage temperatures above 60°F.[15]

Evaporated milk is preserved by sufficient heat to sterilize milk from which about one half of the water had been removed. The use of this ingredient is not suitable for large-scale commercial operations but may be encountered in small ice cream shops where a ''home style'' ice cream is made and sold. A cooked, custardlike flavor is typical in ice cream made with evaporated milk.

2.2.10 Frozen Concentrated Milk Products

Concentrated sources of MSNF in the frozen state are not practical ingredients for two reasons: the freezing process creates problems in physical stability and NDM is an alternative that offers substantial advantages over them. On the other hand, concentrated sources of fat may be successfully preserved by freezing and are useful in geographic areas where fresh cream is unavailable or only seasonally available. The three forms of frozen concentrated fat ingredients—frozen cream, butter, and butter oil—share some precautions in their preparation. As would be suspected, fat oxidation is the principal concern. Preparation steps, therefore, must include provisions to delay or discourage oxidation, that is, avoiding contamination with copper and iron, pasteurization at high temperature (≥170°F for 30 min), and packaging to exclude air. Success in the use of these frozen ingredients also depends on the quality of milk and cream from which they were made. For best results, only perfectly sweet cream with no objectionable off-flavors should serve as the starting material. Addition of 12% sucrose may improve stability of frozen cream. In the case of butter, the unsalted variety is preferred. Correctly made anhydrous butter oil, with only a trace of retained moisture, has been found to resist oxidation and keep quite well. In all cases, however, the frozen storage temperature must be uniformly low to prevent flavor deterioration. Butter oil and NDM have been used as the only dairy ingredients for ice cream in remote areas separated by thousands of miles from their source.

2.2.11 Substitutes for Dairy Products

Mellorine and unstandardized ice cream ''analogs'' may incorporate vegetable fat. In addition, vegetable protein, protein concentrates, and other vegetable derivatives may be found in unstandardized products. A definition and standard of identity for Mellorine may be found in Part 135 of the Code of Federal Regulations. The product is essentially analogous to ice cream or ice milk except that milkfat is replaced with vegetable or animal fat. The language for the requirements of the MSNF simply directs that the product must contain 2.7% milk-derived protein having a protein efficiency ratio (PER) not less than that of whole milk protein, 108% of casein.

Quality as well as functionality are criteria in selecting milkfat substitutes.[16] Physical properties of importance include the melting point (or melting region), rate of crystallization, type of crystal structure, mixed-crystal formation, and extent of super-cooling before crystallization begins. The chemical makeup of the fat, that is, fatty acid composition and their distribution within the triglycerides, is largely responsible for these properties. Milkfat does not have a sharp melting point but rather melts over a range of temperatures, so that at various points of the melting cycle different proportions of solid and liquid fat are present. To emulate this behavior with other fats, one would have to use a blend of fats with different melting points.

Liquid oils, depending on the concentration used, affect the consistency (hardness) as well as whipping ability of the finished product. The literature[16] indicates that their globular structure is unstable during freezing as evidenced by the lack of fat globules when examined by electron microscopy. The flavor of the nondairy fat used should be bland, free of absorbed soapy and oxidized flavors, and it must not contribute to a greasy mouthfeel in the finished product. Close cooperation with a reputable supplier of fat should lead to identifying a product that satisfies both functional and quality requirements.

When nonmilk products are desired in place of MSNF, a good deal of developmental work should precede introduction of the product to ensure that consumer acceptance criteria are met with regard to flavor, body, and texture of the frozen analog. The flavor concerns address both the flavor of the ingredient, for example, soy protein isolate, and the compatibility of the flavoring used. A few products of this type may be found in the marketplace and the quality of some has been rated by the authors as very good.

2.2.12 Sweetening Agents

The functions of sweetening agents are to provide the desired level of sweet taste; as a source of food solids that contribute to the total solids content of the ice cream; as a means of controlling the freezing point and hence the stiffness of ice cream when discharged from the freezer and at any given storage temperature; and as a water binding agent to promote a smooth textured ice cream and one that resists excessive growth of ice crystals as a result of high and fluctuating storage temperatures (heat shock). Because not all of the sweetening agents contribute to all of the functions equally, the technologist must exercise good judgment in selecting the most appropriate sweetener combinations.

The sweetening power of cane or beet sugar (sucrose, which is common table sugar) has become the standard to which the sweetening power of other sugars is compared. To express sweetness numerically, sucrose may be given a value of 100 and the sweetening power of other sweeteners (e.g., corn syrup) can be experimentally compared to it. The determined numerical sweetening value represents a comparison to the sweetening power of sucrose. A sweetener whose sweetening value is 50 has only half the sweetening power of sucrose and twice as much of it would have to be used in ice cream to equal the sweetness level imparted by sucrose. The experimental determination of sweetening power is somewhat complicated by the presence of other flavor notes in some sweetening agents. Further complications are that the results are affected by the concentration of sweetener at which the comparison is made, the background flavor, temperature, and the proportion of different sweeteners present. Turnbow[13] has reported that the sweetening power of dextrose and corn syrups is greater in ice cream than in water solution. The taste comparisons should be made directly in the ice cream at the desired sweetness level. Because of all the complicating factors, the sweetening power data in Table 2.6 should be used only as preliminary guidelines in the search for the appropriate usage level of different sweeteners.

Table 2.6 COMPOSITION AND SOME PROPERTIES OF SWEETENERS (NUMERICAL VALUES ARE APPROXIMATE)[f]

Sweetener	Sweetening[a] Power in Ice Cream (%)	Theoretical[b] Freezing Point Reduction Factor	Mean[b] Molecular Weight	DE	Total Solids (%)	Dextrose (% dry basis)	Maltose (% dry basis)	Triose (% dry basis)	Higher saccharides (% dry basis)	Fructose (% dry basis)
Sucrose[c]	100	1	342							
Dextrose[c]	75	1.9	180							
Fructose[d]	115	1.9	180		77	0.5				99.5
High-fructose corn syrup (42%)	100	1.77	193		71	50	1.5	trace	5	42
High-fructose corn syrup (55%)	110	1.85	185		77	41			4	55
Maltodextrins	0–10	<0.31	>1100	<1–<20		0.3–1.6	0.1–6	0.2–8	remainder	
Corn syrup[d]	25	0.49	700	26	77.5	5	8	11	76	
Corn syrup[d]	45	0.61	557	36	80	13	10	11	66	
Corn syrup[d]	50	0.77	447	42	80.3	19	14	13	54	
Corn syrup	70	1.18	289	64	81.6	37	29	9	25	
High-maltose corn syrup	55	0.8	430	42	80.4	9	34	24	33	
High-maltose corn syrup	55	0.92	374	50	80.7	10	42	22	26	
Invert syrup	110	1.9	180		76.5	50				50
Lactose	20	1	342							
NutraSweet®	(150–200)[e]									
Polydextrose		0.6–0.75								

[a] On dry matter basis, but approximate. Sweetening power may not be the same in different applications. Complications encountered in determining the sweetening power are discussed in the text.

[b] Mean molecular weight and the theoretical freezing point reduction factor are a function of the actual concentration of the saccharides. The theoretical freezing point reduction factor may be somewhat more complex than indicated. More refined values for specific products may be available from the products' suppliers.

[c] Also available in syrup form.

[d] Also available in dry form.

[e] Sweetness intensity is 150–200 times that of sucrose.

[f] Analytical data on corn derived sweeteners courtesy of A. E. Staley Mfg. Co. Decatur, IL 62525.

Certain ingredients used in frozen desserts, such as maltodextrins, are included in the sweetening agent category, although they provide little or no sweetening properties. However, they share a common origin with some sweetening agents and they fulfill some of the same functions.

The fact that some sweeteners are not as sweet as sucrose provides the opportunity for increasing the total solids content of the ice cream without imparting excessive sweetness. The information in Table 2.6 reveals that some of the corn-derived sweeteners not only are less sweet than sucrose but they also lower the freezing point to a lesser extent and bind more water (in an inverse relationship to their sweetening power). These facts must be judiciously applied in formulating ice cream and other frozen desserts, a subject that will be discussed under a separate heading (Section 2.4).

Storage facilities for sweetening agents at the ice cream plant should be given careful consideration. Dry sweeteners must be packaged in sound containers that provide protection against contamination and moisture intrusion. Some dry sweeteners are very hygroscopic. The warehouse should be clean, dry, cool, and free of both insect and animal pests. Good housekeeping practices must not be confined to the warehouse; when bags of sweeteners are moved to the processing area, good housekeeping and sanitary precautions should guide the opening and emptying of the bags so as to avoid spilling and contamination of the mix with fragments of paper bags or any personal articles that the workers may accidentally drop into the vat. Unused portions of sweetener should be carefully sealed, identified, and returned to the warehouse.

Larger plants generally use bulk liquid sweeteners whenever they are available. To do so, they must have liquid sugar (syrup) tanks for each of the different sweeteners that they employ. These tanks must maintain the syrups at the correct temperature for easy handling and contain protective mechanisms against the growth of yeasts and molds. Before accepting a shipment of bulk sweetener, it is prudent to ensure that it complies to specifications, particularly with regard to color (should be free of any browning), composition, and freedom from microbiological fermentation. Dedicated tank-trucks for the transport of syrup are in common use. The authors are aware of a case where the syrup was delivered in a milk tank-truck, which unfortunately had transported rancid milk before picking up the syrup. The syrup absorbed the residue of the rancid flavor and imparted it to the ice cream. This simply emphasizes the requirement of clean and odor-free transport vehicles.

All nutritive sweeteners are carbohydrates, that is, a combination of carbon and water. The simplest sugars are called monosaccharides. When two monosaccharides combine, they form a disaccharide. Sucrose is a disaccharide consisting of dextrose (glucose) and fructose (levulose). In the presence of acid, heat, or specific enzymes, sucrose splits (hydrolyzes or is inverted) into the two monosaccharides. Sucrose is also known as a nonreducing sugar, which makes it more resistant to the browning reaction than dextrose and fructose. The latter are strong reducing sugars that brown readily. There are other differences. The monosaccharides depress the freezing point to a greater extent than disaccharides and their sweetening power varies. Complex carbohydrates are polysaccharides because they are made up of long chains, both straight and branched, of simple sugars (monosaccharides). Invert sugar is the name applied to the mixture of dextrose and fructose that is formed by the hydrolysis of sucrose.

Corn starch is a polysaccharide which when completely broken down to its building blocks yields only dextrose. Both anhydrous and monohydrate dextrose are commercially made by the complete hydrolysis of starch. Other corn sweeteners, which include maltodextrins and various corn syrups, are products of incomplete hydrolysis of starch. Depending on the actual process, they contain varying proportions of dextrose and its oligosaccharides; maltose (a disaccharide), maltotriose (a trisaccharide), and a number of higher saccharides (sometimes called dextrins). To obtain the different saccharide combinations, a solution of the starch is treated with acid, enzyme or both to catalyze the hydrolysis. The extent of hydrolysis in a given syrup is expressed as its dextrose equivalent (DE), which is a measure of the total reducing sugars calculated as dextrose and expressed as a percentage on a dry basis. Anhydrous dextrose has a DE of 100; the hydrated form has a DE of 92. A high DE signifies a substantial conversion to dextrose and maltose and a relatively low conversion to the seven or higher unit oligosaccharides (maltoheptaose, -octaose, etc.). By the use of selected enzymes, corn syrups are manufactured that have a specifically designed composition of saccharides. High-maltose syrup, for instance, may have 40% of its saccharides in the form of maltose, but be designated as having a relatively low 42 DE because its dextrose content is significantly lower (e.g., 8% as opposed to 20%). High-fructose corn syrup is made from either dextrose or a very high DE syrup by the action of a specific enzyme, isomerase. The enzyme catalyzes the conversion of dextrose into fructose, a process called isomerization.

Although theoretically possible, corn syrups are seldom used as the sole source of sweetness in ice cream. Under circumstances of a severe sugar shortage and high sugar prices, as was experienced during war years, corn sweeteners would certainly be used as a replacement for more or all of the sugar. Some combination of low-DE syrup and high-fructose or high-dextrose syrup can be designed to provide satisfactory freezing properties and sweetness level. Under normal circumstances, corn syrups commonly contribute 20 to 50% of the ice creams' sweetening solids. This is the case for essentially all members of the frozen dessert family except for some high-butterfat products and products that already have a high solids content. Sucrose is usually the sole sweetening agents in such products.

2.2.13 Sucrose

In general parlance, the word sugar has come to mean the common table sweeteners, cane and beet sugar. In their pure, refined form, both are chemically identical and identified by the name sucrose. The highly refined, standard white sugar is the common type of sucrose used in dry form. The substance can be expected to be very pure and contain 99.9% solids. Dry, granulated sugar is principally used by small ice cream plants and those in locations where liquid sugar is not available.

Sweetness is the only sensory response to sucrose. In pure form, it is odorless and devoid of any other taste. It complements the flavorings commonly used in ice cream products very well. Being a disaccharide, it lowers the freezing point to a lesser extent than monosaccharides but more than some of the low-DE corn syrups. These are the favorable properties that make sucrose an efficacious sweetening agent. Other properties are summarized in Tables 2.1 and 2.6.

A typical liquid sucrose may test 67.5° Brix, weigh 11.104 lbs/gal at 20°C, and contain 7.495 lbs of sugar per gallon. Degree Brix is merely a measure of percent sucrose; 7.495 is 67.5% of 11.104. Additional criteria addressed by specifications are color, ash content, heavy metal content, yeasts and molds, pH, maximum invert sugar present, and flavor. The invert sugar limit is important because an ice cream formulated to contain a certain concentration of sucrose can acquire different characteristics in the presence of significant quantities of monosaccharides (e.g., effect on freezing point, browning, sweetness etc.). Liquid sugars may also be obtained as blends of sucrose and dextrose or sucrose and one of the corn syrups. The desired proportion of each is a matter of choice and should reflect the actual ratio of the sweeteners in the frozen product. However, one should also consider flexibility in the use of sweeteners for all of the products made and their requirements when deciding on blends versus separate syrup supplies. The sweeteners needed for sherbet, premium ice cream, soft-serve, etc. may differ in amount, type of corn syrup, or proportion of sucrose to corn syrup.

Because sucrose is commonly used in combination with other sweeteners, it is difficult to define its level of usage precisely. It may be stated that plain ice cream (e.g., vanilla) generally contains the sweetness equivalent of 13 to 16% sucrose. The desired sweetness level also varies with the type of flavoring used. A chocolate ice cream may contain 17 to 19% of sucrose. The actual level chosen may be dictated by economic considerations, but should also be a function of consumer preference and acceptance. In any case, excessive as well as inadequate sweetness are flavor defects worthy of management considerations.

2.2.14 Dextrose

The sweetening power of dextrose is 60 to 80% that of sucrose and, theoretically, one should be able to use somewhat more of it to increase the solids content without imparting excessive sweetness. However, because of its effect on the freezing point, the practical limit is defined by the stiffness of the ice cream at the usual freezer discharge and storage temperature. The use of dextrose by itself would yield a very soft product. In combination with sucrose, some ratio, probably in the 10 to 20% range, of sucrose replacement for ice cream, sherbets, and ices is possible. The replacement level may be higher in products that are purposely designed to be softer, such as some gelatos served in the "traditional" way.

Although the sweet taste imparted by dextrose is similar to that of sucrose, dextrose is not a common sweetener in ice cream. In the usual case, ice cream makers are looking to sucrose replacers as a means of improving the body and texture and heat shock resistance. On comparison, corn syrups prove to be more effective in this regard and, therefore, have come into general use. However, dextrose is an optional sweetener, especially when circumstances preclude the use of other sources.

2.2.15 Corn Syrups

As can be seen in Table 2.6, sweeteners derived from corn provide a whole spectrum of products with different properties. Older textbooks suggest that the high-DE syrups were preferred during and after the period of World War II, but now the situation has certainly changed. Syrups of 36 DE and 42 DE have become common ice cream ingredients and have performed a useful function. Much of the credit for the increased acceptance of low-DE corn syrups is due to quality improvement which has made it possible to obtain colorless, bland-tasting syrups that can be used at relatively high levels of sucrose replacement. The methods of marketing ice cream have also changed over the years. Ice cream is expected to withstand considerable temperature abuse before and after it reaches the home refrigerator. This has caused ice cream makers to look for ways to better stabilize the body and texture, and low-DE corn syrups have provided a practical and economical approach to the problem.

The sweet taste of a corn syrup is determined by the concentration of dextrose, its sweetest component, and to a much lesser extent, maltose. The sweetening power of the saccharides containing several units of dextrose (the higher sugars) is negligible by comparison to that of dextrose. Because high-DE corn syrups contain the highest concentration of dextrose and maltose, they are sweeter than the low-DE corn syrups. On the other hand, water binding properties of the higher sugars are greater than those of dextrose and maltose. Because low-DE syrups are not so sweet, a greater quantity of them is needed as a sucrose replacement to maintain the same or a similar level of sweetness. The resulting increase in total solids may also have a beneficial effect on body and texture.

Ingredients with enhanced water binding properties generally have a desirable effect on the body of ice cream and, at the same time, they assist in creating conditions that protect the texture against rapid deterioration during heat shock. Water that is held firmly by physical and chemical forces to other molecules (bound water) behaves as though it were a solid; it loses its ability to freeze or act as a solvent. In simplified language, the ice cream behaves as though it had a higher solids content because there is less water to freeze and the soluble solids become more concentrated. No reliable estimates are readily available of the amount of bound water that conforms to this definition in frozen ice cream.

Maltodextrins are used primarily as bulking agents. They have a very low DE (<1 to <20) and, thus, very little or no useful sweetening properties. Their employment in frozen desserts is of a relatively recent origin and roughly coincides with the surging interest in lower fat and nonfat products. As fat is reduced or eliminated from an ice cream formulation, other solids are needed to take its place, at least in part. In this role, maltodextrins assume the function of a fat replacer that is capable of restoring some of the ice cream body building properties normally provided by the fat. However, one cannot expect them to completely emulate the action of fat and additional benefits may need to be derived from other ingredients. Unfortunately, the sources of food solids that could be appropriately used in this role are rather limited. Of the sweeteners, only very low DE corn syrups and maltodextrins are helpful. Because of the wide range of maltodextrin DEs, the selection of the desired type should be given careful consideration (see Table 2.1).

The DE designation is not employed for identifying high-fructose corn syrups. The products are largely a mixture of dextrose and fructose with only small percentages of maltose and higher sugars. The proportion of fructose in different syrups may range from about 40% to nearly 100%. Fructose is sweeter than sucrose, but the sweetening power of the selected syrup should be determined in the frozen dessert at the intended concentration. The actual sweetening power and the character of the sweetness may be found to vary at different levels. With regard to its effect on body and texture, much of what has been said about dextrose also applies to fructose. The syrup does not contain the higher saccharides that possess the water binding properties and because it is made up largely of monosaccharides, it depresses the freezing point to a greater extent than sucrose. When the proportion of dextrose to fructose approaches 50/50, the product becomes similar to invert sugar, a sweetener resulting from the complete hydrolysis of sucrose, which has the same proportion of monosaccharides. Obviously, the effects on sweetening, body and texture, and freezing point depression would also be similar.

2.2.16 Honey

Although not commonly used, honey is both a sweetener and a flavoring agent. The sweetening power is due largely to invert sugar (dextrose and fructose) which may constitute nearly 75% of the honey (as is). The taste and composition vary between different varieties of honey but many types, particularly the light-colored ones, impart a pleasant flavor to ice cream. Because of its high monosaccharide content, honey has a similar effect on lowering of the freezing point and softening of the ice cream as dextrose and fructose. To impart a honey flavor to ice cream, a concentration of 8 to 10% honey (as is) is needed, which also accounts for approximately 50% of the desired sweetness. The remaining sweetness can be supplied by sweeteners other than monosaccharides so as to minimize the softening effect.

2.2.17 Stabilizers

In physical and chemical terms ice cream stabilizers are colloidal substances called hydrocolloids or simply colloids. They are not soluble in water in the strict chemical sense, but at the same time, they remain dispersed in a stable colloidal (larger than molecular) suspension and thus appear to be dissolved. This is not a unique property of stabilizers; milk proteins and milkfat are also dispersed in a colloidal suspension both in milk and in ice cream. Ice cream has a very complex structure consisting of two liquid phases (water and lipid), each containing soluble substances (particularly the water phase); a colloidal dispersion of lipid in water (but may also include some water in lipid dispersion, especially when shear-induced churning occurs); colloidal dispersion of solids such as proteins, minerals (e.g., calcium phosphate), and stabilizers; and dispersed air. The literature on fundamental aspects of food emulsions, colloidal chemistry, rheology, and the physical and chemical properties of gums is quite extensive.[17–24] Current research reports may be found in food and chemistry journals and any specialized journals such as the following: *Food Hydrocolloids*; *J. Colloid Interface Science*; *Kolloid Z.*; *J. Texture Studies*; *Rheologica Acta*; *J. Rheology*; *Food Microstructure*; and *Colloid and Polymer Sci.*

With the exception of gelatin, which is a protein derived from the connective tissues of skin and bone (collagen), organic substances used as ice cream stabilizers are specific forms of polysaccharides. Chemically, they differ from each other in internal structure; the identity or proportion of the monosaccharide units; the presence, type, and number of acidic groups along the chain; and the presence of inorganic components. A food grade form of calcium sulfate, an inorganic compound, also has stabilizing properties. Several components of commercial stabilizers are described in Table 2.7.

As the name implies, the most important function of these substances is to "stabilize" (i.e., protect against deterioration) the texture of ice cream during storage and distribution. The need for some form of stabilizing action was recognized by early ice cream makers as witnessed by the inclusion of arrowroot flour in ice cream recipes dating to the 18th century.[13] Home recipes included starch in the past and some homemakers may still be using it. Over the years, substances that are more effective at a much lower concentration than starch have been developed. They provide a means for both "shaping" the type of body envisioned for the ice cream and for contributing to the stability of the body and texture under the detrimental effect of heat shock.

It is possible to make an ice cream without stabilizers, but unless its total solids content is quite high (e.g., high fat content), its body is commonly characterized as lacking resistance, being quick to melt, or lacking "chewiness." Of course, the degree to which these characteristics manifest themselves also depends on how much overrun (incorporated air) the ice cream contains. Some manufacturers may purposely refrain from using stabilizers because they desire a light bodied ice cream or when they cannot justify the inclusion of these substances in products designated as "all natural." In any case, without stabilizers, the ice cream is more vulnerable to becoming coarse-textured on storage and especially when heat shocked.

Stabilizers used in present day ice cream manufacture are commonly proprietary blends of two or more stabilizing components along with one or more emulsifiers. Although some stabilizing components are less expensive than others, economy should not be the principal guide in the selection of a commercial stabilizer. Generally, a stabilizer should assist in producing and maintaining an ice cream with a smooth texture, but additional criteria should also be considered. An additional objective is to impart a body (which refers to such properties as firmness, resistance to bite, and cohesiveness) that the ice cream manufacturer perceives as approaching the "ideal" within the constraints of such fixed parameters as composition and overrun of the frozen dessert. The fixed parameters may be dictated by economics or market positioning of the product. A combination of gums may provide the desired

Table 2.7 SOME CHARACTERISTICS OF STABILIZERS AND EMULSIFIERS

Stabilizer or Emulsifier	Properties	Comments
Gelatin	Protein of animal origin; on aging forms a "brush heap" structure that traps water—gels; viscosity of mix is substantially reduced by agitation; available in different Bloom (gel strength) grades, e.g., 250 Bloom; disperses in cold mix but requires heat for activation.	Mix should be aged, preferably 24 h; less is needed as Bloom strength increases—typically 0.5–0.3%; relatively high cost; has lost popularity since World War II in USA; Rate of cooling affects mix viscosity; Interacts with mix constituents at high temperatures.
CMC	Sodium carboxymethyl cellulose, (cellulose gum) obtained by chemical modification of cellulose (a polysaccharide); imparted viscosity is a function of average chain length; does not gel; hydrates at low temperatures; causes whey separation; has excellent absorptive properties.	Usually used in combination with other gums; wheying off problem is controlled by combining with a gel forming stabilizer; depending on viscosity grade and total solids content in mix, 0.1–0.2% needed to fully stabilize mix; imparts excellent body and texture.
Algin	A mixed polymer of anhydro-D-mannuronic acid with anhydro-L-guluronic acid; used as sodium alginate or as an ester alginate; the esterified form does not gel; phosphate helps control reaction of sodium alginate with calcium (gelation); obtained from ocean kelp.	Sodium alginate is added to mix at 160°F to control gel formation; cannot be used with high-acid products; ester alginate disperses in the cold and may be used with acid products; commercial preparations used at 0.25–0.4% level.
Carrageenan	Salt of sulfate esters of polymers of galactose; very strongly charged anionic polyelectrolyte; forms a gel at very low concentration; derived from a marine plant; properties are affected by relative proportion of the three types—kappa, lambda, and iota.	Commonly used in combination with other gums; prevents wheying off due to other gums at concentrations as low as 0.01% of the weight of the mix; disperses in the cold.
Locus bean gum (Carob bean gum)	Polymer of galactose and mannose; a linear chain of D-mannopyranosyl units with every 4th or 5th unit substituted by a D-galactopyranosyl unit; does not gel in ice cream; disperses in the cold, but must be heated to hydrate; causes whey separation; synergistic with xanthan gum; of vegetable origin.	Effective at 0.1–0.15% level; imparts a resistant, chewy body; mix viscosity increases as heat treatment increases; body may be somewhat shorter than with guar gum.

(Continued)

Table 2.7 *(Continued)*

Stabilizer or Emulsifier	Properties	Comments
Guar gum	Similar to locust bean gum, but hydrates better in the cold; the polymer contains a higher proportion of galactose to mannose (~1–2) than locust bean gum (1–4); of vegetable origin.	Body is similar to that imparted by locust bean gum although it may be somewhat "stickier" or more "gummy"
Calcium sulfate	Some proprietary stabilizer blends contain calcium sulfate. It appears to have functional characteristics which may be due to its water binding properties or possibly a specific ion effect. It has an effect on ice cream body.	
Microcrystalline cellulose	Manufactured from purified wood pulp and codried with CMC; imparts excellent bodying properties and texture stability (resistance to heat shock); has fat sparing properties.	Functional at 0.25–0.75%, depending on requirements and total solids content; used in addition to other stabilizers; activated by the shearing action of homogenization.
Xanthan gum	Polymer of glucose, mannose, and glucuronic acid; has pseudoplastic (shear-thinning) properties; provides immediate temperature and pH stable viscosity; a microbial fermentation product (*Xanthamonas campestris*).	Acts synergistically with locust bean and Guar gum; effective at low levels, especially with other gums that act synergistically; may find application in a broad range of products, including still frozen ices.
Mono- and diglycerides	Compounds of glycerol and either one (monoglyceride) or two fatty acids (diglyceride), usually used in combination, although monoglycerides are more effective. Several fatty acids may be present but one may predominate (e.g., stearic or oleic). Used to promote dryness at draw, improved whipping properties and creamier mouthfeel. Those containing unsaturated fatty acids are more effective drying agents but flavor deterioration may be a problem.	Less apt to cause churning, but not as effective a drying and whipping agent as polysorbates; used in the range of 0.1–0.2%; commonly used along with polysorbates; usually obtained and used as part of a proprietary combination of stabilizer(s) and emulsifier(s).

(Continued)

Table 2.7 *(Continued)*

Stabilizer or Emulsifier	Properties	Comments
Polyoxyethylene (20) sorbitan monooleate (Tween 80 or Polysorbate 80)	Chemical compound of sorbitol, oleic acid, and a chain of opened ethylene oxide units; powerful drying agent; aids in improving texture and heat shock resistance; the unsaturated oleic acid tends to become oxidized (this places a limit on the upper level of usage).	Effective in the range of 0.02–0.06%; may promote churning under conditions such as excessive levels, excessive shear (agitation) as is possible in soft-serve and high-fat mixes; usually obtained and used as part of a proprietary combination of stabilizer(s) and emulsifier(s); also available in liquid solution.
Polyoxyethylene (20) sorbitan tristearate (Tween 65 or Polysorbate 65)	Chemical compound of sorbitol, 3 molecules of stearic acid and a chain typically 20 units long of opened ethylene oxide units; an excellent whipping agent but somewhat less effective than Polysorbate 80 as a drying agent.	Choice of polysorbates depends on the requirements of specific products; for equivalent dryness, a slightly higher level is needed than with Polysorbate 80; use of higher levels (around 0.1% is normally not detrimental to flavor).
Lecithin	Chemically a group of diglycerides, also containing a phosphate ester of choline, ethanolamine, etc. (phospholipids); widely distributed in nature, including milk; soybeans are a common source of commercial lecithin.	Potential flavor problems are likely to limit usage level to about 0.1%; unsaturated fatty acids in lecithin are vulnerable to oxidation; similar in effectiveness to mono- and diglycerides; present in high concentration in dry buttermilk and especially in egg yolk solids.

functionality because individual stabilizer components differ in their effect on body, that is, the type of body that they help to impart. Heat shock resistance is also affected in a significant manner by the choice of stabilizing components. The proper level of stabilizer usage is an important consideration because excessive levels may give rise to a gummy body, poor meltdown, and possibly interference with flavor release. Inadequate levels, on the other hand, may not provide the benefits sought from the stabilizer.

Cooperation should be established between the ice cream manufacturer and the stabilizer supplier. To meet the stabilizer requirement of a specific ice cream usually encompasses one or more trial runs in the plant in which the product is made. A mix of the same composition may not yield identical results under all plant conditions because of differences that may exist in equipment performance and procedures for mix processing, freezing, ice cream hardening, and product handling. After actually observing the stabilizer's functionality and performance in a production run, the requirements may be fine tuned to get the desired effects. For their part, suppliers of stabilizers have to ensure that their products perform uniformly from batch to batch. In addition to the obvious quality criteria, such as microorganisms present, physical and chemical properties, and freedom from extraneous matter, their quality assurance program should include appropriate tests to monitor the functional properties and sensory qualities of the individual gums used in their blends.

Commercial stabilizer blends are available in a number of combinations of gums, with or without emulsifiers, with different levels of dispersing agents (e.g., dextrose) which makes them more or less concentrated, and with varying ease of dispersibility in the cold (for application in high temperature–short time pasteurization). Single components are also available. Some blends may be designed for specific application, such as in sherbets (acid compatibility), soft-serve, etc. A given plant is likely to have on hand a number of stabilizers to be used in different products. Obviously, care must be exercised that the right stabilizer is used in the right proportion. Because stabilizers deteriorate on storage, opened containers should be resealed to avoid contamination and to protect the contents from the effects of moisture and high humidity. Prominent labeling and good warehousing practices should help in avoiding some of the problems. The use of stabilizers that have become old and, therefore, could be in a deteriorated condition may be unwise and prove to be false economy.

2.2.18 The Mode of Stabilizer Action

The most obvious manifestation of stabilizer action is an increase in mix viscosity which becomes apparent even by visual observation. One of the criteria in selecting the type of stabilizer and deciding on the level of usage is the degree of viscosity increase. Development of excessive viscosity may create flow problems, slow heat exchange, excessive pressure buildup, and other difficulties derived from these problems. Low viscosity, on the other hand, may signal inadequate stabilization. The relationship between desired stabilization and the viscosity per se does not appear to be a direct one, but once it is established that a given formulation provides both an acceptable mix viscosity and the desired level of stabilization, any departure from the expected viscosity should be investigated.

In addition to the stabilizer, mix viscosity is affected by the fat, MSNF, and total solids content; type of sweetener solids; emulsifier content; homogenization pressures and temperatures; fat globule clumping; "salt balance"; previous heat history of the ingredients; presence of developed acidity; pasteurization methods; rate of cooling; aging period; and many interactions. Because heat treatment is an integral part of mix processing, the contribution of the stabilizer should be viewed as the sum of all of the interactions in a given mix processing system. A simple illustration is presented in Table 2.8 showing the changes in basic viscosity of various model systems with and without added stabilizer components. The data show that in the absence of milk solids, the viscosity of gum solutions was relatively unaffected by heat treatment. The simulated milk salts solution actually exerted a viscosity depressing effect on the corresponding gum-containing solutions at all temperatures studied. At increasing total solids concentrations, the contribution of the gums to the viscosity became more pronounced, partly due to the hydration of the milk constituents on heating. However, the increase in viscosity was greater than expected without

Table 2.8 EFFECT OF HEAT TREATMENT ON BASIC VISCOSITY[a,f–h]

	Basic Viscosity (cp) After Heating to											
	200°F			230°F			260°F			290°F		
System Stabilizer	None	LBG[b,d]	CMC[c,d]	None	LBG	CMC	None	LBG	CMC	None	LBG	CMC
Water		7.5	14.4		8.4	14.1		8.4	14.1		8.0	13.9
5% Lactose	1.7	7.2	15.5	1.8	8.6	15.8	1.7	8.8	15.5	1.6	9.6	15.3
15% Sucrose	2.6	10.3	21.8	2.7	11.6	22.2	2.7	12.4	22.5	2.7	12.9	22.6
Milk salts[e]	1.6	6.4	9.6	1.6	7.4	10.2	1.6	7.3	10.0	1.6	7.6	10.2
Skim milk	3.1	14.0	10.6	3.1	13.8	11.5	3.1	14.2	11.9	3.2	15.4	12.1
12% MSNF	3.8			3.9			4.0			4.1		
16% MSNF	6.3	13.2	18.4	6.5	17.2	21.7	6.9	19.7	26.0	7.7	24.8	39.5
20% MSNF	8.5			9.3			9.5			11.2		
11% MSNF + 28% sucrose	14.7	19.9	24.0	13.8	20.2	26.7	14.0	24.9	25.1	16.7	36.2	25.4
Milk	3.6	13.9	12.7	3.6	14.3	12.8	3.7	14.7	13.3	4.3	16.6	14.3
Ice milk mix	13.7	31.6	37.8	13.8	32.2	37.8	18.2	39.5	43.0	25.4	53.0	54.4
Ice cream mix	19.8	39.1	44.6	18.7	40.5	47.9	26.4	45.3	49.0	31.5	51.3	53.9

[a] Heated in a small tube heat exchanger (Mallory heater) with a 6-s heat-up time and no holding time.
[b] LBG, locust bean gum.
[c] CMC, sodium carboxymethyl cellulose.
[d] Concentration of the stabilizer in the water portion was the same in all samples.
[e] A simulated solution having approximately the same composition as milk salts.
[f] Any structure was broken down by passage through a hand emulsifier.
[g] Viscosities were determined at 40°F, 24 h after heating.
[h] Data taken from a thesis submitted by G. A. Muck to the Graduate College at the University of Illinois in partial fulfillment of the requirements fot the MS degree (1961).

assuming synergism or interaction. It appears that all mix constituents contribute to increased water binding by the action of heat.

Direct evidence of stabilizer–protein interaction may be observed when locust bean gum, Guar gum, carboxymethyl cellulose (CMC or cellulose gum), and certain other gums are incorporated into an ice cream mix. When the mix is allowed to stand undisturbed for about 24 h, whey separation occurs, that is, a clear liquid separates, creating a more concentrated mixture of protein and fat. Depending on the fat content and its effect on the specific gravity, the clear liquid may be observed on top or on the bottom. The separation can be also observed after the ice cream is frozen and then allowed to melt. A mix that has separated can be made uniform by agitation and it will freeze normally; but if someone fails to remix it, the ice cream will not be of uniform composition. A practical solution is to use the gums in combination with a gel-forming stabilizer that prevents the separation. Carrageenan, known for its reactivity with milk proteins, is very effective for this purpose at a very low concentration (0.01%). It is almost invariably used in combination with stabilizers that promote whey separation. Other gel formers, such as gelatin, may also work but would have to be used at a higher concentration than carrageenan. Whey separation (syneresis) is much less of a problem when gel-forming stabilizers are used.

In the hardened ice cream the concentration of gums may increase six- to eightfold due to the low level of remaining unfrozen water (e.g., 10% of the original water content, depending on temperature and mix composition). The low temperature and high gum concentration would be expected to substantially reduce diffusion and curtail mobility of the remaining liquid, which is already a saturated solution of sugars and some of the salts. Temperature fluctuations are inherently quite damaging to the structure of the ice cream due to the tendency for crystals to grow in size as they recrystallize. By impeding the movement of any melted water, it is hoped that the rate of crystal growth can also be reduced. The validity of this assumption is borne out by the general observation that an ice cream formulated with gums (and other ingredients that hydrate readily) usually has a more stable structure in storage. More recent ideas have been advanced in explaining this phenomenon and will be discussed in Section 2.11.

Under normal conditions of rapid freezing and high viscosity, the carbohydrate component of milk solids, lactose, is believed to adopt an amorphous, or glassy, state rather than a crystalline form. Crystallization may be also impeded by the gum stabilizers in common use.[25] However, crystallization may be encountered due to a combination of factors such as: the mix composition includes a high proportion of lactose; the storage temperature fluctuates; and crystal nuclei are available possibly from certain added flavors. As the crystals increase in size they may be perceived as hard, sandy particles that do not readily dissolve in the mouth. Their detection threshold is when their size approaches approximately 15 μm.

The mechanism by which various gums bind water is known to differ in some aspects. Some are capable of forming a gel either through their own structural orientation (e.g., gelatin) or by forming calcium bridges (e.g., sodium alginate); some act synergistically with other gums such as Xanthan which develops higher viscosity with Guar gum, and, depending on total gum level, higher viscosity or a gel with locust bean gum[26]; they hydrate at different rates; and some do not form a gel but act as effective thickening agents (e.g., Guar gum).

The gums used in ice cream also differ in their effect on rheological properties of the mix such as pseudoplasticity (thinning with increase in shear, followed by recovery when the shearing action is reduced or discontinued); thixotropy (time-related viscosity reduction after shear stress); yield value (minimum shear stress before flow is initiated); maximum viscosity production; rate of viscosity development; etc.

Ice cream manufacturers generally rely on their stabilizer suppliers for a product that has been optimized in functional properties for their specific application. Along with actual product trials, rheological tests find practical application in formulating the needed blends of gums that help meet the users' criteria.

The manner in which gums contribute to the sensory perception of body in the frozen ice cream could be related to their molecular structure and orientation as well as their gel forming or viscosity development capability. However, it is difficult to extrapolate results from model solutions to ice cream because the gums in ice cream perform their function at a very low temperature, in a highly concentrated solution with respect to salts, sugars, and oligosaccharides all interacting with other macromolecules.

The body of ice cream may also be modified by the inclusion of emulsifying agents into the blend of gums. The properties of emulsifiers, however, should be understood and will be discussed in the following section.

2.2.19 Emulsifiers

In a physical and chemical sense, an emulsion is a suspension of small particles or globules of one liquid in another liquid. The suspension of milkfat globules in milk is an example of a natural emulsion. To produce a stable emulsion requires the presence of an emulsifying agent that orients (positions) itself at the interface of the two liquids in question and is partially soluble in both. The molecule of the emulsifier is said to have a hydrophilic (water loving) portion and a hydrophobic (water hating, or in this case lipophilic or fat loving) portion. The stability of an emulsion is also affected by the size of the globules. In milk, the emulsion is stable, but because of a difference in specific gravity and other physical and chemical forces, the globules rise and become concentrated in a cream layer. When the milk is homogenized, the size of the globules is reduced and additional protein is deposited on the surface of the globule. Because the specific gravity of protein is much higher than that of fat, the new smaller globule no longer experiences the strong forces of gravity and cream does not separate in homogenized milk.

The naturally occurring emulsifying agent in milk is actually a class of substances called phospholipids (also referred to as lecithin, one of the major components). These substances are widely distributed in both plant and animal matter. Lecithin of plant origin finds use as an emulsifier in a number of foods. By using eggs, early ice cream makers discovered the beneficial effects of emulsifiers indirectly. Egg yolks have had a long history as an ingredient in ice cream due both to the flavor that they impart and their emulsifying properties. They are rich in phospholipids. The benefits derived or hoped for from the use of emulsifiers include the following:

A dry appearing product as it emerges from the freezer
Improved whipping properties
Improved body and texture
Richer mouthfeel sensation
Smaller air cells
Improved heat shock resistance

In the presence of added emulsifiers, ice cream appears drier when it is drawn from the ice cream freezer as compared to an identical ice cream at the same drawing temperature but without added emulsifier. The dryness appears to be the result of an induced fat globule clustering phenomenon at the liquid–air interphase. With proper conditions, these changes are observable under the microscope. A dry, stiff product is essential in the manufacture of extruded novelty items such as sandwiches and stickless bars. Packaging of all types of products is facilitated by a dry ice cream that does not drip as it is filled into containers, especially if the ice cream has to be pumped some distance to the packaging equipment. A dry appearance at the freezer is also associated with desirable effects on body and texture and resistance to heat shock. Soft-serve products usually have a low overrun but should have a dry appearance to maintain the shape of the serving and prevent drippage even on a hot summer day.

Some effects of emulsifiers are predictable from the known properties of emulsifying agents. Because they are surface-active agents that measurably reduce the surface tension, one would expect them to improve whipping properties and promote the development of smaller but more numerous air cells. More air cells provide more surface with a finite quantity of available liquid. This should promote a drier appearance because the liquid is spread over a larger area. However, this is not the only mechanism, and possibly not the predominant one, for the drying effect of emulsifiers.

Attention must be given to both the concentration of the emulsifier and the fat content. As their concentration is increased, emulsifiers acquire a measure of deemulsifying properties displayed by fat (butter) separation in the freezer and a greasy mouthfeel when tasted. As the fat content increases, the deemulsification action is magnified. The objective is to achieve a certain degree of deemulsification because that is how the desired drying effect is produced. However, one should not use more emulsifier than needed to provide just the correct amount of incipient "churning." Thus, less emulsifier is needed in high-fat than in low-fat products. The actual amount of the emulsifier also depends on the specific type of emulsifier used. Emulsifier molecules with a large hydrophilic component promote churning to a greater extent than those with a large fat-soluble component. This fact must be considered when formulating special products such as high- and low-fat ice cream and in soft-serve items. Prolonged agitation in the soft-serve freezer by the action of the dasher tends to promote churning by itself and an improper choice of emulsifier only aggravates the problem. Close cooperation with the supplier of this ingredient should help in identifying the right emulsifier or, as is commonly the case, a combination of emulsifiers for specific purposes. Several emulsifiers are described in Table 2.7. Additional discussion is presented in Section 2.11.1.

2.2.20 Miscellaneous Ingredients

Other than flavoring ingredients, which will be discussed separately, some others find application in special situations. They include artificial food colors (check legality and labeling), ordinary table salt, so-called protein stabilizing salts such as citrates and complex phosphates, acidulants for sherbets and ices (most commonly citric acid), fats other than butterfat, ingredients intended to replace fat (trademarked products of proprietary composition), nonnutritive sweeteners and sweetener substitutes, bulking agents (e.g., polydextrose), and sources of vegetable protein (e.g., soybean protein isolate). Although all of these ingredients must be scrutinized for quality and functionality, they must also satisfy the criteria of safety, appropriateness, and use within the constraints of any legal limitations or prohibitions. Checking local, state, and federal regulations is a prudent approach.

2.3 Calculations and Mix Standardization

To produce an ice cream mix of the desired and consistently uniform composition pertinent analytical data for all ingredients must be available. The accuracy of standardization is completely dependent on these data. Procedures for calculating the required quantities of ingredients are based on arithmetic and algebraic procedures whose principles and application will be illustrated by examples in this section.

To reduce time-consuming calculations, ice cream manufacturers may develop computer programs for mix standardization of their own, or may purchase commercially available programs. A practical option is also provided by the availability of inexpensive hand-held calculators that are capable of solving simultaneous equations involving up to three unknowns in 1 min or less. Their use may be found ideal by beginners and students who want to learn the principles involved in setting up the equations. The authors used a Texas Instruments Model TI-68 calculator for solving the simultaneous equations presented in this section.

2.3.1 Calculating MSNF in Skim Milk and Cream

The fact that the MSNF content of milk is related to the fat content has been illustrated in Table 2.3. Therefore, the original composition of milk must be known before the MSNF content of skim milk and cream can be estimated. Actual analysis of all ingredients could be performed in place of a calculation or to confirm it. When the composition of the original milk is not available, an analysis is required if the mix is to be accurately standardized.

To illustrate the process, let us assume that we have 100 lbs of milk containing 3.5% fat and 8.5% MSNF. Therefore 100 lbs of milk contains 3.5 lbs of fat and 8.5 lbs MSNF. Because all of the MSNF are contained in the nonfat portion of the milk, 100 lbs of milk minus 3.5 lbs of fat = 96.5 lbs nonfat portion and 8.5 lbs MSNF divided by 96.5 × 100 = 8.8% MSNF.

The nonfat portion of milk is actually skim milk, although in practice, skim milk contains 0.05 to 0.09% fat as determined by ether extraction or 0.01 to 0.03% as determined by the Babcock test. A fat content higher than this reflects a reduced efficiency of separation. In this example, the skim milk obtained from this particular milk supply may be assumed to contain 8.8% MSNF.

To calculate the MSNF content of cream, we must know the fat content. As an illustration, let us consider a cream made from the same milk supply as the skim milk in the above example and standardized to contain 40% fat. One hundred pounds of the cream can be visualized as containing 100 lbs cream = 40 lbs fat + 60 lbs skim milk. As this particular skim milk contains 8.8% MSNF, 0.088 × 60 = 5.28 lbs MSNF in 100 lbs of cream, which is another way of saying that this cream contains 5.28% MSNF.

2.3.2 Standardization of Ice Cream Mixes—The Simplest Case

When a single concentrated source of fat and MSNF is to be used, the calculations are simple as illustrated by Example 1:

Example 1

Desired composition	Ingredients
12% Fat	Cream, 35% fat, 5.72% MSNF
11% MSNF	NDM, 97% MSNF (3% water)
15% Sugar	Liquid sugar, 67.5% sucrose (67.5 Brix)
0.35% Stabilizer/emulsifier	Commercial stabilizer

One approach is to calculate the requirements for 100 lbs of mix and then use multiples to obtain the desired weight.

100 lbs of mix must contain:

12 lbs Fat
11 lbs MSNF
15 lbs Sugar
0.35 lbs Stabilizer/emulsifier

Pounds of cream needed to supply 12 lbs of fat:

$$\frac{12}{0.35} = 34.3 \text{ lbs } 35\% \text{ cream}$$

The weight of MSNF provided by the cream is obtained by multiplying the weight of the cream by the determined MSNF content of the cream, in this example 5.72%:

Table 2.9 EXAMPLE 1

Ingredients	Weight	Fat	MSNF	Sugar (lbs)	Stabilizer/Emulsifier
Cream	34.3	12.0	1.96	0	0
NDM	9.32	0	9.04	0	0
Liquid sugar	22.2	0	0	15	0
Stabilizer/emulsifier	0.35	0	0	0	0.35
Water	33.81	0	0	0	0
Total	100.00	12.0	11.00	15.0	0.35
Desired	100.00	12.0	11.00	15.0	0.35

$$34.3 \times 0.0572 = 1.96 \text{ lbs of MSNF}$$

The nonfat dry milk must supply the remainder of the needed MSNF:

$$11 - 1.96 = 9.04 \text{ lbs MSNF}$$

$$\frac{9.04}{0.97} = 9.32 \text{ lbs of NDM}$$

The weight of liquid sugar is obtained by dividing the needed weight of sugar by the percent of sucrose in the liquid sugar:

$$\frac{15}{0.675} = 22.22 \text{ lbs of liquid sugar}$$

The weight of the needed stabilizer/emulsifier is simply 0.35 lbs. As the sum of the weights of the ingredients is <100 lbs, the difference is made up by the addition of water. A convenient way is to prepare a table, as shown in Table 2.9 in which all of the figures can be double checked.

In constructing Table 2.9, the weight of each ingredient is multiplied by its percent composition of fat, MSNF, and sugar. In this example, the weight of cream was multiplied by 0.35 to obtain the weight of fat, and by 0.0572 to obtain the weight of MSNF contributed by the cream. The weight of the NDM was multiplied by 0.97 to obtain the weight of contributed MSNF (the fat content of NDM was assumed to be negligible). The sugar content was obtained by multiplying the weight of liquid sugar by its sucrose content (°Brix). The stabilizer/emulsifier was assumed to be at full strength. The table serves as proof of the correctness of the calculations. If the tested composition of the finished mix made according to the calculated formula is found to be in error, the difficulty may be due to poor sampling, incorrect analysis of the ingredients, errors or malfunction in weighing, dilution with water or another mix, or some other human error.

2.3.3 The Serum Point Method of Mix Standardization

The individual steps in this procedure may be summarized as follows:

1. Add the weights of all nondairy products ingredients.

Table 2.10 EXAMPLE 2

Ingredient	Weight	Fat	MSNF (lbs)	Sugar	Stabilizer
Milk	27.84	1.11	2.35	0	0
Cream	36.28	10.88	2.23	0	0
Condensed skim milk	21.38	0	6.41	0	0
Sugar	14.00	0	0	14.00	0
Stabilizer	0.35	0	0	0	0.35
Total	100.00	11.99	10.99	14.00	0.35
Desired	100.00	12.00	11.00	14.00	0.35

2. Calculate the serum in 100 lbs of mix. Serum is defined as the sum of the weights of MSNF and water contributed by the dairy products. [100 − (wt nondairy ingredients + wt fat)]
3. Calculate the required weight of the concentrated MSNF ingredient to supply the shortage between MSNF needed and the normal MSNF in the serum. The percent normal MSNF is equivalent to the percent MSNF in the skim milk obtained from the available milk supply. In this illustration it is assumed to be 8.8%.

   For 100 lbs of mix, the weight of the concentrated MSNF ingredient may be calculated by the following equation:

$$\begin{matrix}\text{Wt concentrated} \\ \text{MSNF ingredient}\end{matrix} = \frac{\begin{matrix}\text{wt MSNF needed} - \\ (0.088 \times \text{wt of serum in mix})\end{matrix}}{\begin{matrix}\text{\% MSNF in conc. ingred.} - \\ (0.088 \times \text{\% serum in conc. ingr.})\end{matrix}} \times 100$$

4. Add the weights of nondairy and the concentrated MSNF ingredients and subtract from 100 to get the weight of milk and cream needed.
5. Calculate the percent fat of the mixture of milk and cream as follows:

$$\text{\% Fat} = \frac{\text{total fat needed} \times 100}{\text{wt. milk} + \text{cream}}$$

6. Calculate the weight of milk and cream individually by the Pearson Square or other appropriate method.
7. The process is illustrated in Example 2 and proof is presented in Table 2.10.

Example 2

Desired composition	Ingredients
12% Fat	Milk, 4% fat, 8.448% MSNF
11% MSNF	Cream, 30% fat, 6.16% MSNF
14% Sugar	Condensed skim milk, 30% MSNF
0.5% Stabilizer	Granulated sugar
	Stabilizer

For 100 lbs of mix:

Sum of weights of nondairy ingredients = 14.5 lbs

Weight of serum in mix = 100 − (12 + 14 + 0.5) = 73.5 lbs

Normal MSNF = 8.8%

MSNF in serum = 73.5 × 0.088 = 6.468 lbs

Serum in condensed skim milk = 100%

Weight of condensed skim milk needed
= (11 − 6.468)/(30 − 8.8) × 100 = 21.38 lbs

Weight of milk and cream = 100 − (14.5 + 21.38) = 64.12 lbs

Percent fat in mixture of milk and cream = 12/64.12 × 100 = 18.71%

By Pearson square:

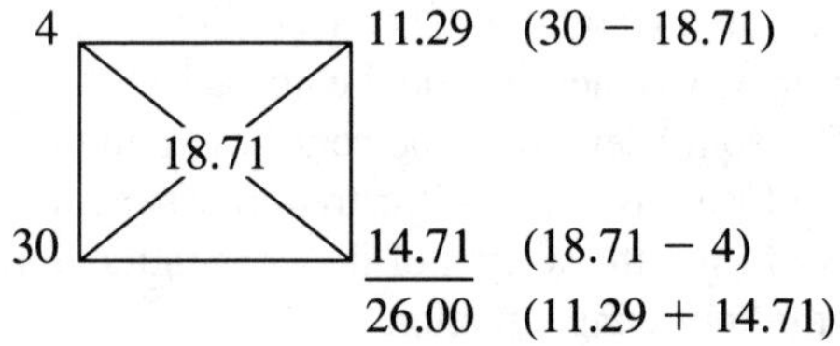

Note: The Pearson Square results indicate that a mixture of 11.29 lbs of 4% milk and 14.71 lbs of 30% cream will yield 26.00 lbs of an 18.71% fat mixture. Since in this example we need 64.12 lbs of the mixture, we can calculate each needed amount by proportion.

Weight of cream needed = (14.71 × 64.12)/26 = 36.28 lbs
Weight of milk needed = 64.12 − 36.28 = 27.84 lbs

The weight of milk and cream needed can also be calculated by one of two alternate methods. Illustrated below is an algebraic procedure and a formula derived from the algebraic method.

Algebraic solution:

x = lbs of cream
y = lbs of milk

$$\begin{array}{lrcl} \text{Fat equation} & 0.3x + 0.04y & = & 12 \\ \text{Weight equation} & x + \quad y & = & 64.12 \\ \hline \text{Solving:} & x & = & 36.29 \text{ lbs of cream} \\ & y & = & 27.83 \text{ lbs of milk} \end{array}$$

Formula method:

$$\text{Lbs of milk needed} = \frac{\left(\text{lbs milk and cream needed} \times \frac{\%\text{ fat cream}}{100}\right) - \text{lbs fat needed}}{\frac{\%\text{ fat cream}}{100} - \frac{\%\text{ fat milk}}{100}}$$

$$= \frac{\left(64.12 \times \frac{30}{100}\right) - 12}{\frac{30}{100} - \frac{4}{100}} = 27.83 \text{ lbs milk}$$

Weight of cream needed = 64.12 − 27.83 = 36.29 lbs cream

The procedure for calculating the required ingredient quantities when liquid sweeteners are used is illustrated in Example 3, and the proof is presented in Table 2.11.

Table 2.11 EXAMPLE 3

Ingredient	Weight	Fat	MSNF	Sugar	CSS (lbs)	Stabilizer/Emulsifier
Milk	45.85	1.60	3.89	0	0	0
Cream	20.99	8.40	1.11	0	0	0
NDM	5.21	0	5.00	0	0	0
Liquid sugar	15.15	0	0	10	0	0
Corn syrup	12.5	0	0	0	10	0
Stabilizer/emulsifier	0.3	0	0	0	0	0.3
Total	100.00	10.00	10.00	10.00	10.00	0.3
Desired	100.00	10.00	10.00	10.00	10.00	0.3

Example 3

Desired composition	Ingredients
10% Fat	Cream, 40% fat, 5.28% MSNF
10% MSNF	Milk, 3.5% fat, 8.49% MSNF
10% Sucrose	NDM, 96% MSNF (4% moisture)
10% Corn syrup solids (CSS)	Liquid sugar, 66% sucrose
0.3% Stabilizer/emulsifier	Corn syrup, 80% CSS
	Stabilizer/emulsifier

Weight liquid sugar in 100 lbs of mix = 10/0.66 = 15.15 lbs

Weight of corn syrup in 100 lbs mix = 10/0.8 = 12.5 lbs

Total weight of nondairy ingredients = 15.15 + 12.5 + 0.3 = 27.95 lbs

Weight of serum in mix = 100 − (10 + 27.95) = 62.05 lbs

Weight of normal MSNF in serum = 62.05 × 0.088 = 5.46 lbs

Weight of MSNF that must be supplied by NDM
= 10 − 5.46 = 4.54 lbs

$$\text{Weight of NDM needed} = \frac{4.54}{[96 - (0.088 \times 100)]} \times 100 = 5.21 \text{ lbs}$$

Weight of milk and cream needed = 100 − (27.95 + 5.21) = 66.84 lbs

Percent fat in mixture of milk and cream = 10/66.84 × 100 = 14.96

By Pearson Square:

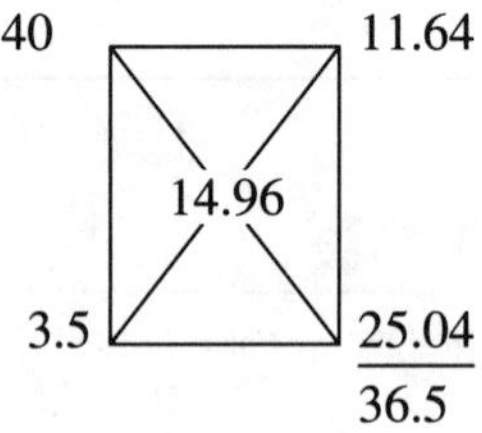

Weight of cream needed = (11.46 × 66.84)/36.5 = 20.99 lbs

Weight of milk needed = 66.84 − 20.99 = 45.85 lbs

The inclusion of sweetened condensed milk introduces another complication which is illustrated in Example 4 and the proof is presented in Table 2.12.

Example 4

Desired composition	Ingredients
10% Fat	Milk, 3.5% fat, 8.49% MSNF
10% MSNF	Cream, 40% fat, 5.28% MSNF
12% Sugar	Sweet condensed milk 8% fat, 22% MSNF, 45% sugar
6% CSS	Liquid sugar, 67% sucrose
0.3% Stabilizer/emulsifier	Corn syrup, 80% CSS
	Stabilizer/emulsifier

Weight liquid sugar in 100 lbs mix = 12/0.67 = 17.91 lbs

Weight of corn syrup = 6/0.8 = 7.5 lbs

Total weight of nondairy ingredients = 25.71 lbs

Weight of serum in mix = 64.29 lbs

Weight of normal MSNF in serum = 5.66

Weight of MSNF to be supplied by sweet condensed milk = 4.34 lbs

Weight of sweet condensed milk needed
(serum = 100 − [8(fat) + 45(sugar)] = 47)

$$\frac{4.34}{22 - (0.088 \times 47)} \times 100 = 24.3 \text{ lbs}$$

Weight of milk and cream needed
= 100 − (wt nondairy + wt sweet. cond. − wt sugar in sweet. cond.)
= 100 − [25.71 + 24.3 − (0.45 × 24.3)] = 60.93 lbs

Weight of fat supplied by sweetened condensed milk
= 24.3 × 0.08 = 1.94 lbs

Therefore milk and cream must supply 10 − 1.94 = 8.06 lbs of fat

% fat in milk and cream mixture = 8.06/60.93 × 100 = 13.23

Weight of cream by Pearson Square = 16.24 lbs

Weight of milk needed = 44.69 lbs

Weight of sugar supplied by sweetened condensed milk
= 24.3 × 0.45 = 10.94

Therefore must reduce liquid sugar by weight that provides 10.94 lbs of sugar:

$$\frac{10.94}{0.67} = 16.33 \text{ lbs}$$

Weight of liquid sugar needed = 17.91 − 16.33 = 1.58 lbs

Weight of water that must be added to compensate for the amount not added by the syrup = 16.33 − 10.94 = 5.39 lbs

Table 2.12 EXAMPLE 4

Ingredient	Weight	Fat	MSNF	Sugar	CSS	Stabilizer/Emulsifier
	(lbs)					
Milk	44.69	1.56	3.79	0	0	0
Cream	16.24	6.5	0.86	0	0	0
Sweet condensed milk	24.3	1.94	5.35	10.94	0	0
Liquid sugar	1.58	0	0	1.06	0	0
Corn syrup	7.5	0	0	0	6.0	0
Stabilizer/emulsifier	0.3	0	0	0	0	0.3
Water	5.39	0	0	0	0	0
Total	100.00	10.00	10.00	12.00	6.00	0.3
Desired	100.00	10.00	10.00	12.00	6.00	0.3

2.3.4 Algebraic Method of Mix Standardization

In this method, the weights of the ingredients are treated as unknowns x, y, and z. With three unknowns, three equations are required which are then solved simultaneously. The process is illustrated in Examples 5 to 8 along with Tables 2.13 through 2.16.

Example 5

Desired composition of Soft-Serve mix	Ingredients
6% Fat	Cream, 36% fat, 5.7% MSNF
12% MSNF	Milk, 3.5% fat, 8.8% MSNF
13% Sugar	Condensed skim milk, 30% MSNF
0.5% Stabilizer/emulsifier	Granulated sugar
	Stabilizer/emulsifier

x = lbs of cream
y = lbs of milk
z = lbs of condensed skim milk

Total weight of milk products in 100 lbs = $100 - (13 + 0.5) = 86.5$

(1) Fat equation	$0.36x + 0.035y + 0z = 6$
(2) MSNF equation	$0.057x + 0.088y + 0.3z = 12$
(3) Milk products equation	$x + y + z = 86.5$
(4) Eq. (1) × 100	$36x + 3.5y + 0 = 600$
(5) Eq. (2) × 100	$5.7x + 8.8y + 30z = 1200$
(6) Eq. (3) × 30	$30x + 30y + 30z = 2595$
(7) Eq. (6) − Eq. (5)	$24.3x + 21.2y + 0 = 1395$
(8) Eq. (4)	$36x + 3.5y + 0 = 600$
(9) Eq. (7) × 3.5	$85.05x + 74.2y + 0 = 4882.5$
(10) Eq. (8) × 21.2	$763.20x + 74.2y + 0 = 12{,}720.0$
(11) Eq. (10) − Eq. (9)	$678.15x + 0 + 0 = 7837.5$

Table 2.13 EXAMPLE 5

Ingredient	Weight	Fat	MSNF	Sugar (lbs)	Stabilizer/Emulsifier
Milk	52.52	1.84	4.62	0	0
Cream	11.56	4.16	0.66	0	0
Condensed skim milk	22.42	0	6.72	0	0
Sugar	13.00	0	0	13	0
Stabilizer/emulsifier	0.5	0	0	0	0.5
Total	100.00	6.00	12.00	13	0.5
Desired	100.00	6.00	12.00	13	0.5

$$x = \frac{7{,}837.5}{678.15} = 11.56 \text{ lbs of cream}$$

(12) Substitute 11.56 for x in Eq. (4)

$$3.5y = 600 - (36 \times 11.56) = 600 - 416.16 = 183.84$$

$$y = \frac{183.84}{3.5} = 52.52 \text{ lbs milk}$$

(13) Substitute 11.56 for x and 52.52 for y in Eq. (3)

$$z = 86.5 - (11.56 + 52.52) = 22.42 \text{ lbs cond. skim milk}$$

Example 6

Desired composition	Ingredients
10% Fat	Cream, 35% fat, 5.27% MSNF
8.5% MSNF	Skim milk, 8.5% MSNF
1.5% Whey solids	Liquid sugar, 67.5% sucrose
8.5% Sucrose	Corn syrup, 80% CSS
8.5% CSS	Dry whey, 97% why solids
0.3% Stabilizer/emulsifier	Stabilizer/emulsifier
	Condensed skim milk, 30% MSNF

Weight of dry whey needed in 100 lbs of mix = 1.5/0.97 = 1.55 lbs

Weight of liquid sugar in 100 lbs of mix = 8.5/0.675 = 12.59 lbs

Weight of corn syrup in 100 lbs of mix = 8.5/0.8 = 10.63 lbs

x = lbs cream

y = lbs skim milk

z = lbs condensed skim milk

Table 2.14 EXAMPLE 6

Ingredient	Weight	Fat	MSNF	WS	Sucrose (lbs)	CSS	Stabilizer/Emulsifier
Skim milk	32.17	0	2.73	0	0	0	0
Cream	28.57	10.00	1.51	0	0	0	0
Condensed skim milk	14.19	0	4.26	0	0	0	0
Dry whey	1.55	0	0	1.5	0	0	0
Liquid sugar	12.59	0	0	0	8.5	0	0
Corn syrup	10.63	0	0	0	0	8.5	0
Stabilizer/emulsifier	0.3	0	0	0	0	0	0.3
Total	100.00	10.00	8.5	1.5	8.5	8.5	0.3
Desired	100.00	10.00	8.5	1.5	8.5	8.5	0.3

$$0.35x = 10$$
$$0.0527x + 0.085y + 0.3z = 8.5$$
$$x + y + z = 100 - (12.59 + 10.63 + 1.55 + 0.3) = 74.9$$

$$x = \frac{10}{0.35} = 28.57 \text{ lbs cream}$$

Substitute 28.57 for x in the remaining equations:

$$0.085y + 0.3z = 8.5 - 1.51 = 6.99$$
$$y + z = 74.93 - 28.57 = 46.36$$

$$0.085y + 0.3z = 6.99$$
$$0.085y + 0.085z = 3.94$$
$$0.215z = 3.05$$

$$z = \frac{3.05}{0.215} = 14.19 \text{ lbs condensed skim milk}$$

$$y = 74.93 - (28.57 + 14.19) = 32.17 \text{ lbs skim milk}$$

Example 7

Desired composition	Ingredients
10% Fat	Cream, 35% fat, 5.27% MSNF
12% MSNF	Skim milk, 8.5% MSNF
13% Sucrose	Sweet condensed skim milk, 30% MSNF, 42% sugar
4% CSS	Corn syrup, 80% CSS
0.3% Stabilizer	Granulated sugar
	Stabilizer

x = lbs cream

y = lbs skim milk

z = lbs sweetened condensed skim milk. Only 58% of it is a milk product as seen in the third equation.

Table 2.15 EXAMPLE 7

Ingredient	Weight	Fat	MSNF	Sucrose (lbs)	CSS	Stabilizer
Skim milk	39.30	0	3.34	0	0	0
Cream	28.57	10.0	1.51	0	0	0
Sweet condensed skim	23.85	0	7.15	10.02	0	0
Sugar	2.98	0	0	2.98	0	0
Corn syrup	5.00	0	0	0	4.00	0
Stabilizer	0.3	0	0	0	0	0.3
Total	100.00	10.00	12.00	13.00	4.00	0.3
Desired	100.00	10.00	12.00	13.00	4.00	0.3

Note: The sugar contributed by the sweetened condensed skim milk is determined by multiplying its weight by 42%. 23.85 × 0.42 = 10.02. The balance of the needed sugar is supplied by the granulated sugar.

$$
\begin{aligned}
0.35x \quad &= 10 \\
0.0527x + 0.085y + 0.3z &= 12 \\
x + \quad y + 0.58z &= 100 - (13 + 5 + 0.3) = 81.7
\end{aligned}
$$

$$
\begin{aligned}
x &= 28.57 \text{ lbs cream} \\
y &= 39.30 \text{ lbs skim milk} \\
z &= 23.85 \text{ lbs sweet condensed skim milk}
\end{aligned}
$$

Occasions may arise when small quantities of certain ingredients are to be used up. Since the weights and compositions of these materials are known, they can be easily accommodated in the calculations. A simple problem has been designed as an illustration:

Example 8

Desired composition	Ingredients	Total weight of mix
10% Fat	150 lbs cream, 20% fat, 7.2% MSNF	4000 lbs
11.5% MSNF	50 lbs condensed skim milk, 30% MSNF	
13% Sucrose	500 lbs milk, 4% fat, 9% MSNF	
4% CSS	Cream, 35% fat, 6% MSNF	
0.3% Stabilizer	NDM, 96% MSNF	
	Skim milk, 9% MSNF	
	Granulated sugar	
	Dry corn syrup solids (use 4% as is)	
	Stabilizer	

150 lbs 20% cream provides 30 lbs fat and 10.8 lbs MSNF
500 lbs milk provides 20 lbs fat and 45 lbs MSNF
50 lbs condensed skim milk provides 15 lbs MSNF

Table 2.16 EXAMPLE 8

Ingredient	Weight	Fat	MSNF	Sugar (lbs)	CSS	Stabilizer
20% Cream	150.00	30.00	10.80	0	0	0
4% Milk	500.00	20.00	45.00	0	0	0
Condensed skim milk	50.00	0	15.00	0	0	0
35% Cream	1000.00	350.00	60.00	0	0	0
NDM	212.05	0	203.57	0	0	0
Skim milk	1395.95	0	125.63	0	0	0
Sugar	520.00	0	0	520.00	0	0
Corn syrup solids	160.00	0	0	0	160.00	0
Stabilizer	12.00	0	0	0	0	12
Total	4000.00	400.00	460.00	520.00	160.00	12.00
Desired	4000.00	400.00	460.00	520.00	160.00	12.00

Total provided by these ingredients
Weight, 150 + 500 + 50 = 700 lbs
Fat, 30 + 20 = 50 lbs
MSNF, 10.8 + 45 + 15 = 70.8 lbs

Total needed in the mix
Weight, 4000 lbs
Fat, 4000 × 10% = 400 lbs
MSNF, 4000 × 11.5% = 460 lbs

x = lbs 35% cream
y = NDM
z = lbs skim milk

$$\begin{aligned} 0.35x \quad\quad\quad\quad\quad\quad &= 400 - 50 = 350 \\ 0.06x + 0.96y + 0.09z &= 460 - 70.8 = 389.2 \\ x + \quad y + \quad z &= 4000 - (680 + 12 + 700) = 2608 \end{aligned}$$

$$\begin{aligned} x &= 1000 \text{ lbs cream} \\ y &= 212.05 \text{ lbs NDM} \\ z &= 1395.95 \text{ lbs skim milk} \end{aligned}$$

Note: Total fat supplied by the miscellaneous ingredients (50 lbs) was subtracted from the total needed in Eq. (1). The same was done in the case of MSNF and total weight. [The 680 in Eq. (3) is the sum of sugar and dry corn syrup solids, and 12 is the weight of stabilizer.]

2.3.5 Restandardizing a Mix of Erroneous Composition

Every plant needs a system that will ensure highest accuracy in testing of mix and ingredients, weighing, and mixing to provide a correct and uniform product composition. An old axiom in chemistry holds that the results of an analysis are reliable only if a representative sample was correctly analyzed. Paraphrased into an ice cream maker's language, it simply says that the samples of ingredients and the mix must be representative and that a mix cannot be accurately formulated if the ingredients are not of a known and uniform composition. Unfortunately, even with all precautions seemingly in place, there may be instances when the mix composition is sufficiently off to require restandardization.

The process of restandardization must comply with all existing regulations and dictates of appropriate enforcement agencies. Compositional imperfections discovered by tests performed right after all ingredients have been thoroughly blended can be corrected prior to pasteurization. This is the most opportune time to make such adjustments. Should restandardization of a pasteurized mix be required, the process becomes more complex. Additional mix with a composition calculated to correct the deficiency needs to be prepared (pasteurized, homogenized, and cooled in an approved manner) and combined with the original mix. The capacity of the available equipment (pasteurizer, storage tank, etc.) will affect the minimum batch size that can be effectively processed for this purpose. Restandardization is a sufficiently sensitive operation that all precautions must be taken to protect the public health qualities of the product. Advance consultation with enforcement agencies on procedures should help in avoiding unpleasantness.

There are several possible scenaria that may necessitate restandardization of the mix. In all cases it may be prudent to recheck the accuracy of the composition and weights, because if these are in error, restandardization may still not accomplish a full correction. Besides, the analysis furnished by the plant laboratory is likely to show only the percent fat and percent total solids. If the weights of the sweeteners is incorrect, the estimate of the MSNF would also be incorrect [MSNF = total solids − (fat + sweeteners + stabilizer)]. These facts point to the necessity of accurate record keeping for every batch of mix made in a format that makes a recheck of all data possible. Following are the various situations that may be encountered:

1. Mix is high in fat and correct in MSNF.
2. Mix is high in fat and high in MSNF.
3. Mix is high in fat and low in MSNF.
4. Mix is low in fat and correct in MSNF.
5. Mix is low in fat and high in MSNF.
6. Mix is low in fat and low in MSNF.

Generally, whenever the fat content is found to be too high, correction is made by determining how much additional mix could be made with the excess fat. For this additional weight, the needed quantities of stabilizer, sweeteners, MSNF (including any that may be deficient in the original mix), and water are calculated to provide the same composition as the original mix was supposed to have. When the fat and MSNF are both high, the MSNF in surplus are subtracted from the total needed in the additional mix. The process is illustrated in Example 9 and the answers are confirmed in Table 2.17.

Table 2.17 EXAMPLE 9

Ingredient	Weight	Fat	MSNF	Sugar (lbs)	CSS	Stabilizer
Original mix	4500.00	517.5	508.5	540.00	270.00	13.5
NDM	9.38	—	9.00	—	—	—
Liquid sugar	36.3	—	—	24.5	—	—
Corn syrup	15.4	—	—	—	12.3	—
Stabilizer	0.61	—	—	—	—	0.61
Water	142.81	—	—	—	—	—
Total	4704.50	517.5	517.5	564.5	282.3	14.11
Desired	4704.50	517.5	517.5	564.5	282.3	14.11

Example 9

Desired	Actual: 4500 lbs mix	Ingredients
11% Fat	11.5% Fat	NDM 96% MSNF
11% MSNF	11.3% MSNF	67.5 °Brix sucrose
12% Sucrose	12% Sucrose	Corn syrup, 80% CSS
6% CSS	6% CSS	Stabilizer
0.3% Stabilizer	0.3% Stabilizer	

Excess lbs of fat $= 4500 \times 0.005 = 22.5$ lbs
Weight of additional mix $= 22.5 \div 0.11 = 204.5$ lbs
Excess lbs of MSNF $= 4500 \times 0.003 = 13.5$ lbs
Weight of MSNF needed in 204.5 lbs of additional mix $= 204.5 \times 0.11 = 22.5$ lbs
Weight of MSNF to be supplied by NDM $= 22.5 - 13.5 = 9$ lbs
Weight of NDM needed $= 9 \div 0.96 = 9.38$ lbs
Weight of sucrose needed in 204.5 lbs $= 204.5 \times 0.12 = 24.5$ lbs
Weight of 67.5 Brix syrup needed $= 24.5 \div 0.675 = 36.3$ lbs
Weight of CSS needed in 204.5 lbs $= 204.5 \times 0.06 = 12.3$ lbs
Weight of corn syrup needed $= 12.3 \div 0.08 = 15.4$ lbs
Weight of stabilizer needed $= 204.5 \times 0.003 = 0.61$ lbs
Weight of water needed $= 204.5 - (9.38 + 36.3 + 15.4 + 0.61) = 142.81$ lbs

When the fat content in the finished mix turns out to be low, a small quantity of additional mix can be made that includes the needed weight of the fat to correct the deficiency. This can be accomplished arithmetically or algebraically, as illustrated in Examples 10 and 11 and Tables 2.18 and 2.19 with confirmation of the calculated results.

Table 2.18 EXAMPLE 10

Ingredient	Weight	Fat	MSNF	Sugar (lbs)	CSS	Stabilizer
Original mix	5000.00	475.00	515.00	600.00	300.00	15.00
Cream	157.14	55.00	8.64	—	—	—
Condensed skim milk	21.2	—	6.36	—	—	—
67.5 Brix sucrose	53.33	—	—	36.00	—	—
Corn syrup	22.5	—	—	—	18.00	—
Stabilizer	0.9	—	—	—	—	0.9
Water	44.93	—	—	—	—	—
Total	5300.00	530.00	530.00	636.00	318.00	15.9
Desired	5300.00	530.00	530.00	636.00	318.00	15.9

Example 10

Desired	Actual: 5000 lbs	Ingredients
10% Fat	9.5% Fat	Cream 35% fat, 5.5% MSNF
10% MSNF	10.3% MSNF	Condensed skim milk 30% MSNF
12% Sucrose	12% Sucrose	or
6% CSS	6% CSS	Skim milk 8.5% MSNF
0.3% Stabilizer	0.3% Stabilizer	67.5 Brix sucrose
		Corn syrup 80% CSS

For the arithmetic solution, the ingredients needed for an additional 300 lbs of mix will be calculated with cream and condensed skim milk as the dairy ingredients.

Shortage of fat in original mix = 5000 × 0.5% = 25 lbs
Weight of fat needed in 300 lbs of additional mix = 300 × 10% = 30 lbs
Total weight of fat needed in 300 lbs of additional mix = 25 + 30 = 55 lbs
Weight of cream needed to supply 55 lbs of fat = 55 ÷ 0.35 = 157.14 lbs
Surplus weight of MSNF in original mix = 5000 × 0.3% = 15 lbs
Weight of MSNF needed in 300 lbs of additional mix = 300 × 10% = 30 lbs
Additional weight of MSNF needed = 30 − 15 = 15 lbs
Weight of MSNF contributed by the cream = 157.14 × 0.055 = 8.64 lbs
Weight of MSNF to be supplied by condensed skim milk = 15 − 8.64 = 6.36 lbs
Weight of condensed skim milk needed = 6.36 ÷ 0.3 = 21.2 lbs
Weight of liquid sugar needed = 300 × 0.12 ÷ 0.675 = 53.33 lbs
Weight of corn syrup needed = 300 × 0.06 ÷ 0.8 = 22.5 lbs
Weight of stabilizer needed = 300 × 0.003 = 0.9 lbs

For the algebraic solution, the cream will be needed to make up the deficiency in fat, but since the MSNF are high, skim milk should be the appropriate choice of MSNF.

Table 2.19 EXAMPLE 11

Ingredient	Weight	Fat	MSNF	Sugar (lbs)	CSS	Stabilizer
Original mix	5000.00	475.00	515.00	600.00	300.00	15.00
Cream	152.50	53.37	8.39	—	—	—
Skim milk	58.66	—	4.98	—	—	—
67.5 Brix sucrose	50.44	—	—	34.05	—	—
Corn syrup	21.28	—	—	—	17.02	—
Stabilizer	0.85	—	—	—	—	0.85
Total	5283.73	528.37	528.37	634.05	317.02	15.85
Desired	5283.73	528.37	528.37	634.05	317.02	15.85

Example 11

The same ingredients and the same mix as in Example 10.

x = new weight of the mix after correction
y = lbs of cream
z = lbs of skim milk

Fat equation $475 + 0.35y = 0.1x$
MSNF equation $515 + 0.055y + 0.085z = 0.1x$
Milk products Eq. $0.7442 \times 5000 + y + z = 0.7442x$

Note:
475 = lbs fat in original mix (5000 × 0.095)
515 = lbs MSNF in original mix (5000 × 0.103)
0.7442 is the percentage of milk products in the mix, obtained by subtracting the weights of nondairy ingredients needed in 100 lbs from 100. In this example, 17.78 lbs of 67.5 °Brix liquid sugar would be needed to supply the required 12 lbs of sugar; 7.5 lbs of corn syrup would supply the needed 6 lbs CSS; and 0.3 lbs of stabilizer must be provided. 17.78 + 7.5 + 0.3 = 25.58. 100 − 25.58 = 74.42. Therefore, the mix contains 74.42% milk products.

When the three simultaneous equations are solved employing the normal rules of algebra, the following results are obtained:

x = 5283.73 lbs (the new weight of the mix)
y = 152.50 lbs (weight of additional cream)
z = 58.66 lbs (weight of additional skim milk)

2.3.6 Mix Made in a Vacuum Pan

Although no longer commonly encountered, making the mix in a vacuum pan has been a viable process. The only dairy ingredients required are cream and milk or

Table 2.20 EXAMPLE 12

Ingredient	Weight	Fat	MSNF	Sugar	CSS	Stabilizer/Emulsifier
				(lbs)		
Milk	106.02	3.17	9.01	—	—	—
Cream	17.97	6.29	0.99	—	—	—
Sugar	10.00	—	—	10.00	—	—
Corn syrup	12.50	—	—	—	10.00	—
Stabilizer/emulsifier	0.30	—	—	—	—	0.3
Total	146.79[a]	10.00	10.00	10.00	10.00	0.3
Desired	100.00	10.00	10.00	10.00	10.00	0.3

[a] Evaporate 46.79 lbs of water.

skim milk. In standardizing a mix to be made in this manner, it is only necessary to bring the fat and the MSNF into the desired ratio. During the vacuum pan operation, enough water is evaporated to yield the desired concentration.

Example 12

Desired composition	Ingredients	
10% Fat	Milk 3.5 fat, 8.5% MSNF	x = lbs cream
10% MSNF	Cream 35% fat, 5.5% MSNF	y = lbs milk
10% Sucrose	Granulated sugar	Fat: MSNF = 10:10
10% CSS	Corn syrup 80% CSS	
0.3% Stabilizer/emulsifier	Stabilizer/emulsifier	

Fat equation $0.35x + 0.035y = 10$

MSNF equation $0.055x + 0.085y = 10$

$x = 17.97$ lbs cream

$y = 106.02$ lbs milk

The results indicate that for every 106.02 lbs of milk, 17.97 lbs of cream, 10 lbs of granulated sugar, 12.5 lbs of corn syrup (10 ÷ 0.8), and 0.3 lbs of stabilizer/emulsifier must be added. When the sum of these ingredients is concentrated to 100 lbs, the resultant product will have the desired composition. The calculations are confirmed in Table 2.20.

2.3.7 Calculating Density and Degrees Baume (Be)

Because of the natural variation in milk composition and differences in mix composition, only approximate density values can be obtained by simple calculation. However, the values provide a reasonable starting point and can be refined by experience and actual measurements. Following are applicable formulas:

$$\text{(1) Specific gravity (60°F, 15.6°C)} = \frac{100}{\dfrac{\%\ \text{fat}}{0.93} + \dfrac{\%\ \text{remaining solids}}{1.601} + \dfrac{\%\ \text{water}}{1}}$$

$$\text{(2) Density (lbs./gal) (60°F, 15.6°C)} = \text{Specific gravity} \times 8.34$$

$$\text{(3) °Be} = 145 - \frac{145}{\text{Specific gravity (60°F)}}$$

$$\text{(4) °Be temperature correction} = 0.03\ \text{°Be per °F}$$

Example 13

Calculate the specific gravity at 60°F and the °Be at 140°F of a mix having the following composition:

14% Fat
10% MSNF
15% Sugar
0.2% Stabilizer

$$\text{Specific gravity (60°F)} = \frac{100}{\dfrac{14}{0.93} + \dfrac{25.2}{1.601} + \dfrac{60.8}{1}} = 1.0918$$

$$\text{°Be (60°F)} = 145 - \frac{145}{1.0918} = 12.19$$

$$\text{°Be (140°F)} = 12.19 - [(140 - 60) \times 0.03] = 9.79$$

The actual weight per gallon and °Be should be determined by physical measurement at the temperature of interest after the mix has been analyzed and found to be of the correct composition. The corrected readings can be used in subsequent runs.

2.4 Formulation

In formulating ice cream mixes, the principal objective is to create a product with physical, chemical, and sensory characteristics perceived by the manufacturer as desirable based on favorable consumer acceptance. Compliance with legal requirements and standards of identity is essential, but within these constraints, the manufacturer is allowed considerable latitude in the choice of formulation. Following is a list of some criteria that may affect the adoption of a particular formulation:

1. Lowest possible price
2. Product and price positioning
3. Competitiveness against market leader

4. Adaptability to available equipment
5. Target sensory characteristics of product
6. Natural label
7. Shelf life
8. Heat shock resistance
9. Meltdown characteristics
10. Flavor release and character
11. Body and texture characteristics
12. Type of product
    a. Sherbet, frozen yogurt, soft-serve, etc.
    b. Super-premium, premium, economy grade, reduced or low-fat, etc.
    c. Reduced calorie, special diet, nondairy, etc.
    d. novelties, molded, extruded type, stick type, still frozen, etc.
13. New product concept
14. Product improvement
15. Emulation of another product.

There are other factors that also have a major impact on the properties of the finished product. Formulation is a major step, but selection of ingredients, flavoring type and quantity, overrun (amount of air whipped into the product), and quality of packaging play a significant role and should complement the selected formula. All processing and freezing steps also contribute in an important way.

In any frozen dessert, the frozen shelf life and resistance to damage caused by fluctuating temperatures (heat shock) are key concerns of the manufacturer. Factors that provide some degree of control over these problems begin with formulation. The total solids content and its components play an important part. When choosing a formula, consideration is given to the desired content of:

1. Fat
2. MSNF
3. Total solids
4. Sweetness level (expressed as sucrose)
5. Stabilizer/emulsifier.

The basic formulas for the various frozen desserts must be considered by product type because of existing standards of identity. In the case of ice cream, the minimum fat and total milk solids content is predetermined by the standard of identity as 10% and 20%, respectively. However, the number of possible formulations with the restricted fat content of 10% is exceedingly high. Some of the possibilities are illustrated in Table 2.21.

Among further variations of the formulations presented in Table 2.21 are: intermediate levels (between 0 and 25% of the MSNF) of whey solids; different percentages of sucrose replacement by corn syrup solids; corn syrups of higher or lower DE; the inclusion of microcrystalline cellulose into formulations other than those indicated; the inclusion of egg yolk solids at different concentration levels (below 1.4% to avoid requirement of labeling the product as frozen custard); inclusion of

Table 2.21 SOME VARIATION IN THE COMPOSITION OF AN ICE CREAM CONTAINING 10% FAT

Constituent	Percent Concentration								
Fat	10	10	10	10	10	10	10	10	10
MSNF	10	7.5	10	8	11	9	7.5	11.5	12
Whey[a]	—	2.5	—	2	—	2	2.5	—	—
Sucrose	15	15	10	10	12	13	—	15	15
CSS[b]	—	—	10	—	6	4	10	—	—
Fructose[c]	—	—	—	5	—	—	8	—	—
Microcrystalline cellulose	—	—	—	—	—	0.25	—	0.3	—
Stabilizer/emulsifier[d]	0.3	0.3	0.2	0.3	0.25	0.25	0.3	0.3	—
Egg yolk solids	—	—	—	—	—	—	—	—	1
Total solids	35.3	35.3	40.2	35.3	39.25	38.5	38.3	37.1	38

[a] Up to 25% of the MSNF may be in the form of whey solids or solids from one of the approved modified whey products.

[b] The CSS were assumed to possess 50% of the sweetening value of sucrose.

[c] High-fructose corn syrup with a sweetening value of at least that of sucrose.

[d] The actual concentration depends on the particular proprietary product used. The supplier's directions should be followed.

egg yolk solids into formulations other than the one indicated; use of stabilizers and emulsifiers made up of different components, etc. Although not all variations produce significant changes, differences in ice cream properties may certainly be brought about by varying the formulations. Products resulting from the illustrated compositions would vary in body and texture, in the freezing point and hardness at any given temperature, in resistance to heat shock, in ingredient cost, in the perception of being "all natural," and possibly in flavor and flavor release. The properties affecting the body and texture of the ice cream can be further modified by the manner of processing and freezing of the mix, the amount of overrun incorporated, the manner in which the flavors are added, and the speed of hardening. The great variety of possibilities testifies to the fact that our federal standards are not a "recipe" forcing everyone to make the same product.

The suggested formulas presented in subsequent pages are intended as starting points in helping ice cream makers to develop formulations with their own specific requirements.

2.4.1 Premium and Superpremium Products

There is no simple or single definition of premium type products and there are several perceptions as to their image and characteristics. The term "premium" cannot be divorced from quality. Therefore, one perception of premium ice cream is that it is made from high-quality ingredients, including the flavorings. The consumer may also expect such a product not to be excessively whipped (not to have a high overrun). Milk fat certainly makes a contribution to the eating quality of ice cream, one that is difficult to emulate with substitutes. Thus a high fat content is compatible with premium eating quality. The visual impact of flavoring is expected to be immediate and positive. This implies that the fruit, nuts, candy, variegating syrups, etc. be distributed in a pattern that is pleasing both for its uniformity and correct quantity. Finally, packaging must receive its due emphasis in conveying the "premium" image. The important criterion of packaging address both attractiveness (eye appeal) and product protection. In summary, a premium or superpremium product may be distinguished from the ordinary product by one or more of the following characteristics:

1. High-quality dairy ingredients
2. High fat content
3. Low overrun
4. High-quality flavoring at optimum level
5. Pleasing visual impact of flavoring
6. Well-balanced formulation for optimum flavor release and body and texture characteristics
7. Attractive, high-quality packaging.

Unfortunately, premium products sometimes fail to live up to their billing. The product may be mishandled during distribution to such an extent that its texture deteriorates, it may shrink, and it may develop off-flavors. In these cases, the manufacturers may or may not be at fault and a reasonable course of action for them is to remove the substandard products from the market and monitor the distribution system in an effort to locate the problem area. In some cases, the fault may be found in the manufacturer's own production and quality assurance program due to human error.

Although a high fat content and low overrun have been commonly associated with the composition of premium ice cream, products with a reduced fat content may also bear premium characteristics. The eating quality of these products can be made outstanding by careful formulation and the judicious choice of the type and amount of flavoring used.

2.4.2 The "All-Natural" Designation

This or similar designations are sometimes perceived as being tantamount to a premium product. They may fall in that category but they do not have to necessarily meet the criteria discussed in the previous paragraphs. Because there is no legal definition of what constitutes an all-natural ice cream, the interpretation is essentially left up to the manufacturers. When questioned, they must be willing and able to defend the label against anyone's objections. In question is the definition of the term "natural." One can argue, on the one hand, that ice cream per se cannot be a natural product. It is not found anywhere in nature in this form and in its manufacture it is heated and homogenized. On the other hand, a great deal of the food that we consume is made up of a number of ingredients that are blended together and heated. Only fruits, nuts, and some vegetables are commonly consumed in their natural state

Table 2.22 WHITE ICE CREAM MIX FORMULATIONS

Fat (%)	10	12	14	16	18
Total milk solids (%)	20–22	20–23	20–24	22–25	24–26
Sweetness as sucrose (%)	13–15	13–15	14–16	14–16	14–16
Stabilizer/emulsifier	Depending on proprietary product used				
Total solids (%)	37–40	38–40	39–41	40–41	40–42

without any processing. Thus, ice cream may approach the natural state only when it is made up of ingredients that are perceived as natural.

When rendering judgment on whether an ingredient may appropriately be designated as natural, the following criteria may be considered: degree of processing; chemical modifications during processing of the ingredient; is it a synthetic ingredient; is it used for cosmetic reasons (e.g., colors); and does it contain chemical additives and preservatives (e.g., in flavoring substances). These points emphasize that the natural designation is a function of ingredient selection rather than formulation. A natural product may be high or low in fat, high or low in total solids, and high or low in overrun.

2.4.3 Formulations for a Plain (White) Ice Cream Mix

The white mix is used for vanilla ice cream and for all other flavors that do not require a chocolate background. The white and the chocolate mixes are usually the only two mixes needed, although a special mix for fruit ice cream can be formulated if desired. The suggested formulations in Table 2.22 provide basic compositional guidelines without addressing the subtleties, discussed earlier, by which the properties of the product made by each of the formulations can be further affected.

2.4.4 Formulations for a Chocolate Ice Cream Mix

A chocolate ice cream may be produced by freezing a white mix flavored with an appropriate quantity of chocolate syrup. In larger plants, however, the general practice is to prepare a separate chocolate mix. The flavor is imparted by the addition of cocoa, chocolate liquor, or both. The suggested formulations do not address the various nuances provided by different types of cocoas and liquors, as these should be considered in the choice of ingredients. Thus, the same formulation may yield different flavor characteristics depending on the choice of chocolate flavoring.

The federal standards of identity contain a provision that allows for a reduction in the minimum fat and total milk solids content due to bulky flavors depending on the amount of flavoring used. In the case of chocolate, the weight of the ingredient is multiplied by 2.5 (this factor is an allowance for the additional sweetener needed) and subtracted from 100. The result is the weight of the mix which must comply with the minimum composition standards. For example, if 5% chocolate liquor is to be used, $5 \times 2.5 = 12.5$; $100 - 12.5 = 87.5$. This weight (87.5 lbs) of a 10%

Table 2.23 CHOCOLATE MIX FORMULATIONS

Milkfat (%)	8.75–10	8.88–10	9.25–10	11	11	12	12
Total milk solids (%)	17.5–19	17.76–19	18.5–19	18–20	18–20	19–20	19–20
Sweetness as sucrose (%)	17–18	17–18	17–18	17–18	17–18	17–18	17–18
Chocolate liquor[a] (%)	5	3.5	0	5–5.5	3	5–6	3–4
Cocoa[a] (%)	0	1	3	0	1–1.5	0	1.5
Stabilizer/emulsifier	Depending on proprietary product used						
Total solids (%)	40.5–42	40–41.5	39–40	40–42.5	40–42.5	41–44	41–43.5

[a] The concentration may be slightly increased or lowered depending on consumer acceptance.

milkfat mix contains 8.75 lbs milkfat. Therefore, this particular mix must contain as a minimum 8.75% milkfat. The milk solids reduction is obtained in a similar way and the new minimum turns out to be 17.5%. In no instance can the milkfat and total milk solids content be reduced below 8% and 16%, respectively. Formulations are given in Table 2.23.

2.4.5 Fruit Ice Cream

Fruits, in various forms, used for flavoring ice cream commonly contain added sugar. The standards of identity recognize this fact by authorizing a dilution factor of 1.4 in calculating the permissible fat and total milk solids reduction in fruit ice cream. If 14% fruit (i.e., pure fruit not including added sweeteners) is used in a 10% fat ice cream, the calculation is as follows: $14 \times 1.4 = 19.6$; $100 - 19.6 = 80.4$. Because 80.4 lbs of mix must minimally contain 8.04 lbs of fat and 8.04% MSNF, the ice cream composition after addition of the fruit cannot be lower than 8.04% fat and 16.08% total milk solids. This is essentially at the limit of the maximum permissible reduction to 8% and 16%, respectively. The factor of 1.4 implies that the flavoring contains 2.5 parts of fruit and 1 part of sugar ($2.5 \times 1.4 = 3.5$), although this does not preclude the use of flavorings with a different fruit to sugar ratio (e.g., 3 + 1 or 4 + 1).

When the quantity of fruit flavoring (with added sugar) exceeds 20% by weight, it can no longer be added to a mix of minimum legal composition (10% fat and 20% total milk solids). To avoid reducing the milk solids below the permissible level, should this situation arise, two options may be considered. One possibility is to start with a mix that has a higher fat and solids content. Another option is to prepare a special mix so designed that it will have the desired composition after addition of the flavoring. Following is an illustration.

To every 70 lbs of mix, 30 lbs of peaches (2.5 + 1) are to be added and the resulting ice cream is to have the following composition:

9% Fat
9% MSNF
17% Sweetness as sucrose
0.3 Stabilizer/emulsifier.

Table 2.24 FORMULATIONS FOR REDUCED FAT PRODUCTS

	Soft-Serve		Hard Frozen[a]	
Fat (%)	2–3	4–7	2–3	4–7
Total milk solids (%)	15–17	18–19	15–17	17–19
Sweetness as sucrose (%)	13	13	15	15
Stabilizer/emulsifier	Depending on proprietary product used			
Total solids (%)	30–32	32–34	35–36	36–38

[a] Depending on the desired results, the formulation may include microcrystalline cellulose in the stabilizer system and bulking agents such as very low DE corn syrups as components of the sweetener solids.

The peach flavoring contains $(1/3.5) \times 100 = 28.6\%$ sucrose and yields $30 \times 0.286 = 8.57$ lbs sucrose. Therefore, 70 lbs of mix must contain:

9 lbs fat or 12.86% fat
9 lbs MSNF or 12.86% MSNF
$17 - 8.57 = 8.43$ lb sweetener or 12.04% sweetness as sucrose
0.3 lbs stabilizer/emulsifier or 0.43% stabilizer/emulsifier.

The peach flavoring must be uniformly distributed and in the correct proportion to yield the targeted composition in the ice cream. The desired sweetness level may be affected by the sensory characteristics of the fruit preparation actually used. Fruit flavors are often complemented by a somewhat higher sweetness level than is usual in vanilla ice cream. Attention should be focused on the stabilizer used to guard against an excessive mix viscosity which could create processing problems. In the freezing process, applicable regulations such as those pertaining to weight per gallon of finished product and the weight of total solids in a gallon of ice cream should not be overlooked.

2.4.6 Products Containing 2 to 7% Fat

Known as ice milk, these reduced fat products may be hard frozen like ice cream or may be sold as soft-serve directly from the freezer. The required properties of the hard frozen product are essentially the same as those of ice cream. The total solids content, stabilization, and emulsification must be adequate to yield a body similar to that of ice cream and to protect the texture against the effects of heat shock. On the other hand, requirements for the soft-serve product chiefly address its appearance as it is drawn from the freezer. It should be dry appearing, stiff, and readily capable of shaping into an attractive serving at the discharge temperature (about 19°F). Although reduced fat products are the ones most commonly encountered in soft-serve form, ice cream may also be frozen in this manner. By comparison to the hard frozen ice cream, the formula for the soft-serve product will likely have a reduced sugar content and a different emulsifier system to prevent churning. Some illustrations are given in Table 2.24.

2.4.7 Products Containing 0 to 2% Fat

Interest in low-fat and no-fat frozen desserts may be seen as a counterattack against a possible market deterioration due to the continuing stream of adverse health claims leveled against the consumption of cholesterol and saturated fat. As this is not the proper place to address the rationale of these assertions, only the available formulation options will be considered. The obvious problem that must be solved is the source of the solids. Attention is usually directed toward fat sparing and bulking agents, fat substitutes, normal sweeteners, and MSNF. In addition to skim milk, a concentrated source of skim milk solids (MSNF), and conventional sweeteners, the list of possible ingredients includes microcrystalline cellulose, maltodextrins, very low DE corn syrups, sodium caseinate, whey protein concentrates, and proprietary products consisting of egg whites, soy proteins, and other vegetable sources treated in a special way to act as fat substitutes or fat sparing agents. Because of the diversity of possible ingredients, a difficulty arises in presenting a model formula. The MSNF content may be between 10% and 15% and the total solids between 30% and 35%, although either of them need not be restricted to these ranges. Care must be exercised in selecting flavorings. Ordinary chocolate, for instance, could raise the total fat content sufficiently to violate the no-fat or less than 2% fat label.

2.4.8 Sherbets and Ices

The requirements for milk products in sherbets are given in the standards of identity as 1% (minimum) to 2% (maximum) milkfat, and 2 to 5% (maximum) total milk solids. Incorporation of the maximum permissible milk solids has beneficial effect on the body and texture but may result in the masking of some fruit flavors. The selected composition should reflect the manufacturer's concept of a high-quality sherbet, which should preferably be based on the interpretation of any feedback received from consumers. Water ices cannot contain any milk solids and thus have an unobstructed flavor release for many fruit flavors. The sugar content of both sherbets and water ices is substantially higher than that of ice cream both in sweetness level and total sweetener solids. The choice of sweetener solids has a definite effect on body and texture but the importance of sweeteners goes beyond that. Sweetener solids, particularly corn syrups, and certain types and amounts of stabilizer are depended on to prevent bleeding (syrup separation and settling), surface crustation (sugar crystallization on the surface), and ice separation in the freezer.

The body and texture of sherbets and ices is significantly affected by sweeteners, stabilizers, and whipping agents. Monosaccharide sweeteners usable at no more than about 25% replacement provide protection against the above problems and impart an excellent flavor release. Corn syrups, however, can be used at higher replacement levels, yield a smoother texture and a firmer although somewhat stickier body. Emulsifiers and whipping agents team up with stabilizers in guarding against a crumbly body which may be a problem particularly with certain orange flavors. The emulsifiers and whipping agents may also permit drawing the product with a somewhat higher overrun (but must weigh at least 6 lbs/gal). The stabilizer should be of a type

Table 2.25 FORMULATIONS FOR SHERBET AND ICES

	Sherbets[a] (%)	Ices (%)
Milkfat	1–2	0
Total milk solids	2–5	0
Sucrose	20–23	23–25
Corn syrup solids[b]	7–11	7–11
Total sweetness as sucrose	26–28	27–29
Flavor[c]		
Stabilizer[d]	Depending on proprietary product used	
Citric acid[e]		

[a] Nonfruit sherbets have a similar composition except for the absence of an acidulant and possibly a lower total sweetness level (more corn syrup and less sucrose). All must contain not less than 1% MSNF. See CFR in Appendix.

[b] The choice of corn syrup to provide these solids is a matter of preference. Criteria to be considered are flavor release, desired sweetness level, total solids content, and hardness when frozen.

[c] The standards of identity specify the following minimum quantities of fruit flavoring on a weight basis: 2% for citrus flavors; 6% for berries; and 10% for all other fruits. Fruit flavoring obtained from proprietary sources should comply with these and other requirements imposed by the standards.

[d] The stabilizer and emulsifier should be appropriate for use in sherbets or ices. The concentration needed depends on the specific product used.

[e] To prevent curdling of the milk solids, the citric acid solution is best added to the cold mix in the flavor tank just before freezing. The standards of identity for fruit sherbets and ices require that enough acid be used to give a titratable acidity of at least 0.35% expressed as lactic acid.

that is stable and effective in an acid solution. Sherbets and ices that are not fruit flavored may now be manufactured without added acid. Most commonly, however, sherbets and ices are fruit flavored and acidified to an acidity between pH 3 and pH 4 (titratable acidity of 0.4 to 0.6% expressed as lactic acid). The amount of acidulant, usually 50% citric acid solution, depends on the concentration of milk solids and the tartness of the fruit flavoring. A taste test should confirm that the appropriate quantity of acid has been selected or added (roughly 8 to 10 oz of 50% citric acid solution per 10 gal, but must be fine tuned). Formulations are given in Table 2.25.

2.4.9 Direct-Draw Shakes

Selection of a formulation for these products should follow consideration of their desired properties, for example, the freezing point, smoothness of texture, body characteristics, whether frozen flavored or unflavored, etc. High total solids and increasing levels of corn syrup favor a smooth textured shake. Some stabilizers and the level of stabilizer also affect the smoothness as well as the body of the shake. The stabilizer and the MSNF may affect the whipping properties if the shake is whipped on a spindle after removal from the freezer. A difference may be encountered in the legal definition of "shake" and "milk shake." Generally, with "milk" in the name, compliance with the existing standards for milk composition is required. Applicable regulations must be checked. Formulations are given in Table 2.26.

Table 2.26 FORMULATIONS FOR DIRECT-DRAW SHAKES

	Unflavored (%)	Chocolate Flavor (%)
Milk fat	2–4	2–4
MSNF	10–14	10–12
Sucrose	7–9	9–11
Corn sweetener	0–4	0–4
Cocoa	0	1–1.75
Stabilizer	Depending on proprietary product used	
Total solids	23–27	25–30

Table 2.27 FORMULATIONS FOR FROZEN YOGURTS

	Hard-Frozen (%)	Soft-Serve (%)
Milkfat	0–3.5	0–2
MSNF	10–14	10–14
Sucrose	9–11	9–10
CSS	9–12	6–8
Sweetness (as sucrose)	15	13–14
Stabilizer	Depending on proprietary product used	
Total solids	34–36	30–31
Acidity	As required by existing regulations or higher, if desired	

2.4.10 Frozen Yogurt

These products are perceived as having either a low or no fat content. They differ from ordinary low fat products by the presence of viable microorganisms used in yogurt fermentation, *Lactobacillus bulgaricus* and *Streptococcus thermophilus*. Because a prescribed level of developed acidity must be produced by fermentation, without subsequent pasteurization (which would destroy the viable microorganisms), a fermentation step must be included in the process. Unfortunately, sweeteners, at the level used in frozen desserts, inhibit the growth of yogurt bacteria and, thus, the required or desired level of acidity may not be obtained by culturing the whole mix. One option is to prepare a sweet (uncultured) yogurt base to which a cultured yogurt of known acidity is added. The composition of the base depends on the composition and acidity of the cultured yogurt, the amount of flavoring used, and the final composition and acidity desired. The sample formulas in Table 2.27 are for the finished yogurt mix without the flavoring. If the flavoring causes excessive dilution, the solids content may be increased accordingly.

2.4.11 Other Frozen Desserts

A federal standard of identity has been established for a product designated as Mellorine. Its fat content is not less than 6% and is usually of vegetable origin. Except for the source of fat, the formulation for Mellorine is essentially the same as that for an all-dairy product frozen dessert of the same composition (e.g., 6% fat). The quality, flavor, and melting point of the substitute fat have a strong effect on the flavor and consistency of the frozen product. See CFR in the Appendix.

2.4.12 Nonstandardized Products

Because of the many forms that they may assume, these products cannot be readily defined. Some may contain dairy ingredients, and others may be entirely nondairy. They may be sweetened in the conventional manner with sucrose and corn syrups and stabilized with the commonly used gums. The fact that they are not standardized leads to a variety of possibilities.

Attention will be focused on a group identified as artificially sweetened frozen desserts. Before marketing products of this type, manufacturers must carefully study and comply with all regulations pertaining to reduced and low-calorie foods, foods useful to diabetics, and all applicable labeling regulations. From the technological standpoint, the main problem is to find acceptable and effective substitutes for sucrose and corn sweeteners to provide solids, maintain the desired freezing point, and to ensure absence of off-flavors. Some of these ingredients contribute the same caloric input of 4 Cal/gram as sucrose (maltodextrin, sorbitol, glycerol, and fructose); others contribute fewer calories (polydextrose 1 Cal/g). Some may be acceptable to diabetics (e.g., sorbitol and fructose) and some may impart off-flavors at relatively low levels of usage (e.g., glycerol). To a varying extent, some are sweet (sorbitol, fructose, and glycerol). Some depress the freezing point to a greater extent than sucrose (sorbitol, fructose, glycerol), some to a lesser extent (polydextrose), and some little or not at all (maltodextrin). A freezing point between 27 and 28°F is desirable. A suggestion has been made that the lactose in MSNF be hydrolyzed by the enzyme lactase to produce the monosaccharides dextrose and galactose. This could assist in lowering the freezing point and producing a frozen dessert with an acceptable consistency at normal serving temperatures. It should be noted that a product made with skim milk solids cannot be made sugar-free because lactose (which is hydrolyzed to galactose and dextrose) is a normal component of MSNF.

Depending on the choice of bulking agents, part or all of the desired sweetness must be imparted by an artificial sweetener that has been specifically approved for use in this type of product. The latest provisions of Title 21, Code of Federal Regulations are applicable and any individual state regulations must also be complied with. Suggested formulations for the use of specific bulking agents and other technical assistance may be obtained from their suppliers and the suppliers of other ingredients including the artificial sweetening agent. Depending on the type of product desired, one possible combination of bulking agents may be 6% sorbitol, 6% polydextrose-K, and 4% maltodextrin. The fat content may be from 2 to 4%, MSNF from 12 to 13%. An effective stabilization system should be chosen for optimum body and texture characteristics.

2.5 Mix Processing

2.5.1 Pasteurization

All of the mix ingredients have to be assembled in the correct proportion, blended to achieve uniform dispersion, pasteurized by one of the legally accepted methods, homogenized, and cooled. The method of pasteurization, batch versus continuous, affects the manner in which some of the steps are accomplished and the type of equipment required. Thus, there are multiple ways in which the mix may be processed at the option of the manufacturer, although good manufacturing practices must be followed in all cases. The major criteria on which a choice of method is based include the ubiquitous economic factors, product characteristics, available equipment, size of operation, plant efficiency, etc.

2.5.1.1 Assembly of Ingredients

The available options for assembling exact quantities of ingredients include scales (or tanks on scales), load cells, and liquid flow meters. All must be carefully and frequently calibrated. The sole reliance on volume measurements of liquids of known density (lbs/gal) may come close, but is lacking in full control (e.g., entrained air could be a problem). Ideally, a printout record of the weight of each ingredient should be available to assist in addressing compositional problems, should any be encountered (may also be useful in inventory control).

With the batch pasteurization system, the desired weights of all ingredients are delivered to the pasteurization vat. Powders (e.g., NDM, whey powder, granulated sugar, and stabilizers) may be preliquified by passage through a funnel pump or other means (e.g., a blender) if difficulties are encountered in properly solubilizing them during the pasteurization process. Stabilizers must be handled carefully to ensure good dispersion, hydration, and complete ''solubility.'' Generally, few difficulties are encountered in batch pasteurization when the supplier's recommendations for a specific stabilizer are followed. Careless addition of some stabilizers may result in undissolved ''lumps'' which may deposit downstream. Checks on the composition of the mix (fat and total solids) should be made after all ingredients have been assembled and the batch has been sufficiently agitated to provide a uniform sample. If necessary, adjustments to the fat and total solids content should be made before the start of pasteurization.

When continuous pasteurization is employed, the systems for the assembly of ingredients may vary in detail, but all must satisfy certain requirements. The ingredients, in the cold, must be brought into uniform suspension and remain so while the product is pumped through the continuous pasteurizer. A batching tank of sufficient size is needed to hold all of the ingredients that are finally assembled prior to continuous pasteurization. Some means must also be provided for the more difficult dispersion of dry ingredients and very viscous syrups. Blenders or liquifiers provide this function. Recirculating milk and cream (from the batching tank or the supply tanks) as well as water, if the mix formulation calls for water, act as the dispersing medium. When all of the ingredients are finally assembled in the batching tank and after sufficient agitation, which from experience has been shown to provide a uniform composition, a sample is taken for analysis to determine if restandardization is necessary. Other means for dispersing ingredients may also be found effective under specific local conditions. A funnel pump may be used to assist in dispersing dry ingredients. A presolution tank may be a heated liquifier (blender) or a small capacity heated and agitated vat in which materials that are very difficult to disperse are "presolubilized." A good example is chocolate liquor and some stabilizers.

Obviously, the system that is selected must accommodate specific requirements as to the size of batches, number of different mixes made, and ingredients used. Based on consideration of these requirements, a determination is made as to the type and size of equipment and the number of batching tanks needed for most efficient operation. There are some quality considerations that may be affected by this step of mix processing. Severe agitation in the presence of raw dairy ingredients may cause the development of a rancid flavor. This danger is magnified if any of the other ingredients have been homogenized (e.g., reprocessed mix). Under some conditions, the sum of the times needed to assemble and disperse the ingredients and hold them for test results may be sufficient for rancid flavor development. Rancidity is also promoted by foam formation. Air incorporation should be kept to a minimum even when only pasteurized dairy ingredients are used in the mix. During heating any collapsing foam may be a contributing factor to burn-on on the heating surfaces. In addition to causing possible operating and cleaning difficulties, burn-on may also contribute to the development of a scorched flavor. Air interferes with effective homogenization later in the process and may contribute to whey separation in the mix and finished product after melting.

2.5.1.2 Pasteurization

Pasteurization requirements may vary in different localities, but the minimum time–temperature combination employed should coincide with (or exceed) the conditions specified in 21 CFR Part 135.3. The requirements set forth there are:

155°F for 30 min
175°F for 25 s

This section also includes the statement "or other time–temperature relationship which has been demonstrated to be equivalent thereto in microbial destruction." A check with a regional office of the Food and Drug Administration[27] has revealed that in the opinion of the FDA, the following time–temperature relationships are equivalent for the pasteurization of frozen desserts:

180°F for 15 s
191°F for 1 s
212°F for 0.01 s

There are additional requirements that address the design, installation, and operation of the equipment. All time–temperature relationships listed may be considered as minimum requirements, so that a higher temperature or a longer holding time may be employed as long as both minimum conditions have been met. The main purpose of pasteurization is the protection of human health, the importance of which certainly warrants the requirement that each installation of a pasteurizer be thoroughly checked out by competent regulatory authorities.

2.5.1.3 Batch Pasteurization

The minimum requirement of heating to 155°F and holding for 30 min at 155°F is commonly exceeded. The holding time of 30 min is retained (actually, some of the mix is held considerably longer depending on the time required to empty the vat) but the final temperature may be in the range of 160 to 170°F. The aim is not just an improvement in keeping quality of the mix (in an ice cream plant the mix may be frozen the same or the next day), but the beneficial effects that the heat treatment has on the body and texture of the ice cream. It promotes increased hydration of the proteins and stabilizers, and other interactions between the mix constituents which all combine to yield body and texture characteristics that some manufacturers desire.

The system may be quite simple, consisting of a pasteurizing vat of approved design with connections downstream to a homogenizer. When two or more pasteurization vats are operated simultaneously, the operation may be timed to be essentially continuous providing all the equipment downstream is designed to handle the flow at the needed rate. The batch pasteurization method has a number of positive attributes. Small quantities of various milk products that need to be used up and any reprocess product can easily be handled in the pasteurizing vat; the system is effective in removing dissolved air from the mix; and intermixing of different successive batches can be kept to a minimum. For many years, this was the standard method for processing mix and some manufacturers choose to use it to this day.

2.5.1.4 Continuous Pasteurization

There are three principal versions of continuous pasteurization: high temperature–short time (HTST, 175°F for 25 sec), the originally approved process (more recently, another version of the HTST process gained acceptance, 180°F for 15 s); and higher heat–short time pasteurization (HHST of either 191°F for 1 s or 212°F for 0.01 s, the holding times in both cases being calculated). Processing temperatures or holding times above their minima for each process may be encountered in industry as manufacturers strive for various objectives in product properties and characteristics.

Another process that has been in use around the world is termed ultra-high-temperature (UHT) processing. The product may be ultrapasteurized (minimum of 280°F for 2 s) to achieve extended shelf life at refrigeration temperature, or sterilized to provide room temperature stability. Shake mixes and soft-serve mixes are UHT processed by some manufacturers.

The primary advantage of HTST pasteurization is the saving in energy, both in heating and cooling, achieved by the use of the regeneration principle. With 70 to 90% regeneration feasible, one would expect that more product can be processed with the same boiler and refrigeration compressors than would have been possible without employing regeneration. Maximum energy savings are realized when operating as close to the mandated minimum temperature as possible but sensory properties are affected by the heat treatment and should be monitored. If the body and texture fall short of expectations and no fault can be found in the freezing, hardening, and distribution process, remedial measures may be sought along the following lines: check formulation; check ingredients; evaluate stabilizer system for type and quantity; check on hydration and solubilization of dry ingredients during the assembly step; consider increasing holding time, raising pasteurization temperature, or both; and make trial runs to evaluate corrective steps and confirm that desired improvements have been achieved.

2.5.1.5 Effect of Heat Treatment

Progressively higher heat treatment increases the amount of bound water in the mix. Although viscosity measurements do not directly assess the quantity of bound water, under controlled conditions they provide indirect evidence for water binding. This is illustrated in Table 2.8 which shows the basic viscosities obtained in various systems after almost instantaneous heating and cooling. The observed changes in viscosity with increasing heat treatment were magnified as the solids content increased. The increase in viscosity as a result of heat treatment is commonly described as the superheating effect, after the process that leads to the production of superheated condensed milk. Several facts concerning it are known. The effect is influenced by the previous heat history of the ingredients, the total solids content, the protein content, and the ratio of casein to whey proteins. Any developed acidity in the mix can enhance the viscosity development to the point that processing problems are encountered. Milk from different localities and obtained during different seasons of the year may vary in composition and salt balance (the balance between calcium and magnesium ions versus the citrate and phosphate ions). This, in turn, affects the stability of proteins toward heat. When raw milk is contaminated with psychrotrophic bacteria (bacteria that grow in the cold), the proteins may be attacked by proteolytic enzymes. Any change in the proteins may influence heat stability.

When an ice cream mix is heated, interactions apparently occur between all of the mix constituents.[28] Because fat globules are coated with a layer of protein on their surface, they are no exception. The gist of this discussion is merely to illustrate that there are many factors that may be involved in mix behavior fluctuations. Over a period of years, manufacturers have used a number of time–temperature relationships to capitalize on the superheating effect (e.g., 220°F for 25 s), but presently processing temperatures are largely between 176 and 190°F, and holding times between 25 and 50 s. The exceptions are ultrapasteurized and sterile mixes which are heated to much higher temperatures.

2.5.2 Homogenization

The mix is homogenized to prevent churning of the fat in the freezer. When the size of all but a few of the fat globules is reduced to between <1 and 2 μm and the globules are evenly distributed without clumping, the tendency to churn is drastically reduced. The efficiency of the homogenizer should be checked by examining a diluted sample of the mix under a microscope. Some of the factors that affect homogenization and its efficiency are condition of the homogenizing valve; condition of suction and discharge valves; temperature of homogenization; homogenization pressure; type of homogenizing valve; number of pistons; single- versus two-stage homogenization; fat-to-protein ratio; salt balance; location of the homogenizer relative to the pasteurization system; presence of air in the mix; etc.

There is no conflict between the functions of homogenization and the deemulsification properties of emulsifiers, as may appear at first glance. The fat destabilizing effect of emulsifiers should be accomplished under controlled conditions to avoid excessive churning with such consequences as progressive buildup of fat in the freezer, a greasy mouth coating sensation when the product is consumed, and poor meltdown characteristics of the ice cream. The fat in unhomogenized mixes would be very apt to separate or churn in the ice cream freezer. Faulty homogenization, due to such conditions as defective or poorly fitting valves and fluctuating pressures, provides an opportunity for some of the fat globules to escape homogenization and hence to be more susceptible to churning. Thus, a measure of control is provided by the fat globules which have been reduced in size by homogenization and stabilized by their newly acquired membrane.

2.5.2.1 Homogenization Temperature

To be effectively homogenized, the fat must be completely in the liquid state and preferably at a temperature well above its melting point. The lowest acceptable temperature is about 140°F. Below this temperature, and depending on the type of mix being processed, fat globule clumping and excessive viscosity may be encountered. The upper temperature limit for homogenization does not appear to be firmly established but is affected by the design and construction of the homogenizer. Temperatures between 150 and 185°F are usually found satisfactory.

2.5.2.2 Location of the Homogenizer

In the batch pasteurization system, the homogenizer is located immediately downstream from the pasteurizer. Thus, the mix can be homogenized at the pasteurization temperature (155 to 165°F) and discharged from the homogenizer directly to the cooling system.

With a continuous pasteurization system there are several options to consider. Because the homogenizer is a positive displacement pump it can assume the function of a timing pump to ensure that the mix has been held at the pasteurization temperature for at least the legally required holding time. Alternatively, another positive displacement pump may be used as the timing pump. Because it would be extremely difficult to operate two positive pumps at exactly the same rate, the design of the system must provide for a relief mechanism that meets both engineering and public health requirements.

The several possible locations for the homogenizer in the HTST system include: between the raw regenerator and the heater; after the heater but before the holding tube; and between the end of the holding tube and the pasteurized regenerator. Some units may be equipped with a split regenerator. In this case, the mix could be homogenized between the first and the second pasteurized regenerator section.

When the mix enters the homogenizer, it is desirable to have all of its constituents fully in solution to avoid damage to the homogenizer valves by hard crystalline materials. This consideration implies that it may be wise to locate the homogenizer at some point after the final heating section. Whether it is installed before or after the holding tube depends on whether or not the homogenizer is used as the timing pump. The temperature of the mix increases as it passes through the homogenizer. With the homogenizer located just ahead of the holding tube, the increase in temperature may constitute a part of the heating process as the mix will be held at this final temperature in the holding tube. A prediction of which arrangement will provide the best results is difficult to render without carefully weighing all locally applicable conditions. A reasonably safe general statement is that the homogenizer should be located as far downstream as is necessary to ensure that the fat and emulsifiers are melted; to control mix viscosity; to obtain full hydration of the stabilizer(s); to provide complete solution of all crystalline materials (e.g., lactose); and to ensure the least emulsion destabilizing effect by subsequent flow through the system.

2.5.2.3 Homogenizing Pressures

Due to the existence of a variety of homogenizers and homogenizer valves, one cannot identify a single pressure that would prove satisfactory for all of them. The aim is to achieve effective homogenization at the lowest possible pressure, for economic reasons, or at a pressure that provides the desired results intended for the characteristics of the finished product. There are indications that homogenization pressure affects mix viscosity, whipping properties, and body and texture characteristics of the ice cream. For an average white mix, some two-stage homogenizers yield good results with pressures of 2500 and 500 lbs/square inch on the first and second valve, respectively. Optimum pressures for a particular unit should be determined on the basis of microscopic examination and product performance.

High-fat mixes generally cannot be homogenized at the same high pressure as low-fat mixes because their viscosity may increase beyond a manageable level. Chocolate mixes tend to behave in a similar manner. Experience should reveal how much the pressure can be reduced and still maintain satisfactory homogenization and an acceptable viscosity. A reduction of about 1000 lbs in the first stage pressure may be necessary in some units for a mix with 16 to 18% fat (in the present example, a reduction to 1500 to 500 lbs/square inch). Homogenization of a high fat mix creates such crowding of small globules that clustering is difficult to control. Usually, high-fat mixes also have a lower protein content which may contribute to clustering of fat globules. The problem may be further compounded by an unfavorable salt balance due to season or origin of the milk solids.

2.5.2.4 Condition of the Homogenizer

Difficulties that are traceable to homogenization are very often due to wear and pitting of the homogenizer valves and the suction and discharge valves. Periodic inspection of the valves should be a routine procedure. The equipment manufacturer can be consulted for specific guidance. When homogenization problems due to defective valves are encountered, increasing the homogenization pressure may do little to correct the difficulty. Entrained air in the product or leakage on the suction side of the homogenizer will cause the pressure to fluctuate and the homogenizer to "move around." Subsequent problems may be encountered in whey separation and freezing. When problems are encountered, the equipment manufacturer may be requested to validate the accuracy of the readings on the pressure gauges.

2.5.3 Mix Cooling and Storage

The mix should be cooled rapidly to a temperature within the range of 32 to 40°F. With the batch method of pasteurization, the flow continues from the pasteurizer to the homogenizer and the cooler (surface cooler, plate cooler, etc.). In the continuous pasteurization system, the cooling segment is generally an integral part of the equipment. In a typical plate pasteurizer, the mix exits the holding tube through a flow diversion valve and enters the pasteurized regenerator where it is partially cooled by the incoming cold, raw mix. From there it flows directly to the cooling section and on to the pasteurized mix holding tank. Although there are effects on the mix attributable to the rate of cooling, the overriding criterion is one of public health significance that requires that the mix be cooled as rapidly as possible.

2.5.3.1 Aging of the Mix

After a mix has been cooled to its storage temperature physical changes occur that are beneficial to freezing properties. The process of holding the mix prior to freezing is called aging. When the mix is stabilized with gelatin, improvements in the body and texture of the ice cream may be experienced over an aging period of up to 24 h. The performance of mixes with gum stabilizers is less dependent on aging, although a short aging period is still beneficial. As the temperature of the mix is rapidly reduced from the pasteurization temperature, some of the physical changes fail to reach an immediate "equilibrium." Perhaps the most obvious is fat crystallization (hardening) which continues over several hours after cooling.

In addition to fat crystallization other changes that may be expected include continued hydration of proteins, stabilizers, and other mix constituents. These are usually associated with an increase both in apparent viscosity due to a progressive gel structure formation, and in basic viscosity in response to increased hydration, fat crystallization, and, possibly, continuation of the interaction of various mix constituents. Changes at the surface of the fat globule membrane appear to depend on the presence and type of emulsifier. The heat treatment may have resulted in considerable whey protein denaturation (e.g., in ultrapasteurization), but one can only speculate whether any casein–whey protein complexes formed,[29,30] along with other heat-induced interaction products, actually participate in defining the outer layer of the fat globule membrane. The presence of casein micelles surrounding the fat globule has been demonstrated by electron microscopy.[31,32] In the absence of an emulsifier, the coating of the fat globules may continue. In the presence of emulsifiers, the protein layer at the surface is progressively replaced by the more surface active lipid emulsifiers.[16,30,31,33] The implication is that with time (aging) the fat globules become more completely covered with more hydrophilic emulsifiers and, therefore, more subject to destabilization (see Section 2.2.19).

Unfortunately, variables encountered in different commercial ice cream plants may introduce an element of uncertainty into predictions of expected behavior derived from experiments performed under carefully controlled conditions. However, an awareness that the changes discussed previously are taking place during aging should assist in addressing certain problems (e.g., products of the same composition may be found wetter in appearance when drawn from the freezer than expected, or in another case, to be excessively churned). Obviously, these problems may also be due to mix processing, homogenization, freezing, flavoring, and air incorporation variables (e.g., pitted homogenizer valves; fluctuating pressures; insulating film of oil in the refrigerant side of the freezer barrel; changes in the mix temperature on entering the freezer; changes in drawing temperature and in the rate of throughput; poor mechanical condition of the freezer including the dasher, blades, pumps, and controls; percent overrun, etc.). In spite of these complexities, experience indicates that improvements in mix performance in the freezer can be attributed to mix aging, although the optimum time and the time needed before a point of diminishing returns is reached depends on local conditions and should be established under those circumstances. A general suggestion to allow mixes to age at least about 4 h appears reasonable.

When the production schedule in a plant is not encumbered by a lack of available mix storage capacity, the age of the mix when frozen into ice cream is controllable and can include an aging period that has been found to be optimal. When mix is manufactured for sale to other freezer operators, it is bound to be adequately aged under most circumstances, but its age when frozen is unpredictable and largely limited by its useful storage life (keeping quality). Aseptically packaged sterile mix may be held for periods of weeks or months before freezing. It would appear that the performance of mixes manufactured for sale should be checked at several levels of their expected useful storage lives. In the field, the temperature abuse and postpasteurization contamination of the mix (beginning at the plant) should not be overlooked as possible sources of uncontrolled variables.

2.5.3.2 *Mix Packaging*

When sold in liquid form, mix is packaged in containers of the desired size ranging from ½ to 5 gal. The filling operation is analogous to that of fluid milk, that is, the product may be filled into paper cartons, or plastic bags for larger volumes that, in turn, are loaded into boxes or crates. Equipment designed to fill products with extended shelf life is gaining in popularity as manufacturers strive for greater efficiencies in product distribution.

When the entire batch of mix has been transferred to the pasteurized mix storage vat, a sample of it should be analyzed for fat and total solids. If the composition is off, additional mix may have to be prepared of such composition as to bring the total batch within specifications. Hopefully, all of the compositional corrections were made prior to pasteurization so none will be required at this point. Other quality control tests may be performed at this time on the mix including bacterial and taste tests. Tests for potential pathogens (e.g., *Salmonella* and *Listeria*) are performed by outside laboratories on a contractual basis. Mixes sold in liquid form should also be checked for keeping quality in their final container.

2.6 Flavoring of Frozen Desserts

Some ice cream plants may prepare one or more of their own flavorings, but in the usual case, flavorings are purchased from suppliers specializing in their manufacture. Obviously, close cooperation between the supplier and the ice cream manufacturer is needed to ensure that flavors with the desired characteristics and highest possible quality are furnished and properly handled. Because the success of any flavor in frozen desserts depends on how well it is accepted by the consumers, both the supplier and the ice cream manufacturer have a vested interest in the performance of a specific product. There are many variations in the type of flavor that may be imparted to frozen desserts. The word ''type'' is used here to mean different characteristics of the same flavor. This is why not all vanilla ice creams taste the same. The same is true for chocolate, fruit, nut, and other flavors. The factors that affect the selection of a flavoring include perceived or demonstrated consumer acceptance; cost; available equipment to handle a particular flavor or flavor combination; ''all-natural'' considerations; customer requests; rate of ''movement'' of an item; whether seasonal or year around; packaging considerations; flavor stability; etc.

Flavorings may be either concentrated or bulky. Concentrated flavorings may be derived from natural sources, be synthetic (artificial flavor), or a combination of both. Some concentrated flavors may be derived from a single source (e.g., strawberry); others may contain two or more natural derivatives in which case they carry the designation WONF (with other natural flavors). Labeling requirements are quite explicit in the choice of language to differentiate between naturally and artificially flavored products and products that contain both natural and artificial flavor. These provisions may be found in Title 21 of the code of Federal regulations, reprinted here in the Appendix.

The chemical constituents responsible for the aroma portion (which is the principal portion) of flavor in a given food are present in minute quantities (parts per million and parts per billion). A concentrate of these flavor imparting compounds may be known as an essential oil (e.g., an essential oil of orange). An alcoholic solution of the essential oil becomes an extract, in this example, an orange extract. Water dispersions of the essential oils can also be prepared and are available under the name of flavor emulsions. The chemical compounds of flavor significance in the essential oils vary from one specific flavor to another but generally include organic compounds of the types known as ketones, acids, alcohols, aldehydes, esters, lactones, etc. Many of the specific flavors have an ''impact constituent,'' which is a compound that makes the major contribution to that particular flavor. This constituent also may be the principal component of an artificial flavor simulating the natural one. A familiar example is vanillin, the main flavor component of vanilla extract. Imitation vanilla usually contains methyl vanillin. However, the natural vanilla flavor also contains other compounds that complement the flavor and, therefore, an imitation vanilla cannot be expected to fully emulate true vanilla.

Besides the most popular flavoring, vanilla, the large variety of frozen dessert flavors includes those derived from sugar caramelization; the browning (Maillard) reaction (the flavor of chocolate, coffee, tea, nutmeats, maple, and baked goods produced by the action of heat, i.e., boiling, roasting, or baking); fruits; candy; ground spices; liqueurs; etc. The manner in which some flavorings are used is rather straightforward. When the label of the product identifies it as strawberry ice cream, for instance, there is a reasonable expectation that the flavor be recognizable in a blindfold test and that it be reminiscent of high-quality strawberries (and not be over- or underflavored). The visual impact of fruit particles is also significant, although some of the best looking berries may be severely lacking in strawberry flavor. Other flavors may require a considered judgment on how they should be presented. In chocolate chip ice cream, for instance, how large should the chips be and should they be sweet, semisweet, or bitter? How large should the pieces of nutmeats be? What percent weight of candy, nuts, fruit pieces, variegating syrups, etc. should be incorporated? What should be the background flavor in products such as butter pecan and how strong should it be? It may appear that there are more questions than answers, but actually there are more answers than questions because ice cream manufacturers may solve each problem in their own way. There obviously are some wrong answers, but there are also several correct answers to each question. As always, consumers, right or wrong, exercise the ultimate authority by determining the degree of the product's acceptance.

With few if any exceptions, at least some components of the total flavorings are added to the pasteurized mix in the flavor tank just before freezing. The potential of product contamination by bacteria and foreign substances must be dealt with by rigid sanitation, hygienic personal habits, and good housekeeping. Whether added to the flavor tank or later in the process, the flavorings must not serve as a vehicle for contamination. Nothing should be left to chance and specifications for flavoring materials should include both flavor quality criteria and criteria addressing bacterial content and product purity. There can be cases when sanitary considerations will preclude the use of an otherwise desirable flavoring. For example, frozen fruit packs can be an excellent source of fruit flavor, but in some cases their bacterial and coliform count may make their use unwise or illegal. Flavor extracts and solutions of colors may also be a source of bacterial contamination.

Some flavors may be incorporated entirely in the flavor tank; others require addition in part both before and after freezing. Materials that are incorporated into the flavor tank must be completely dispersible and contain no particulate matter of the type that would cause wear to parts of the continuous freezer. Concentrated flavors added in small quantity must be agitated sufficiently to ensure uniform distribution. Particulate materials that are intended to make a "showing" in the ice cream (and those unsuitable to run through the freezer) are fed through a fruit or ingredient feeder directly into the ice cream as it exits the freezer. Syrups for variegating are introduced into the exiting ice cream by means of a syrup pump.

When the flavoring is added directly to the mix, both the mix and the bulk of the flavoring can incorporate air during the freezing process. The whipped-in air is uniformly distributed throughout the ice cream. This is not the case with flavorings that are introduced as the ice cream leaves the freezer. In the latter case all of the air is contained in the mix portion. An example will illustrate the achieved overrun in the mix portion when the ice cream is drawn at 4.5 lbs/gal and the same amount of flavoring is added either before or after freezing.

Assume: weight of mix 9 lbs/gal
Weight of flavoring 10.7 lbs/gal
Weight of ice cream 4.5 lbs/gal
Flavoring added at the rate of 20% by weight
4.5 lbs of ice cream contains 80% × 4.5 = 3.6 lbs mix and
20% × 4.5 = 0.9 lbs flavoring

Case I Flavoring added directly to the mix

3.6 lbs of mix occupies a volume of 3.6/9 = 0.4 gal
0.9 lbs of flavor occupies a volume of 0.9/10.7 = 0.084 gal
Total volume occupied by 4.5 lbs = 0.4 + 0.084 = 0.484 gal[a]
Weight per gallon of flavored mix = 4.5/0.484 = 9.3 lbs/gal

$$\% \text{ Overrun} = \frac{\text{wt of 1 gal mix} - \text{wt of 1 gal ice cream} \times 100}{\text{wt of 1 gal of ice cream}}$$

$$= \frac{9.3 - 4.5}{4.5} \times 100 = 107\%$$

[a] No change in volume due to mixing of two liquids is assumed

Case II Flavoring added after freezing

Volume occupied by the flavoring $= 0.9/10.7 = 0.084$ gal
Volume occupied by mix $= 3.6/9 = 0.4$ gal
Volume occupied by ice cream minus flavor $= 1 - 0.084 = 0.916$ gal

$$\%\ \text{Overrun} = \frac{\text{Volume of ice cream} - \text{Volume of mix}}{\text{Volume of mix}} \times 100$$

$$= \frac{0.916 - 0.4}{0.4} \times 100 = 129\%$$

The same answer can be obtained by using the weight formula for overrun as long as the weights are for the same volume product.

Weight of ice cream minus flavor $= 3.6$ lbs/0.916 gal
Weight of mix/0.916 gal $= 9 \times 0.916 = 8.244$

$$\%\ \text{Overrun} = \frac{8.244 - 3.6}{3.6} \times 100 = 129\%$$

The results show that when the flavor is added after freezing, the ice cream portion carries a disproportionate amount of the air. At 4.5 lbs/gal, which is the minimum weight permitted for ice cream, the overrun on the unflavored portion of the product in this example is certainly high enough to cause some concern. One should be on the alert for body and texture problems and stability toward heat shock.

2.6.1 Flavor Character and Intensity

Flavors used in frozen desserts must be compatible with a sweet background; those used in sherbets, ices, and frozen yogurts must also be complemented by the acid present in those products. The optimal intensity of sweetness applies to its presence in the background (the mix) as well as to that of some of the flavoring materials. Formulations for a white mix generally provide for a sweetness level equivalent to 13 to 15% sucrose which is within the optimal range for vanilla ice cream and a number of other flavors that use vanilla as the background. Vanilla extract contains no additional sweeteners and is used in such low levels that its effect on the mix composition is negligible. Although this is true for many other concentrated flavors, bulky flavors may contain significant quantities of sugar. This may be illustrated by examining the sugar content in a fruit pack.

Fruit preparations may be packed in a ratio of two parts of fruit to one part of sugar (2 + 1) and up to 5 + 1. In the first instance, the concentration of sugar is 33.3%, but the 5 + 1 pack has a concentration of 16.7% of added sugar. The fruit itself contains some natural sugar. Flavor formulators may differ in their opinions, but some feel that a fruit-flavored product should have a higher sweetness level than vanilla. The 5 + 1 pack may not quite accomplish that, but a simple calculation will illustrate the effect of a 3 + 1 pack on the final sugar content. When 20% of a 3 + 1 pack (by weight) are used for flavoring:

20 lbs of 3 + 1 pack contains 15 lbs fruit and 5 lbs sugar
80 lbs of mix (15% sucrose) contains 12 lbs sugar
100 lbs of flavored ice cream contains 17 lbs sugar.

The additional sugar may not be uniformly distributed depending on how the fruit preparation is handled. When the juice is strained out for direct addition to the mix and the pulp is fed into the ice cream leaving the freezer, the sugar concentration may be quite uniform. However, if the fruit and pulp are ''gelled'' together, the entire mixture may be added through the fruit feeder and contrasting high flavor areas are created at the points of injection.

The freezing point of the water in the fruit should be compatible with the consumption temperature of the ice cream. If it is not, the fruit will be icy and appear to lack flavor. The sugar content, and to some degree the acid and the mineral content of the fruit, provide the additional functions of lowering the fruit's freezing point and assisting in controlling the fruit pulp consistency.

Of great importance to the ice cream maker is the flavor quality of the original source of ice cream flavor (fruit, nuts, coffee, chocolate, mint, etc). The flavor character varies considerably between different varieties of fruit, stage of ripeness, growing season conditions, and handling (e.g., bruising of fruits which causes enzymatic browning and a ''bruised flavor''). The manner of processing of the flavoring may also affect the flavor character (e.g., Dutch process chocolate, jamlike flavor in fruit preparations, etc). Flavorings may deteriorate on storage and develop off-flavors, lose intensity, change appearance, or simply ''spoil.'' Flavorings should be checked for quality when received; nuts, for instance, may lack crispness or be rancid at the time of receipt. In all cases, storage conditions for flavorings should follow recommended practices.

The flavor intensity imparted by the flavoring should be delicate but definite, pleasing but not overpowering. Concentrated flavorings used in excess can impart a flavor that is too high. Non-acid-type flavorings, such as vanilla and chocolate, create a problem in frozen yogurts with a substantial acid content. In some cases, special preparations of vanilla flavoring may be formulated by the flavor chemist that may prove satisfactory. Chocolate may be difficult to formulate for products beyond a certain acid level.

When a combination of flavors is used, there is an additional requirement that the blend be pleasing. Many fancy flavors are blends or combinations that incorporate a background flavor, added materials introduced through the ingredient feeder, and possibly a variegating syrup. Other products, such as Neapolitan ice cream (three flavors), spumone, and rum and raisin have been long-time favorites.

2.6.2 Quantity of Flavoring

Recommendations on the usage rate must be limited to the particular flavoring in question and both the flavor and visual impact intended. Suppliers of proprietary flavorings offer recommendations that may be followed, or at least serve as a starting point. Quantities will obviously vary when used in conjunction with concentrated flavorings. For different fruits, the actual fruit content, not including the added sugar, may be in the range of from 2 to 20% (consult applicable CFR in the Appendix for permissible minima and labeling requirements). As pointed out earlier, the composition of the mix affects the dilution limit with flavoring; the aim is not to dip below the minimum permissible fat and milk solids content. Some proprietary flavorings used at a relatively high level may be stabilized to help prevent iciness and improve heat shock resistance. Nuts may be added at levels from 2 to 10%; variegating syrups at 5 to 20% depending on type and whether or not they are the sole flavoring source; other added materials including candy (e.g., chocolate chip) can be used at the rate of 5 to 10% of the weight of the mix. Confections with a honeycomb structure, making them light in weight, may be added at a lower rate. Because of their high flavor impact, the quantity of concentrated flavorings should be near the supplier's recommended level based on a standardized strength of the extract. When the concentrate is used in conjunction with the actual fruit, each will furnish a percentage of the total flavor. The desirability of all flavored products should be confirmed by taste test.

2.6.3 Proprietary Flavorings

In addition to concentrated flavorings, which may be natural (e.g., true fruit extract) or imitation, commercial flavor preparations are available ready for use in the frozen dessert. They may be preserved by heat, freezing, sugar, and acid, and hot packed or aseptically packaged. The advantage of using them is the reasonable assurance of uniformity of subsequent batches once a particular flavor has been chosen. The technical staff of the supplier also provides a valuable consultative service on questions relating to flavorings. The flavoring particulates may be reduced in size to the form of a puree, and if there are no hard particles or seeds in it, the supplier may recommend that the material is suitable for incorporation into the mix prior to freezing. Optimum storage recommendations will also be provided. Other flavorings may be obtained in the frozen state and require thawing prior to use. These may be citrus juices or purees, strawberries, peaches, bananas, etc. The flavor imparted by them can be excellent but their sanitary quality must be carefully monitored. When they are packed without heat treatment, they could carry serious contamination of *E. coli* and other bacteria. Dehydrated fruits (e.g., raisins) and candied fruits also find their way into some flavor preparations. Artificial flavors by themselves are generally used only in the "economy" grades of frozen desserts. In combination with natural flavors, and when no claim of an all natural product is made, some imitation flavors may be found in intermediate grade products, but probably never in premium products (see 21 CFR in the Appendix for labeling products containing imitation flavors).

2.6.4 Vanilla Flavor

Vanilla may be a complete flavor by itself or it may serve as a background flavor for fruits, nuts, variegating syrups, and candy. It may also be used as a modifying flavor for chocolate. In the days of Montezuma's Mexico, chocolate and vanilla were combined to prepare a favorite beverage that was subsequently introduced to Europeans. Since then, vanilla has made it on its own and in the United States has become the most popular flavor of ice cream.

The ice cream manufacturer usually obtains vanilla flavoring in the form of an alcoholic extract which, when in single strength, contains the soluble extractives of one-tenth its weight of vanilla beans (13.35 oz/gal). More concentrated extracts, two- up to tenfold, are also available and proportionately contain the extractives of larger weights of the beans. The impact-producing component in the extract is vanillin but a chromatographic analysis reveals the presence of other components that contribute to the flavor character. Some vanilla extracts are fortified with synthetic vanillin. For this purpose, one ounce of vanillin per gallon is considered equivalent in intensity (not flavor character) to a single -fold of pure vanilla extract. Thus a single strength extract with 1 oz of vanillin is roughly equivalent to a twofold extract. The fortified extracts impart a less delicate vanilla bouquet but one that makes a more immediate and stronger impact. The flavor release, although not of the same character as that of pure vanilla, is less hindered by the flavor of the ingredients in the background.

For high-quality vanilla ice cream, both the flavor of the unflavored mix and that of the vanilla extract should be free of criticism. Vanilla flavoring does not ''cover up'' any off-flavors that may be present in the mix due to poor-quality ingredients. Flavoring extracts have their own quality criteria. Those made from Tahiti vanilla beans have a completely different flavor, which may or may not be desired. A large proportion of the vanilla extracts are made from Bourbon vanilla beans grown in Madagascar, although Bourbon beans are also grown in Indonesia and other places. The quality of the beans is best assessed by the quality of the extract made from them. The technology employed in making the extract and subsequent aging and handling may also affect quality. The extract that is used should contain the desired typical components in the proper proportion, it should impart the desired intensity of flavor, and should not have any ''fermented'' or other types of off-flavors. Labeling of the ice cream must conform to the required language when vanilla alone, vanilla–vanillin mixtures, or vanillin alone is used for flavoring (21 CFR).

2.6.5 Chocolate Flavor

A general consensus is that the best chocolate ice cream is made from a special chocolate mix that is processed with all of the chocolate flavor ingredients. If it is desired to include vanilla flavoring to make the chocolate flavor somewhat more mellow, the vanilla extract can be added in the flavor tank (usually one half to two thirds of the quantity needed for flavoring a vanilla ice cream). The chocolate mix also provides a foundation for other flavors such as chocolate marshmallow, various chocolate–nut combinations, candy and variegating syrup combinations, mocha, Neapolitan, etc. When the volume does not justify making a special mix, chocolate ice cream can be madc by adding a chocolate syrup to a vanilla mix. Syrups especially formulated for this purpose are available from chocolate flavor suppliers. Flavoring for chocolate ice cream is in the form of cocoa, chocolate liquor, or both. Chocolate liquor is the primary product obtained after the seeds of the roasted cocoa beans have been dehulled and degermed and the remaining ''nibs'' have been finely ground. It contains about 50% cocoa butter. Under pressure, some of the cocoa butter can be removed from the liquor and the residue becomes cocoa. Its fat content may vary roughly between 10 and 25%. The bulk of the chocolate flavor is contained in the cocoa but there are some delicate, complementary, fat-soluble flavor notes that are retained by the cocoa butter. Thus, the flavor character of cocoas may vary with the fat content of the cocoa.

Basically, the chocolate flavoring may be made by one of two processes, the natural or the Dutch process. Treatment with alkali in the Dutch process yields a darker chocolate with an altered flavor. There are still other variables in the flavor of chocolate. As is true with any food commodity, the quality of cocoa beans varies depending on the source of the beans, growing conditions, and handling procedures including the important fermentation step carried out in the area of the beans' origin. The intensity of the roasting process is also related to flavor characteristics.

After this brief review of chocolate basics, the emerging conclusion is that there are many variants of chocolate flavor. The ice cream maker must decide on the types of cocoa, liquor, or cocoa–liquor combinations, level of usage, light or dark, with or without modifying flavor, and other variables that affect the chocolate flavor character. An opportunity exists to ''individualize'' a flavor by blending several types of cocoa or liquor products to obtain a unique combination. For soft-serve application, a low-fat cocoa may be desirable to prevent separation of dark, flaky granules during the freezing process. If the chocolate mix is to be sterilized, the chocolate flavoring should be checked for highly heat-resistant bacterial spores, sometimes present, that could survive the heating process.

2.7 Freezing of the Mix

Many of the ultimate properties of ice cream are predetermined by the selection of ingredients, the formulation, and the selected flavoring. However, the product must be frozen properly to take full advantage of any benefits derived from the earlier processing steps. Except for soft-serve products, freezing is a two-step operation. During the first stage, the liquid mix at a temperature as close to 30 to 32°F as conditions permit, is frozen in an ice cream freezer at a rapid rate to its discharge temperature (about 21 to 22°F, but may differ depending on the freezing point of the mix). The second stage occurs after the ice cream has been packaged, when its temperature is lowered until it becomes very hard. This phase of freezing, called hardening, will be discussed later.

The principles of operation of the various types of ice cream freezers are discussed in Chapter 5 (Vol. III), but some points are relevant here. A continuous ice cream freezer is depended on to discharge a steady flow of ice cream at the desired rate and temperature and with the desired amount of air incorporated in it. The freezer operator performs a function of utmost importance to the freezing process. Every freezer tends to have characteristics of its own; even different barrels of the same freezer may behave differently. The freezer operator learns from daily observations what to expect from each freezer, so that, when he notes an unusual behavior he must initiate immediate corrective action by whatever procedures have been established for the purpose. This does not apply only to major problems, such as a shutdown, but to any deviation from the norm, even subtle day-to-day changes. The appearance of the product as it leaves the freezer may be wetter or shinier than expected; the product may be softer; gauge readings may be erratic or abnormal; adjustments may not produce the desired response; there may be excessive overrun fluctuation; etc. Some of the problems may relate to compositional errors (e.g., incorrect quantity of stabilizer/emulsifier); processing errors (e.g., mix not properly homogenized); incorrect assembly of freezer parts; breakdown in freezer maintenance (e.g., worn, damaged, or improperly sharpened blades; worn or damaged pumps; worn shaft bushings); oil deposit in the refrigerant side of the freezer barrel; insufficient refrigeration; overcrowding the freezer; faulty gauges; etc. The freezer may still be turning out a product but the deviations from norm presage a definite possibility that profits, legal weight of product, or product quality in this particular run are in jeopardy. Training of freezer operators must include this aspect of their responsibilities. Other workers who handle the freezer, including the clean-up crew, also must be aware of the serious consequences of mishandling or dropping freezer parts, particularly the pumps and blades. Failure to recognize the need for timely maintenance, early, can have an immediate effect on the quality of product and lead to expensive repairs later.

The batch freezer is a less complicated machine, at present used mainly in small plants and product development laboratories. The drawing temperature is somewhat higher (23 to 25°F) and depends on when the proper overrun has been attained. Typically, the overrun goes up rather rapidly when freezing begins but as the product becomes stiff, some of the overrun is lost. At this point, the refrigeration is turned off and the ice cream is allowed to whip to the desired overrun as its temperature increases slightly. The temperature at which the desired overrun is attained, the maximum overrun attainable, and the time required to obtain it are a function of mix composition, particularly the emulsifier content, all other factors being equal. Ice cream made in the batch freezer is generally somewhat coarser textured than that produced in a continuous freezer. This is because of the higher drawing temperature and slower freezing rate. The frozen ice cream must be removed from the batch freezer rapidly because the dasher continues its beating action, causing the overrun to fluctuate. Some variation during discharge is unavoidable. Depending on the type, flavoring may be added directly to the freezer at the beginning of the run or later as the product is being discharged to preserve particle identification. Some flavorings may also be stirred into the ice cream after discharge from the freezer. Excessive warm-up of the product must be avoided and sound sanitary precautions must be taken to avoid contamination. Large containers may be filled directly from the freezer discharge. Unless filled manually after discharge from the freezer (watch sanitation and product warm-up!), small containers may be filled by a packaging machine of sanitary construction fed through a hopper.

The soft-serve and shake freezers normally dispense their product in individual serving sizes on demand, which may be in rapid succession or at a very slow rate depending on business volume. On slow days, the product may remain in the freezer barrel for a long period of time and be subjected to agitation during successive refrigeration cycles (well insulated freezer design can keep these cycles to a minimum). The mix must be formulated and processed to help it stand up under these unfavorable conditions. Some of the common difficulties are churning (emulsifier system and homogenization) and progressive softening with the product temperature actually going down. This is probably due to emulsion and protein destabilization or freeing of bound water. Churning problems increase with higher fat content but the softening phenomenon can occur with any mix. When churning has progressed to the point of being troublesome, the only available option to the freezer operator is to empty the freezer, clean and sanitize it, and start all over again with fresh mix. Under certain conditions, lactose crystallization can also occur over a period of time in the freezer barrel. Soft-serve products are usually drawn at a temperature of 19°F and shakes at 27°F, but in both cases, the actual drawing temperature depends on the freezing point of the mix.

Some stick novelties are frozen without agitation in molds that are partially immersed in a refrigerated brine bath. In this case, the complete freezing process occurs here and the finished product only needs to be stored at a sufficiently low temperature to maintain its quality (− 15 to − 25°F). For other stick novelties which are frozen under agitation and with air incorporation, the product exits an ice cream freezer and from there is filled into molds for hardening in a brine tank. To ensure a complete fill of the molds, the product has to be on the soft side when the molds are being filled.

When the frozen dessert is discharged from the continuous ice cream freezer, it may flow through an ingredient feeder for the incorporation of fruits, nuts, or candy. Variegating syrups may also be introduced downstream in addition to, or in the absence of, other particulate flavor ingredients, and the flow of the product is then directed to the filling machine. The distance of the flow from the freezer to the filler should be as short as conditions permit to hold down temperature increases and minimize the back pressure against which the freezer must operate. Some freezers may have difficulty handling the back pressure. During startup and changeovers, provisions must be made for the sanitary handling of any sound but unusable product for refreezing or reprocessing. (The product may be too soft, wrong overrun, a mixture of two products, or good product wasted while adjustments were being made on the packaging machine.) The final overrun (after packaging) should be checked by weight. With all systems operating properly, the variation in weights should be in the range of 1 to 3% (not exceed ± ½ oz to ± ¾ oz for ½-gal containers).

2.7.1 Amount of Water Frozen

The freezing point of the mix has been alluded to in a number of places in the text but no actual values have been given. An ice cream mix has an initial freezing point at which ice begins to form, but as soon as some ice forms, the soluble solids become more concentrated and lower the freezing point further. The initial freezing point may be determined experimentally or may be calculated with reasonable accuracy.

It corresponds to the lowest temperature at which no ice is present. As the temperature is lowered, ice begins to form but under commercial conditions, it is unlikely that all of the water is ever frozen, even after hardening. A method for calculating the amount of frozen water in ice cream at various temperatures was developed by researchers at the U.S. Department of Agriculture in the 1920s.[34] Results obtained by this calculation are quite useful, although some assumptions must be recognized. The calculation of the freezing point and the method for estimating the amount of frozen water at any temperature are carried out in the following steps:

1. The percent lactose in MSNF is obtained by assuming that 54.5% of the MSNF is lactose. The actual percent lactose should be used if it is known.
2. The percent lactose in whey solids (WS) is obtained by assuming that 76.5% of the WS is lactose. The actual percent lactose should be used if known.
3. Calculate the sucrose equivalent (SE) of lactose and all sweeteners used in the formula. The assumption here is that all sweeteners respond in relation of their molecular weight to that of sucrose. This obviously ignores all interactions between sweeteners and between sweeteners and other mix constituents. Other complicating effects are hydration and changes in bound water occasioned by the mediating effect of processing (heat treatment, homogenization). However, a useful approximation is obtained as follows: percent each sweetener in the formula × MW factor from Table 2.6.
4. Calculate total percent SE by summing percent SEs of all sweeteners:

$$\text{Total \% SE} = \text{\% lactose} + \text{\% sucrose} + \text{SE of all other sweeteners used}$$

5. Calculate percent SE in the aqueous portion of the mix

$$\text{\% SE} = \frac{\text{Total \% SE}}{\text{\% Unfrozen water} + \text{Total \% SE}}$$

6. Determine the freezing point depression due to the sweeteners by reference to Table 2.28.
7. Determine the freezing point lowering due to milk salts by the Leighton formula.[34] The assumptions in this formula are that the salt content in MSNF and WS is the same, their "effective" molecular weight is the same, and their assumed concentration is correct. The "effective" molecular weight is assumed to be 78.6 and salts concentration as 10% of MSNF and WS. It is further assumed that the formulation contains no other added salts.

$$\begin{array}{l}\text{Freezing point depression}\\ \text{due to milk salts (°F)}\end{array} = \frac{\text{\% MSNF} + \text{\% WS}}{\text{\% unfrozen water}} \times 9/5 \times 2.37$$

(9/5 converts the results to °F from °C)

8. Add the freezing point lowering due to sweeteners and milk salts and subtract from 32°F to get the calculated freezing point in °F.
9. By substituting progressively smaller values for percent unfrozen water in steps 5 and 7 and calculating the corresponding freezing point, a freezing curve may

be constructed from which the percent frozen or unfrozen water may be estimated at different temperatures.

The calculations will be illustrated for a mix of the following composition:

10% Fat
8% MSNF
2.5% WS
12% Sucrose
7% 36 DE corn syrup solids
0.3% Stabilizer
39.8% Total solids
60.2% Water

At 0% water frozen:

1. % Lactose − (8 × 0.545) + (2.5 × 0.765) = 6.27
2. % Sucrose = 12.00
3. % SE of 36 DE corn syrup solids (7 × 0.61) = 4.27

Total % SE (step 4) = 22.54

4. % SE in aqueous portion of mix (step 5)

$$\frac{22.54}{60.2 + 22.54} \times 100 = 27.24\%$$

5. Freezing point depression due to sugars (from table) 4.14°F
6. Freezing point lowering due to milk salts (step 7)

$$\frac{8 + 2.5}{60.2} \times 9/5 \times 2.37 = 0.74°F$$

Total freezing point depression = 4.88°F
Freezing point = 32 − 4.88 = 27.12°F

Other points on the freezing curve of this mix are obtained in the same manner except that the percent unfrozen water in steps 5 and 7 is reduced. Assuming that 10% of the water is frozen, then 90% of the original 60.2% would remain unfrozen. The freezing point is calculated using 90% of 60.2 or 54.18 as the percentage of unfrozen water. Additional calculations are made for 80% of 60.2, 70%, etc. The freezing points so obtained are plotted on a graph as temperature against percent water frozen and the points when connected become the freezing curve of this particular mix. Examples of freezing curves are shown in Figure 2.1.

There are several different ways of interpreting the freezing point data. From the information now available one can determine the percent unfrozen water, percent ice, and percent solids in the unfrozen water at the drawing temperature and any other temperature of interest. To assess the potential seriousness of heat shock, the amount of water that actually melts and refreezes when the storage temperature

Table 2.28 RELATIONSHIP OF SUCROSE EQUIVALENT CONCENTRATION TO FREEZING POINT DEPRESSION[a]

Percent SE in Water	Freezing Point Depression (°F)	Percent SE in Water	Freezing Point Depression (°F)
2	0.22	46	10.55
4	0.42	47	11.00
6	0.67	48	11.50
8	0.91	49	12.17
10	1.18	50	12.79
12	1.47	51	13.48
14	1.78	52	14.22
16	2.10	52.5	14.45
18	2.40	53	14.81
20	2.78	54	15.7
22	3.14	55	16.45
24	3.51	56	17.3
26	3.87	56.5	17.69
28	4.31	57	18.05
30	4.79	58	18.95
32	5.28	59	19.72
34	5.85	60	20.54
36	6.46	61	21.40
38	7.12	62	22.55
40	7.89	63	23.45
42	8.64	64	24.40
44	9.51	64.5	24.84

[a] Original data by Pickering,[19] not shown. Values were interpolated from Keeney and Kroger,[9] where the original data were quoted. Interpolation may result in some error because the slope of the data when plotted is not uniform.

fluctuates may be calculated (e.g., between 0 and 10°F). The data in Table 2.29 illustrate some of these relationships. Note that at the drawing temperature of 21°F, ice cream B has essentially the same solids in unfrozen water as ice cream A, but the latter has considerably more ice frozen due to a lower total solids content and higher freezing point. The same is true at 5°F. The higher total solids content, nearly the same proportion of unfrozen water, and the water binding properties of the sweetener type used, tend to favor ice cream B over A from the standpoint of body and texture and resistance to heat shock. Ice creams C and D would be progressively softer. Although consistency can be modified by stabilizers, ice cream D would probably be too soft for most purposes. Freezing point data do not provide all the answers, but they contribute an important element to the total picture.

Bulky flavors added directly to the mix may affect the freezing point as well as the drawing temperature and appearance at draw. If their sugar content is very high, they may cause the ice cream to be softer at any given temperature than its vanilla counterpart.

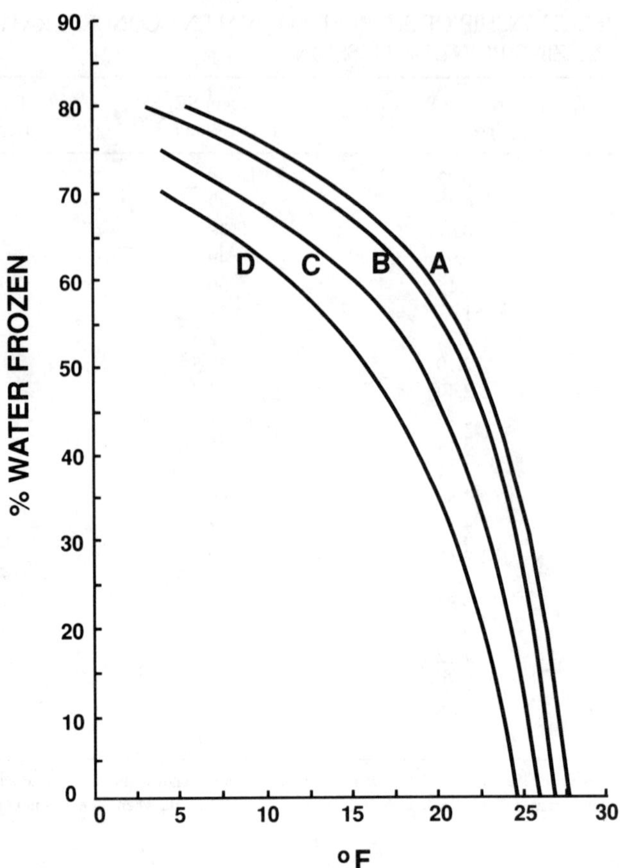

Figure 2.1 Examples of ice cream freezing curves. Composition of the mixes is given in Table 2.29 as a footnote.

2.8 Ice Cream Hardening

A clear distinction must be made between the process of hardening and storage of the ice cream after it is hardened. In the hardening process, the aim is to reduce the temperature of the product to at least 0°F in the center of the package as quickly as possible. After the ice cream reaches this point, it is only necessary to store it at a uniformly low temperature to prevent ice melting and recrystallization.

Several factors affect the rate of hardening, such as size of container; whether several containers have been bundled together and the nature of the wrapping material (paper or plastic); the manner of stacking of the containers; the temperature and velocity of the circulating air; obstructed versus unobstructed exposure of the containers to the cooling medium (one side as opposed to several sides); etc. Figure 2.2 illustrates how the ice cream temperature drops in half-gallon containers

Table 2.29 UNFROZEN WATER IN ICE CREAMS OF DIFFERENT COMPOSITIONS[a]

Ice Cream[b]	Temperature (°F)	Solids (%)	Unfrozen Water (%)	Ice (%)	Solids in Unfrozen Water (%)
A	21	36.3	27.4	36.3	57.0
B	21	38.3	29.0	32.7	56.9
C	21	38.3	35.8	25.9	51.7
D	21	36.3	43.9	19.8	45.2
A	5	36.3	12.4	51.3	74.5
B	5	38.3	13.3	48.4	74.3
C	5	38.3	16.0	45.7	70.5
D	5	36.3	19.8	43.9	64.8

[a] Approximate values obtained by calculations which involve a number of assumptions.

[b] A = 10% fat, 11% MSNF, 15% sucrose, 0.3% stabilizer. B = 10% fat, 7.5% MSNF, 2.5% WS, 12% sucrose, 6% CSS, 0.3% stabilizer. C = 10% fat, 7.5% MSNF, 2.5% WS, 6% sucrose, 6% monosaccharide, 6% CSS, 0.3% stabilizer. D = 10% fat, 11% MSNF, 15% monosaccharide, 0.3% stabilizer.

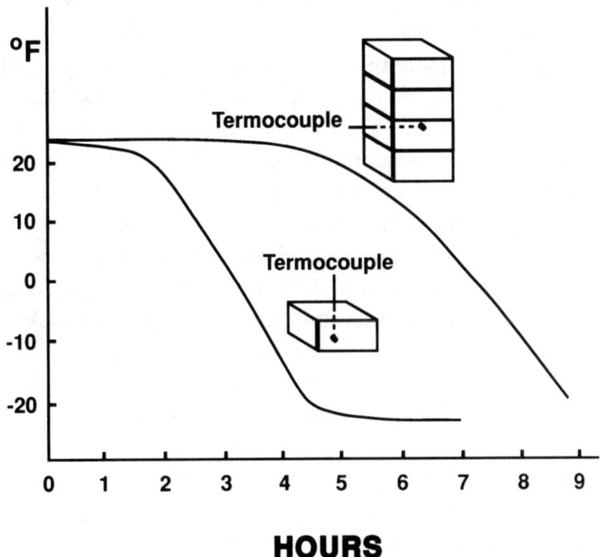

Figure. 2.2 Rate of convection hardening as observed under conditions existing in one commercial plant. Air temperatures was −30°F (velocity was not measured). When bundled, as illustrated, a brown paper overwrap was used. Thermocouples were located in the center of the containers as indicated.

when subjected to one system of air convection hardening. Local conditions at other plants may yield better or worse cooling rates depending on hardening room design and some of the factors enumerated above.

The size of the container makes a significant difference on how fast the product hardens. Very small containers harden quickly (assuming there is adequate air movement), but they also warm up quickly when removed from the freezing temperatures. For that reason, they suffer severe body and texture damage as a result of heat shock. This also applies to novelties (stick bars, small cups etc.) and constitutes one of their most serious quality problems. Large containers (e.g., 3 gal size) harden much slower in the interior (where cooling is largely by conduction) and must be given ample time to reach 0°F in the interior (the actual time obviously depends on the hardening room temperature). If containers are stacked before they are adequately hardened, deformation may occur and some overrun may be squeezed out causing surface discoloration.

Direct refrigerated contact plate hardening provides very effective heat transfer but requires that all containers be of the same size and geometry. Because square half-gallon containers constitute a major portion of the volume in many plants, the contact plate hardeners can be dedicated to this line of products, while the remainder of the production is hardened by some other systems.

Hardening of the half-gallon containers to a temperature of 0°F at the core (center of container) may be accomplished within a period of 1 to 2 h, but not necessarily with all hardening systems. Figure 2.2, for which the data were collected in a commercial setting, provides one illustration of this. It can be seen that in the bundled units of four containers, those on the inside hardened much slower. They required about 7 h to reach 0°F at core as opposed to about 3 h for the unbundled half-gallon containers. In this particular plant the results were acceptable as judged by existing production and quality criteria, and actually represented a substantial improvement over previous hardening performance. Obviously, further improvements would be possible if deemed necessary. The introduction of fast hardening systems constitutes one of the most significant improvements in ice cream technology.

After the ice cream has been hardened, subsequent steps are dictated by local requirements. The containers may be palletized and stacked according to a plan designed to facilitate load-out operations. In some cases, the fully hardened ice cream may be loaded directly onto trucks for transfer to distribution points. Whether during warehousing or the transportation and transfer phase, a constant and low temperature (in the range of −15 to −25°F) should be maintained to minimize heat shock. Maintaining a frost-free environment is also important and should not be disregarded just because it presents a challenge.

Good inventory control should include products coded to reveal date of production, location in storage, and destination in shipment. Maximum storage time which does not inflict an unacceptable degree of quality deterioration obviously depends on local conditions and the management's concept of what constitutes unacceptable quality. Invariably, there are some items that move rather slowly, but for one reason or another must be kept in the inventory. These items may stay in storage longer than desired. The production and storage of faster moving items should be so synchronized that they are still at the peak of their quality when they are moved out. Sensory evaluation should provide an indication of an acceptable storage time for every item manufactured.

Vehicles used for transporting ice cream must be maintained and monitored so they will not become a source of heat shock. Ideally, a vehicle is used only for rapid transport, not for hardening or storing of the product. Several trips on a truck can be very damaging to the product.

2.9 Defects of Ice Cream

Ice cream defects are generally traceable to some identifiable cause which should be included in the surveillance and control measures assigned to quality assurance. However, gross abuse of the product may occur beyond the sphere of a plant's control (possibly in the hands of the ultimate consumer), in which case little can be done other than to attempt to educate those involved. There are several criteria which may render a product unacceptable:

- Failure to meet legal composition
- High standard plate count and/or coliform count (above legal maximum)
- Weight below the legal minimum
- Serious flavor defect(s)
- Serious body and texture defect(s)
- Serious defect(s) in appearance (both product and container)
- Contamination with any harmful substance (e.g., bacteria, chemicals)
- Inadequate pasteurization
- Presence of ''foreign'' substance(s)
- Product mislabeled
- Failure to meet company's own specifications
- Food solids content below legal minimum (e.g., federal standards require that 1 gal of ice cream contains not less than 1.6 lbs of food solids)
- Damaged or unsealed container.

Most of these criteria are self explanatory, but those pertaining to sensory quality merit further elaboration. One aspect of sensory quality is the hedonic component, that is, the degree of like or dislike for a particular product. A hedonic evaluation requires no special training because all individuals know best what they like or dislike. This is a judgment rendered by the consumers of the product and one that should serve as a guide during the various stages of new product development. Once a particular item has achieved a significant level of consumer acceptance to justify its production, it becomes the responsibility of the quality control people to ensure that the sensory qualities do not change. This type of evaluation requires trained experts who can identify any sensory notes that deviate from the product's design, regardless of their own personal preference. Inherent to the training of the quality control sensory evaluator is the ability to identify the defects of dairy products caused by bacteria, enzymes, chemical mechanisms, contamination, etc. which may be the source of quality problems. Corrective measures can be initiated most effectively when defects are identified and their causes are known.

2.9.1 Defects Identified by Sight

Some of these defects are the first to be observed by the consumer and, if serious, may lead to rejection of the product. Eye appeal is an important attribute of the product as well as its container. A review of dairy products evaluation was published by Bodyfelt et al.[35]

2.9.2 Defective Container

Numerous problems may be identified, including soiled containers, with either dirt or ice cream on the exterior of the package; dented, torn, or otherwise damaged containers; unsealed or improperly sealed container; improperly or illegibly coded; inferior packaging material; misshapen container; etc.

2.9.3 Product Appearance

Packages may be over- or underfilled, which are defects traceable to the filling operation in the plant. However, the product may also be bulging due to changes in atmospheric pressure (when product is transported from a low to a high elevation); or it may be pulling away from the sides and top of the container and appear to be "shrunken." High overrun and heat shock accentuates both problems, although shrinkage may occur for no apparent reason and, just as mysteriously, go away. It seems to be related to some subtle condition in the milk proteins because changing the source of MSNF sometimes stops an outbreak of shrinkage.

The color and appearance are largely defined by the ice cream manufacturers who make the product. They decide whether to use artificial colors, and their type and intensity. They also select the fruit, nuts, candy, and variegating syrups and control the concentration of each to be used. The appearance should conform to the manufacturer's design from one run to the next. If an illustration of the product appears on the container, the ice cream should look reasonably the same. Generally, the color should be appealing, compatible with the flavor, and not artificial-looking. Most common color defects are too light, too intense, uneven, and unnatural. The last implies that the color is not compatible with the flavor (e.g., a lemon color in a peach flavored ice cream). Added ingredients (fruit and nut particles, syrups, etc.) should be of desired size, uniformly distributed at the desired density, not icy, and their color should not be bleeding into the surrounding ice cream.

2.9.4 Meltdown Characteristics of Ice Cream

These are observed by the consumer when a serving is not completely consumed. Products may vary in the rate of meltdown and the appearance of the melted portion. Ideally, ice cream should melt to a liquid of the consistency of the mix from which it was made. An old ice cream or one that has been highly stabilized tends to melt slower. Stabilizers and emulsifiers also affect the appearance of the meltdown, which may be curdy, foamy, or actually separated into clear whey. A "buttery" meltdown may result when the ice cream has churned in the freezer. Whey separation may also be observed in the undisturbed mix due to the same causes and when air has been incorporated during processing. The addition of approved food grade protein stabilizing salts (various citrates and phosphates) may affect the meltdown.

2.9.5 Defects of Texture

The aim is to produce an ice cream with a smooth, "creamy" texture consistent with an internal structure made up of small ice crystals and small air cells. There should be no discontinuity of the internal structure perceptible to the consumer as excessive coldness, ice crystals, sugar crystals, or relatively large masses of churned butter. Sugar crystals (lactose) large enough to be perceptible do not melt as rapidly as ice in the mouth and thus impart a "sandy" texture. Smaller undissolved particles may be perceived as chalkiness or astringency. Many steps in the manufacture of ice cream are aimed directly at promoting a smooth textured product (e.g., use of stabilizers and emulsifiers, high solids content, fast freezing, fast hardening, etc.). However, the ice crystals begin to grow in size as soon as the ice cream is made and it is the rate of growth that must be controlled by the choice of proper ingredients and the avoidance of heat shock.

Defects in texture due to ice crystals are described as cold, coarse, and icy. The presence of other undissolved particles produces a chalky or sandy texture. The use of excessive emulsifier or ineffective homogenization gives rise to a buttery texture.

2.9.6 Defects in Body

The type of body desired in the ice cream is an option that the manufacturer can exercise. The principal contributors to the body are the solids content (both type and level), stabilizer and emulsifier, and overrun. An ice cream body may be too heavy (excessively "chewy" or resistant to bite); too weak (quick disappearance in the mouth due to low solids, high overrun, or inadequate stabilization); crumbly (lacking cohesiveness due to high overrun, low solids, or ineffective stabilization); short (similar to crumbly and usually caused by high overrun; when scraped, the ice cream lifts up in relatively thin layers, and thus lacks cohesiveness); too dry both in appearance and mouthfeel (solids content and certain stabilizers and emulsifiers); or gummy (due to overstabilization).

2.9.7 Flavor Defects

Flavor defects may be imparted by any of the ingredients, but some may also develop in the mix or the ice cream. A logical division of the various defects is based on their source, because it is along these lines that corrective measures must be sought. Some defects will appear under more than one source.

2.9.8 Defects Contributed by the Dairy Ingredients

Any off-flavor present in the milk products may be reasonably expected to appear in the ice cream, although mild defects such as slight feed, slight cooked, or slight flat would be of little consequence or undetectable. More serious off-flavors to be guarded against are:

High acid (sour). This is one of the defects caused by bacteria when due to favorable temperature and length of storage they are given an opportunity to multiply. Depending on the specific bacteria present, the acid development may be accompanied by other off-flavors of an unpleasant and generally unclean character. Some acid producing bacteria also produce a malty flavor.

Old ingredient. There are several types of old ingredient flavor. Dehydrated products may become stale due to chemical changes. Fluid dairy products may become subject to bacterial action as in the high acid flavor or when psychrotrophic bacteria (those growing at refrigeration temperature) are active. These bacteria produce off-flavors described as fruity, unclean, bitter, putrid, rancid, etc.

Unclean. When the flavor suggests unsanitary conditions or has a barny character, its generic description is unclean. The term is aptly chosen because of the unpleasant aftertaste which persists after the sample has been tasted.

Oxidized. Cardboardy, tallowy, and stale-metallic are other terms used to describe this off-flavor. Fat oxidation that leads to the development of oxidized flavor proceeds more rapidly in the presence of copper or iron contamination. Products intended to have a long storage life (dehydrated products, butter, sweetened condensed milk) are also susceptible. Once the off-flavor develops, it continues to get worse, which makes it even more serious. Some milk supplies are particularly susceptible to oxidation. Dry-lot feeding of cows has been shown to be one responsible factor. Another form of oxidation occurs when milk is exposed to sunlight and fluorescent light of low wavelength. It is caused by the oxidation of a protein component and is identified as a cabbage or burnt featherslike flavor.

Rancid. The enzyme lipase is normally found in milk and under conditions of excessive agitation, foam formation, and alternate warming and cooling, catalyzes the breakdown of the fat. The free fatty acids that are liberated (butyric, caproic, caprylic, capric, and lauric acids) produce the off-flavor which has been variously described as soapy, goaty, bitter, stale coconutlike, and perspirationlike. Pasteurization inactivates the enzyme. Mixing of raw milk with homogenized products can initiate the off-flavor production. Homogenization of a product containing active lipase may produce rancidity in a very short time.

Cooked. There are several variants of the cooked flavor. The milder form is simply described as cooked or custardlike. The more unpleasant variants are caramelized, scorched, burnt, or scalded. High-heat NDM, sweetened condensed milk, evaporated milk, ingredients that have turned brown (due to caramelization or Maillard reaction), or ingredients processed at high temperatures when considerable ''burn-on'' occurred on the heating surfaces are the possible causes of the defect.

Whey. When whey is used as an ingredient, its flavor quality should be carefully checked. Any off-flavors present will very likely appear in the ice cream.

Foreign. This represents a serious category of defects caused by contamination of the ingredient by a substance completely foreign to food material. The substances may be sanitizers, detergents, pesticides, paints, lubricants, etc. Quality surveillance of ingredients must discover such problems and reject the ingredient from use.

2.9.9 Defects Due to Mix Processing and Storage

During processing, the mix is susceptible to the development of a cooked flavor (see previous section). Foreign flavors may also gain access to the mix from the equipment, carelessness on the part of the plant workers, or from the plant environment. Under certain conditions, rancidity may be promoted if the factors discussed in Section 2.9.8 under rancidity are not controlled. If the mix is stored any length of time, it may deteriorate in much the same manner as milk, cream, and other perishable products. Off-flavors may be caused by bacterial action, oxidation, or absorption of odors from the surroundings, including foreign odors.

2.9.10 Defects Due to Flavoring Materials

The quality of flavoring materials must be constantly monitored to ensure that it conforms to the products' design. Difficulties may be encountered with comingling of different flavors when one flavored ice cream follows another in freezing and packaging. The flavoring material may have the desired characteristics, but the imparted flavor may lack perfection due to an excessive or inadequate intensity. The flavor may be slightly lacking in "blend" or be a little harsh, in which case one may criticize it as "lacking fine flavor." If the flavor is uncharacteristic or artificiallike, it can be labeled as unnatural. Other specific shortcomings may be identified by descriptive terminology. For instance, fruits may lack tartness, chocolate may be too bitter, nuts may be rancid, and citrus may have a peel flavor.

2.9.11 Defects Due to Sweetening Agents

In addition to being excessive or deficient, sweetness can also be uncharacteristic. A syrupy flavor suggests caramelization. It may detract from the fine flavor of the flavoring ingredients, particularly vanilla. Defective syrups may also impart a fermented flavor to the ice cream.

2.9.12 Defects Due to Storage of Ice Cream

On storage, the flavor of ice cream may undergo chemical changes and the product may absorb odors from the surrounding atmosphere. The flavor may lack the luster of the fresh product, in which case it is criticized as lacking freshness. On further storage, a staleness may become evident and the criticism becomes storage flavor. Oxidation may also take place giving rise to an oxidized flavor. When the frozen storage facility experiences an ammonia leak, the consequences generally lead to the product being pulled from distribution and discarded.

2.9.13 Defects of Frozen Dessert Novelties

Depending on their type, novelties are subject to specific defects in addition to those encountered in packaged frozen desserts. Two defects appear to head the list—coarse texture due to heat shock and coliform contamination. The severity of the damage due to heat shock is accentuated by the small size of the individual items which encourages rapid temperature fluctuations throughout the product. Coliform contamination may come from conveyor belts or moisture condensation.

Many novelty items have an exterior coating, most commonly of chocolate. The coating may be defective in several different ways:

Incomplete coverage of the bar
Coating deposited too far down the stick
Coating too thick
Coating too thin
Coating cracked or slipping
Off-flavored coating
Unnatural or undesired flavor of coating
Product bleeding through coating.

The coating contributes significantly to the appearance of the items, but the bars may also be defective for other reasons. Following are some examples:

Bars with voids
Misshapen bars
Incorrect volume
Incorrect weight
Improper pattern or proportioning of constituents of composite bars (those containing two or more constituents).

Additional defects due to various causes include:

Empty wrappers
Torn wrappers
Wrappers sticking to the bars
Soiled wrappers
Unsealed wrappers
Broken sticks
Improperly inserted sticks
Wafers, cookies, or cones "soggy" or lacking in crispness
Contamination with brine
Comingling of flavors and colors
Body and texture defects
Flavor defects
Color defects.

Some defects may be corrected by a mechanical adjustment on the equipment; others require a wider quality assurance effort. Constant observations should anticipate and, hopefully, prevent potential problems. Among the process control steps that should be monitored are the following:

Temperature going into the hopper
Bar temperature
Coating temperature
Dwell times
Product weight before and after enrobing
Volume of bar
Overrun
End of day inventory
Specific gravity of brine
Formulation of product—composition
Coliform counts and other bacterial tests
Sensory properties of ingredients and finished product
Temperature monitoring to prevent heat shock
Product rotation

2.10 Plant Management

Simply stated, the objectives of a commercial ice cream operation are to achieve a desirable product, an efficient and cost-effective production and distribution system, and successful sales. To do so requires astute management and a competent, responsive work force. An attempt is made here to summarize the significant issues requiring executive decisions in the management of an ice cream plant.

1. Personnel
   a. Job descriptions
   b. Wage and salary administration
   c. Selection and hiring
   d. Training
   e. Discipline (including chemical dependency abuse and testing)
   f. Performance measurement
2. Engineering
   a. Buildings—both exterior and interior
   b. Utilities—steam, cold and hot water, refrigeration, compressed air, electricity
   c. Process equipment
   d. Dry storage
   e. Cold and frozen storage
   f. Maintenance facilities and procedures
   g. Process control
   h. Special equipment such as computers, measuring devices, and instrumentation

   i. Maintenance of up-to-date diagrams and flowsheets of all piping and equipment used in processing, cleaning, and sanitizing
3. Environmental: air, water, waste, and noise management
4. Product line
   a. Maintenance of core business line
   b. Modifications to existing products
   c. Introduction of new products
   d. Competitive planning
5. Pricing: price–value relationship
6. Packaging
   a. Sizing
   b. Single vs. bundling and type of overwrap
   c. Package graphics and coding
   d. Case coding, product identification, and tracking
7. Quality assurance
   a. Raw material specifications
   b. Finished goods product specifications
   c. Laboratory procedures including tests contracted to outside laboratories
   d. Plant sanitation procedures
   e. Testing requirements including critical control point surveillance
   f. Housekeeping procedures
   g. Uniforms and personal hygiene requirements
   h. Product recall management
   i. Handling of product to be reprocessed
   j. Handling of returns
   k. Temperature control from ingredient to finished product
   l. Records and documentation
8. Production planning and scheduling
   a. Purchasing
   b. Inventory control—raw materials
   c. Inventory control—finished goods
   d. Coordination with sales and marketing
   e. Production scheduling
9. Production
   a. Trained personnel
   b. Proper equipment
   c. Properly maintained, cleaned, and sanitized equipment ready for use
   d. Adequate supply of raw materials
   e. Adequate supply of packaging material
   f. A production plan
   g. Production records
10. Storage (dry and cold)
   a. Suitable space and location for edible and nonedible materials
   b. Humidity and temperature control
   c. An inventory control system

   d. Efficient handling system
11. Distribution
   a. Drivers
   b. Vehicles
   c. Temperature and frost control
   d. Effective organization of distribution management
   e. Controls and records
12. Others
   a. Accounting
   b. Cost control
   c. Regulatory compliance management
   d. Safety management
   e. Insurance against loss and stability
   f. Sales and marketing
   g. Research and development

2.11 Active Areas of Research

Technological advances in ice cream should be made possible by an understanding of the principles governing the interaction between its components. Optimal functionality from each ingredient selected in the production of ice cream is achieved through careful formulation and processing. Although past research has greatly increased our knowledge, there are still many circumstances when it is difficult to accurately predict the effects that changes in ingredients, formulation, and other variables will have on the finished product. One of the major objectives of research in foods is a detailed understanding of the interactions and changes of the different components and to apply that knowledge in product development and improvement.

Among the active areas of study are those exploring the role of ingredients such as proteins, emulsifiers, stabilizers, fat, sweeteners, fat replacers, and fat mimetics. The basic resources available to the ice cream technologists are the physical and chemical studies of emulsions and foams; the chemistry of the ingredients; and the effect that processing such as heating, homogenization, whipping, and freezing may have on them. Due to the scope and space limitations of this chapter only some selected research areas will be highlighted under the headings of (1) ice cream mix, (2) ice cream structure, and (3) processing and freezing. This is not to imply that research not specifically addressed is of lesser importance.

2.11.1 Ice Cream Mix

The ice cream mix, as discussed in Section 2.5, consists of ingredients such as cream, milk condensed skim milk, nonfat dry milk, sugars such as sucrose or corn syrups, stabilizers, emulsifiers, in some cases flavors, etc. All of these ingredients are then blended, pasteurized, homogenized, cooled, and aged. The sum of these processes results in an ice cream mix, which may be sold as such (e.g., soft-serve for fast food restaurants). The principal unit operations directly relevant to the ice cream mix are pasteurization, homogenization, and aging.

The applicable areas of research include studies addressing development of the emulsion structure, identification of the mechanism of emulsifier and stabilizer action, and investigations into the nature of molecular interactions between stabilizers and other ice cream components.

The formation of a new fat globule membrane as a result of homogenization and its subsequent reactions have been studied in great detail. Oortwijn and Walstra[36] studied the properties of cream by combining milkfat with different sources of protein. The amount of protein available, the composition of the membrane, and fat crystallization were important factors in controlling the stability of the emulsions. The nature of the proteins involved in the process of fat globule membrane formation has been found to have implications on the tactile properties in frozen dairy products. For example, increasing concentrations of whey proteins were found by Goff et al.[37] to have fat destabilization properties. However, they were not able to provide guidelines for conclusive predictions, partly due to the diverse processing conditions encountered.

Better understanding of the function and performance of emulsifiers in ice cream has been provided recently by a number of studies.[16,30,38–42] As pointed out in Section 2.2.19, emulsifiers are not needed in ice cream mix to stabilize the fat emulsion. There are many components, mainly proteins, available in the ice cream mix to perform this function. In ice cream mixes homogenized without emulsifiers, the new fat globule membrane will be formed by caseins and whey proteins. However, the surface-active character of emulsifiers, when present in the mix, allows them to be preferentially adsorbed at the surface of the fat globule replacing the proteins. As the interfacial tension is lowered due to the action of the adsorbed emulsifiers, the fat globules are more readily destabilized. Due to their size and structure, proteins at the interface form a more complex membrane than one made up of emulsifiers.[30] Incorporation of air into the mix results in the adsorption of fat globules at the air/serum interface due to a differential in the created surface forces. The shear forces resulting from freezing concentration, agitation, and whipping in the freezer barrel cause the emulsion to partially destabilize with the formation of clusters of fat globules and with possibly some coalescence. Both aggregation and coalescence of globules is facilitated by the creation of the weaker emulsifier-containing membranes. These clusters and possibly some free fat are responsible for stabilizing the air cells and creating a matrix throughout the product. Matrix formation, partially coalesced fat globules at the air–cell interfaces, and stable air cells result in a dry appearance, smooth texture, and resistance to melting.

Not all of the emulsifiers work in a similar manner. An excellent review on emulsion stability is presented by Friberg et al.[43] They describe two methods for classifying emulsifiers. In the first, the surfactant per se is characterized by a value for the hydrophilic and lipophilic balance (HLB) of the molecule (water loving and lipid loving parts of the molecule). The second approach combines the surfactant with oil and water and the whole system is characterized by a number. Generally, emulsifiers with a higher HLB number have a higher affinity for water, and are more effective as "deemulsifiers" at a given concentration than those with a lower HLB. However, the unsaturation of the fatty acid components of monoglycerides and polysorbates has also been found to be significant. Emulsifiers containing predominantly unsaturated fatty acids (e.g., glyceryl monooleate and polysorbate 80) are more effective destabilizing agents than their counterparts containing predominantly saturated fatty acids (e.g., glyceryl monostearate and polysorbate 65).[16,33]

2.11.2 Ice Cream Structure

Human senses are the ultimate evaluation tool of ice cream body, texture, and taste. Thus, product acceptance, which is the necessary goal of any producer, is based on sensory perception. However, much useful information regarding the structure of ice cream and the ways in which different ingredients and processes may change its tactile properties have been studied by objective physical methods. One of the important tools in analyzing ice cream structure has been the electron microscope.[16,31,38,44–46] Excellent electron micrographs showing details of mix and ice cream structure have been published by Buchheim[31] and Berger[16] using techniques of freeze *etching* and freeze fracturing in their sample preparation. Continuing efforts to find new techniques in microscopy and sample preparation are likely to yield further enlightment of ice cream structure and its development. One of these techniques is low-temperature scanning electron microscopy (LT-SEM). In this technique the samples are stabilized by quench-freezing in liquid nitrogen (−210°C). This provides an opportunity to examine intact biological materials in a fully hydrated frozen state. Samples so prepared for LT-SEM are stable because below −130°C the vapor pressure of the components nears zero and the ice recrystallization process is halted. This avoids introduction of artefacts through chemical fixation and structural collapse.[47]

Body and texture of ice cream are affected by the use of stabilizers. The mode of action and the importance of stabilizers in ice cream were discussed in Sections 2.2.17 and 2.2.18. Some current research focuses on the basic aspects of the particular molecular characteristics and their effects on the structure of hydrocolloids. One example is the work reported on the direct measurement of forces in the strands of Xanthan gum.[48] Stress measurements of forces between molecular helices of Xanthan were performed using a method that correlates these forces to the osmotic pressure of the polysaccharide in solution. This method provides the opportunity to relate the functionality of a polymer solution to the microscopic properties that underlie them.

Other examples of basic or fundamental research that may contribute to the understanding of ice cream structure are rheological tests. Experiments are being developed that correlate liquid and semisolid texture to rheological and frictional properties of foods. This work has direct implication to ice cream structure due to the semisolid nature of ice cream. In addition, relationships may be established to correlate texture–taste interactions to diffusion coefficients. An excellent review on these experiments is presented by Kokini.[49] The possibilities of these contributions are exemplified in the development of a model based on theoretical calculations and practical, sensory data. Kokini presented a model for testing the melting action of ice cream in the mouth and how it generates a layer of lubricant between the solid ice cream and the mouth.[49] This model suggests that the shear stress on the tongue is the mechanism of texture perception even in the presence of a melting layer. Another interesting model correlated viscosity of a solution to taste intensity. In summary, it can be said that in ice cream, as well as in other foods, considerable work is being done to relate textural attributes to physical quantities from a basic understanding of perception mechanisms.

2.11.3 Processing and Freezing

Stabilizers, sweeteners, and glass transitions are subjects of very active research in ice cream freezing. The action of stabilizers and carbohydrate sweetening agents in ice cream results from their ability to bind water or to form gellike structures. These properties greatly increase the viscosity of the serum phase during freezing and freeze concentration (Section 2.2.17). Efforts to elucidate the mechanism of action of stabilizers on rates of recrystallization have not correlated well with increases in mix viscosities before freezing. Budiaman and Fenema[50] concluded that stabilizers do not have a significant effect on (1) the amount of ice that forms in ice cream mix, (2) the size and shape of the ice crystals existing soon after freezing, and (3) the rate at which recrystallization of ice occurs after a 2-week period at − 15°C.[50] Their data do not confirm the usual mechanism by which stabilizers are thought to retard ice crystal growth initially and during storage. They stated that their data neither disprove or support the common contention that one function of hydrocolloids in frozen desserts is to limit crystal size. Apparently, the mechanism of ice crystal control is related to mass diffusion and the factors that control its rate, rather than the fundamental thermodynamics of ice nucleation and ice-crystal growth. Further research is needed, perhaps at the molecular level, to elucidate the mechanism by which stabilizers exert their function.

A glass is characterized as an amorphous (not crystalline) solid. Glass transitions are phase changes with defined temperatures of transitions for different materials. At the glass transition temperature Tg', polymeric materials change from a viscoelastic fluid to a glass (very high viscosity). In foods, the Tg' is defined as that temperature at which a solution reaches its maximum freeze concentration. In ice cream it has been calculated[51–53] that at temperatures of − 30°C or below, the superconcentrated solutes should be present in a glass state. In this case, the unfrozen water is in the glass state (characterized by an extremely high viscosity) and unable to diffuse to the surface of an existing water-crystal nucleus. Above this glass transition temperature, or at lower viscosity than that corresponding to the glass state, water would be able to migrate with the concomitant result of crystal growth.[46,51,53] Ingredient formulation can elevate Tg', thus increasing the stability of ice cream or alternatively, one could store the product at temperatures lower than the Tg'. The overall viscosity of a solution does not correlate well with the observed increase in ice cream structure stability. However, it is possible that the interaction of polysaccharides with sugar and other solids in ice cream increases the local viscosity of the unfrozen serum, thereby increasing the viscosity of the serum phase surrounding the ice crystals to above the viscosity corresponding to Tg'. This would result in the physical resistance to recrystallization and structural collapse. Future research should provide the answers.

Viscosity in the serum can also be modified by the interaction between partially denatured proteins, or between proteins with extremely different isoelectric points.[54] Poole et al.[55] have found that basic proteins such as lysozyme (with isoelectric point pI = 10.7) or clupeine (pI = 12) enhance the surface activity of acidic proteins such as whey proteins (pI ~ 5) resulting in extremely stiff foams after whipping. Sucrose was found to further enhance the interaction between the proteins. One may speculate that further studies may be designed to uncover appropriate protein–protein and protein–carbohydrate interactions which may be useful in the substitution of fat in ice cream.

To conclude this section, it can be said that much of what can be learned about the ice cream making process is highly dependent on the theoretical tools and equipment used. However, as more studies come to light, it may be possible to establish some general principles on which to base technological advancement. Empirical experiments (trial and error) in product development and improvement have been historically very important. Hopefully they may be supplemented in the future by scientific knowledge that will provide a predictable basis for further advances in ice cream technology.

2.12 References

1. Anonymous. 1951. 1851–1951. Ice cream centennial. *Ice Cream Trade J.* **47:**222.
2. Anonymous. 1955. A 50 year history of the ice cream industry. *Ice Cream Trade J.* **51:**1–270.
3. Arbuckle, W. S., 1986. *Ice Cream*, 4th edit. AVI, Westport, CT.
4. Burke, A. D. 1947. *Practical Ice Cream Making.* Olsen, Milwaukee, WI.
5. Fisk, W. W. 1919. *The Book of Ice Cream.* Macmillan, New York.
6. Frandsen, J. H., and E. A. Markham. 1915. *The Manufacture of Ice Creams and Ices.* Orange Judd, New York.
7. Frandsen, J. H., and D. H. Nelson. 1950. *Ice Cream and Other Frozen Desserts.* Frandsen, Amherst, MA.
8. Keeney, P. G. 1960. *Commercial Ice Cream and Other Frozen Desserts*, p. 50. The Pennsylvania State University, College of Agriculture, Extension Service.
9. Keeney, P. G., and M. Kroger. 1974. Frozen dairy products. *In* B. H. Webb, A. H. Johnson, and J. A. Alford (eds.), *Fundamentals of Dairy Chemistry.* AVI, Westport, CT.
10. Lucas, P. S. 1956. Ice Cream Manufacture (commemorating 50 years of progress). *J. Dairy Sci.* **39:**833.
11. Sommer, H. H. 1951. *Theory and Practice of Ice Cream Making.* Sommer, Madison, WI.
12. Tobias, J., and G. A. Muck. 1985. Ice cream and frozen desserts. *J. Dairy Sci.* **64:**1077.

13. Turnbow, G. D., P. H. Tracy, and L. A. Raffeto. 1956. *The Ice Cream Industry*. John Wiley & Sons, New York.

14. American Dry Milk Institute. 1971. *Standards for Grades of Dry Milk including Methods of Analysis*. Vol. Bulletin 916 (Revised). American Dry Milk Institute, Chicago, IL.

15. Hunziker, O. F. 1946. *Condensed Milk and Milk Powder*. Hunziker, La Grange, IL.

16. Berger, K. G. 1990. Ice Cream. *In* K. Larsson and S. Friberg (eds.), *Food Emulsions*, pp. 367–444. Marcel Dekker, New York.

17. Sherman, P. 1978. *Food Texture and Rheology*. Vol. UFST Symposium. Academic Press, New York.

18. Larsson, K., and S. E. Friberg. 1990. *Food Emulsions*, 2nd edit. Marcel Dekker, New York.

19. Pickering, S. V. 1891. The freezing point relationship of cane sugar. *Berichte Deutsch. Chem. Gensellschaft*, **24**:333.

20. Okos, M. R. 1986. *Physical and Chemical Properties of Food*. American Society of Agricultural Engineers, St. Joseph, MI.

21. Whistler, R. L., and J. R. Daniel. 1985. Carbohydrates. *In* O. Fennema (ed.), *Food Chemistry*, pp. 69–137. Marcel Dekker, New York.

22. Whistler, R. L. 1973. *Industrial Gums: Polysaccharides and Their Derivatives*. Academic Press, New York.

23. Dikinson, E., and G. Stainsby. 1982. *Colloid in Food*. Elsevier London.

24. Dickinson, E. 1987. *Food Emulsions and Foams*. Royal Society of Chemistry, London.

25. Nickerson, T. A. 1962. Lactose crystallization in ice cream. IV. Factors responsible for reduced incidence of sandiness. *J. Dairy Sci*. **45**:354.

26. Schappner, H. R. 1986. British Patent GB-1,108,376.

27. Price, C. 1990. *Time–Temperature Requirements for Ice Cream Mix*. Midwest Region, Public Health Service, Office of the Regional Food and Drug Director, Chicago, IL.

28. Muck, G. A., and Tobias, J. 1962. Effect of high heat treatment on the viscosity of model milk systems. *J. Dairy Sci*. **45**:481–485.

29. Tobias, J., M. Whitney, and P. H. Tracy. 1952. Electrophoretic properties of milk proteins II; Effect of heating to 300°F by means of the Mallory small-tube heat exchanger on skimmilk proteins. *J. Dairy Sci*. **35**:1036–1045.

30. Walstra, P., and R. Jenness. 1984. *Dairy Chemistry and Physics*. John Wiley & Sons, New York.

31. Buchheim, W. 1978. Mikrostruktur von geshlagenem Rahm. Microstructure of whipped cream. *Gordian*, **78**:184–188.

32. Brooker, B. E., M. Anderson, and A. T. Andrews. 1986. The development of structure in whipped cream. *Food Microstruct*. **5**:277–285.

33. Goff, H. D., Liboff, M., Jordan, W. K., Kinsella, J. E. 1987. The effects of Polysorbate 80 on the fat emulsion in ice cream mix: evidence from transmission electron microscopy studies. *Food Microstruct*. **6**:193–198.

34. Leighton, A. 1927. On the calculation of the freezing point of ice cream mixes and of the quantities of ice separated during the freezing process. *J. Dairy Sci*. **10**:300.

35. Bodyfelt, F. S., J. Tobias, and G. M. Trout. 1988. *The Sensory Evaluation of Dairy Products*. Van Nostrand Reinhold, New York.

36. Oortwijn, H., and P. Walstra. 1982. The membranes of recombined fat globules 4. Effects on properties of the recombined milks. *Netherlands Milk Dairy J.*, **36:**279–290.

37. Goff, H. D., J. E. Kinsella, and W. K. Jordan. 1989. Influence of various milk protein isolates on ice cream emulsion stability. *J. Dairy Sci.* **72:**385–397.

38. Buchheim, W., and Dejmeck, P. 1990. Milk and Dairy-Type Emulsions, pp. 203–246. *In* K. Larsson and S. Friberg (eds.), *Food Emulsion* Marcel Dekker, New York.

39. Goff, H. D. 1988. Emulsifiers in ice cream: How do they work? *Modern Dairy* **67:**15–16.

40. Goff, H. D., and W. K. Jordan. 1989. Action of emulsifiers promoting fat destabilization during the manufacture of ice cream. *J. Dairy Sci.* **72:**18–29.

41. Keeney, P. G. 1982. Development of frozen emulsions. *Food Technol.* **36:**65.

42. Lin, P. M., and J. G. Leeder. 1974. Mechanism of emulsifier action in an ice cream system. *J. Food Sci.* **39:**108.

43. Friberg, S. E., R. F. Goubran, and I. K. Kayali. 1990. Emulsion Stability. *In* K. Larson and S. E. Friberg (eds.), *Food Emulsions.* Marcel Dekker, New York.

44. Berger, K. G., Bullimore, B. K., White, G. W., Wright, W. B. 1972. The structure of ice cream. *Dairy Industries* **37:**419, 493.

45. Brooker, B. E. 1985. Observations on the air serum interface of milk foams. *Food Microstruct.* **4:**289.

46. Goff, H. D., and K. B. Caldwell. 1991. Stabilizers in ice cream: How do they work? *Modern Dairy* **70:**14–15.

47. Caldwell, K. B., H. D. Goff, and D. W. Stanley. 1992. A low-temperature SEM study of ice cream. II. Influence of selected ingredients and processes. *Food Struct.* **2:** (in press).

48. Rau, D. C., and V. A. Parsegian. 1990. Direct measurement of forces between linear polysaccharides Xantan and Schizophyllan. *Science* **249:**1278–1281.

49. Kokini, J. L. 1987. The physical basis of liquid food texture and texture–taste interactions. *J. Food Engin.* **6:**51–81.

50. Budiaman, E. R., and O. Fenema. 1987. Linear rate of water crystallization as influenced by viscosity of hydrocolloid suspensions. *J. Dairy Sci.* **70:**547.

51. Eisenberg, A. 1984. The glassy state and the glass transition. *In* J. E. Mark et al. (eds.), *Physical Properties of Polymers.* American Chemical Society, Washington, D.C.

52. Levine, H., and L. Slade. 1988. Principles of cryo-stabilization technology from structure property relationships of carbohydrate/water systems. A review. *Cryo-Lett.* **9:**21–63.

53. Levine, H., and L. Slade. 1990. Cryostabilization technology: thermoanalytical evaluation of food ingredients and systems. *In* C. Y. Ma and V. R. Harlwaker (eds.), *Thermal Analysis of Foods.* Elsevier Applied Science, London.

54. Dickinson, E., and G. Stainsby. 1987. Progress in the formulation of food emulsions and foams. *Food Technol.* **41:**74–81.

55. Poole, S., S. I. West, and J. C. Fry. 1986. Lipid tolerant protein foaming systems. *Food Hydrocolloids* **1:**45.

Note: In order to provide the most recent standards for frozen desserts, this legal document is reproduced as an appendix to this volume instead of this chapter.

CHAPTER

3

Cheese

K. Rajinder Nath

3.1 Introduction

Cheese is one of mankind's oldest foodstuffs. It is nutritious. It was Clifton Fadiman—epic (and Epicurean) worksmith—who coined the phrase that best describes cheese as "milk's leap to immortality."[1] The first use of cheese as food is not known, although it is very likely that cheese originated accidentally. References to cheeses throughout history are widespread: "Cheese is an art that predates the biblical era."[2] The origin of cheese has been dated to 6000 to 7000 B.C. The worldwide number of cheese varieties has been estimated at 500, with an annual production of more than 12 million tons growing at a rate of about 4%.[3]

Cheesemaking is a process of dehydration by which milk is preserved. There are at least three constants in cheesemaking: milk, coagulant, and culture. By introducing heating and salting steps in cheesemaking, a potential for numerous varieties has been realized.

The techniques employed by early cheesemakers varied geographically. A cheese made in a given region with the available milk and prevailing procedures acquired its own distinctive characteristics. Cheese made in another locality under different conditions developed other properties. In this way specific varieties of cheese originated, many of which were named according to the town where produced, for example, Cheddar, England. Although varieties of cheese are known by more than 2000 names, many differ only slightly, if at all, in their characteristics.[4]

About 1900, the following five developments in cheese technology contributed to the rapid growth of commercial cheesemaking[4]:

- The use of titratable acidity measurements to control acidities
- The introduction of bacterial cultures as "starters"
- The pasteurization of milk used in cheesemaking which destroys harmful microorganisms
- Refrigerated ripening
- The appearance of processed cheese

3.1.1 Classification

Cheeses have been classified in several ways. Several attempts to classify the varieties of cheese have been made. One suggestion consists of a scheme that divides cheeses into the following superfamilies based on the coagulating agent.[3]

1. *Rennet cheeses.* Cheddar, Brick, Muenster
2. *Acid cheeses.* Cottage, Quarg, Cream
3. *Heat-acid.* Ricotta, Sapsago
4. *Concentration-crystallization.* Mysost

A more simple but incomplete scheme would be to classify cheeses as follows:

1. *Very hard.* Parmesan, Romano
2. *Hard.* Cheddar, Swiss
3. *Semisoft.* Brick, Muenster, Blue, Havarti
4. *Soft.* Bel Paese, Brie, Camembert, Feta
5. *Acid.* Cottage, Baker's, Cream, Ricotta

Natural cheese types can be classified according to the distinguishing differences in processing[4] as shown in Table 3.1.

Another broad look at cheeses might divide them into two large categories, ripened and fresh.

3.1.1.1 Ripened

Cheeses can be ripened by adding selected enzymes or microorganisms (bacteria or molds) to the starting milk, to the newly made cheese curds, or to the surface of a finished cheese. The cheese is then ripened (cured) under conditions controlled by one or more of the following elements: temperature, humidity, salt, and time.

Depending on the style of cheese, the ripening can be principally carried out on the cheese surface or the interior. The selection of organisms, the appropriate enzymes, and ripening regime determine the texture and flavor of each cheese type.

Table 3.1 DISTINCT TYPES OF NATURAL CHEESE CLASSIFIED BY DISTINGUISHING DIFFERENCES IN PROCESSING

Distinctive Processing	Distinctive Characteristics	Typical Varieties of Cheese
Curd particles matted together	Close texture[a], firm body	Cheddar
Curd particles kept separate	More open texture	Colby, Monterey
Bacteria ripened throughout interior with eye formation[b]	Gas holes or eyes throughout cheese	Swiss (large eyes), Samsoe, Edam, Gouda (small eyes)
Prolonged curing period	Granular texture; brittle body	Parmesan, Romano
Pasta filata (stretched curd)	Plastic curd; threadlike or flaky texture	Provolone, Caciocavallo, Mozzarella
Mold ripened throughout interior	Visible veins of mold (blue-green or white). Typical piquant, spicy flavor	Blue, Roquefort, Stilton, Gorgonzola
Surface ripened principally by bacteria and yeasts	Surface growth: soft, smooth, waxy body, typical mild to robust flavor	Bel paese, Brick, Limburger, Port du salut
Surface ripened principally by mold	Edible crust: soft creamy interior, typical pungent flavor	Camembert, Brie
Curd coagulated primarily by acid[c]	Delicate soft curd	Cottage, cream, Neufchatel
Protein of whey or whey and milk coagulated by acid and high heat	Sweetish cooked flavor of whey	Gjetost, Sap sago, Primost, ricotta

Source: Ref. 4. *Newer Knowledge of Cheese,* Courtesy of NATIONAL DAIRY COUNCIL.®

[a] Close texture means no mechanical holes within the cheese; open texture means considerable mechanical holes.
[b] In contrast to ripening by bacteria throughout interior without eye formation.
[c] In contrast to coagulation by acid and coagulating enzymes, or in whey cheese, by acid and high heat.

3.1.1.2 Fresh

These cheeses do not undergo curing and are generally the result of acid coagulation of the milk. The composition, as well as processing steps, provide the specific product texture, while the bacteria used to provide the acid usually generate the characteristic flavor of the cheese.

3.1.2 Cheese Production and Composition

Per capita consumption of cheese is highest in Greece, at 47.52 lbs per year compared to 21.56 lbs per year in the U.S.A., which ranks sixteenth.[3] Production and composition of cheese in the United States is growing steadily.

Manufacturer's sales of cheese and projections[5] for the United States are shown in Tables 3.2 and 3.3

Unless otherwise indicated on the label, the basis of cheese is cow's milk which may be adjusted by separating part of the fat or by adding certain milk solids. The composition of cheese and related cheese products for interstate commerce is gov-

Table 3.2 MANUFACTURERS' SALES OF CHEESE

Year	Total ($, Millions)	Annual Percent Change
1967	1,751.8	—
1972	3,094.6	12.1[a]
1973	3,644.4	17.8
1974	4,504.7	23.6
1975	4,900.5	8.8
1976	5,764.1	17.6
1977	6,073.6	5.4
1978	6,688.5	10.1
1979	7,903.6	18.2
1980	9,415.9	19.1
1981	10,188.0	8.2
1982	10,170.0	−0.2
1983	10,561.7	3.9
1984	10,492.1	−0.7
1985	10,707.5	2.1
1986	11,378.3	6.3
1987	11,232.5	−1.3
1988[b]	11,388.8	1.4
1997[c]	17,644.8	4.6[a]

Source: Ref. 5.

[a] Average annual growth.
[b] Estimate.
[c] Projection

erned by the definitions and standards of identity developed, promulgated, and revised by the Food and Drug Administration (FDA) of United States Department of Health, Education, and Welfare. Cheese regulations assure the consumer of constant cheese characteristics and uniform minimum composition.[4] Federal standards of identity concerning cheese and cheese products[6] where established are given in Table 3.4. Typical analysis of cheeses[7] is given in Table 3.5.

Cheesemaking, as an artform, has been around for thousands of years. In earlier times cheese had been less than uniform and often with blemishes. The cheesemakers of the past worked diligently to learn intuitively the causes of and ways to avoid cheese failures. The discovery in 1935 by Whitehead in New Zealand, that bacteriophage(s) caused the milk acidification problem and gassy cheese,[8] was the first step toward more uniform and mechanized cheesemaking. The intervening 57 years of intensive research on milk and its conversion to cheese has brought a great deal of understanding and knowledge of milk composition—proteins, fat, lactose, and minerals—and their interaction as it affects cheesemaking. A great deal is being learned about the causes and metabolic behavior of starter organisms and their proteinases and peptidases, and their ability to cope with bacteriophages in the environment. There is considerable information in the published literature that has been recently arranged and compiled into reviews and books.[9–11]

Table 3.3 MANUFACTURERS' SALES OF CHEESE BY TYPE

	Natural Cheese		Process Cheese and Related Products		Cottage Cheese		Other Cheese[a]	
	Sales ($, Millions)	Percent Change	Sales ($, Millions)	Percent Change	Sales ($, Millions)	Percent Change	Sales ($, Millions)	Percent Change
1967	829.2	—	562.5	—	218.0	—	142.1	—
1972	1,400.0	11.0[b]	1,134.1	15.1[b]	340.9	9.4[b]	219.6	9.1[b]
1973	1,705.9	21.9	1,363.5	20.2	405.6	19.0	169.4	−22.9
1974	2,458.7	44.1	1,496.6	9.8	456.0	12.4	93.4	−44.9
1975	2,668.7	8.5	1,654.4	10.5	508.7	11.6	68.7	−26.4
1976	3,267.9	22.5	1,859.7	12.4	530.7	4.3	105.8	54.0
1977	2,727.2	−16.5	2,518.5	35.4	545.6	2.8	282.3	66.8
1978	3,104.1	13.8	2,681.4	6.5	588.5	7.9	314.5	11.4
1979	3,949.3	27.2	2,822.0	5.2	729.3	23.9	403.0	28.1
1980	4,821.1	22.1	3,303.4	17.1	840.9	15.3	450.5	11.8
1981	5,225.6	8.4	3,567.9	8.0	856.5	1.9	538.0	19.4
1982	5,625.6	7.7	3,194.3	−10.5	683.2	−20.2	666.9	24.0
1983	5,824.0	3.5	3,325.4	4.1	693.8	1.6	719.5	7.9
1984	5,617.3	−3.5	3,390.1	1.9	748.3	7.9	736.4	2.3
1985	5,664.6	0.8	3,552.6	4.8	738.3	−1.3	752.0	2.1
1986	6,289.8	11.0	3,548.9	−0.1	725.1	−1.8	814.5	8.3
1987	6,208.0	−1.3	3,463.7	−2.4	731.6	0.9	829.2	1.8
1988[c]	6,294.9	1.4	3,529.5	1.9	722.8	−1.2	841.6	1.5
1997[d]	9,826.9	4.7[b]	5,482.8	4.7[b]	1,083.9	3.9[b]	1,251.2	4.2[b]

Source: Ref. 5.

[a] Includes cheese substitutes.

[b] Average annual growth.

[c] Estimate.

[d] Projection.

Table 3.4 CODE OF FEDERAL REGULATIONS CHEESE COMPOSITION STANDARDS

Cheese Type	Legal Maximum Moisture, %	Legal Minimum Fat (Dry Basis), %	Legal Minimum Age
Asiago fresh	45	50	60 days
Asiago soft	45	50	60 days
Asiago medium	35	45	6 months
Asiago old	32	42	1 year
Blue cheese	46	50	60 days
Brick cheese	44	50	—
Caciocavello Siciliano	40	42	90 days
Cheddar	39	50	—
Low-sodium Cheddar	(Same as cheddar but less than 96 mg of sodium per pound of cheese)		—
Colby	40	50	—
Low-sodium Colby	(Same as cheddar but less than 96 mg of sodium per pound of cheese)		—
Cottage cheese (curd)	80	0.5	—
Cream cheese	55	33	—
Washed curd	42	50	—
Edam	45	40	—
Gammelost	52	(skim milk)	—
Gorganzola	42	50	90 days
Gouda	45	46	—
Granular-stirred curd	39	50	—
Hard grating	34	32	6 months
Hard cheese	39	50	—
Gruyère	39	45	90 days
Limburger	50	50	—
Monterey Jack	44	50	—
High-moisture Monterey Jack	44–50	50	—
Mozzarella and Scamorza	52–60	45	—
Low-moisture Mozzarella and Scamorza	45–52	45	—
Part-skim Mozzarella and Scamorza	52–60	30–45	—
Low-moisture, part-skim Mozzarella	45–52	30–45	—
Muenster	46	50	—
Neufchatel	65	20–33	—
Nuworld	46	50	60 days
Parmesan and Reggiano	32	32	10 months
Provolone	45	45	—
Soft-ripened cheese	—	50	—
Romano	34	38	5 months
Roquefort (sheep's milk)	45	50	60 days
Samsoe	41	45	60 days
Sapsago	38	(skim milk)	5 months
Semisoft cheese	39–50	50	—
Semisoft, part-skim cheese	50	45–50	—
Skim-milk cheese for manufacturing	50	(skim milk)	—
Swiss and Emmentaler	41	43	60 days

Source: Ref. 6.

In this chapter, effort is made to select and interpret information that is current and germane to the topic of cheese. Milk composition, cheese yield, starter proteinases and peptidases, and bacteriophage are not discussed because of space limitation. The subjects of fresh cheese, cheese defects, and pathogens in cheese are also not discussed. Some aspects of milk composition and casein micelle assembly and rennet coagulation are discussed in Chapter 1.

Although much is known about in vitro chymosin-induced proteolysis of casein(s) little is truly understood about the augment of changes and microbiological shifts in vivo that occur in cheese as a result. The efforts to accelerate cheese curing and to harness ultrafiltration of milk to produce superior Cheddar cheese and Swiss cheese have largely failed, indicating the lacuna in our understanding of cheese as an entity. It is ironic that most studies dealing with starter organisms and rennet reactions deal with optimum conditions, but most of cheesemaking and cheese curing is done under suboptimal conditions as they relate to starter or adventitious bacteria found in cheese. Wherever applicable, comments are made to provoke thinking in the unexplored facets of cheesemaking, curing, and longevity of cheese as a good food.

3.2 Heat Treatment of Milk for Cheesemaking

The bacterial flora in raw milk can vary considerably in numbers and species depending on how the milk is soiled. Major types of microorganisms found in milk are listed in Table 3.6.[12] Raw milk may also contain microorganisms pathogenic for man. Some of the more important ones are *Mycobacterium tuberculosis*, *Brucella abortus*, *Listeria monocytogenes*, *Coxiella burnette*, *Salmonella typhi*, *Campylobacter jejuni*, *Clostridium perfringens*, and *Bacillus cereus*. All of these pathogens with the exception of *C. perfringens* and *B. cereus* are destroyed by pasteurization because of their ability to sporulate.[12] Pasteurization of milk involves a vat method of heating milk to 62.8°C for 30 min or by a high temperature–short time (HTST) method, 71.7°C for 15 s. Originally most cheese was made from raw milk, but currently most manufacturers use heat-treated or pasteurized milk. Cheeses such as Swiss and Gruyère may be produced from heat-treated or pasteurized milk, but they are ripened or cured for at least 60 days for the development of eyes. In those instances where unpasteurized milk is used in the making of cheese, the cheese must be ripened for a period of 60 days at a temperature of not less than 1.7°C to ensure safety against pathogenic organisms.[4,13]

The pasteurization of milk for cheesemaking is not a substitute for sanitation. The advantages of pasteurization include:

- Heat treatment sufficient to destroy pathogenic flora
- A higher quality product due to destruction of undesirable gas and flavor-forming organisms
- Product uniformity
- Higher cheese yield[14]
- Standardized cheesemaking—there is easier control of the manufacturing procedure, especially acid development. The disadvantage of pasteurization is the dif-

Table 3.5 TYPICAL ANALYSIS OF CHEESE

Type	Cheese	Moisture (%)	Protein (%)	Total Fat (%)	Total Carbohydrate (%)	Fat in Dry Matter (%)	Ash (%)	Calcium (%)	Phosphorus (%)	Sodium (%)	Potassium (%)
Soft unripened low fat	Cottage (dry curd)	79.8	17.3	0.42	1.8	2.1	0.7	0.03	0.10	0.01	0.03
	Creamed cottage	79.0	12.5	4.5	2.7	21.4	1.4	0.06	0.13	0.40	0.08
	Quarg	72.0	18.0	8.0	3.0	28.5		0.30	0.35		
	Quarg (highfat)	59.0	19.0	18.0	3.0			0.30	0.35		
Soft, unripened high fat	Cream	53.7	7.5	34.9	2.7	75.4	1.2	0.08	0.10	0.29	0.11
	Neufchatel	62.2	10.0	23.4	2.9	62.0	1.5	0.07	0.13	0.39	0.11
Soft, ripened by surface bacteria	Limburger	48.4	20.0	27.2	0.49	52.8	3.8	0.49	0.39	0.80	0.13
	Liederkranz	52.0	16.5	28.0	0	58.3	3.5	0.30	0.25		
Soft, ripened by external molds	Camembert	51.8	19.8	24.3	0.5	50.3	3.7	0.39	0.35	0.84	0.19
	Brie	48.4	20.7	27.7	0.4	53.7	2.7	0.18	0.19	0.63	0.15
Soft, ripened by bacteria, preserved by salt	Feta	55.2	14.2	21.3	4.1	47.5	5.2	0.49	0.34	1.12	0.06
	Domiati	55.0	20.5	25.0		55.5					
Semisoft, ripened by bacteria with surface growth	Brick	41.1	23.3	29.7	2.8	50.4	3.2	0.67	0.45	0.56	0.14
	Munster	41.8	23.4	30.0	1.1	51.6	3.7	0.72	0.47	0.63	0.13
Semisoft, ripened by internal molds	Blue	42.4	21.4	28.7	2.3	49.9	5.1	0.53	0.39	1.39	0.26
	Roquefort	39.4	21.5	30.6	2.0	50.5	6.4	0.66	0.39	1.81	0.09
	Gorganzola	36.0	26.0	32.0		50.0	5.0				
Hard, ripened by bacteria	Cheddar	36.7	24.9	33.1	1.3	52.4	3.9	0.72	0.51	0.62	0.09
	Colby	38.2	23.8	32.1	2.6	52.0	3.4	0.68	0.46	0.60	0.13
Hard, ripened by eye-forming bacteria	Swiss	37.2	28.4	27.4	3.4	43.7	3.5	0.96	0.60	0.26	0.11
	Edam	41.4	25.0	27.8	1.4	47.6	4.2	0.73	0.54	0.96	0.19
	Gouda	41.5	25.0	27.4	2.2	46.9	3.9	0.70	0.55	0.82	0.12

(Continued)

Table 3.5 *(Continued)*

Very hard, ripened by bacteria	Parmesan (hard)	29.2	35.7	25.8	3.2	36.5	6.0	1.18	0.69	1.60	0.09
	Romano	30.9	31.8	26.9	3.6	39.0	6.7	1.06	0.76	1.20	
Pasta filata (stretch cheese)	Provolone	40.9	25.6	26.6	2.1	45.1	4.7	0.76	0.50	0.88	0.14
	Mozzarella	54.1	19.4	21.6	2.2	47.1	2.6	0.52	0.37	0.37	0.067
Low-fat or skim milk cheese (ripened)	Euda	56.5	30.0	6.5	1.0						
	Sapsago	37.0	41.0	7.4							
Whey cheese	Ricotta	71.7	11.3	13.0	3.0	45.9	1.0	0.21	0.16	0.08	0.10
	Primost	13.8	10.9	30.2	36.6	35.0					
Processed Cheese	American pasteurized processed cheese	39.2	22.1	31.2	1.6	51.4	5.8	0.62	0.74	1.43	0.16
	American cheese food, cold pack	43.1	19.7	24.5	8.3	43.0	4.4	0.50	0.40	0.97	0.36
	American pasteurized processed cheese spread	47.6	16.4	21.2	8.7	40.5	6.0	0.56	0.71	1.34	0.24
	Pimento pasteurized processed cheese	39.1	22.1	31.2	1.7	51.2	5.8	0.61	0.74	1.42	0.16
	Swiss pasteurized processed cheese	42.3	24.7	25.0	2.1	43.3	5.8	0.77	0.76	1.37	0.22
	Swiss pasteurized processed cheese food	43.7	21.9	24.1	4.5	42.8	5.8	0.72	0.53	1.55	0.28

Source: Hargrove and Alford (1974), Posati and Orr (1976).
Source: Ref. 7. Reproduced with permission.

Table 36 TYPES OF AEROBIC MESOPHILIC MICRO-ORGANISMS IN FRESH RAW MILK AND FORMING COLONIES ON MILK COUNT AGARS

Micrococci	Streptococci	Asporogenous Gram + Rods	Sporeformers	Gram − Rods	Miscellaneous
Micrococcus	*Enterococcus*	*Microbacterium*	*Bacillus* (spores or	*Pseudomonas*	Streptomycetes
Staphylococcus	([c]fecal)	*Corynebacterium*	vegetative cells)	*Acinetobacter*	Yeasts
		Arthrobacter		*Flavobacterium*	Molds
	Group N	*Kurthia*		*Enterobacter*	
				Klebsiella	
	Mastitis streptococci			*Aerobacter*	
	S. agalactiae			*Escherichia*	
	S. dysgalactiae			*Serratia*	
	S. uberis			*Alcaligenes*	

Source: Ref. 12. Reproduced with permission.

Note: Special media or incubation conditions are needed for isolation or detection of species of *Clostridium*, *Lactobacillus* and lactic acid bacteria, *Corynebacterium*, and certain pathogens.

ficulty of developing the full typical flavor in some cheeses such as Cheddar, Swiss, and hard Italian type cheeses.[4,15]

Higher than normal pasteurization temperatures were evaluated in Danish danbo cheese. The protein recovery ratios were 73.5%, 77.5%, and 78.5% when the milk was pasteurized at 66.7°C, 87.2°C, and 95°C respectively. The advantages of greater protein recovery and cheese yield by higher heat treatment were tempered by the lower quality of cheese made from milks heated at the two higher temperatures. Eye formation was not typical compared to the control cheese, and flavor and body defects were more prevalent in cheeses made from milk heated at 95°C.[16]

When cheese was made from milk pasteurized for 16 s at 73.3°C, 75.5°C, and 77.75°C, no significant differences in flavor preference or intensity of off-flavors were noted between the cheeses during ripening, although differences in body characteristics were evident. As the pasteurization temperature increased, the resulting cheeses were firmer and more rubbery and did not break down as readily when chewed.[17]

In another study, it was demonstrated that during aging, Cheddar cheese from pasteurized milk showed decreased proteolysis of α_s- and β-casein and production of 12% trichloracetic acid (TCA)-soluble nitrogen compared to the raw milk cheese. It is explained that the pasteurization of milk caused heat-induced interaction of whey proteins with casein and resulted in greater than normal retention of whey proteins in cheese. It is suggested that heat-denatured whey proteins affect the accessibility of caseins to proteases during aging.[18] The concentration of sulfhydryl (–SH) groups in cheese decreased as the temperature of milk heat treatment was increased. Kristoffersen believed that the concentration of –SH groups ran parallel to the intensity of characteristic Cheddar cheese aroma.[19–21]

The use of heat-treated milk is preferred for ripened cheeses such as Cheddar, Swiss, and Provolone to preserve a more typical cheese flavor.[4] Heat-treated milk is usually heated to 63.9 to 67.8°C for 16 to 18 s.

The heat treatment of raw milk can exert a significant role in producing microbiologically safe cheese. Recent thorough research has affirmed that milk heat treatment at 65.0 to 65.6°C for 16 to 18 s will destroy virtually all pathogenic microorganisms that are major threats to the safety of cheese.[13,22–24] For further discussion on heat treatment of milk for cheesemaking the reader should consult an excellent three-part review by Johnson et al.[13,15,25]

3.3 Cheese Starter Cultures

Starter cultures are organisms that ferment lactose in milk to lactic acid and other products. These include lactococci, leuconostocs, lactobacilli, and *Streptococcus salivarius* subsp. *thermophilus*. Starter cultures also include propionibacteria, brevibacteria, and mold species of *Penicillium*. These latter organisms are used in conjunction with lactic acid bacteria for a particular characteristic of cheese, for example, the holes in Swiss cheese are due to propionibacteria, and the yellowish color and typical flavor of Brick cheese is due to *Brevibacterium linens*. Blue cheese and Brie cheese derive their characteristics from the added blue and white molds, respectively.

Acidification of cheese milk is one of the essentials of cheesemaking. Acidification of milk is realized by the addition of selected strains of bacteria that can ferment lactose to lactic acid. Both the extent of acid production and the rate of acid production are important in directed cheese manufacture.[26] Mesophilic cultures (lactococci) are used in cheese where curd is not cooked to more than 40°C, for example, Cheddar cheese. Those cheese types that are cooked to 50 to 56°C (Swiss and Parmesan) use thermophilic cultures.

Acid production is the major function of the starter bacteria. During cheesemaking starter bacteria increase in numbers from about 2×10^7 cfu/g to 2×10^9 cfu/g in the curd at pressing.[27] During cheese ripening the added starter bacteria die off,[28] releasing their intracellular enzymes in the curd matrix which continue to act on components of the curd to develop desirable flavor, body, and textural changes. There are other incidental changes in milk and cheese and they come about as a result of acid production by lactic acid bacteria.

Lactic acid producing bacteria have several functions[3]:

1. Acid production and coagulation of milk.
2. Acid gives firmness to the coagulum which affects cheese yield.
3. Developed acidity determines the residual amount of animal rennet affecting cheese ripening; more acid curd binds more rennet.
4. The rate of acid development affects dissociation of colloidal calcium phosphate which in turn impacts proteolysis during manufacture and affects rheological properties of cheese.
5. Acid development and production of other antimicrobials control the growth of certain nonstarter bacteria and pathogens in cheese.
6. Acid development contributes to proteolysis and flavor production in cheese.
7. Growth of lactic acid bacteria produces the low oxidation–reduction potential (E_h) necessary for the production of reduced sulfur compounds (methanethiol, which may contribute to the aroma of Cheddar cheese).

3.3.1 Types of Cultures

Mesophilic cultures have their growth optimum at around 30°C and are used in cheeses where curd and whey are not cooked to over 40°C during cheesemaking. These starters are propagated at 21 to 23°C. These cultures along with their new and old names and some pertinent characteristics are listed in Tables 3.7 and 3.8. Culture compositions used for different cheese types are shown in Table 3.9.

Lactococcus lactis subsp. *lactis* belongs to Lancefield group N. Some strains isolated from raw milk produce nisin, a bacteriocin. Nisin is heat stable.[32] Its production is linked to a plasmid ranging from 28 to 30 MDA.[33,34] The plasmid also codes for sucrose fermenting ability and nisin resistance. Steel and McKay believe Suc^+, Nis^+ phenotypes are plasmid encoded but could not find physical evidence linking this phenotype to a distinct plasmid.[35]

Table 3.7 CHARACTERISTICS OF MESOPHILIC STARTER LACTIC ACID BACTERIA

Old Name	*Streptococcus lactis*	*Streptococcus cremoris*	*Streptococcus diacetylactis*	*Leuconostoc lactis*	*Leuconostoc cremoris*
New Name	*Lactococcus lactis* subsp. *lactis*	*Lactococcus lactis* subsp. *cremoris*	*Lactococcus lactis* subsp. *lactis* biovar *diacelylactis*	*Leuconostoc lactis*	*Leuconostoc mesentroides* subsp. *cremoris*
Optimum temp. (approx.)	30°C	30°C	30°C	30°C	30°C
Growth at 10°C	+	+	+	+	+
Growth at 40°C	+	−	+	+	−
Growth at 45°C	−	−	−	−	−
Survive 72°C/15 s	−	−	−	−	−
Growth in 2% salt	+	+	+	−	−
Growth in 4% salt	+	−	+	−	−
Growth in 6.5% salt	−	−	−	−	−
Production of NH_3 from arginine	+	−	+/−	−	−
Metabolize citrate	−	−	+	+	+
CO_2 production	−	−	+	+	+
Isomer of lactate produced	L	L	L	D	D
Lactic acid % in milk	0.8	0.8	0.4–0.8	0.2	0.2
Production of bacteriocin	Nisin[a]	Diplococcin[a]	—[a]	−	−
Lactose	+	+	+	+	+W
Glucose	+	+	+	+	+
Galactose	+	+	+	+	+

Source: Refs. 29–31.

[a] All strains do not produce bacteriocins

+ = Positive; +W = weakly positive; − = negative.

Table 3.8 CHARACTERISTICS OF LACTOBACILLI ASSOCIATED WITH CHEESE MANUFACTURE AND CHEESE RIPENING

	Percent Lactic Acid in Milk	Lactic Acid Isomer	Growth 15°C	Growth 45°C	Growth 50°C	Sensitivity to Salt	Glucose	Galactose	Lactose	Ammonia from Arginine	Bacteriocin
L. delbrueckii subsp. *bulgaricus*	1.8	D	−	+	+	<2%	+	−	+	±	+
L. delbrueckii subsp. *lactis*	1.8	D	−	+	+	<2%	+	−	+	−	+
L. helveticus	3.0	DL	−	+	+	<2%	+	+	+	−	+
L. casei subsp. *casei*	0.8	L	+	−	−	8% +[a], 10% −[a]	+	+	±	−	
L. casei subsp. *pseudoplantarum*		DL	+	−	−	6% +[a], 8% −[a]	+	+	+	−	
L. casei subsp. *rhamnosus*		L	+	+	−	8% +[a], 10% −[a]	+	+	+	−	
L. plantarum		DL	+	−	−	6% +[a], 8% −[a]	+	+	+	−	+
L. curvatus		DL	+	−	−	8% +[a], 10% −[a]	+	+	±	−	
L. fermentum[A]		DL	C	+	+	4% +[a], 6% −[a]	+	+	+	+	+
L. brevis[A]		DL	+	−	+		+	±	±	+	
L. buchneri[A]		DL	+	−	+		+	±	±	+	
L. bifermentans[A,B]		DL	+	−	−		+	+	−	−	

Source: Ref. 31.

[a] Unpublished: growth in MRS broth containing sodium chloride, 4 days at 35°C, + = growth, − = no growth.

A = Produce gas in cheese.

B = Ferments lactate in cheese with the production of CO_2, ethanol, and acetic acid.

C = Can grow in cheese at 15°C.

Table 3.9 STARTER CULTURES FOR CHEESE

Cheese	Culture Organisms Added
Cheddar, Colby	*Lactococcus lactis* subsp. *lactis*, *L. lactis* subsp. *cremoris* *Leuconostoc mesentroides* subsp. *cremoris*,* *L. lactis* subsp. *lactis* var. *diacetylactis** (*optional)
Swiss	*Streptococcus salvarius* subsp. *thermophilus*, *Lactobacillus helveticus* or *lactobacillus delbrueckii* subsp. *bulgaricus* or *L. delbrueckii* subsp. *lactis and Propionibacterium*
Parmesan, Romano	*Streptococcus salivarius* subsp. *thermophilus*, *L. helveticus* or *L. delbrueckii* subsp. *bulgaricus* or *L. delbrueckii* subsp. *lactis*
Mozzarella, Provolone	*S. salivarius* subsp. *thermophilus*, *L. delbrueckii* subsp. *bulgaricus* or *L. helveticus*
Blue, Roquefort and Stilton	*S. salivarius* subsp. *thermophilus*, *L. lactis* subsp. *lactis/ cremoris*, *L. lactis* subsp. *lactis* var. *diacetylactis*, *Penicillium roqueforti*
Gorgonzola	*S. salivarius* subsp. *thermophilus*, *L. delbrueckii* subsp. *bulgaricus*, *Penicillium roqueforti*, *L. lactis* subsp. *lactis* biovar. *diacetylactis* or yeast
Camembert	Lactococcus culture *Penicillium camemberti*
Brick, Limburger	Mixture of lactococcus culture and *S. salivarius* subsp. *thermophilus* Smear of *Brevibacterium linens* and yeast
Muenster Gouda and Edam	*L. lactis* subsp. *lactis* *L. lactis* subsp. *cremoris* With B or BD flavor cultures
Cream cheese	*L. lactis* subsp. *lactis* *L. lactis* subsp. *cremoris* With B or BD flavor cultures
Cottage cheese	*L. lactis* subsp. *lactis* and *L. lactis* subsp. *cremoris*

B = *Leuconostoc mesentroides* subsp. *cremoris/Leuconostoc lactis*.
D = *Lactococcus lactis* subsp. *lactis* var. *diacetylactis*.
BD = Where both leuconostocs and *L. lactis* subsp. *lactis* var. *diacelylactis* are included.

Nisin is active against *Clostridum botulinum* spores and several other Gram-positive organisms. Many of the isolates of *L. lactis* subsp. *lactis* from raw milk produce a malty odor. These strains metabolize leucine to produce 3-methylbutanol which is highly undesirable,[36] and as little as 0.5 ppm is sufficient to give milk this malty defect.

Lactococcus lactis subsp. *cremoris* also belongs to Lancefield group N. To date it has not been isolated from raw milk and its origin is not known. Some strains produce a narrow range bacteriocin diplococcin.[37–39] These organisms do not grow at 40°C and are more sensitive to salt. Many commercial cultures contain predominantly strain(s) of this specie.

Mixtures of these two lactococci are used as starters for Cheddar, Colby, and cottage cheese, where gas production in cheese and open texture are undesirable.

Lactococcus lactis subsp. *lactis* var. *diacetylactis* is used in combination with other starters to produce mold-ripened cheese, soft ripened cheese, Edam, Gouda, and cream cheese. It is capable of producing CO_2, diacetyl, acetoin, and some acetate from citrate in milk.[40]

3.3.2 Leuconostoc

The leuconostocs are heterofermentative, and ferment glucose with the production of D-(−)-lactic acid, ethanol, and CO_2. Leuconostocs are found in starter cultures and are considered important in flavor formation due to their ability to break down citrate, forming diacetyl from the pyruvate produced. The leuconostocs are less active than *Lactococcus lactis* subsp. *lactis* var. *diacetylactis*, attacking citrate only in acidic media.[29] *Leuconostoc* form only 5 to 10% of the culture population. Addition of a larger inoculum does not change their proportion of the population in a mixed lactic culture.[41] When the lactococci culture contains leuconostoc as a flavor producer, the mixed culture is called B or L type. When the flavor component is *Lactococcus lactis* subsp. *lactis* var. *diacetylactis*, it is called D type. The cultures designated as BD or DL contain both the leuconostocs and the *L. lactis* subsp. *lactis* var. *diacetylactis*. The lactococci without flavor components are called N or O type.[42]

3.3.3 *Streptococcus salivarius* subsp. *thermophilus*

This organism is a Gram-positive, catalase-negative anaerobic cocci and it is largely used in the manufacture of hard cheese varieties, Mozzarella, and yogurt. It does not grow at 10°C but grows well at 40 and 45°C. Most strains can survive 60°C for 30 min. It is very sensitive to antibiotics. Penicillin (0.005 Iu/ml) can interfere with milk acidification.[43] It grows well in milk and ferments lactose and sucrose. Two percent sodium chloride may prevent growth of many strains. These streptococci possess a weak proteolytic system. It is often combined with the more proteolytic lactobacilli in starter cultures. Most streptococci grow more readily in milk than lactococci and produce acid faster. These streptococci strains possess β-galactosidase (β-gal) and utilize only the glucose moiety of lactose and leave galactose in the medium.[31]

In a recent study,[44] proteolytic activities of nine strains of *Streptococcus salivarius* subsp. *thermophilus* and nine strains of *Lactobacillus delbrueckii* subsp. *bulgaricus* cultures incubated in pasteurized reconstituted NFDM at 42°C as single and mixed cultures were studied. Lactobacilli were highly proteolytic (61.0 to 14.6 μg of tyrosine/ml of milk) and *S. thermophilus* were less proteolytic (2.4 to 14.8 μg of tyrosine/ml of milk). Mixed cultures, with the exception of one combination, liberated more tyrosine (92.6 to 419.9 μg/ml) than the sum of the individual cultures. Mixed cultures also produced more acid (lower pH). Of 81 combinations of

L. bulgaricus and *S. thermophilus* cultures, only one combination was less proteolytic (92.6 μg of tyrosine/ml) than the corresponding *L. bulgaricus* strain in pure culture (125 μg of tyrosine/ml).

3.3.4 Lactobacilli

The lactobacilli are Gram-positive, catalase-negative, anaerobic/aerotolerant organisms. *Lactobacillus helveticus*, *L. delbrueckii* subsp. *lactis*, and *L. delrueckii* subsp. *bulgaricus* and homofermentative thermophiles are used in combination with *S. salivarius* subsp. *thermophilus* as starter culture for Swiss type cheeses, Parmesan, and Mozzarella. The phenotypic properties of these along with other lactobacilli commonly found in ripening cheese are given in Table 3.8. Premi et al. (1972)[45] screened strains of a number of species and found β-gal to be the dominant enzyme in *L. helveticus*, *L. delbrueckii* subsp. *lactis*, and *L. delbrueckii* subsp. *bulgaricus*.

Lactobacillus casei did not have β-gal, but some β-P-gal activity was recorded, and no galactosidase was found in *L. buchnerii*, which does not ferment lactose. There are several implications of this fermentation pattern to cheese quality. Cultures with β-gal use the glucose moiety of lactose and release galactose in the medium. An excess of galactose in Mozzarella can cause browning of cheese pizza, or galactose may serve as an energy source for undesirable fermentations by resident populations in cheese. It is recommended that *L. helveticus*, which is able to ferment galactose, be used in conjunction with *S. salivarius* subsp. *thermophilus*.[46] A symbiotic relationship exists between *L. delbrueckii* subsp. *bulgaricus* and *S. salivarius* subsp. *thermophilus*[47]; CO_2, formate, peptides, and amino acids are involved. In a mixed culture, associative growth of rod–coccus cultures results in greater acid production and flavor development than using single culture growth.[48,49] It has been established that numerous amino acids liberated from casein by proteases from *lactobacillus bulgaricus* stimulate growth of *S. thermophilus*.[50,51] In turn, *S. thermophilus* produces CO_2 and formate which stimulates *L. bulgaricus*.[51–54] During the early part of the incubation *S. thermophilus* grows faster and removes excess oxygen and produces the said stimulants. After the growth of *S. thermophilus* has slowed because of increasing concentrations of lactic acid, the more acid-tolerant *L. bulgaricus* increases in numbers.[55,56] For a one-to-one ratio of rod and coccus, inoculum level, time, and temperature of incubation must be controlled and bulk starter should be cooled promptly. Many strains *L. bulgaricus* continue to produce acid when in the cold and it is likely that some degree of population imbalance will occur.

3.3.5 Lactobacilli Found During Cheese Ripening

Lactobacilli occupy a niche in the ripening cheese.[57] A number of lactobacilli have been isolated from cheese and identified in the author's laboratory. The more common ones are subspecies of *L. casei*, *L. fermentum*, and *L. brevis*.

The presence of heterofermentative organisms, *L. fermentum* and *L. brevis* ($>10^6$ cfu/g), caused open texture defect in Cheddar cheese.[58] The addition of homofermentative lactobacilli affected cheese positively by accelerating the curing process.[59] The phenotypic traits of these are given in Table 3.8.

3.3.6 Propionibacteria

Propionibacteria are Gram-positive, catalase-positive anaerobic/aerotolerant organisms.[31] The cell can be coccoid, bifid, or even branched. Four species—*P. freudenreichii, P. jensenii, P. thoenii*, and *P. acidipropionici*—are associated with milk and Swiss cheese. Fermentation products include large quantities of propionic acid, acetic acid, and CO_2. These organisms can tolerate 125°F or higher temperatures in Swiss cheese manufacture. *P. thoenii* and *P. acidipropionici* can cause red, brown, and orange-yellow pigmentation in cheese which is not desirable. Some strains form curd in milk without digestion. Glucose, galactose, and glycerol are utilized by all species, and lactose utilization is not universal. These can grow in 20% bile. Glucose is fermented according to the following reaction[29]:

$$3 \text{ Glucose} \rightarrow 2 \text{ Acetate} + 4 \text{ Proprionate} + 2\ CO_2 + 4\ H_2O$$

3.3.7 Pediococci

Pediococci are associated with plant materials. These are Gram-positive, catalase-negative, or weakly positive, grow in 6.5% salt, grow at 45°C, and produce ammonia from arginine. These can be confused with micrococci. Pediococci are not used in any dairy cultures, though they may grow in some maturing cheese and ferment residual lactose over a long period. Only two species, *P. pentosaceus* and *P. acidilactici*, are found in dairy products; neither ferments lactose actively.[29]

Pediococci were first reported in New Zealand[60,61] and later in English cheese[62,63] and were thought to enhance flavor. They produce DL-lactate from lactose and racemize L-lactate. Their effect is negligible until the population exceeds 10^6 to 10^7 cfu/g. Their growth in cheese is temperature dependent.[64]

Pediococci occur in very insignificant numbers in Canadian Cheddar[65] and in Cheddar cheese or other cheeses in the United States (personal observations). There is a renewed interest in pediococci because some strains possess antimicrobial activity against *Listeria monocytogenes, Staphylococcus aureus*, and *Clostridium perfringens*.[66] In an examination of 49 strains of *P. pentosaceus*, valine and leucine amino peptidases, weak lipase or esterase, α-glucosidase, β-glucosidase, and *N*-acetyl-β-glucosamidase were found in all strains.

These studies were done with the API ZYM system.[67] In a more thorough investigation, Bhowmik and Marth[68] found intracellular aminopeptidase, protease, dipeptidase, and dipeptidyl aminopeptidase in six strains of *P. pentosaceus* and two of *P. acidilactici*. They also noted that purified α_{s1}- and β-casein fractions as well as skim milk were hydrolyzed. These authors could not detect esterase activity in any of the *P. acidilactici* strains studied.[69]

Utilization of lactose is poor in these organisms and varies from strain to strain.[69] Recently it was demonstrated that all strains of *P. pentosaceus* and *P. acidilactici* had intercellular β-galactosidase which was greater in cells grown in the presence of lactose rather than glucose, indicating the inducible nature of β-gal synthesis.[69] The enzyme was induced fully by galactose and lactose. Glucose failed to induce the enzyme in the strain (*P. pentosaceus* ATCC. 25745). Although these organisms are considered homolactic with the production of lactate, production of ethanol and acetate was observed when *P. pentosaceus* PC 39 was grown on different hexoses and pentoses.[70] The molar ratios of lactate and acetate were higher with ribose as substrate.

3.3.8 Molds

3.3.8.1 *Penicillium Roqueforti*[43,71]

It is used in the manufacture of Roquefort, Stilton, Gorgonzola, and other blue-veined cheeses, and usually produces blue-green spreading colonies changing to a dark green. A white mutant of this mold was developed for use in Nuworld cheese. These mutants form white rather than blue mycelia, but otherwise the mold produces a cheese of typical flavor. Spore preparations, dried form or suspension in saline solution, are added either to the vat milk or sprayed onto the curd. Air passages must be provided in cheese to permit aeration of the cheese and growth of the mold. Strains of *Penicillium roqueforti* can grow in an atmosphere containing 5% oxygen and 8% salt, although slowly.[71,72]

Its optimum temperature is 20 to 25°C with a range from 5 to 35°C. Production of mycelium is abundant at pH from 4.5 to 7.5, although it can tolerate pH 3.0 to 10.5.[72] Five strains isolated from cheeses and cultures showed differences in their salt tolerance.[71] The germination of spores of all five strains was inhibited by >3% NaCl in water and agar. In cheese, *P. roqueforti* could tolerate 6 to 10% salt.[71]

3.3.8.2 *Penicillium Camemberti*[72–74]

This grows on the surface of Brie and Camembert cheese. Due to its biochemical activity in conjunction with other flora on the cheese surface the mold produces its typical aroma and taste. *P. caseicolum* is a white mutant of *P. camemberti*[73] that forms a fluffy mycelium that turns gray-green in color from the center outward with aging. The white mutants may have short ''hair,'' rapid growth with white, dense, close-napped mycelium. Another white mutant has long hair and grows more slowly, producing a tall mycelium with loose nap. The Neufchatel form grows vigorously, producing a thick white-yellow mycelium. It has stronger lipolytic and proteolytic activities; only the white forms of the mutant are used as starters.

It has been shown that spores of *P. camemberti* do not grow well at the pH (4.7 to 4.9) and salt content present at the surface of fresh Camembert.[75] Maximum development of mold takes place in 10 to 12 days. *P. camemberti* possesses aspartate proteinases (acid proteinases) with a pH optimum of 5.5 on casein.[74,76]

3.4 Growth of Starter Bacteria in Milk

Milk is a suitable medium for the growth of lactic acid bacteria. In fact, *Lactobacillus delbrueckii* subsp. *bulgaricus*, *L. helveticus*, and *Streptococcus salivarius* subsp. *thermophilus* find milk a preferred medium for growth and utilize the abundant lactose found in milk. The lactic acid production of starter depends on the milk itself. Auclair and Hirsch were the first to point out that a balance exists between growth promoting and inhibitory factors in milk.[77] It is generally recognized that the ability of a starter to multiply in milk partly depends on its proteolytic activity. *Lactococcus lactis* subsp. *lactis* grew in a medium with caseinate as the sole source of nitrogen, whereas *L. lactis* subsp. *cremoris* required amino acid supplementation.[78] All dairy lactic acid bacteria either require or are stimulated by amino acids. The free amino acids available in milk are not adequate and the lactic acid bacteria use their proteinases, peptidases, and transport systems to meet their nutritional requirements.[79] Minimum concentrations of amino acids required by some lactic acid bacteria for maximum growth in a defined medium have been calculated. The data are not extensive and should be considered as directional. The amino acids Glu, Leu, Ile, Val, Arg, Cys, Pro, His, Phe, and Met are considered important in the nutrition of lactococci.

It is not uncomon that on continued transfers and propagation, organisms lose activity and ferment milk slowly. This is due to accumulation of slow variants in the culture. This was traced to the loss of one or more plasmids that control protein and lactose metabolism; phenotypic evidence for this was presented.[80,81]

3.4.1 Inhibitors of Starter Bacteria

3.4.1.1 Bacteriocins

Bacteriological quality of milk and the length of storage before it is used is important. Milk always contains organisms that can grow and utilize the amino acids and peptides in milk and produce inhibitors (bacteriocins) that can be inhibitory at very low concentration.[82]

Mattick and Hirsch[83] isolated an inhibitor, nisin, from *S. lactis*, that was active against Gram-positive organisms including starters, lactobacilli, and sporeformers. Oxford[84] isolated a bacteriocin from *S. cremoris* and called it diplococcin. Diplococcin has a very narrow spectrum of activity.[38]

3.4.1.2 Lipolysis

In stored raw milk psychrotrophs can grow and can cause lipolysis if the population exceeds 10^6 to 10^7 cfu/ml. Fatty acid C_4 to C_{12} and sorbic acid in cheese are inhibitory to starter bacteria. Cells accumulate free fatty acids on the cell surface and are not metabolized.[85–89] Resting cells of Group N lactococci at pH 4.5 metabolized pyruvate with the formation of acetate (volatile acids) acetoin + diacetyl and CO_2. In the presence of oelic acid the utilization of pyruvate was maximal at pH 6.5 and completely inhibited at pH 4.5.[90]

3.4.1.3 Hydrogen Peroxide

Hydrogen peroxide is metabolically produced by Group N lactococci through the action of reduced nicotinamide adenine dinucleotide (NADH) oxidase which catalyzes the oxidation of NADH by molecular oxygen. The enzyme is activated by flavine adenine dinucleotide (FAD). Some of the hydrogen peroxide formed is removed by NADH peroxidase.[91]

The reaction is:

$$NADH + H^+ + O_2 \xrightarrow{\text{(NADH) oxidase}} NAD^+ + H_2O_2$$

$$NADH + H^+ H_2O_2 \xrightarrow{\text{(NADH) peroxidase}} NAD^+ + 2H_2O$$

Milk is agitated during filling of the vat and addition of starter and during addition of rennet in the course of cheese manufacture, and sufficient hydrogen peroxide can be formed in milk. Addition of trace amounts of H_2O_2 had a deleterious effect on the rate of acid production by lactococci.[92] In milk, cultures of lactococci and lactobacilli produced hydrogen peroxide in the early period of acid production, followed by a drastic reduction in the accumulation of H_2O_2 as the acid production increased. Addition of ferrous sulfate and catalase prevented or reduced the accumulation of H_2O_2 and stimulated the rate of acid production.[93] Addition of a capsular preparation from a *Micrococcus*[94] and the addition of *Micrococcus* reduced the amount of H_2O_2 in the medium and stimulated acid production through multiple effects.

3.4.1.4 Lactoperoxidase/Thiocyanate/H_2O_2 System

Hydrogen peroxide produced metabolically can also inhibit some strains of lactococci indirectly in milk cultures by oxidizing the thiocyanate present in milk to an inhibitory product, a reaction catalyzed by lactoperoxidase.[91] Small concentrations of hydrogen peroxide form a complex with lactoperoxidase (LP) which stabilizes the oxidizing power of H_2O_2, catalyzing the oxidation of thiocyanate (SCN^-) according to the reaction:

$$H_2O_2 + SCN^- \xrightarrow{\text{Lactoperoxidase}} OSCN^- + H_2O$$

$$OSCN^- + H_2O_2 \xrightarrow{\text{LP}} O_2SCN^- + H_2O$$

$$O_2SCN^- + H_2O_2 \xrightarrow{\text{LP}} O_3SCN^- + H_2O$$

The end products of the oxidation of thiocyanate are CO_2, NH_4^+, SO_4^{2-}, which are inert but the intermediate oxidation product ($OSCN^-$) is inhibitory to Gram-positive organisms (starter organism) and bacteriocidal to coliform, pseudomonads, salmonellae, and other Gram-negative organisms. Under aerobic conditions $OSCN^-$ affects the inner membrane and other cell wall components.[95]

Wright and Tramer noted that some starter cultures show inhibition by the presence of milk peroxidase which can be prevented by the addition of cysteine or

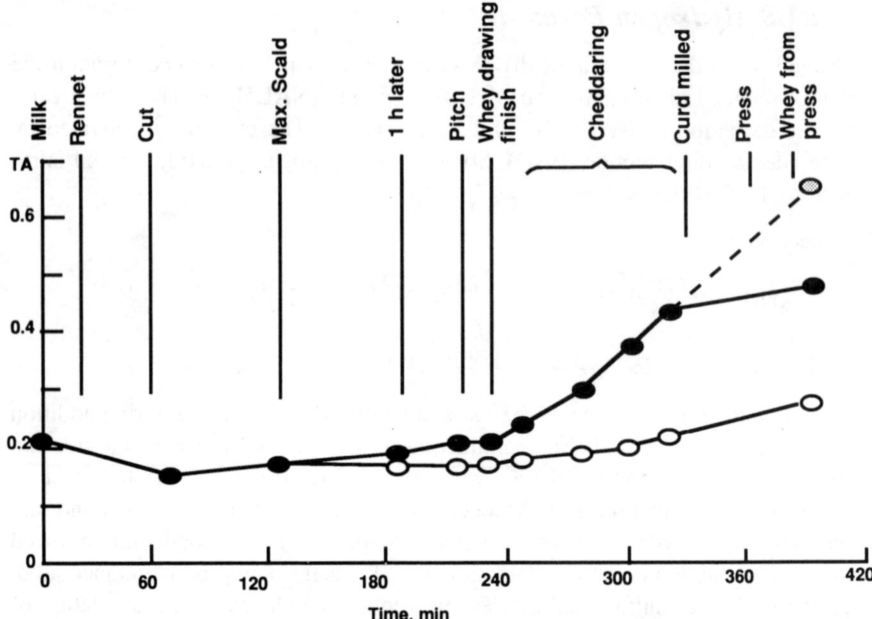

Figure 3.1 Lactic acid development in Cheddar cheese made with peroxidase/SCN^--sensitive starter in the presence of SCN^- and after removal from milk by ion-exchange treatment. (●——●), SCN^- removed from milk; (○——○), untreated milk; (●- - -●), control (lactic acid production with peroxidase resistant starter *Strep. cremoris* 803).

Source: Ref. 97 (This figure is reproduced by kind permission of the Society of Dairy Technology, Crossley House, 72 Ermine Street, Huntington, Cambs PE18 6EZ, UK and is taken from a paper 'Some Thoughts on Cheese Starter' by Bruno Reiter published in the Society's Journal Vol **26** no. 1, January 1973.)

generation of –SH groups by heating.[96] The effect of peroxidase/thiocyanate on cheesemaking was demonstrated (Fig. 3.1). Thiocyanate was removed from milk with ion-exchange resins and it is shown that the peroxidase-sensitive strain *S. cremoris* 972 was not inhibited, and lactic acid production rate was normal during cheesemaking. The addition of thiocyanate prevented any appreciable acid development, similar to the behavior of phage-infected starter culture.

Stadhouders and Veringa[98] noted that inhibition of lactococci and the prevention of inhibition of lactic streptococci by cysteine were related. They explained that in a mixture of H_2O_2, cysteine, and milk peroxidase, cysteine is oxidized and acts as an H-donor. If cysteine and the milk peroxidase are incubated together without H_2O_2, the cysteine and the enzyme form an irreversible compound. If H_2O_2 then is added the cysteine acts as an inhibitor of the enzyme.

They theorized that peroxidase-sensitive variants of lactic streptococci probably had an absolute requirement for free cysteine but the cysteine was complexed with peroxidase. Peroxidase in milk is inhibited by the presence of very small amounts of hydrogen sulfide which is produced during heating of milk.[99]

The susceptibility of dairy starter cultures to lactoperoxidase/hydrogen peroxide/ thiocyanate system (LPS) inhibition is dependent on[100–102]:

1. Strain sensitivity
2. Ability of the strain to generate H_2O_2 which activates the LPS system
3. The presence of nonspecific enzymes, for example, xanthine oxidase, or hypoxanthine that generate H_2O_2.

This inhibitory system is heat labile and destroyed by heat treatment of the starter culture milk.

Inhibitory substances can also be produced by lactic streptococci during their propagation; D-leucine was formed in mixed-starter cultures during growth at controlled pH in broth and had an autoinhibitory effect.[103]

3.4.1 Heat Treatment

Milk is given a heat treatment to preserve it and to make it safe for consumption. The extent of heat treatment is dependent on the product and its intended use. Many workers have studied the effect of heat on starter culture activity. It is generally recognized that different cultures show varied activity when propagated in milk that has received a certain heat treatment. Olson and Gilliland[104] and Speck[105] noted that the rate of acid production by lactococci was highest in the lots of milk pasteurized at 71.1°C for 30 min followed in order by that in milk sterilized at 121.1°C for 15 min, 61.6°C, 82.2°C, and 98.8°C for 30 min, respectively. Those cultures that produced acid rapidly in milk pasteurized at 61.6°C for 30 min or 71.7°C for 16 s were called ''low-temperature cultures.'' The cultures that produced acid rapidly in high-heat-treated milk were called ''high-temperature cultures.'' Of 37 commercial lactic cultures tested, 49% were classified as low heat, 35% as high heat, and 16% as indifferent cultures.

For thermophilic cultures such as *S. salivarius* subsp. *thermophilus* and *L. delbrueckii* subsp. *bulgaricus*, heat treatment of milk at various time–temperature combinations ranging from HTST pasteurization to 180°C for 10 min was studied. It had no observable effect on the growth of *S. salivarious* subsp. *thermophilus* but stimulated *L. delbrueckii* subsp. *bulgaricus*; the effect increased with the severity of heat treatment. At heat treatments up to 95°C/10 min, the stimulation occurred only in mixed culture.[106] The stimulatory factor could be replaced by formic acid.[53] The production of formic acid by *S. thermophilus* was confirmed.[107]

3.4.1.6 Agglutination

The inhibitory property of agglutinating antibodies is of minor importance in bulk starters as the heat treatment employed or by the formation of rennet coagulum during cheesemaking destroys this inhibition.[108–110] However, agglutinins are important and impact negatively in cottage cheese production where a sludge is formed at the bottom of the vat and culture activity is slowed.

Table 3.10 ACTIVITY OF SINGLE STRAIN BULK-STARTER GROWN IN AUTOCLAVED SKIM MILK WITH DIFFERENT LEVELS OF PENICILLIN

	Bulk-Starter Analysis[a]		
Bulk-Starter Sample	PH	Plate Count per ml	Activity Test pH[b]
Control 104	4.45	5.9×10^8	5.18
104 + 0.025 IU penicillin ml	4.44	4.3×10^8	5.49
104 + 0.05 IU penicillin ml	4.55	1.2×10^8	5.87
104 + 0.1 IU penicillin ml	4.60	2.5×10^7	6.39
Control 134	4.48	8.7×10^8	5.85
134 + 0.025 IU penicillin ml	4.44	6.2×10^8	5.85
134 + 0.05 IU penicillin ml	4.46	6.2×10^8	5.88
134 + 0.1 IU penicillin ml	4.57	6.2×10^8	6.30

Source: Ref. 111. reproduced with permission.

[a] After incubation for 18.5 h at 22°C.

[b] Averaged pH results from two separate trials.

3.4.1.7 Antibiotics

The presence of a low level of antibiotics can cause slow culture activity and cheese-making to be more difficult. Heap[111] demonstrated that given time, lactococci could grow and produce acid in reconstituted skim milk containing different levels of penicillin; acid production looked normal but the culture had poor activity. The data are shown in Table 3.10. Starter culture activity must be performed to verify culture activity. Sensitivity of cheese and dairy-related organisms to antibiotics is presented[112] in Table 3.11.

3.4.1.8 pH

One of the common causes of observed variation in starter activity in the cheese vat is the difference in the ability of the culture to retain activity when held for long periods in the high acid concentrations existing in overripe bulk starters.[113] Olson[114] demonstrated that fully ripened starter cultures survive better under less acid conditions; addition of calcium carbonate increased the survival. When lactococci were allowed to grow below pH 5.0, cells were damaged and a period of growth above pH 5.0 was required to correct this damage.[115] Growth at low pH could result in direct inactivation of a number of enzymes or in loss of control of the differential rates of synthesis of individual enzymes. The cells stopped growing when the pH reached 4.9, even though lactic acid continued to be produced until the pH had fallen to about 4.6. Neutralization of the acid permitted resumption of growth and glycolysis by the cell.[116]

Of all the factors studied, bacteriophages are the most important enemy of cheese starter bacteria. These will be discussed in a later section.

Table 3.11 CRITICAL PENICILLIN LEVELS IN MILK FOR BACTERIA

Bacteria	Penicillin Concentration Significantly Inhibiting Growth (IU per ml)
S. cremoris	0.05–0.10
S. lactis	0.10–0.30
Streptococci starter	0.10
S. thermophilus	0.01–0.05
S. faecalis	0.30
L. bulgaricus	0.30–0.60
L. acidophilus	0.30–0.60
L. casei	0.30–0.60
L. lactis	0.25–0.50
L. helveticus	0.25–0.50
L. citrovorum	0.05–0.10
Proprionibacterium shermanii	0.05–0.10

Source: Compiled from K. E. Thomé. *Refresher Course on Cheese*. Poligny, France, 1952; Overby, A. J. *J. Dairy Sci. Abstr,* **16**:2–23, 1954; and F. V. Kosikowski, Unpublished, 1954.
Source: Ref. 112. Reproduced with permission of FAO of the United Nations.

3.5 Starter Culture Systems

As stated earlier, the primary function of starter bacteria is to ferment lactose in milk to lactic acid and other products. It is also important that rate of acid development be such that cheese of proper composition is made within the limits of manufacturing parameters. This has become more critical where automated cheesemaking is practiced in large plants pumping milk at 120,000 lbs/h. The major problem associated with the commercial use of starters is inhibition of acid production by bacteriophage (phage).

Researchers all over the world have tried to understand the etiology of phage-mediated lack of milk acidification and have developed considerable understanding and various strategies to combat phage in cheese and dairy plants. The work done in New Zealand for the past 55 years had a major impact on culture selection, culture composition, culture handling, and bacteriophage control. Various culture systems are operative today and these are described briefly.

In the 1930s mixed cultures used in New Zealand produced gas and caused open texture in cheese. Whitehead isolated pure strains of non-gas-producers and used them as single strains. The rate of acid production with these strains was virtually uniform from day to day. Eventually these strains also failed due to phage.

In 1934 Whitehead and Cox[117] noted that sudden failure of the starter resulted from aeration of the cheese milk. It was proposed that their failure was due to disrupting phage present in the starter.

In 1935, they proposed that phage are present in very small amounts in the culture and may exist in an ‘‘occluded’ state.[118,119] These phage may then be ‘‘triggered’’ by aeration and liberated into the culture, where they would multiply and inhibit acid production.

In 1943 Whitehead introduced a 4-day rotation of non-phage-related single starters.[120] Subsequently, single strains were paired as a precaution against failure of one of the members. Pairing also tended to even out differences in the rate of acid production and any tendency to produce bitter flavors by the individual members. Lawrence and Pearce[121] noted good flavor cheese made with slower starters. However, the use of slower starters took longer for cheesemaking. This was overcome by pairing a ''slower'' starter with a ''fast'' starter in a ratio of about 2:1. It was also noted that a combination of slow and fast strains not only improved the quality of cheese but also reduced the number of phage particles produced; faster acid producing strains propagated phage to the highest level. Perhaps the level of lysin (cell wall degrading enzyme) produced by phaged out starter was also reduced, thereby helping the viability of the bulk starter. It was emphasized that stock cultures must be replaced regularly with strict observance to procedures.[122] In 1976 Heap and Lawrence published a test procedure where a projected viability of a new strain in a plant environment could be established.[123] It involved growing the culture for successive growth cycles in the presence of bulked plant whey. Any difference in 5-h pH between successive growth cycles was an indication of phage against the strain. Only strains that were not attacked by phage in at least ten growth cycles were used. Based on the above selection criteria, a multiple starter consisting of six carefully selected strains was introduced for continued use in cheese factories.[124] Whey samples were monitored using the strains as host. Strains showing high levels of phage were replaced with less sensitive strains. This seemed to have worked well. The multiple starter concept is only an extension of the paired starter system, as the single strains are not mixed until the mother culture stage.[125] Suitability criteria of a strain for use in multiple starter is given in Table 3.12. In the past few years, the number of strains in the starter has been reduced from six strains to two without any reported problems.[126]

3.5.1 Culture Systems

1. Defined culture system requires good starter tanks, and proper air flow and plant layout along with trained people to do simple culture activity testing with and without filtered whey. This system is operative in New Zealand, some plants in Australia, and in many large factories in the United States, United Kingdom,[127,128] and Ireland.[129] The defined strains may be grown as a mixed culture or strains propagated singly and mixed after harvesting.

Exclusive use of defined-strain cultures was reported to yield significant savings ($1 million for a cheese plant producing 11.35 million kg of cheese/year) with no reported cheese vat failures due to phage. Because the starter activity was uniform and predictable, cheesemaking could be standardized.[127]

2. Mixed strain mesophilic cultures containing undefined flora, some containing leuconostocs or *L. lactis* subsp. *lactis* var. *diacetylactis*, are still in use in United States and in Europe. These cultures are propagated as mixed cultures without regard

Table 3.12 PROCEDURES TO DETERMINE THE SUITABILITY OF A STRAIN FOR USE IN MULTIPLE STARTERS; CHARACTERISTICS OF A GOOD STRAIN

1. Colony appearance on bromcresol purple medium.
2. Ability to coagulate sterile reconstituted skim milk (rsm) at 22°C in 18 h.
3. Activity in simulated cheesemaking test (using both rsm and pasteurized factory milk).
4. Viable cell counts after simulated cheesemaking test.
5. Temperature sensitivity.
6. Salt sensitivity.
7. Tolerance to antibiotics.
8. Survival in wheys from cheese plants.
9. Host/phage relationships.
10. Multiplication factors of phages attacking strain.
11. Phage adsorption.
12. Induction of phage from strain by ultraviolet light.
13. Compatibility with other strains.
14. Small-scale cheesemaking trials.
15. A suitable strain should have the following characteristics for producing good flavor in Cheddar cheese:
    - Poor survival both in cheese matured at 13°C and in pasteurized skim milk (PSM) containing 4–5% NaCl at ~pH 5.0.
    - A low rate of cell division at 37.5–38.5°C resulting in low starter population in the cheese curd.
    - Low proteolytic activity at 13°C and pH 5.0 in PSM containing 4–5% NaCl.
    - High acid phosphatase activity after growth to pH 5.2 in PSM at 35°C.

Source: Refs. 28, 125.

to the component strain balance. The cultures may be concentrated and then frozen. Many small factories and some large factories use these cultures with rotation recommendations from culture suppliers.

Because most cultures sold are mixed, phage profiling is not practicable, and the recommended rotations are useless because plants use cultures from different suppliers which may use strains of the same phage type. Many plants have suffered considerable lack of milk acidification and cheese quality losses.[127]

3. Bacteriophage-carrying starter cultures are widely used in the Netherlands.[130,131] The cultures are called P-(Practice) cultures. These cultures are in equilibrium with the phages in their environment and normally contain phages that do not affect culture activity. When a phage emerges against the dominant strain, a slight weakness in culture activity may be noticed but the culture activity recovers quickly due to the presence of a large phage-insensitive population. The Netherlands Institute of Dairy Research maintains a supply of P-starters that it had collected and preserved in a concentrated frozen state. These cultures are provided to the plants. This system appears to work almost flawlessly. When the P-starters are propagated in the laboratory without phage contamination (L-starter), they become sensitive to phages. This is attributed to the domination of one or of a small number of strains in the so called L-starters.

4. Direct-to-vat (DVS) set cultures had become popular in the late 1970s. These are highly concentrated (10^{11} cfu/g)[132,133] cell suspensions of defined strains in milk

along with cryoprotective agents such as glycerol or lactose,[134] quick frozen in liquid nitrogen, and held frozen at $-196°C$. For the shipping to plants, frozen culture containers are packed in dry ice in Styrofoam boxes.

One culture container is added to 5000 lbs of cheese milk which is roughly equivalent to 1.0% bulk starter addition.[135] DVS cultures are mixtures of three or four defined strains propagated mixed together or propagated separately and blended in a proprietary manner.

Use of these cultures is supposed to eliminate phage infection related problems associated with bulk starter propagation and make cheesemaking easier.

Several advantages are claimed[136]:

1. *Convenience*. The cultures can eliminate the need for bulk starter facilities including tanks, laboratory, and expensive sterile air systems. They can supplement the conventional system at weekends or during holidays and can be used as a backup in the event of a bulk starter failure.
2. *Culture reliability*. Because the cultures are pretested for activity, the cheesemaker can standardize the cheese make for each blend used.
3. *Improved daily performance*. The pretested cultures afford the same strain balance day after day and should result in a more uniform cheese production.
4. *Improved cheese yield.*

Disadvantages:

a. Use of DVS cultures necessitates a large dependable freezer. The cost of a freezer is claimed to be offset by savings in labor and starter preparation in antibiotic-free milk.
b. Due to lower acid development at the time of setting, some coagulants containing porcine pepsin may have to be used at a higher level. To increase firmness of the curd, vats need to be set at 90 to 91°F instead of 86°F.[136]

Although DVS cultures are still in use, many of the claims made a few years ago are not fully realized for the following reasons:

1. Lack of enough strains with discretely different phage types to support a large cheese factory reliably.
2. Many strains are difficult to concentrate 50- to 80-fold by centrifugation.
3. Activity of the frozen cultures inoculated in vat milk is slow[132,137] during the cutting and cooking stages of cheese. Cheesemaking steps had to be modified to accommodate slow wet-acid production and fast acid development in dry state (cheddaring).
4. DVS culture cost to cheese is high; this view is not without opposition.
5. Due to availability of easy-to-maintain electronics and automation, plant propagation of starter cultures with internal pH control was introduced in the United States in the last decade. In this propagation, the cell concentration is 10 to 15 times higher than the conventional bulk starter cultures. Now many of the well maintained large cheese plants have adopted pH-controlled propagation of defined strains with exellent success.

Richardson et al. were largely responsible for bringing external pH-controlled starters to cheese factories,[138] recommending a whey-based medium for greatest economic return because of high cheese yield and a lower medium cost, one third the cost of internal pH-control-buffered media.[139]

3.6 Culture Production and Bulk Starter Propagation

3.6.1 History

Traditionally cultures were carried from seed to intermediate mother cultures to inoculate the bulk tank. These were propagated in 10% or 12% nonfat dry milk heat treated at 90°C for 45 min or more to render it bacteriophage- and cell-free. Stadhouders found that 95°C/55 s was required to inactivate phage.[130] Such cultures were dispensed in sterile glass bottles and sent by post to reach cheese factories within 72 h. These were subcultured for further propagation by cheese plants.[140] For long-distance shipment, cultures were made into powder form by blending with lactose, followed by neutralization with calcium carbonate and vacuum drying. Cultures produced in this manner needed several transfers for full activation due to only 1 to 2% survivors in the powder.

Freeze-dried cultures showed 42 to 80% survival for different cultures; these cultures grew slowly with a long lag phase.[135] In order to reduce the lag phase, addition of stimulants to the culture before freeze-drying or to the substrate in which the culture was reactivated were practiced.[140]

In 1963, frozen, nonconcentrated, 1-ml vial cultures were made available commercially to cheesemakers. These could be stored in liquid nitrogen over a longer period without much loss in activity and produced a good active mother culture in the first transfer.[135]

3.6.2 Concentrated Cultures

Work on concentrated cultures began in the late 1960s and was commercialized in 1973. This development eliminated the chores of preparing mother culture and intermediate cultures. This practice minimized starter handling in the phage-contaminated atmosphere of the cheese factory and paved the way for DVS cultures.[135–141]

For a conventional bulk starter, the heat-treated (90°C/45 min) milk tempered to 21 to 27°C is inoculated and incubated at ~27°C until it reaches a pH of 4.6. At this point the culture may contain 5 to 8 $\times$ 10^8 cfu/ml and has good activity. However, if the cells are held at pH 5.0 for extended periods of time, the culture activity is reduced.[115] The final population of lactococci can be greatly increased by controlling the pH of the growth medium at 6.0 to 6.5.[132,142–147] When culture was propagated in a medium (2% tryptone, 1% yeast extract, 2.5% lactose, and 2.5% glucose) at a constant pH of 6.0 (maintained by the addition of NaOH), the cell population was 15 times that of non-pH-controlled propagation.[132] At this pH both the rate and the total amount of growth were optimum. When mixed species of starter bacteria containing aroma bacteria were grown in skim milk (9.1% solids), whey medium, and tryptone medium at a constant pH with continuous culturing, relative lactic acid production activity (%), aroma bacteria (%), and diacetyl production were highest in milk at pH 5.9.[148] Specific growth rate and productivity were found to be affected by both the medium and the pH value. Continuous culturing below pH 5.9 to 6.1 was not recommended.[148] Batch culture was considered preferable to continuous culture and the best yield, approximately 10^{10} cfu/ml, was obtained at 30 to 32°C with pH maintained between pH 6.0 and 6.3.[146] The maximum cell density and culture activity were affected by the neutralizer; higher cell densities were obtained with NH_4OH than with NaOH.[132,145] Culture concentrate prepared using NH_4OH had a reduced rate of acid production compared to the milk cultures. This was traced to a lower proteinase activity in the NH_4OH-neutralized cell preparations.[132] Lactic acid or lactate salts[144,147] accumulation and secretion of D-leucine[103] in the medium limit growth of lactic acid bacteria.

3.6.3 Bulk Starter Propagation

For bulk starter preparation inoculation, about 10^6 to 10^7 cfu/ml are required. For a properly prepared culture concentrate containing 10^{11} cfu/g of culture, 25 g should be sufficient for 500 gal.[149]

Starter organisms grow well in milk of normal composition. In the past it was difficult to keep bacteriophage out of the bulk starter and at times culture activity was affected. The most important aspect of starter production is the preparation of the growth medium, and the protection of the culture from phage attack. Several approaches, singly or in combination, are in practice. These are:

1. Aseptic technique
2. Specially designed starter vessels to prevent phage entry from without
3. Phage inhibitory media that prevent phage multiplication in the medium
4. DVS cultures—frozen or freeze-dried

3.6.3.1 Aseptic Techniques

These involve separate starter room, chlorination, and steam sterilization of the starter vessel and chlorine fogging of the starter room before inoculation. These techniques are helpful but not entirely satisfactory for keeping phage out of the starter if it is present in the environment.

3.6.3.2 Specifically Designed Starter Tanks

Specially designed starter tanks aim at preventing post heat treatment contamination of the starter medium. Some of these are described:

The Lewis System

The technique involves the use of polythene bottles for mother and feeder cultures. These bottles are fitted with Astell rubber seals. The medium is sterilized and cooled in the bottle and culture is transferred by means of two-way hypodermic needles.

The Lewis system requires a pressurized starter vessel; no air enters or leaves the vessel during heating and cooling. This system is detailed in a recent book.[150]

The Jones System

In this system, the tank is not pressurized. The tank openings are protected by water seals. The air is forced out during the heating of the medium and sterile air (heated and filtered) reenters the tank during cooling. The system is used in New Zealand and described in detail by Heap and Lawrence.[151]

A starter vessel combining the Lewis and Jones System has been developed in the United Kingdom.[152]

The Alfa-Laval System

In this system the mother and intermediate cultures are propagated in a viscubator and the culture is transferred to a large tank using filter-sterilized air under pressure. The system is described by Tamime.[152]

Systems Using High-Efficiency Particulate Air Filters (HEPA)

Dutch cheese manufacture utilizes P-starters prepared by NIZO. Every care is exercised to prevent phage contamination during inoculation and cultivation of the starter. Milk is heated to 95°C or higher for 1 min, and during cooling, inoculation and cultivation tanks are pressurized with sterile air. Absolute filters (Pall Enflonfilter Type ABI FR7PV) permit penetration of less than one per 2.5×10^{10} phages, ensuring that the pressurized tank is always free of phage.[153] Recently, depth filters have been made available that have a pore size of 0.015 μm that can filter out bacteriophage from air. These are in-line filters and can be steam sterilized in place up to 50 cycles.[154]

Recently, Bactosas, an ultra-clean room with 12 filtered air changes/h, has been designed and patented.[155] In this system, any number of pressurized vessels are grouped together in such a manner that the entry ports of the vessels—and only the ports—are accessible to the operator from inside a large and carefully controlled enclosed area. The vessels are CIPable and the service units, the valves, pumps, pipe work, instruments, electrical wiring, etc. are separated.

3.6.3.3 Phage Inhibitory Media

That bacteriophage require divalent ions, particularly calcium, for adsorption and subsequent proliferation is established.[156] Reiter[157] removed calcium from the medium by ion exchange and noticed inhibition of bacteriophage.

Addition of 2% sodium phosphate ($NaH_2PO_4 \cdot H_2O/Na_2HPO_4$ in a ratio of 3:2), to sequester calcium, prevented phage growth in skim milk bulk starter.[158] The bacteria grew normally and the cheese made with starter challenged with homologous bacteriophage had normal texture and flavor. Other formulations[159] were developed where media containing nonfat dry milk, dry blended mono- and dibasic phosphate, yeast extract, and electrodialyzed whey could prevent the growth of most phages while permitting culture growth. These media were called phage inhibitory media (PIM) or phage-resistant media (PRM). Numerous such media were made available in the marketplace and contain milk solids, carbohydrate, growth promoting factor(s), and buffering agents such as phosphate and citrate. However, it was noticed that all phage active against lactococci were not restricted in Ca^{2+}-reduced media.[160] In a comprehensive study, seven commercial PIM were compared for their buffering capacity, ability to support lactococci growth, and extent of suppression of bacteriophage replication.[161] Only two of the seven media were adequate in preventing phage proliferation; the effectiveness was linked to the buffering capacity. Such media contained sufficient nutrients to overcome the effects of high phosphate or citrate concentrations which depressed growth. The most effective media also contained citrate buffer and cereal hydrolyzate as a stimulant. Ledford and Speck[162] clearly demonstrated that PIM caused metabolic injury to starter bacteria and their proteinase activity was diminished. Addition of 1 or 2% phosphate to reconstituted nonfat dry milk reduced about 30% of proteinase activity as measured by tyrosine release.

The development of PIM was an important step and brought some relief from phage-mediated lack of milk acidification. These media serve a useful function where physical protection against phage, that is, proper bulk tank design, provision of pressurization with sterile air, and inoculation and other general procedures, are not adequate. However, these media are not suitable for cultures containing lauconostocs[163,164] because they promote culture imbalance, which may lead to flavor defects in products. Also, these media add substantially to the cost of cheese production and counteract addition of Ca^{2+} to cheesemilk to aid rennet coagulation. LaGrange and Reinbold in 1968 documented that the cost of PIM was 10 to 15¢/lb more than the low-heat NFDM which cost 20 to 25¢/lb and that the starter media cost was 70% of the cost of starter.[165] Many changes and developments have come about in starter cultures handling and culture media in the last 20 years. In a later study,[166] LaGrange found that starter costs per 100 lbs of milk converted to cheese ranged from 13.66¢ for DVS to 3.47 to 6.09¢ for external pH control systems used by four large plants.

3.6.4 pH-Controlled Propagation of Cultures

Considerable information regarding culture concentrate production[132,142,145] and injury to starter cells kept at pH <5.0[116] has accumulated in literature. Recently, cessation of starter culture growth at low pH was explained by Nannen and Hutkins.[167] They found that a gradient of 0.6 to 1.44 pH units was achieved in early log phase, and a noticeable decline in ΔpH between the extracellular medium and the cell cytoplasm occurred during the late log phase of growth, corresponding to pH_{in} of 5.0 to 5.5 or pH_{out} <5.0. The critical or minimum pH compatible for cell growth was similar for the three different media tested, with slightly different buffering capacities. Cessation of growth appears to occur when pH_{out} of 5.0 is reached and this was linked to a dissipation of ΔpH resulting in a low pH_{in}.

3.6.4.1 External pH Control

Due to the cost of commercial starter PIM and due to the availability of easy to maintain automated starter propagation operation, Richardson's group pioneered the development and introduction of whey-based phage inhibitory media to the cheese industry. These compositions included fresh whey (Cheddar/Swiss/Parmesan), phosphate, and yeast extract. Propagation was carried out at pH 6.0 using ammonia as a neutralizer. Starter culture produced in this manner was very active even when held for several days and only 20 to 30% culture inoculum was required compared to nonfat milk culture.[168] Good quality Cheddar and cottage cheese was produced with said medium. Compared to milk cultures, PIM culture addition increased the clotting time of milk by rennet at 30°C. Soluble calcium in the phosphated whey medium was lower than PIM at pH >5.7 because of removal of calcium during cheesemaking.[170] Because the soluble calcium was low, a reduced level of phosphates could be employed to achieve phage inhibition equal to or better than PIM that contained high levels of phosphates.[170] The composition of whey-based or nonfat dry milk-based media for pH-controlled propagation was further optimized to include 5.2% whey solids, 0.71% yeast autolysate, and 0.43% casein hydrolysate. This formulation permitted 36% more cells and 38% higher activity over the control whey medium. Nonfat dry milk-based media with stimulants proved superior in activity and phage protection compared to commercial PIM.[171]

3.6.4.2 Internal pH Control

In the external pH control propagation, pH of the medium is controlled by the addition of ammonia or sodium hydroxide in response to acid production. In contrast, in internal pH control systems, the medium contains a very sophisticated buffering agent that solubilizes in response to acid production in the medium. Phase 4 is an example of such a medium developed by Sandine's group at Oregon State University.[172] This medium contains sweet dairy whey, autolyzed yeast, and phosphate–citrate buffer. The pH of the medium does not drop below 5.1 to 5.2, thus avoiding acid injury to the culture. The insoluble buffering salts are solubilized as the pH drops below 5.1. It is claimed that the pathogens do not grow in the medium at this pH. It is also claimed that the cell population is about four to eight times higher than the conventional media; cheese yield and starter activity were also higher. Phage proliferation was vigorously controlled and in some cases it showed some decline in numbers. Its superior performance with cultures used for Italian and Swiss cheese was also reported.[173] There are other numerous small modifications of these basic starter propagation systems to meet particular needs.

3.6.4.3 Temperature Effect

After the bulk starter is propagated, it should be cooled to a temperature below 10°C to preserve maximum culture activity during holdover.[149] The effects of temperature and holding time on the activity of liquid culture are shown in Figure 3.2.

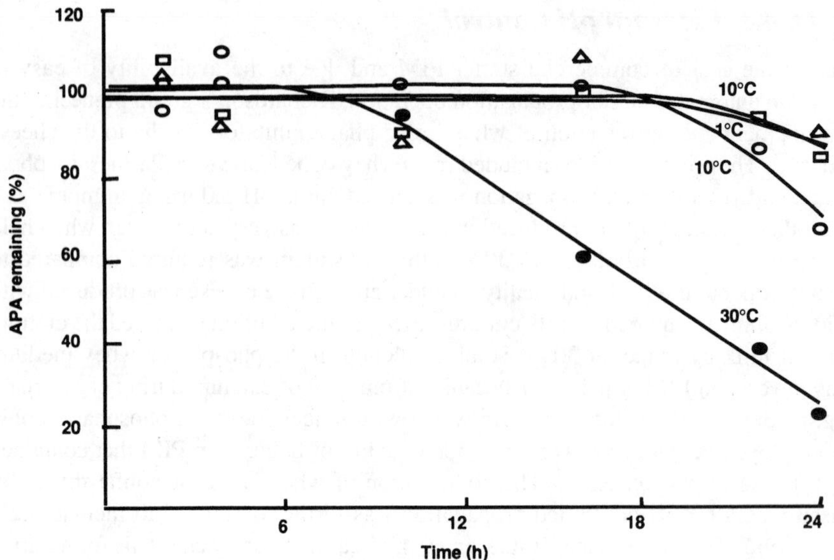

Figure 3.2 Effect of temperature and holding time on the activity of liquid fermenter cultures. After growth had ceased (zero time), cultures of *S. cremoris* 134 were held at various temperatures in the lactose-depleted medium and APAs (acid producing abilities) determined at intervals (●—●), 30°C; (○—○), 22°C; (△—△), 10°C; (□—□), 1°C.
Source: Ref. 149 (reproduced with permission).

3.6.5 General Comments

It should be understood that control and elimination of bacteriophages in a cheese plant is imperative to the viability and business success of an operation. Central to this theme is the total understanding of the cheese plant layout and its operation. Use of inhibitory media alone is never sufficient to prevent phage-related lack of acidification. Phages are restricted in PIM but not destroyed. When this medium containing phage is inoculated in cheesemilk, the viable phages multiply and contaminate the plant personnel and plant environment.

In large and small plants a continued effort in training and education of plant personnel is needed. Phage monitoring and daily starter activity are needed to ensure phage-free bulk starters.[174].

3.6.6 Helpful Points to Phage-Free Starters[127,174]

1. Use as few starter strains as possible.
2. The ratio of the strains that make up a culture should stay constant.
3. Use frozen blends for starter inoculation and avoid subculturing. Subculturing upsets the strain balance.
4. Monitor whey for phages and remove cultures that show progressive increase in whey phage titer.

5. Ensure that air, water, people, and product movement through the plant are known and recognized as potential channels in phage attack.
6. The starter room should be away and completely separated from the cheese-making room and from whey separators. As the phage-laden whey droplets dissipate in air, phage is concentrated in the atmosphere.
7. The starter room should have 15 to 20 air changes of 100% fresh air that is HEPA filtered. The sterile starter tanks should be pressurized with sterile air (.015 μm depth filters) when under operation so that contamination cannot get in.
8. Avoid opening the tank after it has been heat processed.
9. All affluent and washings from the tanks should be piped to the closed drains.
10. The person dedicated to starter making should not do other chores in the plant and no other plant personnel should be allowed in the starter room.
11. It is imperative that plant personnel thoroughly understand and conceptualize the phage phenomenon and be obsessive in hygiene and the production disciplines associated with starter production and usage.
12. Much attention should be given to the cheese vat layout with respect to air movement and flow pattern and their location with respect to the whey side.
13. Source and quality of incoming air are important and should be critically planned.

3.7 Manufacture of Cheese

Cheese manufacture is essentially a process of dehydration of milk in which casein, fat, and minerals of milk are concentrated 6- to 12-fold. About 90% of the water in milk is removed and it carries with it almost all of the lactose.[175] Addition of rennet, acid development by starter culture, and a degree of heat treatment applied to curd after it has been cut into small pieces constitute the cheesemaking constants. It is the modulation of these constants coupled with different microorganisms and curing regimens that result in different cheese types.

General steps are as follows[4,43,175–178]:

1. Milk is clarified by filtration or centrifugation.
2. Dependent on the composition of final cheese, the fat content is standardized using a special centrifuge (separator).
3. Depending on the variety of cheese, milk is either pasteurized at 71.8°C/15 s or heat treated at 62.8 to 68.3°C/16 to 18 s.
4. For some cheese types, milk may be homogenized.
5. Starter culture is added to cheese milk tempered to 30 to 35°C at 0.5 to 1.5% of milk. The milk is generally ripened for 30 to 60 min. In modern plants, starter is injected into the milk line going from the pasteurizer to the vat. It takes about 40 to 60 min to fill a vat and this filling time then serves as the ripening time. During this time, fermentation of lactose to lactic acid by the added starter bacteria begins.

6. At the end of the ripening period, a milk coagulant is added to milk to effect a coagulum in 25 to 30 min.

   The coagulant (70 to 90 ml/1000 lbs of milk) is diluted (1:40) with clean water and evenly distributed throughout milk by stirring milk for 3 to 5 min. Calcium chloride may be added to milk to accelerate coagulation and to increase curd firmness. Its addition to milk should not exceed 0.02%.

   Distinct differences in texture and physical characteristics can be affected by variations in the coagulating temperature. The combination of the temperature of coagulation, the starter culture, the coagulating enzyme, and the acid produced affect the rate of formation; the firmness, elasticity, and other physical properties of the resulting curd; and the degree of whey expulsion.

   The curd produced by acid and a coagulating enzyme is a gel. Variations in the manner in which the curd is treated primarily affects the moisture and secondarily the body and texture which ascertain the characteristics of the finished cheese.
7. When the coagulum is firm enough to be cut, a horizontal-wired stainless steel knife is drawn through the cord followed by a vertical knife in a rectangular vat. If an automatic enclosed circular vat is used, the cutting is programmed to ensure a curd size range by the speed and timing of the automatic knives. The purpose of this step is to increase the surface area of the curd particles which in turn permits whey expulsion and more uniformly thorough heating of the equal sized smaller curd. Cutting the curd into comparatively small cubes reduces the curd moisture. The curd particles should be cut to the similar size.
8. After the curd is cut, it is allowed to sit undisturbed for 5 to 15 min. This period is called ''heal time.'' This allows the newly cut surface to form new intramolecular linkages and firm up the curd while expelling whey. To help make a firm, low-moisture cheese, the curd should be stirred for 30 min after cutting before heat is applied. This also prevents formation of tough skin around the curd cube.
9. Following the ''heal'' period, heat is applied to the jacket of the vat and gradual stirring is initiated. For most ripened cheeses the curds are cooked in whey until the temperature of the curds and whey reaches 37 to 41°C, depending on the variety. For Parmesan and Swiss cheese, the cooking temperature of curd may be as high as 53°C.

   The temperature should be raised slowly to the desired cooking temperature, taking from 30 to 40 min but never less than 30 min for Cheddar cheese. The temperature of the whey should be raised slowly at first and then more rapidly as cooking progresses. The cooking should accompany stirring slowly when curd is fragile and more vigorously when curd firms up. Fresh cheeses, such as cottage, cream, and Neufchatel, are cooked at temperatures as high as 51.5 to 60°C to promote syneresis and provide product stability.
10. Generally, when the cook temperature is reached, a 15 to 60 min period for ''stir out'' is allowed. During this period, contents of the vat are agitated somewhat vigorously.

Agitation during cooking or removing some whey increases the pressure on cheese particles and the frequency of their collision with each other and with the container walls, and promotes syneresis. Syneresis is also promoted by increasing temperature. Syneresis is, initially, a first-order reaction because the pressure depends on the amount of whey in the curd; holding curd in whey retards syneresis due to back pressure of the surrounding whey, whereas removing whey promotes syneresis.[175]

During healing, cooking, and stir out, acid is being produced by lactic starter bacteria which helps syneresis of rennet curd. Approximately 65% and 55% of the calcium and phosphate, respectively, in milk are insoluble and associated with the casein micelles as colloidal calcium phosphate (CCP).[175] The solubility of the CCP increases as the pH of milk decreases (it is fully soluble at pH 4.9). As acid is produced in cheese curd during manufacture, CCP dissolves and is removed in the whey; thus, the pH at curd whey separation determines the calcium content of cheese which in turn affects cheese texture:

Fast acid development → low pH → low calcium → crumbly texture, for example, Cheshire.

Slow acidification → high pH → high calcium → elastic, rubbery texture, for example, Swiss.[175,179]

While curd remains in the whey there is an equilibrium between the lactose in the curd and that in the whey. The whey provides a reservoir of lactose that prevents any great decrease in lactose concentration in the curd. After the whey is removed, the remaining lactose in the curd is depleted rapidly as the fermentation proceeds. Curd that has been left in contact with the whey for a longer period has a higher lactose content than curd of the same pH from which the whey has been removed earlier.[180,181] When the high acidity is reached quickly in the vat, sufficient calcium is removed to alter the physical properties of the curd but insufficient phosphate is lost to seriously affect the buffering capacity of the cheese.[180,181] When high acidity is a consequence of an increase in the time between cutting and draining of whey, a high loss of calcium and phosphate occurs. The loss of phosphate is sufficient to reduce the buffering capacity of the cheese significantly and the pH of the cheese is consequently lowered. Such a cheese develops an acid flavor and a weak, pasty body and texture.[181]

11. When proper acidity has developed, the whey is permanently separated from curd. Many techniques are used to perform this simple but important step. These are

- Let the curd drop to the bottom and let clear whey flow out.
- The curd and whey are pumped to an automatic curd and matting machine where whey is quickly separated from curd and the curd mats in a ribbon form under controlled conditions of temperature and curd depth.
- In an automatic version of the above, curd and whey are pumped onto a draining and matting conveyer under controlled conditions. When the curd has reached the proper pH, the mat is cut and is automatically salted and transferred to another conveyer which takes the salted curd to a boxing station.

For a historic perspective on automation of cheese making see ref. 182. From this point the whey is drained and the new curd is treated differently depending on the nature of the final product.

3.7.1 Cheddar Cheese

Traditionally, for Cheddar cheese, the curd left in the vat after whey drainage is allowed to sit for 10 to 15 min when it is trenched in the middle of the vat, lengthwise. The curd is hand cut, turned over, and then piled at intervals, one slab upon another. During this time acid development continues and syneresis of the curd also continues. This process is called cheddaring and is important in controlling the moisture of cheese. If the curd is piled too soon or too high, it will retain moisture. The slabs should not be piled until the curd is sufficiently firm. It is believed that hydrophobic interactions within the casein network are probably responsible for the advanced stages of syneresis.[175] When the acidity of whey is 0.55 to 0.65% and the curd pH is 5.1 to 5.3, the curd is ready to be milled. Milling is cutting the cheddard slabs into uniform ½-inch × 2-inch particles. The primary purposes of milling curd are to promote further removal of whey and to make it possible to distribute salt quickly and uniformly throughout the curd. Immediately following milling, the curd should be forked for at least 10 min. Too much forking leads to fat loss in whey. Also, the greasy curd may need washing to wash off fat.

The curd is salted at the rate of 2.5 to 3.0 lbs of salt per 1000 pounds of milk. It should be applied in three equal applications. A cheesemake schedule with expected pH/% titratable acidity is shown in Table 3.13.[183]

The salted curd is hooped into 40-lb stainless steel hoops or filled into a 640-lb box. The curd is pressed at 20 psi in the beginning. The 40-lb cheese is dressed after an hour and pressed again at 20 to 25 lbs for the night. Dressing refers to opening the hoop and straightening the wrinkled cheesecloth to obtain a smooth, even, well-knit cheese surface after repressing. The large box is pressed and the resulting free whey in the center of the block is withdrawn with vacuum-operated probes. These blocks may be pressed under vacuum for 45 min to obtain air-free close-knit cheese. The temperature of cheese curds at the time of hooping should be at 30 to 35°C.

3.7.2 Stirred Curd or Granular Cheddar Cheese[43,176–178]

Follow the procedures recommended for milled Cheddar cheese up to draining the whey. Stop draining whey when the curds are just evident through the whey surface. Stir the curds for 10 min, then drain all the whey and stir the curds vigorously for 20 additional minutes. When acidity reaches 0.25 to 0.35%, salt the curd with continuous stirring for 30 min. No cheddaring or milling is done in this cheese.

3.7.3 Colby Cheese[43,176–178]

Follow the directions for making stirred curd Cheddar cheese just before termination of the whey drainage. At the point where curd is just visible through the whey surface, add clean cool water at 15.6°C to the vat with continuous sitrring so that the whey and curd temperature is 26.7 to 32°C. Stir the vat for about 15 min. Drain the watered whey and stir the curd vigorously for about 20 min. Salt the curd when curd has a titratable acidity of 0.19 to 0.24%. Salt and hoop the curds as in the regular Cheddar cheese described. Colby and Monterey Jack cheeses are not placed under vacuum to preserve a sight open texture.

3.7.4 Swiss Cheese[43,184]

For Swiss cheese, milk is heat treated at 62.7 to 67.8°C, then cooled to 31 to 35°C and inoculated with *Streptococcus salivarius* subsp. *thermophilus* (0.5%) and a very small quantity (50 to 100 ml/35,000 lb/milk) of *L. delbrueckii* subsp. *bulgaricus* or *L. helveticus* culture and *Propionibacterium* at 100 to 1000 cfu/ml. The milk is set with animal or microbial rennet at the rate of 2 to 3 oz/1000 lbs of milk. The curd is cut in 30 min with a ¼-inch knife to the size of rice grains. Let the curd sit for about 5 min and then stir it for 30 min without turning on the heat. This is called foreworking. Start to cook the curd slowly to 50 to 53.3°C in 30 min. Then turn off the steam but continue to stir for an additional 30 to 60 min. until the curd is firm and the pH of the whey is about 6.3 to 6.4. At this point the curd is separated from whey by pumping the curd and whey into perforated stainless vats called "universal." The curd is allowed to settle evenly. Large stainless steel plates are used to press the curd at a precise depth of curd mass. This pressing under whey results in a tightly fused cheese required later for eye development. The huge curd block is kept under pressure. During this time acidity continues to develop and should reach a pH of 5.15 to 5.20 in 16 to 18 h.

For smaller operations, the curd is collected in a large coarse cloth bag and pressed in 15- to 20-lb hoops.

The large block of cheese (3000 to 3500 lbs) is cut into 180-lb sections and immersed in saturated brine at 2 to 10°C for 12 to 24 h. The blocks are removed from the brine and the surface is allowed to dry. These are then packaged and boxed. The boxes are stacked and banded to help keep the block shape during the hot room cure. Smaller wheels may be brined for 2 to 3 days and then placed in drying rooms for 10 to 14 days at 10 to 15.6°C with 90% relative humidity to form a rind. This cheese is placed in curing rooms/hot rooms maintained at 20 to 25°C. The eyes in cheese start to develop in 18 to 24 days. Eye formation should take place at a slow uniform rate. Due to the production of gas, the cheese starts to rise a little. Too much rise indicates strong fermentation and must be controlled. After the eye formation is complete, the cheese is transferred to a room about 2°C to prevent further eye development and held for at least 60 days, preferably longer, to develop a fully sweet, nutty, typical flavor before cutting into retail size. A manufacturing schedule is presented in Table 3.14.

3.7.5 Parmesan Cheese[43,185]

Parmesan, or "grana" as it is known in Italy, is a very hard granular bacteria-ripened cheese made from partially skimmed cow's milk. Cheese contains a maximum of 32% moisture and a minimum of 32% fat on a dry basis.

Table 3.13 CHANGES OCCURRING DURING THE MANUFACTURE OF CHEDDAR CHEESE

No.	Step in the Manufacture of Cheese	Duration of Process	Conditions in the Vat: Temp. S^a	Temp. E^b	pH/Acidity S^a	pH/Acidity E^b	Changes Occurring in the Cheese Milk or Cheese Curd	Purpose of the Step in the Manufacturing Process
1	Flash heating of cheese milk	Held for 16 s		147	6.6/ 0.16	6.6/ 0.16	Expulsion of dissolved gases and volatile odors. Partial destruction of microflora, and natural milk enzymes.	To eliminate undesirable flora such as coliforms, psychrophiles, yeasts, and staphylococci.
2	Ripening of milk after addition of 1% starter	60 min	86	86	6.6/ 0.16	6.5/ 0.17	Increase in acidity and shift in salt balance. Growth of starter organisms.	To initiate rapid microbial growth. For liberation of soluble Ca and neutralization of Zeta potential. To facilitate rapid rennet coagulation.
3	"Setting"—addition of rennet and allowing milk to coagulate	20–30 (preferably 30) min	86	86	6.5/ 0.17	6.4/ 0.12	Formation of smooth shrinkable matrix due to uniform coagulation.	To rapidly coagulate milk to form a uniform, smooth coagulum.
4	Cutting of curd	10 min[c]	86	86	6.4/ 0.12	6.4/ 0.12	Large mass of curd cut into small cubical or rectangular pieces with the liberation of whey.	To expel entrapped moisture as whey. Increase surface area for expulsion of whey as the curd matrix shrinks.
5	Cooking with agitation	30–40 min[d]	86	104	6.4/ 0.12	6.2/ 0.13	Shrinkage and firming of the curd. Increase in whey acidity and temperature. Increase in microbial numbers.	To facilitate expulsion of whey from the curd caused by the shrinkage of curd matrix with increasing temperature and acidity.
6	"Dipping" or drainage of whey	Starts 135 min after rennet	104	101	6.2/ 0.15	6.2/ 0.16	Removal of whey. Expulsion of moisture due to compaction as the curd settles. Slight cooling of curd.	To remove whey from the curd.

Table 3.13 *(Continued)*

7	"Packing"—matting of curd	15 min	101	101	6.1/ 0.16	6.1/ 0.18 −0.22	Further expulsion of whey due to compaction, fusion of curd particles.	To fuse curd particles into a solid mass to make convenient for handling.
8	"Cheddaring"—turning and piling of cheese curd slabs	105 min (varies)	99	97	6.1/ 0.18 −0.22	5.3/ 0.55	Further expulsion of whey due to pressure. Increase in acidity. Increase in microbial population. Changes in texture.	To expel whey and gas, and develop meaty body and close texture.
9	"Milling"—cutting curd into small 2″ × 1″ × ½″ strips	10 min	97	95	5.1/ 0.55	5.1/[e]	Reduction in the size of curd slabs. Slight cooling of curd. Increase in bacterial numbers.	To facilitate uniform salting of curd and cooling and further expulsion of whey. Cooling to prevent fat loss.
10	Salting (2.5% by weight of raw curd)	20 min	95	92	5.1/[e]	5.1/[e]	Further cooling of curd.	To allow dissolution of salt in the curd to improve flavor and to cool the curd further to prevent fat loss.
11	Hooping—Filling milled, salted curd into molds	15 min	92	90	5.1/[e]	5.1/[e]		To pack the cheese curd into blocks for easy handling and marketing. Pressure applied to expel moisture and fuse curd pieces.
12	Dressing, etc.				5.1/[e]	5.1/[e]		To protect the cheese from molds etc. during curing period.

Source: Ref. 183. Reproduced with permission.

[a] Start of process; [b]End of process. (Temp. in °F)

[c] Depending on moisture level desired in finished product, up to 30 min of stirring curd in whey may precede cooking.

[d] Following cooking, up to 45 min of stirring curd in whey may precede dipping.

[e] Titratable acidity measurements no longer necessary or applicable.

Table 3.14 PROCEDURE FOR MANUFACTURE OF RINDLESS BLOCK SWISS CHEESE

Operation	Time (min)	Temperature (°C)	pH
Fill stainless			
Vat			
Add starter		31.1–35	6.5–6.7
Ripening	0–30	31.1–35	6.5–6.6
Rennetting	25–30	31.1–35	6.5
Cutting	15–20	31.1–35	6.5
Foreworking	30–60	31.1–35	6.5
Cooking	30–40	48.9–52.8	6.4–6.5
Stir-out	30–70	slight decrease	6.4
Dipping		47.2–51.1 +	6.3–6.4
Pressing	12–18 h	22.2–25.5 room	5.15–5.4
Brine salting	1–2 days	7.2–14.4 tank	
Drying	up to 1 day	7.2–12.7 room	
		35 drying tunnel	
Wrapping			
Cold room	0–10 days	7.2–12.7 room	5.2 ± 5.5
Warm room	2–7 wk	21.1–25.5 room	5.5 ± 5.7
Finished cooler	until sold	2.2–12.7 room	5.5 ± 5.7

Source: Ref. 184.

Standardized (1.8% fat) heat-treated milk (63 to 69°C) is cooled to 32 to 35°C and inoculated with *Streptococcus salvarius* subsp. *thermophilus* and *Lactobacillus delbrueckii* subsp. *bulgaricus* or *Lactobacillus helveticus* at 1% mixed inoculum. These cultures can be propagated mixed or singly. Milk is ripened for 40 to 60 min. Coagulant (3.5 oz/1000 lbs) is added to effect a curd in 20 to 25 min. The curd is cut using ¼-inch wired knife and allowed to sit for 10 min undisturbed before beginning to cook. The curd is cooked to 42.2°C in 15 min and then stirred vigorously for 15 min. It is further cooked to about 51.6°C in 30 min. Start draining the whey while stirring when the acidity reaches about 0.13%. Drain the whey completely when acidity reaches 0.18 to 0.20%. Some salt may be added at this point. The curd is hooped in 20-lb capacity and pressed in a horizontal press immediately. The cheese may be redressed in 1 h and further held at 21.1 to 24.4°C until the following morning. Cheese is brined for several days. Dry salt is added on the surface of the cheese in the brine.

The cheese is then removed from the brine and allowed to dry for several days at 13 to 16°C and form a rind. This also allows the cheese to reach a proper moisture. It is important that the surface of the cheese should be free of cracks or the cheese will get moldy. The cheese may be waxed or vacuum packaged in plastic bags. Parmesan is cured at 10°C for at least 10 months as required by federal regulation.

3.7.6 Mozzarella and Provolone Cheese[43,185,186]

These cheeses are referred to as ''pasta filata'' varieties. These cheeses have traditionally been further cooked after whey drainage in hot water and stretched until they become close knit, elastic masses. The hot curd is molded into forms.

Per capita consumption of Mozzarella cheese in the United States has increased from 0.4 lb in 1960 to 4.1 lbs in 1984.[186] Production of Mozzarella cheese now ranks second to Cheddar cheese.

It is also reognized that physical properties of Mozzarella cheese vary greatly based on cheese age, pH, salt content, and starter cultures used.

Milk composition for Mozzarella is adjusted to suit the type of cheese. Milk is pasteurized and cooled to 32 to 35°C. Milk or whey culture of *S. salivarius* subsp. *thermophilus* and *L. delbrueckii* subsp. *bulgaricus* or *L. helveticus* is used at about 1 to 2% level. It is important that strains used should ferment lactic acid rapidly and tolerate high temperature 48.8 to 54.4°C.

Add 2 to 3 oz/1000 lbs of milk coagulant to set milk in 30 min. Acidity of milk at setting should be about 0.18%. Cut the curd using ⅜-inch wired knives. Let the curd stand for 5 to 10 min and then start to stir gently, turn steam on, and cook slowly, one degree rise during the first 5 min, 1.5°C rise during the second 5 min, and then at the rate of 0.55°C per minute until 43.3 to 46.6°C is reached depending on the culture used. The titratable acidity of whey at the end of cooking should be about 0.13%. After the cooking has ended, it is stirred, first gently and then vigorously for about 40 min. until the whey acidity reaches about 0.19%. The whey is then drained and the curd allowed to mat in a manner similar to Cheddar. When the whey acidity reaches 0.30%, it is milled. The milled curd is molded in hot water at 74 to 82.2°C and then formed and released into cold water to firm up. The cheese is then brined, about 1 day for each 3 to 5 lb of cheese.

The brined cheese is dried and shrink wrapped. It is ready for use right away or it can be cured. Mozzarella cheese is also manufactured by acidification with citric acid or vinegar in place of starter culture. About 1 quart of vinegar is used for about 1000 gal of milk.[187]

3.7.7 Brick Cheese[43,177,188]

Brick cheese originated in Dodge County, Wisconsin, U.S.A. It should have about 42% moisture, 28% fat, and 1.5% salt. The starter culture for this cheese consists of 0.25% mesophilic lactococci and 0.25% of *Streptococcus salivarius* subsp. *thermophilus*. The coagulum at 32°C is cut with ¼-inch wired knives and very gradually heated to 36°C. Whey is drained until 1 inch of whey is left on the curd. While the curd is stirred, water at 36°C is added in 5 min amounting to 50% of milk volume. After 15 min, watered whey equal to the amount of added water is drained. A positive-action pump is used for pumping the curd and whey over to the hoops. The curd in hoops is turned using cover followers. The second turn is made in 1 h and a 5-lb weight is applied. Three additional turns are made, one every hour. During this operation, room temperature should be maintained at 21 to 24°C. Weights are removed after the fourth turn.

The loaves of cheese are placed in brine at 10°C for 24 h. During brining, dry salt is sprinkled on the surface of the loaves. The loaves in brine are turned once after 16 h. The pH of cheese at one day is at 5.2 to 5.3. Higher pH values are obtained by removing more moisture from cheese with longer washing treatments.

The salted loaves are placed on shelves in curing rooms maintained at 15.6°C with 90% humidity. If the wooden shelves in the curing room were never used for ripening Brick cheese, a suspension of *B. linens* is applied to the shelves using a cheesecloth. Each day, the cheese is turned on its new side and rubbed with hands dipped in 5% salt water. The cheese is shelf cured for 5 to 10 days depending on the intensity of growth desired. The smear can then be washed off and the cheese dried in rooms at 15.6°C with 70% humidity. The cheese is then wrapped in plastic film and cured at 4.4°C for 4 to 8 weeks. If more pungent flavor is desired, the cheese is wrapped unwashed.

3.7.8 Mold-Ripened Cheese

Blue Cheese, Gorgonzola, Stilton, Brie, and Camembert are the cheese types where mold is added directly to effect ripening and so determine the characteristics of the cheese.[189] Blue mold is used for Blue Cheese, Gorgonzola, and Stilton, whereas white mold is used for the manufacture of Camembert and Brie cheeses.

Blue veined cheeses are linked together by the common use of *Penicillium roqueforti*. This mold is unique in that it can tolerate low oxygen and high CO_2 tension and is relatively salt tolerant. It is hardier than the white mold used for Camembert production.[189] Roquefort cheese is made from sheep milk in the Roquefort area of France and its cow's milk counterpart is known as Bleu cheese in other areas of France.[4]

3.7.8.1 Blue Cheese[43,177,189,190]

Blue cheese contains not more than 46% moisture, 29.5 to 30.5% fat, 20 to 21% protein, and 4.5 to 5% salt. Blue cheese may be made from homogenized milk (Iowa method) or unhomogenized milk (Minnesota method).[190]

Raw or pasteurized milk is homogenized at 2000 psi at 32.0 to 43.3°C. A mesophilic lactic culture containing *Lactococcus lactis* subsp. *lactis* var. *diacetylactis* is added to milk at 0.5% level at 32.2°C. Mold spores may be added to milk in the vat just before adding rennet at the rate of 4 oz/1000 lbs of milk. The rennet coagulum is cut with ½-inch wire knives. The curd is allowed to heal for 5 min and then stirred gently once every 5 min. Stirring is continued for 60 min while the temperature remains at 31.1 to 32.2°C. The acidity of whey should rise to 0.11 to 0.14%. Just before draining the whey, the temperature is raised to 33.3°C and held for 2 min.

All the whey is drained and the curds trenched. If the mold spores were not added to the milk, these can be added to the curd at this time. Two pounds of coarse salt and 1 oz of spore powder per 100 lb curd are mixed and applied to the curd with thorough stirring. The curd is scooped to perforated stainless steel circular molds placed on drainage mats. The hoops are turned every 15 min for the first 2 h and then left on drainage mats overnight at room temperature at about 22.2°C. The cheese is removed and dry salted liberally. The cheese is placed on its side in a cradle in a room at 15.6°C with 85% relative humidity. Salt is applied four more times, once every day. After the cheese salting is complete, it is pierced with needles ⅛ inch thick on both sides. It is then placed in a room at 10 to 12.8°C with 95% relative humidity. The cheese is turned on its side one quarter every 4 days and wiped with a clean cloth. This process continues for 20 days. Cheese is then wrapped in foil or other appropriate wrapping material and cured at 2 to 4°C for 3 to 4 months.

3.7.8.2 Camembert Cheese[4,43,186]

The Camembert and brie style cheeses are most characteristic of the white soft mold cheeses. *Penicillium camemberti* and the probable biotypes *P. caseioculum* and *P. candidum* are used to ripen the cheese by external growth of the mold.

Whole pasteurized milk at 29 to 33.5°C is ripened with mesophilic lactococci and the mold spores. When the acidity of milk reaches 0.22%, rennet is added and the coagulum cut with ½-inch wire knives. Curd temperature is maintained around 32.2°C and no cooking of the curd is exercised. The curd and whey are transferred to perforated 8-oz stainless steel round molds placed on drainage mats. The curd is allowed to drain at room temperature, 22.2°C for 3 to 4 h. The curds are now firm enough for turning. The curds are turned three to four times at 30-min intervals. If mold is not inoculated in milk, it can be applied to the cheese now. The small wheels are allowed to stay on draining mat for an additional 5 to 6 h. At this time the cheese pH is around 4.6. The rate and extent of acid production are critical for product attributes and product stability.

The cheese is dry-salted on all sides and left at room temperature overnight. The next morning the cheese is transferred to rooms at 10 to 13°C with relative humidity of 95 to 98%. Cheese lies there undisturbed for about a week when white mold emerges on the surface; the cheese is turned over once. After about 14 days in the curing room, the cheese is wrapped and left at 10°C and 95% relative humidity for an additional 7 days. The cheese is now transferred to (4.4°C) and is ready for distribution.

As the cheese ripens, the pH increases rapidly to about 7.2 due to deamination of amino acids and the texture and palatability can begin to change. The detection of a pronounced ammonia aroma indicates that cheese is overripe; also, the white cottony mold starts to turn brown on the surface of cheese.

3.8 Cheese from Ultrafiltered Retentate

Ultrafiltration is a sieving process that employs a membrane with definite pores that are large enough to permit the passage of water and small molecules. When pressure is applied to a fluid, the semipermeable membrane allows small species to pass through as permeate and larger species are retained and concentrated as retentate. In ultrafiltration of milk, nonprotein nitrogen and soluble components such as lactose, salts, and some vitamins pass through the membrane, whereas milk fat, protein, and insoluble salts are retained by the membrane.[191]

During the past 20 years, the use of UF-retentate for cheesemaking has attracted considerable attention. The ''precheese'' technology known as the Maubois, Macquot, and Vassal (MMV) process is used in many dairies in the world to produce cheese varieties such as Camembert, Feta, Brie, cream, Cheddar, Havarti, Colby, Domiati, Brick, and Mozzarella.[191–194]

The principle is that the milk is concentrated by ultrafiltration to a composition very close to the chemical composition of the cheese in question. Then the retentate is coagulated by starter culture and rennet.

The main advantages of this method are:

1. Substantial increase in yield due to whey protein and minerals inclusion.[192]
2. Simple, continuous process open to almost complete automation.[194]
3. Reduction in cheese cost due to reduction in costs of energy, equipment, and labor.[191]
4. The process uses substantially less salt and rennet.[195]

The main disadvantages are[194]:

1. Cheese becomes very homogeneous and has a high bulk density.
2. The acidification is slow due to high buffer capacity; therefore minimum pH might be difficult to obtain.
3. Very viscous retentate is difficult to mix with starter and rennet, etc. and cannot be cooled without solidification.
4. Cheese does not correspond to its definition in properties.

The general conclusion is that the MMV process is not suitable for making cheese of traditional quality. To overcome these problems, Alfa-Laval[194] and others have tried and developed methods for using UF retentate in the production of variety of cheese types.[195–205]

When milk is ultrafiltered and Cheddar-type cheese is made from the retentate (40% total solids) by a modification of conventional cheesemaking procedures, considerable quantities of whey proteins are lost in whey during syneresis of the curd. Heat treatment of retentate before coagulation with rennet has been found to reduce the loss of whey protein and so increase cheese yield.[204,205] Heat treatment (90°C/15 s) of retentate reduced the rate of whey loss and slightly improved the curd structure but did not affect fat losses.[204] Light homogenization slightly reduced heat denaturation of whey protein and fat loss. The structure of the curd from the heated concentrated milks was coarser than those of the control and the curd particles fused poorly. This appeared partly responsible for the crumbly texture in the cheeses from the heated concentrates. The texture was not improved by the addition of a bacterial proteinase.[205]

When Cheddar cheese was made from reconstituted retentates, the pH of cheese rose from 5.2 to 6.0 when cured at 10°C and developed eyes and had a flavor reminiscent of Gouda or Swiss cheese.[197] Bush et al.[200] prepared satisfactory Colby but not Brick cheese from creamed skim milk retentate; reductions in cooking temperature and milk-clotting enzyme and elimination of curd-washing were helpful. A satisfactory Cheddar cheese was made from milk concentrated twofold by ultrafiltration with the following modifications[203]: (1) Use lower setting, cooking, and cheddaring temperatures. (2) Offset the effect of increased buffer capacity of the UF milk by the addition of higher amounts (2%) of starter culture. (3) Overcome the slow ripening rate and flavor development by adding rennet on the basis of the original amount of milk.

A commercial process called "Siro curd process" for cheese manufacture was developed at CSIRO and commercialized with the help of APV Bell Bryant, APV International, Ltd. and the Milk Marketing Board for England and Wales.[195] The process claims a number of benefits and advantages:

1. A Cheddar yield increase of 6 to 8% over conventional processing.
2. The make time is reduced by 1 h.
3. The process uses substantially less salt.
4. Rennet usage is about one third.
5. The process and the starter systems are totally enclosed and greatly reduce the risk of bacteriophage infection.
6. Consistent cheese composition, through accurate automatic control of moisture, salt, and pH.
7. The process is flexible and adaptable to other cheese types.
8. The process can effectively handle seasonal variations in milk composition.

Manufacture of Mozarella cheese with good melting properties from 1.75:1 retentate volume concentration is described by Fernandez and Kosikowski.[202] A commercial process for Mozzarella manufacture achieving 18% cheese yield was developed using Pasilac equipment.[201] In this process the skim milk is acidified to pH 6.0 with acetic acid and allowed to sit for 2 h before ultrafiltration. The excess calcium then follows the permeate phase and the calcium content of the retentate is reduced to effect the stretching properties of cheese. The retentate is diafiltered to remove excess lactose which can cause brown discoloration of cheese in making pizza. The retentate has 38% solids with 34% protein. It is mixed with 82% fat cream to achieve further high solids with 52% dry matter. The whole process is automated. Similarly, production of Gouda cheese from UF retentate has been reported.[199,200] Another process using preacidification of retentate claims traditional cheese qualities.

Alfa-Laval has developed the Alcurd continuous coagulator; process description of blue mold cheese is given.[194]

After years of research with UF retentate, much remains to be understood before ultrafiltered milk can be successfully converted to hard and semihard cheese varieties.

3.9 Salting of Cheese

In natural cheese, salting of curd is traditional and an integral art of the manufacture of most if not all cheese varieties. Salt exercises one or more of the following functions[206]:

1. It modifies cheese flavor. The unsalted cheese is insipid which is overcome by 0.8% sodium chloride. In the unsalted cheese, body breakdown is rapid and cheese flavor is not normal.[178]
2. Salt promotes syneresis and thus regulates the moisture content of cheese.[207]
3. It reduces water activity (A_ω) of cheese.[208]
4. It controls microbial growth and activity. If the salt in the moisture (S/M) value is $<4.5\%$, the starter numbers remain high in cheese and off-flavors due to starters are likely.[209] For Cheddar cheese, a S/M value between 4.5 and 5.5 is desirable.[210,211] At this salt concentration, residual lactose metabolism by starter and nonstarter lactobacilli is controlled.
5. Salt concentration in cheese has an effect on the rate of proteolysis of both α_{s1}- and β-casein.[212] In general β-casein hydrolysis is impeded more by the salt. Salt can be incorporated in cheese by:

   a. Dry salting, for example, Cheddar, Colby, Cheshire
   b. Brine salting, for example, Swiss, Parmesan, and Dutch cheese varieties
   c. Rubbing dry salt on the surface of cheese, for example, Blue, Gorgonzola
   d. Combination of dry salting and brining.

3.10 Cheese Ripening and Flavor Development

The properties of a cheese depend on its original composition, curing conditions, and shape and size of the cheese. The combination is largely governed by physiochemical and bacterial processes during curd making and directly afterwards. Both types of processes affect each other.[213]

The terms "ripening" and "curing" are sometimes used interchangeably and are not defined clearly. The term "curing" was arbitrarily applied to the methods and conditions, that is, temperature, humidity, and other treatment of cheese.[214] "Ripening" denotes the chemical and physical changes during curing of cheese. A young cheese of specific composition is purposely exposed to certain conditions where limited but essential proteolysis of milk by rennet in concert with proteinase, peptidase, and other activities of starter bacteria augment the shift of microbial populations. The emerging starter and adventitious populations in turn are pressed into summoning those metabolic activities that must transform the milk components simply to survive. In doing so, the chemical entities generated interact among themselves and with the microbial population to result in a more flavorful and preserved milk. Both casein and milkfat hydrolysis are needed for cheese flavor development, but the rate and extent of such hydrolysis must be controlled to maintain cheese identity. The distinguishing features of a cheese within a family are a function of smaller but significant deviations from set practices. The following pairs of cheeses represent good examples: Edam and Gouda, Brick and Limburger, Mozzarella and Provolone, and Cheddar and Colby.[215]

Lactic starter cultures are added to vat milk to give about 10^6 to 10^7 cfu/ml. The amount and type of starter may vary significantly depending on the type of cheese and characteristics of cheese desired. In most cheese types an overnight pH range is 4.95 to 5.3. The primary biochemical changes involve glycolysis, proteolysis, and lipolysis. These changes are followed and overlapped by a number of secondary catabolic changes including deamination, amination, decarboxylation, transamination, desulfurization, β-oxidation, and some anabolic changes, such as esterification.[3,216,217]

The number of starter bacteria decline in cheese during ripening. In ripened Cheddar cheese the lactococci constituted only 13% of the total number of bacteria[218] while other Gram-positive bacteria from milk formed the majority of the flora. This indicates large changes in the microflora during ripening.

Proteolysis is critical to the conversion of curd to a well ripened cheese. Casein degradation in cheese can come from residual microbial proteinases (starter and nonstarter) and milk proteinases.[219–221] Proteolysis is probably the most important biochemical event during the ripening of most cheese varieties.[222] During cheese manufacture, early proteolysis and coagulation of milk results from rennet cleavage of Phe_{105}–Met_{106} of κ-casein.[223] This is followed by a general proteolysis in which the caseins are slowly degraded.

3.10.1 Proteolysis of Caseins

The physical form of casein affects the rate of hydrolysis by rennet enzymes. Ledford et al.[224] demonstrated that degradation was faster when casein was dissolved than when in micelles. Dissolved α_s-casein was hydrolyzed first. Both α_s- and β-caseins yielded two fractions on Sephadex G-50. Peak areas for α_s-casein proteolysis were several times that of β-casein. In α_{s1}-casein, Phe_{23}–Phe_{24} is the most sensitive bond for chymosin.[225,226] Hydrolysis of Phe_{24}–Val_{25} was reported by Creamer and Richardson,[227] who also found that α_{s1}-casein A was resistant to chymosin because it lacked the 14–26 peptide segment. In fact, both 23/24/24/25 bonds are hydrolyzed.[228] Proteolytic specificity of first cleavage by chymosin of α_{s1}-casein (i.e., release of α_{s1-1}, residues 24/25–199) is independent of pH, ionic strength, and urea content.[229,230] The subsequent hydrolysis is dependent on pH, ionic strength, and state of aggregation. An increase in NaCl concentration from 0 to 20% (w/v) reduced proteolysis at pH 5.8 whereas at 5% salt level, inhibition was greatest at higher and lower pH values. Hydrolysis of the primary susceptible bond of α_{s1}-casein was slightly stimulated at pH 5.2. A polypeptide α_{s1-VII} was formed at pH 5.8 only in the presence of 5% NaCl. At pH 5.2 α_{s1-1} was hydrolyzed to α_{s1-V} in a salt-free system but was hydrolyzed instead to α_{s1-VII} and $\alpha_{s1-VIII}$ in the presence of 5.0% NaCl.[228] Gel-filtration studies on α_{s1}-casein hydrolysate using Sephadezx G-150 and electrophoresis showed molecular weight species corresponding to α_{s1-1}, α_{s1-V},

α_{s1-VII}, and $\alpha_{s1-VIII}$ and only trace of α_{s1-II}. Cheddar cheese contained high levels of peptides corresponding to these fractions.[231]

In all there are 26 hydrolyzable bonds in α_{s1}-casein B by chymosin.[226]

In β-casein most chymosin susceptible bonds are Leu_{192}–Tyr_{193} and Ala_{189}–Phe_{190} starting from the C-terminal. Hydrolysis of β-casein is significantly decreased by increasing ionic strength and is dependent on pH.[226,232]

3.10.2 Proteolysis in Cheese

In cheese, the coagulant is mainly responsible for the formation of initial large peptides from α_{s1-1} and β_1 caseins, by cleaving the 1–23/24 and 190/193–209 segments respectively,[226] after which starter bacteria produce smaller peptide fragments and free amino acids.[222,233,234] The caseins are hydrolyzed to different extents in the cheese varieties. The α_{s1}-fraction was degraded in every cheese; however, β-caseins appeared largely intact in some varieties, whereas its content was appreciably diminished in others.[233] In addition, alkaline milk proteinase and acid proteinase also play a part.[233] Nath and Ledford[235] first reported that para-κ-casein was not hydrolyzed in cheese. This observation was confirmed by other researchers.[236]

Existence of proteolysis associated with intact metabolizing cells and with preparations from disrupted starter cells has been demonstrated.[237–240] The large polypeptides generated by chymosin trapped in the curd are degraded further by enzymes from starter bacteria and adventitious populations of nonstarter lactobacilli in cheese.[234] Lactobacilli have proteinases that degrade α_{s1} and β-casein.[241] These organisms also possess intracellular aminopeptidases, dipeptidases, carboxypeptidases, and endopeptidases which play a vital role in the production of free amino acids that are precursors of some cheese flavor compounds. Although largely intracellular, membrane-associated peptidases have been noted. Although weak, intracellular lipases and entrases activity are also thought to contribute to cheese flavor.[241–244]

Growth and autolysis of lactic acid bacteria in cheese potentially can release enzymes and cellular constituents that would serve as metabolites for other microorganisms in cheese.[245,246] Some autolysis of culture growing in milk was seen even during the log phase, even though the cell number decline was not detected.[247] The intracellular enzymes must be released by autolysis to be of any importance in peptide degradation and cheese maturation.[247,248] Autolysis of starter bacteria takes place in cheese and a considerable inter- and intraspecies variation in autolytic behavior has been noted.[249–251] Some strains showed temperature-induced lysis when heated to 38 to 40°C, whereas other strains continued to grow.[252] This autolysis difference among starter bacteria apparently reflects the presence of different enzyme systems, different sensitivity toward inhibitory substances, or just varying amounts of available active enzyme.[251] These differences are speculated to affect the rate of cheese curing. The thermoinducible mutants appear tempting for use in accelerated cheese maturation.[245]

3.10.3 Amino Acid Transformations

The presence of free amino acids in the ripening cheese and the fate of these is considered important in cheese flavor. The concentration of free amino acids in cheese correlated well with flavor development and was considered a reliable indicator of ripening.[253] Concentrations of methionine, leucine, and glutamic acid were considered good indicators of proteolysis in cheese.[254,255] Views about the relationship between amino acids level and cheese flavor are opposing.[256,257] Law and Sharpe[258] considered amino acids to be intermediate products in the production of certain aromatic compounds.

Many amino acids are decomposed and rebuilt by microbiological enzyme systems.[216] Free amino acids can undergo a variety of changes as shown below[217]:

A. Side chain alteration
Tryptophan → Indole

B. Decarboxylation
Lysine → Cadavarine
Glutamate → Aminobutyric Acid
Tyrosine → Tyramine
Tryptophan → Tryptamine

C. Deamination
Alanine → Pyruvate
Tryptophan → Indole
Glutamate → α-Ketogluterate
Serine → Pyruvate
Threonine → α-Ketobutyrate

D. Transamination
Aspartate → Oxalacetate

E. Strickland Rection
Alanine → Acetate
Leucine → Isovalerate
Proline → γ-Aminovalerate
Hydroxyproline → γ-Amino-α-hydroxyvalerate

3.10.4 Flavor Development

Lactic acid, acetic acid, formic acid, diacetyl, acetaldehyde, ethanol, and propionic acid are derived from lactose and citrate in milk. Ketones, lactones, aldehydes, and fatty acids are mainly derived from lipids.[9]

Many research groups have sought the chemical basis to answer the riddle of cheese flavor. Many aspects of cheese chemistry and flavor development have, however, been elucidated and described in reviews over the past 30 years. Mulder[259] and Kosikowski and Mocquot[112] proposed that cheese flavor is produced by a blend of compounds, no one of which produced the characteristic flavor. If the proper balance of components was not achieved, then undesirable or defective flavors occurred. This view has held ground. Two approaches have been used to study cheese flavor. One is to isolate and identify flavor contributing components and the other is to determine the factors that affect or control the development of flavor.[260]

Experiments with cheese made in aseptic vats[261] clearly indicated that starter bacteria were needed for cheese flavor and this flavor was intensified by the addition of certain organisms isolated from milk and cheese. The fat is essential to the development of flavor and that the ratio of acetate to total free fatty acids must be in a given range for typical Cheddar cheese flavor was proposed by Ohren and Tuckey.[262] Kristofferson,[263] on the other hand, hypothesized that oxidation of protein sulfur in milk is critical to cheese flavor development and that the ratio of hydrogen sulfide to free fatty acids should fall within certain limits.

Manning proposed that sulfur compounds—methanethiol, H_2S, and dimethylsulfide—contribute to the full Cheddar flavor and methanethiol was the most significant component of flavor.[264]

Methyl ketones[265] 2-pentanone,[266] and a water-soluble, nonvolatile fraction[268] containing free amino acids[269] are also thought to contribute to the Cheddar cheese flavor.[267–269] In old raw-milk Cheddar, methional, phenols, and pyrazines were considered to be significnt in flavor.[260] Other compounds such as ethyl butyrate and ethyl hexanoate were implicated in the fruity defect in Cheddar cheese; lactones (C_{12} and C_{10}, C_{12}, and C_{14} δ lactones) were considered to impact Cheddar flavor directly.[269]

Aston and Douglas[270] noted that H_2S and methanethiol increased until the Cheddar cheeses were approximately 6 months and then decreased. They found that carbonyl sulfide levels increased with age of the cheese. They concluded that none of the volatile sulfur compounds could be considered as reliable indicators of flavor development. Lloyd and Ramshaw[271] used ethanol, propan-2-ol, propan-1-ol, butan-2-one, ethyl acetate, butan-2-ol, hydrogen sulfide, menthanethiol, dimethyl sulfide, acetic acid, lactic acid, and water-soluble nitrogen to objectively characterize several brands of Cheddar cheese. These objective profiles were compared with subjective panel assessment and it was concluded that authors could validate a ''mark'' of cheese quality. The elements found in Edam, Jarlesberg, vintage Cheddar, and soft cheese were different. They found that a profile of Feta contained high levels of volatiles and emphasized the similarity of the starter system to that used for Cheddar cheese. The lack of several components in Edam and Jarlsberg appeared to reflect differences in manufacturing and the enzyme systems at work. Differences in cheese flavor can also come from the feed of cow and the quality of milk used.[272] When raw milk was stored at 2°C and 7°C, some volatile carbonyls were reduced to the corresponding alcohols.[273] Some carbonyls such as acetone were present in fresh milk, whereas others were formed from the corresponding amino acids, for example, 3-methylbutanal from leucine. Ethanol, propan-2-ol, and 3-methylbutan-1-ol found in milk were partially esterified with volatile acid on storage. Sulfur compounds, for example, dimethyl disulfide and 2,4-dithiapentane, were also formed on storage. The bacterial cell count, the off-flavors, and volatile production were much greater at

7°C than at 2°C. Headspace volatiles from cold-stored raw milk and bacterial populations increased in parallel.[274] In the study of aroma compounds in Swiss Gruyère cheese[275–277] it was found that some compounds (benzaldehyde, limonene, camphor, ketoalcohols, ketones, nitrogen-containing volatiles) were found in much higher concentration in the outer zone whereas esters and lactones were found in the middle or central zone of the cheese.

What characterizes a given variety of cheese is not yet fully clear. Some of this is perhaps due to differences in the chemical composition and microbiological flora of milk, and the manufacturing and ripening of cheese. On top of these differences are the data from the varied methodologies (distillation, dialysis, and solvent extraction) used for isolation of cheese flavor compounds. In a recent investigation, techniques of distillation, solvent extraction, and membrane dialysis were compared on three sets of cheeses.[277] The solvent extraction (acetonitrile) method was the fastest, cheapest, and gave the most characteristic flavor isolate. Eighty-six odor-active components were detected while one was characterized as cheesy but could not be identified.[277]

After many studies and chemical constituent measurements and identifications, a single compound or a few compounds characteristic of Cheddar flavor have not been identified. On the contrary, a number of groups of compounds provide correlations with Cheddar cheese flavor scores of a similar magnitude. This reinforces Mulder's theory that Cheddar flavor may result from the contribution of many compounds, which in the correct ratio produce a good flavor.[277,278,280]

3.11 Microbiological and Biochemical Changes in Cheddar Cheese

3.11.1 Fate of Lactose

In the manufacture of Cheddar cheese, uniform starter activity is important. The proper rate of acid development, particularly before the whey is drained from the curd, is essential to attain proper composition and subsequent events in ripening of Cheddar cheese.[26] In Cheddar and Colby types of cheeses, about 30 to 40% of the added culture cells are lost in whey. The cells trapped in the rennet coagulum rapidly multiply and ferment lactose to lactic acid. The population of starter organisms may reach in excess of 5×10^8 cfu/g in curd before salting.[27] The number of starter organisms in the fresh curd depends on the strain of culture used and the manufacturing procedure.

The cultures multiply only slightly during coagulation and cooking, but growth and acid production accelerate after whey is removed and continue through cheddaring as the starter cells are concentrated in the curd. Acid production will continue until the lactose is depleted.[26] At the time of milling, the curd may have a pH of 5.3 and titratable acidity of whey at 0.57% or higher.[279] It is well established that the rate of lactic acid fermentation and the amount of lactic acid formed are critical to the quality of the resulting cheese. Examination of several samples of commercial

Cheddar cheese showed 1.03 to 1.6 g of lactic acid, 72 to 479 mg of lactose, 0.4 to 11 mg of glucose, 2 to 147 mg of galactose, and 0 to 19 mg of succinic acid per 100 g.[280] Turner and Thomas[64] noticed that lactose utilization and L-lactate production in cheese by starter bacteria was a function of salt-in-moisture (S/M) levels between 4% and 6%. In a cheese with low S/M levels (about 4%), lactose was completely utilized in about 8 days and L-lactate was the major end product. In contrast, with high S/M levels (about 6%) lactose concentrations were high after several weeks. This residual lactose was utilized by nonstarter bacteria and D-lactate was a major end product. It is believed that the quality of the resulting cheese may be determined by the "fate" of this residual lactose, as there is the potential for the formation of high concentrations of various end products. A greater role of adventitious nonstarter bacteria in cheese flavor production is recognized than previously acknowledged.[64] In another study[281] low-lactose cheeses developed most flavor after 1 month, whereas high-lactose cheeses developed most flavor after 3 months. The high-lactose and control cheeses had a higher and sharper flavor than low-lactose cheese. The low-lactose cheese with the greatest decrease in lactate contained the highest concentrations of –SH groups and had the highest pH during curing. The authors hypothesized that the hydrogen released by lactate dehydrogenation to pyruvate could be used to reduce –SS– to –SH and thus be detected as an increase in reactive sulfydryls.

3.11.2 Fate of Casein

Salt in the moisture phase not only affects lactose utilization by starter bacteria, it also controls bacterial growth and enzyme activity in the cheese, especially the proteolytic activity of chymosin,[282,283] plasmin,[284] and starter proteinases.[285] Salt concentration had a large effect on the rate of proteolysis of both α_{s1}- and β-casein. In 1-month-old cheese containing 4% S/M, approximately 5% of the α_{s1}-casein and 50% of the β-casein remained unhydrolyzed. Corresponding figures for 6% S/M were 30% and 80%. In Cheddar cheese, S/M value between 4.5% and 5.5% is targeted. In this range, the rate of metabolism of lactose and proteolysis is controlled and further adjusted by lower temperature of ripening. Lower temperature of ripening also controls the growth of nonstarter lactic acid bacteria such as lactobacilli and pediococci.[181] It is now generally recognized that coagulant is primarily responsible for the formation of large peptides whereas small peptides and free amino acids result principally due to starter organisms, possibly from coagulant produced peptides.[237] Ledford et al.[233] first reported that rennet cleaved α_{s1}-casein during the initial stages of ripening of Cheddar cheese, yielding a product of higher electrophoretic mobility. This large peptide was later identified as α_{s1}^{-1} corresponding to the 24/25–199 of C-terminal of α_{s1}-casein.[225,286] During the normal ripening of Cheddar cheese, α_{s1}-casein is the principal substrate for proteolysis with little degradation of β-casein.[233]

Proteolysis of β-casein is more extensive when the level of salt is low.[287] Peptides with mobilities and molecular weight identical to $\alpha_{s1-\mathrm{V}}$ and $\alpha_{s1-\mathrm{VII/VIII}}$ were present in Cheddar cheese and were located between α_{s1}- and β-casein.[231]

A fairly large amount of β-casein remains unattacked by the proteinases at the end of ripening.[233,288] Proteolysis products of β-casein (β-$_{\text{I}}$, β-$_{\text{II}}$, and β-$_{\text{III}}$) by rennet have not been seen in Cheddar, whereas γ-caseins have been noted in most of the cheese varieties examined;[288] these are derived by plasmin activity in Cheddar.[283,289] The overall breakdown of β-casein in Cheddar cheese appears to be small and affected by the salt concentration.[284]

In Cheddar cheese and other cheeses, α_{s1}-casein is always the first to be hydrolyzed and generally extensively degraded. Nath and Ledford[235] noted that α_{s1}-casein in Cheddar cheese was completely hydrolyzed in 35 days, whereas β-casein remained intact. Para-κ casein was not proteolyzed at 170 days of cheese ripening. Creamer and Olson[290] found that the amount of intact α_{s1}-casein in commercial Cheddar cheese was related directly to the yield force in a compression test. This suggests that proteolysis of caseins determines the rheological properties of cheese.

3.11.3 Microbiological Changes

Cheddar-type cheese is internally ripened by chymosin in concert with starter proteinases and adventitious lactobacilli. Franklin and Sharpe[291] noted that lactobacilli may be present in small numbers in curd. They are the only lactic acid bacteria to multiply in the maturing cheese. A number of species of lactobacilli have been isolated from cheese. These include *L. casei* varieties, *L. plantarum, L. fermentum, L. brevis, L. buchneri, L. curvatus,* and many others. Micrococci, aerobic and anaerobic spore formers, and enterococci are also seen in cheese at ~10 to 10^4cfu/g and these numbers generally decline during ripening.

As stated earlier, for normal ripening of cheese, a high starter population must lyse to release proteinases and peptidases to effect cheese flavor, body, and texture development.[292] The intensity of Cheddar flavor was not increased in starter cheeses by the presence of additional lysozyme-treated starter cells and no Cheddar flavor developed in chemically acidified cheese containing the lysozyme-treated cells. It was concluded that the intracellular starter enzymes play no direct part in flavor formation but produce breakdown products from which Cheddar flavor compounds may be formed by other unknown mechanisms.[293] Cheese flavor intensity seems to be closely related to soluble nitrogen compounds, especially amino acids and small peptides.[267,268]

Lactobacilli are the only lactic acid bacteria to increase significantly in number during maturation of Cheddar cheese, except for the less frequently occurring pediococci (most often *P. pentosaceus*), which may multiply at a similar rate and reach levels as great as 10^7 cfu/g.[27] The fact that lactobacilli can multiply in ripening cheese whereas most other bacteria decrease in numbers has caused investigations into the means by which strains of this genus can sustain growth in an environment nearly devoid of fermentable carbohydrates.[294] The subject of lactobacilli in cheese was recently reviewed.[244,294] Following is a brief perspective on the means of survival and growth of lactobacilli in cheese.

Lactobacilli isolated from cheese grew poorly in milk, perhaps from lack of suitable available nitrogen.[295] Serum of mature Cheddar cheese inhibited *Lactobacillus brevis,* whereas sera of 4- to 6-month old cheese supported its growth.[296] Peptides from mature Edam cheese were stimulatory to *L. casei.*[297] Recent studies indicate that cheeseborne lactobacilli possess significant proteinase and peptidase activities.[257,298,299] Perhaps these activities are needed to cope with large concentrations of protein and peptide fractions present in a carbohydrate-depleted cheese matrix held at low temperature. Nath and Ledford[235] demonstrated that aqueous fractions from 120-day- and 180-day-old cheese were stimulatory to *L. casei* growing in milk. In younger cheese there were inhibitory and stimulatory fractions. Evidence was also presented that the stimulatory peptides came from α_s-casein. Other than peptides, common compounds found in the stimulatory fractions were *N*-acetylhexosamine, glutamic acid, and riboflavin. However, riboflavin added to milk was not stimulatory. The essential amino acids are utilized more efficiently from peptides containing them than from an equivalent amount of the essential amino acids in free form.[299–301] Carbohydrates bound to proteins,[302] citrate,[303] and glycerol[304] can serve as a carbon source for lactobacilli in cheese. Thomas[246] showed evidence that dying starter bacteria present in cheese can also serve as carbon source for emerging lactobacilli in cheese. The intracellular contents of starter bacteria may provide small molecular weight (somewhat heat stable) growth promoting substance(s) for lactobacilli.[305]

There is a great deal of interest in shortening the ripening period of cheese by the use of added nonstarter mesophilic lactobacilli.[306–309] It was concluded that even among the homofermentative lactobacilli, only a few, two out of 22, were found suitable for accelerated aged cheese ripening. Strains of *L. casei* subsp. *casei* and *L. casei* subsp. *pseudoplantarum* yielded high quality cheese whereas other strains caused some off-flavors.[307] Strains of *L. casei* subsp. *rhamnosus* contributed to high acidity and low pH. All amino acids increased during ripening and were higher in the *Lactobacillus*-added cheeses than in the control cheese. Glutamic acid, leucine, phenylalanine, valine, and lysine were detected in large quantities. The proteolytic process and accumulation of higher concentrations of free amino acids were affected by higher ripening temperature.[306] In these experiments, hetero- and homofermentative lactobacilli produced similar proteolytic breakdown, but the former resulted in off-flavors and gassy cheese.[306] High levels of γ-amino acid butyrate (0.3 to 19.4 mg/g) were associated with poor quality aged cheese.[310]

3.11.4 Fate of Fat

It is well known that Cheddar cheese from skim milk does not develop full typical flavor, indicating that fat is required for the development of characteristic cheese flavor.[262] Free fatty acids (FFAs) play a major role in flavors of many cheese varieties. They have been considered the backbone of Cheddar cheese flavor by Patton[311] and are thought to contribute cheesiness in Cheddar cheese.[312,313] Acetic acid is found in cheese and its concentration can vary considerably in cheese.[314] It probably adds to the sharp mouthfeel of cheese conferred by lactic acid concentration, but overproduction of acetic acid can lead to a vinegarlike off-flavor.[313] The claim that ratios of acetic acid to other fatty acids are important determinants of Cheddar flavor[262] have not been confirmed.[314,315] Acetic acid in cheese arises through microbial activity whereas other volatile fatty acids increase in cheese due to the weak esterase and lipase activities of the milk and the starter bacteria.[316] It was shown that mesophilic starters hydrolyzed mono- and diglycerides but their activity on triglycerides was very weak. Volatile fatty acids can also arise from amino acids via oxidative demination activity of *Lactococcus lactis* subsp. *lactis* var. *diacetylactis*,[317] but this activity is considered uncertain in cheese.[318] It is widely believed that lipolytic and esterolytic activities of lactic acid bacteria are limited, but the search for these enzymes in lactococci and lactobacilli is continuing,[245,247,319] perhaps to find a suitable replacement for glottal tissues and enzymes[320] which are added to cheese for rapid and increased flavor development.

3.11.5 Flavor of Cheddar Cheese

Many flavor compounds are chemical interactions of microbially derived substrates under conditions of low pH and low oxidation reduction potential. Numerous compounds such as hydrocarbons, alcohols, aldehydes, ketones, acids, esters, lactone, and sulfur are important in cheese flavor. Hydrogen sulfide, dimethyl sulfide methanethiol, diacetyl,[321] phenylacetaldehyde, phenylacetic acid and phenethanol,[322] butanone diacetyl and pentan-2-one[323] terpenes ethyl butyrate,[324] methanol, pentan-2-one, diacetyl, and ethyl butyrate[325] are considered key compounds of good Cheddar flavor. In a recent study,[277] 86 odor-active components were found. Most of these compounds possessed odors characteristic of free fatty acids, ketones, and saturated and unsaturated aldehydes. The researchers also identified 2-propanol, 1,3-butanediol, γ-decalactone, and δ-undecalactone in cheese for the first time. One component that had a weak cheeselike aroma and eluted after butyric acid from the gas chromatograph could not be identified. These authors also support the component theory for Cheddar cheese flavor.

3.12 Microbiological and Biochemical Changes in Swiss Cheese

Swiss, Emmentaler, and Gruyère type cheeses are made with thermophilic streptococci and lactobacilli to which propionibacteria are added for eye formation. During the early phases of cheesemaking *S. salivarius* subsp. *thermophilus* multiplies rapidly and utilizes the glucose moiety of lactose to produce L-lactate, leaving behind galactose.[326] Lactobacilli start vigorous acid production after whey drainage when the curd temperature drops to 46 to 49°C. At 1 day, population of streptococci and lactobacilli reach a little over 10^8 cfu/g.[327] In large blocks of cheese there are temperature gradients. The center of the cheese cools more slowly than the periphery.[328] As a consequence, the lactic acid fermentation starts more rapidly in the outer area where the temperature has dropped compared to the center of the cheese. The growth of the starter streptococci and lactobacilli is greater in the periphery of cheese than in the center. This difference may be as large as one log in population.

The propionibacteria added to cheese milk do not show measurable growth during cheese manufacture. Growth of these organisms starts after the whey is drained and the population may reach 10^6 cfu/g in 24 h.

During the hot room curing, the number of lactic starter bacteria decline by a log or more to 10^6 cfu/g or less whereas the propionibacteria reach a population in excess of 5×10^8 cfu/g.[327] During the hot room curing, growth of enterococci (Group D streptococci) and homofermentative and heterofermentative lactobacilli also takes place. A typical Swiss cheese made from milk heat treated at 64.5°C/18 s contained propionbacteria (6×10^8), total lactobacilli (2×10^8), *L. fermentum* (4×10^7), enterococci (5×10^5), aerobic sporeformers (5×10^2), and presumptive anaerobic sporeformers ($\sim 10^3$) per gram of cheese at 60 days. The lactobacilli population consisted of *L. casei* subsp. *casei, L. casei* subsp. *alactosus, L. plantarum,* and *L. fermentum.* At this point no starter lactic acid bacteria were detected at 10^{-3} dilution.[329]

3.12.1 Fate of Lactose

In milk cultures, *S. thermophilus* metabolizes lactose to L-(+)-lactic acid utilizing only the glucose moiety and leaves the galactose free in the medium.[326] The thermophilic *L. helveticus* can utilize lactose with the production of D- and L-lactic acid and it is galactose positive. *L. delbrueckii* subsp. *bulgaricus* and *L. delbrueckii* subsp. *lactis* generally do not utilize galactose and produce D-lactate in milk.[46] About 1.7% lactose was present in Swiss cheese curd after the curd was pumped.[330] It was rapidly metabolized during the 10 h in the press (<0.1%). At 1 day no lactose was detectable. A low level of glucose (0.15%) was present in the curd at 4 h into press. At 10 h, curd contained 0.7% galactose which fell to 0.3% in 1 day whereas no glucose was detected. Halfway (14 days) through the hotroom curing, galactose was undetectable and the cheese contained 1.2% L-lactate and 0.35% D-lactate. In 35 days during cure, L-lactate decreased from 1.2% to 0.2%, whereas D-lactate increased to 0.4% at 21 days and fell to about 0.1% in 35 days. At 35 days, acetate and propionate were present at concentrations of 0.25% and 0.55%, respectively.[330]

It has been pointed out that carbohydrate fermentation balance will change according to the type and level of lactobacilli used for cheesemaking.[331] Both D- and L-lactate accumulation and the disappearance of galactose in cheese were more rapid with higher inoculum of lactobacilli.[46]

3.12.2 CO_2 Production

Under normal conditions the propionic acid fermentation is the source of CO_2 production and eye formation. The accepted equation of this fermentation is[331,332]:

$$3 \text{ Lactate} \rightarrow 1 \text{ Acetate} + 2 \text{ Proprionate} + CO_2 + H_2O$$

Under certain conditions lactate is also fermentated to butyric acid according to the equation[333]

$$2 \text{ Lactate} \rightarrow 1 \text{ Butyrate} + CO_2 + 2H_2$$

It has been noted that instead of production of 2:1 propionate to acetate ratio from lactate, 1.16:1.00 to 2.15:100 occurred in cheese.[331] Crow and Turner[334] attempted to explain this discrepancy by taking into consideration the production of succinate in Swiss cheese as follows.

The succinate is formed at the expense of an equivalent concentration of propionate and CO_2 which are formed from lactate or carbohydrate. The quantity of acetate produced from lactate is unaffected by succinate formation due to this CO_2 fixation step. Citric acid, malic acid, and fumaric acid are also metabolized to succinic acid.[335] Acetate is also produced from citric acid and lactate by cheese lactobacilli.[336] Aspartic acid is converted to succinate during lactate fermentation by strains of *Propionibacterium freudenreichii* subsp. *shermanii*.[337] This resulted in a greater proportion of the lactate being fermented to acetate and CO_2 rather than to propionate. The CO_2 fixation and aspartate pathways in propionibacteria, although both producing succinate, give rise respectively to a decrease and an increase in CO_2 production from lactate.[334] It was postulated that eye formation in Swiss cheese would be affected by the contribution of both these pathways to succinate production. Experiments with propionibacteria suggest that carbohydrates, when present in cheese, may be used directly by the propionibacteria along with aspartate and lactate.[334] In another study,[337] it was shown that aspartate was cometabolized with lactate by propionibacteria. After lactate exhaustion, alanine was one of the two amino acids to be metabolized according to the following equation[338]:

$$3 \text{ Alanine} \rightarrow 2 \text{ Propionate} + 1 \text{ Acetate} + CO_2 + 3 \text{ Ammonia}$$

Studies with resting cell suspensions of propionibacteria in an amino acid mixture showed that amino acids were potential sources of CO_2 production in Swiss cheese during long-term storage, possibly causing secondary fermentation and split defect in cheese.[339]

3.12.3 Eye Formation

The quality of Swiss cheese is judged by the size and distribution of eyes. Swiss cheese eyes are essentially due to CO_2 production, diffusion, and accumulation in the cheese body.[184] The number and size of eyes depend on CO_2 pressure; diffusion rate; and body, texture, and temperature of cheese. Fluckiger[340,341] followed CO_2 production and eye formation in Emmental cheese for 5 months during ripening. He found a total production of 130 to 150 L of CO_2 per 100 kg of cheese. This volume was composed of 50% dissolved gas, 15% of CO_2 present in the eyes, and 30% lost by diffusion through the paste (cheese). It was noticed that the values of CO_2 from calculations were lower than those that were measured.[341] The difference was 50 to 70 L per 100 kg. It was explained that the calculated gas was based on fermentation and did not take into consideration the decarboxylation of amino acids. Aspartate and alanine catabolism also contributes to CO_2 production.[334,335] In another study, French Emmental cheese was wrapped in gas-tight bags and analyzed for dissolved gas and the gas present in the eyes. The comparison of measured CO_2 to calculated CO_2 from the volatile fatty acids was in good agreement.[333] The increase in the eye volume occurred at the same time as the gas diffusion. This observation opposes the generally accepted hypothesis that CO_2 saturates the cheese before diffusion and forms the eyes at the same time that its release is restrained by the rind.[330] It is believed that CO_2 diffusion, eye formation, and dissolution in the cheese are simultaneous. A scanning electron microscope study of Emmental cheese revealed that casein micelles compacted during manufacturing and ripening. The fat globules lost their integrity and appeared as large masses with diverse forms. A few junctions in the grain and the formation of gas microbubbles were observed which may be responsible for eye formation.[342]

3.12.4 Fate of Proteins

The Swiss cheese curd is cooked to about 52°C and at this temperature the coagulant is rendered inactive. In this type of cheese, bulk of casein proteolysis results from proteolytic enzymes of lactic acid bacteria and milk proteinase.[283] At 42 days, Swiss type cheese had retained 70% of its original plasmin activity. Ollikainen and Nyberg[343] noticed higher than expected plasmin activity in cheese during ripening, due possibly to the increasing pH. They also noted that unclean flavor was associated with low plasmin activity. In Swiss cheese α_s-casein is more proteolyzed than β-casein.[233,329]

3.12.5 Flavor of Swiss Cheese

Cheese flavor is derived in part from cheese milk. Production of Swiss cheese with desirable body, flavor, and texture requires that milk be of low count and properly clarified. Mild heat treatment, 68 to 72°C/15 to 18 s, is recommended.[344] However, the characteristic flavor of Swiss-type cheese comes from microbial transformation of milk components. These contain milk-soluble volatiles (acetic acid, propionic acid, butyric acid, and diacetyl) which give the basic sharpness and general cheesy notes.[322,345] Water-soluble nonvolatile amino acids (especially proline), peptides, lactic acid, and salts provide mainly sweet notes. Oil-soluble fractions (short-chain fatty acids) are also important to flavor.[346] Nutty flavor is attributed to alkylpyrizines.[322] Several compounds, for example, ketones, aldehydes, esters, lactones, and sulfur-containing compounds are also important.[315,331] Due to the activity of certain strains of lactobacilli and fecal streptococci, biogenic amines are sometimes found in Swiss cheese.[347]

3.13 Microbiological and Biochemical Changes in Gouda Cheese

Gouda and Edam cheese are made with mesophilic lactic starters containing citrate fermenting lactococci and leuconostoc. Gouda cheese has slightly higher fat than Edam. Gouda cheese milk is standardized to casein-to-fat ratios of 0.8 to 0.82

whereas Edam cheese milk is brought to a casein-to-fat ratio of 1.06 to 1.08.[348] Edam and Gouda curd is cooked to 35°C using hot water. The curd is pressed under whey to obtain a close texture. The formed cheese is brined in 14% brine at 14°C. The cheese is cured at 12 to 16°C and 85 to 90% relative humidity.[349]

3.13.1 Fate of Lactose

The pH of the cheese should be 5.7 to 5.9 after 4 h from the start of the manufacture, and 5.3 to 5.5 after 5.5 h. The pH of the cheese is 5.1 to 5.2 in about 24 h.[350] Lactose is fermented almost completely and rapidly. Fermentation of citric acid is of particular importance to eye formation. In Edam and Gouda, CO_2 for eye formation comes from the residual citrate in cheese and not from lactate as in Swiss cheese.[349]

3.13.2 Fate of Proteins

The initial proteolysis in cheese is due to the rennet enzymes and further proteolysis is brought about by enzymes of starter bacteria and to a much lesser extent by milk proteinase.[349] Proteolysis in Gouda cheese is much like in Cheddar cheese where virtually all of α_{s1}-casein is degraded leaving behind β-casein.[233,351] In cheese trials using purified calf chymosin and microbiologically produced chymosin, it was demonstrated that proteolysis of α_{s1+2}- and β-casein took place rapidly during the first 3 months of ripening. Subsequently, the protein breakdown occurred more slowly and after 6 months 20% of α_{s1+2}- and 30 to 40% of β-casein remained. A marked increase in γ-caseins was also observed, indicating that milk proteinase was active.[352] In a recent study of water-soluble extracts of Gouda cheese after ripening for 1, 2, and 3 months, three major peaks were isolated which increased in size as the cheese ripened.[256] The amino acid compositions of these peptides were similar to the fragments of α_{s1}-casein (f1–9), α_{s1}-casein (f1–13), and α_{s1}-casein (f1–14). The α_{s1}-casein (f1–23) was hydrolyzed by cellular proteinases of *Streptococcus cremoris* H61 to seven main peptides including the three mentioned above. This study clearly indicates that α_{s1}-casein degradation in Gouda and perhaps other cheeses is caused by lactic acid bacterial proteinases.[256]

Several factors in cheesemaking affect proteolysis by rennet and these are[353]:

1. The quantity of rennet used in cheesemaking.
2. Moisture in cheese.
3. pH of cheese during manufacture; the lower the pH, the more calf rennet is bound to para-κ-casein.
4. The amount of starter used.
5. The rate of acidification and the initial pH of the cheese milk and its composition.
6. Cooking temperature of curd; the higher the temperature, the less active is the rennet.
7. Heat treatment of the milk; the more intensive, the more rennet the curd will contain.

8. The salt level in cheese.
9. Bacteriophages.
10. Inhibitors in milk.

Factors 9 and 10 affect indirectly by affecting culture activity.

3.13.3 Fate of Fat

A limited lipolysis in Gouda cheese is desirable as it adds to its flavor balance, but greater fat acidity is not desirable.[349] In pasteurized milk (72°C/15 s) much of the milk lipase is destroyed which can act on the triglycerides in milk to generate mono- and diglycerides.[349] The starter and lactobacilli esterases and lipases act on the mono- and diglyceride fractions to liberate FFAs.[316] A high count milk may have higher levels of FFAs.[316]

3.13.4 Microbiological Changes

In Gouda cheese a high number of starter bacteria are seen at the time of brining. Their rate of disappearance in cheese depends on the strain of the organism.[354] In addition, a considerable difference was found among lactococci in their ability to liberate amino acids and other flavor compounds. Lactobacilli are always present in cheese. Their numbers can increase dramatically from as low as 1 cfu/g in milk to $>10^7$ cfu/in cheese in several weeks.[355] It is believed that lactobacilli do not add positively to the flavor of Gouda cheese. If anything, their presence in large numbers is believed to cause off-flavors and texture defects.[355]

3.13.5 Flavor of Gouda Cheese

It is generally accepted that lactic acid, diacetyl, CO_2, peptides, amino acids, and FFAs contribute to the flavor of cheese. Several secondary compounds, resulting from transformations of lactic acid, also affect flavor. These are aldehydes, ketones, alcohols, esters, organic acids, and CO_2. Cheese also contains volatile compounds arising from the degradation of amino acids, for example, ammonia, amines, hydrogen sulfide, and phenylacetic acid.[356,357] Anethole, 2,4-dithiopentane, and several alkyl pyrazines and bismethylthiomethane are considered important aroma compounds in Gouda cheese.[358]

3.14 Microbiological and Biochemical Changes in Mold-Ripened Cheese

3.14.1 Blue Cheese

Blue cheese and its relatives—Roquefort, Gorgonzola, and Stilton—are characterized by peppery, piquant flavors produced by the mold *Penicillium roqueforti*. This mold can tolerate low oxygen and high CO_2 tension and is relatively halotolerant.[189]

For these cheese types, acid development is slow and the curd mass is not pressed.[190] This promotes an open texture necessary for the CO_2 to escape and oxygen to gain access.

Cheese is ripened at 8 to 12°C with relative humidity of 95%. Due to high acid and salt, the starter lactococci decline rapidly in 2 to 3 weeks.[279] After salting the surface flora mostly consists of yeast and micrococci.[359] The yeasts start to grow and deacidify the curd on the surface.

In 2 to 3 weeks *Brevibacterium linens* appears on the surface of cheese.[360] Growth of mold in cheese is evident in 8 to 10 days and development is complete in 30 to 90 days. Lactobacilli (*L. casei* varieties and *L. plantarum*) are also present in the cheese.

Due to deacidification of the cheese and extensive proteolysis, the cheese pH rises from 4.5 to 4.7 at 24 h to a maximum of 6.0 to 7.0 at 16 to 18 weeks. The increase is more rapid and pronounced on the surface. Molds are both proteolytic and lipolytic, resulting in extensive proteolysis and lipolysis of cheese.[361] In Blue cheese both α_{s1}- and β-casein are degraded.[329] There is a large accumulation of free amino acids due to extracellular acidic and alkaline endopeptidases.[362,363] Maximum proteolytic activity in cheese occurs during the first few weeks when mycelium has attained full growth.[364] Sodium chloride and free fatty acids depress proteinase activity and prevent excessive proteolysis.[361] The contribution of plasmin,[363] *Geotrichum candidum*[365] yeast, and *B. linens*[366] in proteolysis and flavor production in mold-ripened cheeses is well recognized. The occurrence of citrulline, ornithine, γ-aminobutyric acid, histamine, tyramine, and tryptamine reflects amino acid breakdown products.[362] Amino acid breakdown products can also generate ammonia, aldehyde, acids, alcohols, amine, and methanethiol.[362] Amino acids also enhance methyl ketone production.[361]

The quality of Blue cheese flavor critically depends on the metabolism of lipid substrate in cheese. The unique and dominating flavor of mold-ripened cheeses comes from methyl ketones which are predominantly derived from partial oxidation of FFAs resulting in a ketone with one less carbon atom.[367] Activity of lipase to liberate FFAs is important in the production of methyl ketones.[361] Homogenization of milk promotes lipolysis. Also, there is evidence that some strains of *P. camemberti* possess mono- and diacylglycerol lipase.[368] The enzyme was separated into two forms, A and B. B enzyme was the predominant form and was specific for mono- and diacylglycerol and preferred long-chain monoacylglycerols in the α-position. Of the various methyl ketones, 2-heptanone is usually the most abundant followed by 2-nonanone, 2-pentanone, and 2-undecanone.[361] β-Decarboxylase activity was shown to be present in resting spores, germinated spores, and mycelium; β-ketolaurate was actively decarboxylated to 2-undecanone.[369] Activity of the enzyme was in the order of mycelium > germinating spores > resting spores. Of various β-ketoacid substrates, β-ketolaurate was the preferred substrate for mold decarboxylase.[370] It is believed that in later stages of cheese ripening, spore metabolism is favored where spores can continue to generate methyl ketones in the presence of high fatty acid concentration and at relatively high CO_2 levels.[361] Methyl ketones (2-alkanones) are easily reduced to their corresponding secondary alcohols

(2-alkanols) supposedly to minimize the toxic effects of methyl ketones on the mold.[371] Flavor-simulation studies suggest that δ-tetradecalactone and δ-dodecalactone improved the quality of the cheese flavor.[372,373] Addition of δ-tetradecalactone and δ-dodecalactone improved the quality of Blue cheese flavor. It has been proposed that traces of δ-hydroxyacid in milk glycerides released by lipases during cheese ripening may undergo ring closure to form lactones, or they may be converted enzymically.[374] The production of δ-lactones can also arise by the lipase release of esterified δ-ketoacids in milk glycerides. These ketoacids are reduced to hydroxyacids and then converted into lactones. Such a pathway has been demonstrated in yeasts and molds.[374]

3.14.2 Camembert and Brie Cheese

These are soft white cheeses that are ripened by external mold growth. The mold involved is *Penicillium camemberti* or its biotypes *P. Caseicolum* and *P. candidum.* These cheeses have a relatively high water activity (0.98) and a low pH (4.6) at make time.[189]

Lactose in the exterior of cheese disappears in about 15 days, whereas lactose in the interior and galactose and L-lactate in cheese disappear by 30 days.[375] First to appear on surface of the cheese are yeasts, *Kluyveromyces lactis, Sacchromyces cerivisae,* and *Debaryomyces hanseni* and deacidify the cheese. *Geotrichum candidum* also appears at the same time but growth is somewhat limited.[362] After 5 to 7 days, the surface of cheese is less acid and salt has diffused into the cheese. *P. camemberti* appears on the surface and growth is complete in 15 to 20 days. At this time micrococci and sometimes *B. linens* is seen on the cheese surface.[362] At the end of cheesemaking, curd has about 60% moisture and after 1 month cheese must not lose more than 5 to 7% moisture at the time of packaging.

In Camembert, ripening of the cheese takes place from the surface to the center of the cheese. On the surface, pH of the cheese rises to ~7.0 due to proteolytic activity of the organisms. Due to deamination of amino acids, ammonia is released which contributes to the aroma profile of cheese. Ammonia constitutes about 7 to 9% of the soluble nitrogen.[362]

β-Casein is not degraded extensively and α_{s1}-casein degradation is less than in Cheddar,[233] and the appearance of some γ-caseins suggests plasmin activity in cheese.[363] Free fatty acids found in large quantities, 22.27 ± 13.73 meq acid/100 g of fat, contribute to the basic flavor of cheese and serve as the precursors of methyl ketones and secondary alcohols. Primary alcohols, secondary alcohols, methyl ketones, aldehydes, esters (ethyl esters of C_2, C_4, C_6, C_8, C_{10}, butyrate 2-phenylethyl acetate), lactones (C_9, C_{10}, C_{12}), phenol, *p*-cresol, hydrogen sulfide, methanethiol, methlylsulfide, and other sulfur compounds along with anisoles, amines, and other compounds constitute the volatile compounds of Camembert cheese.[376]

3.15 Microbiological and Biochemical Changes in Bacteria Surface-Ripened Cheese

3.15.1 Brick Cheese

Brick cheese is a representative of a large group of cheeses (Limburger, Muenster, Tilsiter, Bel Paese, and Trappist) that are ripened by growth of bacteria and yeast on the surface. The organisms involved are yeast, micrococci, and *B. linens.*[279]

When *Streptococcus salivarius* subsp. *thermophilus* is used along with *L. lactis* subsp. *lactis* as a starter, it will grow rapidly during cooking and for a few hours after the curd is drained. Growth of this organism stops when temperature of the curd drops to about 32°C or lower. Growth of lactococci continues and pH of the cheese reaches 5.1 to 5.3.[377] After brining for 1 or more days, cheese is held at about 15°C in a room with 90–95% relative humidity (RH). Yeast (*Mycoderma*) appear on the surface in 2 or 3 days followed by micrococci and then *B. linens.* The yeast oxidize the acid on the surface of cheese making it less acid, thus permitting growth of micrococci and *B. linens* and the pH of cheese surface may reach 5.4 in a 2-week period. Sometimes *Geotrichum candidum* may also be present. α_{s1}-Casein is always hydrolyzed but β-casein disappearance was seen in Muenster and not in brick.[233] Yeasts isolated from surface-ripened cheeses also contribute to the proteolysis of cheese.[381] Yeasts found on Limburger cheese synthesize considerable amounts of pantothenic acid, niacin, and riboflavin.[382] Pantothenic acid and *p*-aminobenzoic acid are required by *B. linens.* Liberated free amino acids are much higher on the surface of cheese where *B. linens* is present.[383] It is clear that association among different organisms present on the surface of cheese is essential to the definition of cheese-smear and its role in flavor production. It has been suggested that yeasts and *B. linens* are essential for flavor of Brick cheese but typical flavor is attained only in the presence of micrococci.[378,379] Organisms of the genus *Arthrobacter* are also isolated from surface ripened cheese and appear earlier than *B. linens* in the presence of salt.[384] *B. linens* is very proteolytic and able to convert methionine into methane-thiol.[385,386] Many of the compounds formed on the surface of cheese are absorbed into the cheese and compounds such as methyl mercaptan and 2-butanone were higher at the surface and H_2S, dimethyl disulfide, acetone, and ethanol were higher in the interior of cheese.[387]

3.16 Microbiological and Biochemical Changes in Mozzarella Cheese

Mozzarella cheese is primarily used on pizza, lasagna, and other recipes in cooking. Consequently, good quality of Mozzarella refers to its stretchability, meltability, and shredability with little pronounced flavor. In order to preserve these characteristics some manufacturers freeze Mozzarella after it has been graded.[388] Several factors affect the physical properties of Mozzarella, including salt, pH, fat, moisture, and microbiology of the ripening cheese. Salt (NaCl) concentration between 1% and

2.4% has little effect.[188] Lower concentrations of salt cause softening and a high level of salt promotes firmness. As the moisture and FDB (fat on dry basis) of cheese increases, the cheese becomes soft and less shredable.[389] Cheese made with a mixture of mesophilic starter and *S. salivarius* subsp. *thermophilus* starter tends to have a greater protein and fat hydrolysis during storage. Such cheese is difficult to shred and has atypical flavor when aged. The molded cheese is brinned. Cheese should have pH 5.2 (range pH 5.1 to 5.4) as it ensures sufficient removal of calcium from the caseins to effect proper stretch.[390] When cheese was made with proteinase-deficient and proteinase-positive single strains of *L. delbrueckii* subsp. *bulgaricus,* cheese from proteinase-deficient strains lost its ability to stretch after 7 days. With time stretchability decreased for all cheese.[188]

Cheese made with proteinase-deficient strains melted more easily than cheese made with proteinase-positive cultures. These differences were not dramatic after 28 days of storage.[188] Cheese made with normal starter composed of rod and coccus melted better and was more brown on cooking than proteinase-deficient pairs. It was noticed that as stretch decreased with time, melt increased.[188] α_{s1}-Casein is proteolyzed to a lesser degree by rennet enzymes compared to other cheese types, whereas β-casein is largely intact. This level of rennet proteolysis of milk protein appears sufficient to give the melt and stretch characteristics to cheese during hot water kneading at about 57°C. Creamer[391] suggested that stretching properties may be related to higher concentrations of intact casein and large peptides in the cheese.

There is little lipolysis and fatty acid liberation in traditional Mozzarella cheese.[392]

Mozzarella and Provolone are manufactured in a similar manner. The former is consumed fresh while the latter may be ripened at 12.5°C for 3 to 4 weeks and then stored at 4.5°C for 6 to 12 months for grating.[43] The ripened cheeses have mainly *L. casei* and its subspecies. Provolone has more lipolytic flavors than Mozzarella. Provolone may be molded in pear, cylindrical, or salami shapes. Smoked Provolone is also popular in trade.

3.17 Microbiological and Biochemical Changes in Parmesan and Romano Cheese

Parmesan and Romano cheese are made with *S. salivarius* subsp. *thermophilus, L. delbrueckii* subsp. *bulgaricus,* or other species of thermophilic lactobacilli. In addition to rennet, pregastric estrases, or rennet paste may be added to cheese milk for their lipolytic activity. The curd is cooked to 51 to 54°C, when the whey acidity reaches about 0.2% it is hooped (packed) in round forms. Sometimes salt is added to the curd, which slows the starter and regulates moisture. The cheese at pH 5.1 to 5.3 is placed in 24% brine for several days.

Compared to other cheese types, the starter population in the fresh cheese is low. Throughout cheese ripening, 12 to 24 months, cheese flora seldom exceeds 10^5 cfu/g[393] and fecal streptococci and salt-tolerant lactobacilli predominate. In these hard grating cheese α_s- and β-caseins are not overly proteolyzed compared to other cheese varieties.[394] However, a high concentration of γ-casein in some samples indicates plasmin activity. This is attributed to high cooking temperatures of curd, which inactivate the coagulant, and the high salt in the moisture, which discourages growth of adventitious flora. In ripened cheeses quality varies from location to location. Volatile free fatty acid and nonvolatile free fatty acid (C_4 through C_{18}) concentrations are high in these cheeses, particularly in Romano.[392] Butyric acid and minor branched-chain fatty acids that occur in milk appear to contribute to the piquant flavor of Parmesan.

The total concentrations of methyl ketones in grana cheese are quite low, 0.075 μM/g fat, compared with those in blue (19.14 μM/g), Roquefort (5.18 μM/g), and even Cheddar (0.24 μM/g). The proportions of all methyl ketones, except C_3, in grana were similar to the proportions of β-ketoacids in the cheese fat, suggesting the spontaneous formation of methyl ketones from β-ketoacids in grana cheese.[394] It is claimed that addition of 1-phenylpropionic acid and isovaleric acid to fresh cheese curds imparted Italian cheese flavor.[328] For a more balanced flavor, a concomitant increase in free amino acids (glutamic acid, aspartic acid, valine, and alanine) has been noted. Too high a free fatty acid level in cheese gives a strong, soapy, undesirable flavor.

3.18 Accelerated Cheese Ripening

One of the major costs of cheese is the expense of curing time before desired flavor develops. While some maturation time is inevitable, there are systems available where ripening time is shortened by speeding up proteolysis and lipolysis to generate flavor and modify texture. Elevated temperature (13°C or higher) curing offers the simplest approach to speed up ripening of otherwise normal cheese. Cheese intended for this type of curing must not contain measurable levels of heterofermentative lactobacilli or leuconostocs, because an open-texture defect and off-flavors will develop.[395]

Microbial proteinases and gastric esterases have been used with little success to achieve acceptable cheese with uniformity. Activity of these exogenous enzymes is unregulated and may contribute to the detriment of cheese quality. Several unproven systems are available from culture houses.

Additions of partially inactivated starter organisms have been used with mixed results. Presently, this is not economical. Most of the proprietary systems investigated caused a minor to major deviation from characteristic flavor, body, and texture of cheese.

3.19 Processed Cheese Products

Process cheese is produced by blending several lots of different ages of cheese that are comminuted and mixed together by stirring and heating. Water, emulsifying salts, color, and condiments may be added. The final product is smooth and homogeneous.

Process cheeses were prepared as early as 1895 in Europe, but the use of emulsifying salts was not widely practiced until 1911 when Gerber and Co. of Switzerland invented process cheese. A patent issued to J. L. Kraft in 1916 marked the origin of the process cheese industry in America and describes the method of heating natural cheese and its emulsification with alkaline salts.[215]

Process cheeses in the United States generally fall in one of the following categories.[215,396]

1. Pasteurized blended cheese. Must conform to the standard of identity and is subject to the requirements prescribed by pasteurized process cheese except:
   a. A mixture of two or more cheeses may include cream or Neufchatel.
   b. None of the ingredients prescribed or permitted for pasteurized process cheese is used.
   c. The moisture content is not more than the arithmetic average of the maximum moisture prescribed by the definitions of the standards of identity for the varieties of cheeses blended.
   d. The word process is replaced by the word blended.
2. Pasteurized process cheese.
   a. Must be heated at no less than 65.5°C for no less than 30 s. If a single variety is used the moisture content can be no more than 1% greater than that prescribed by the definition of that variety, but in no case greater than 43%, except for special provisions for Swiss, Gruyère, or Limburger.
   b. The fat content must not be less than that prescribed for the variety used or in no case less than 47% except for special provisions for Swiss or Gruyère.
   c. Further requirements refer to minimum percentages of the cheeses used.
3. Pasteurized process cheese food.
   a. Required heat treatment minimum is the same as pasteurized process cheese.
   b. Moisture maximum is 44%; fat minimum is 23%.
   c. A variety of percentages are prescribed.
   d. Optional dairy ingredients may be used, such as cream, milk, skim milk, buttermilk, and cheese whey.
   e. May contain any approved emulsifying agent.
   f. The weight of the cheese ingredient is not less than 51% of the weight of the finished product.
4. Pasteurized process cheese spread.
   a. Moisture is more than 44% but less than 60%.
   b. Fat minimum is 20%
   c. Is a blend of cheeses and optional dairy ingredients and is spreadable at 21°C.
   d. Has the same heat treatment minimum as pasteurized process cheese.
   e. Cheese ingredients must constitute at least 51%.
   f. A variety of percentages are prescribed.

3.19.1 Advantages of Process Cheese over Natural Cheese

1. Can be kept at room temperature without oil separation.
2. Flavor and other attributes of cheese can be consistantly maintained by proper selection and blending of cheeses.
3. Keeping quality and safety of the product is improved because pathogenic organisms present in cheese are destroyed during heating.
4. Numerous compositions containing fruits, vegetables, meats, smoke, and spices are possible.
5. Offers versatility in use—cooking, dips, sauces, snacks, etc.
6. Process cheese provides a home for off-cuts, cheese with poor maturing properties, and other cheese not suitable for consumption as natural cheese, economically.

3.19.2 Processing

Steps in processing involve[397]:

- Selection of natural cheese
- Blending
- Grinding and milling
- Adding emulsifiers, water, salt, and color
- Processing and packaging
- Homogenization (optional)
- Storage

It is important that cheese with rancid, putrid, and severe microbiological defect be not included in the process cheese blend. The age and proportion of the cheese in the blend depends on the characteristics of the process cheese desired. For the production of slices, 75% of cheese up to 3 months old can be blended with about 25% well-ripened cheese 6 to 12 months old.[178] Generally, young cheese with elastic unhydrolyzed casein lends smooth texture and firm body and good slicing properties. Mature, older cheese tends to give higher flavor and grainy texture. For cheese spreads, slightly larger portions of higher acid cheese and older cheese can be used in the blend.[178]

3.19.3 Emulsifiers

A good emulsifier system should consist of monovalent cations and polyvalent anions. Some salts are better emulsifiers and have poor calcium binding capacity. The ability to sequester calcium is one of the most important functions of the emulsifying agents. Emulsifying agents supplement the emulsifying capacity of cheese proteins to provide unique properties to process cheese. Following are some functions of the emulsifying salts[397,398]:

1. Removing calcium from the protein system

2. Peptizing, solubilizing, and dispersing the proteins
3. Hydration and swelling of proteins
4. Emulsification of fat and stabilization of the emulsion
5. Control and stabilization of cheese pH
6. Structure formation during cooling

To obtain desired body, texture, and spreadability, a number of ingredients such as nonfat dry milk, whey, powder, whey protein concentrate, whey proteins,[397] calcium caseinate, and butteroil[398] can be drawn on to develop a blend for process cheese. For a more detailed review consult refs. 397, 399, and 400.

3.19.3.1 Basic Emulsification Systems for Cheese Processing

Citrates

- Trisodium citrate (most common, used for slices).
- Tripotassium citrate (used in reduced-sodium formulations, promotes bitterness).
- Calcium citrate (poor emulsification).

Orthophosphates

- Disodium phosphate and trisodium phosphate (most common, used for loaf and slices).
- Dicalcium phosphate and tricalcium phosphate (poor emulsification, used for calcium ion fortification).
- Monosodium phosphate (acid taste, open texture).

Condensed Phosphates

- Sodium tripolyphosphate (nonmelting).
- Sodium hexamethaphosphate (used to restrict melts).
- Tetrasodium pyrophosphate and sodium acid pyrophosphate (minimal usage).

As the amount of calcium phosphate in protein is decreased, the solubility of casein in water is increased and so is its emulsifying capacity.[401] Reduction of calcium in the calcium-paracaseinate in the cheese by emulsifiers solubilizes the insoluble paracaseinate and improves the emulsifying capacity of cheese proteins.[400,401] Affinity of phosphates for calcium in process cheesemaking is in the order of monosodium phosphate > disodium phosphate > disodium pyrophosphate > trisodium pyrophosphate > tetrasodium pyrophosphate > sodium tripolyphosphate.[399] Furthermore, polyphosphates possess the peptizing capacity lacking in orthophosphates. Peptizing ability is essential for process cheese production. Peptization rate of casein in the presence of polyphosphates increases with increasing chain length and phosphate concentration. For peptization of casein, three or more P atoms are required and the rate is greatest at pH 6.5.[402,403] Sodium-containing emulsifier salts including trisodium citrate and disodium phosphate are used extensively.[403–406] Manufacture of process cheese in the presence of phosphates tends to increase soluble nitrogen.

Table 3.15 CHARACTERISTICS OF EMULSIFIERS MOST COMMONLY USED IN THE MANUFACTURE OF PROCESS CHEESE AND RELATED PRODUCTS

Emulsifier[a]	Formula	Characteristics
Sodium citrate	$2Na_3C_6H_5O_7 \cdot 11\ H_2O$ $Na_3C_6H_5O_7 \cdot 2\ H_2O$	Versatile; produces firm cheese with good melting properties; inexpensive; best qualities.
Disodium phosphate	Na_2HPO_4	Good firming, buffering, and melting properties; poor creaming properties. Least expensive.
Trisodium phosphate	Na_3PO_4	Highly alkaline; improves sliceability when used in combination with other emulsifiers; good buffering ability; used at low concentrations.
Sodium hexametaphosphate (Graham's salt)	$(Na\ PO_3)_6$	Produces tartish flavor and a very firm body; product does not melt easily; least soluble of all; bacteriostatic.
Tetrasodium disphosphate	Polyphosphates $Na_4P_2O_7$	Good creaming properties; strong buffering capacity; high protein solubility; excellent ion exchange; tartish flavor.

Source: Ref. 43.
Adapted by permission of VCH Publishers, Inc., 220 East 23rd St., New York, N.Y., 10010 from: Kosikowski, Frank. CHEESE AND FERMENTED MILK FOODS. 2nd edition, 1977: Table 66, p. 392.

[a] Other emulsifiers permitted by the U.S. Federal Standards of Identity are: sodium acid pyrophosphate, sodium potassium tartrate, tetrasodium pyrophosphate, dipotasium phosphate, potassium citrate, calcium citrate, and sodium aluminum phosphate.

However, no increase in water-soluble nitrogen was observed when tetrasodium pyrophosphate and sodium citrate were used at the 2 to 4% level.[403]

Of the citric acid salts, trisodium citrate is commonly used. Process cheese made with citrate has a higher melting point than the cheese made with other emulsifying salts. It should not be used at a rate higher than 3% of natural cheese weight. A small proportion of phosphates and citrate works best for cheese of average to high maturity.[397] Some characteristics of the commonly used emulsifiers are listed in Table 3.15.

The melting properties of processed cheese are not governed only by the age of cheese in the blend and the emulsifier, but also by the heat treatment given to the product. Process cheese was prepared from the same lot of Cheddar cheese using sodium citrate, disodium phosphate, tetrasodium pyrophosphate, or sodium aluminum phosphate and cooked at 82°C for different times from 0 to 40 min. All cheeses had different physical properties but in general all cheeses became firmer, more elastic, and less meltable as the cooking time increased from 0 to 40 min.[407] In another study hard and soft process cheeses were prepared by using 2.2% polyphosphate and 1.0% trisodium citrate plus 1.5% polyphosphate, respectively. Electron microscopy of these samples revealed that soft type process cheese had mostly single particles in the protein matrix (20 to 25 nm in diameter), whereas the hard type showed pronounced networklike structures of longer protein strands.[408]

The search for new types of emulsification system(s) for process cheese continues. In Yugoslavia, new emulsifiers, KSS-4 (pH 6) and KSS-11 (pH 11), produced good quality process cheese.[409] In Egypt, Cremodan SE 30 proved to be the best emulsifier as regards organoleptics and texture stability during storage.[410] Japanese workers produced process cheese without emulsifying salts. They used Cheddar cheese of different moisture contents (35.4 to 38.9%) and a twin-screw extruder with screw rotation speeds ranging from 50 to 150 rpm. Continuous emulsification by extrusion heating was demonstrated and a finer emulsion in cheese was produced at faster rotation speed.[411] Studies on the effect of batch and extrusion cooking on lipid–protein interaction have indicated that batch cheese possessed firmer texture with less peptidization than extruded cheeses of identical composition. It is postulated that this may be due to improved protein restructuring as a result of stirring and the use of a lower temperature in batch cooking.[412] Extrusion cooking is claimed to be the way of future processing by the year 2000.[413]

3.19.4 Heat Treatment

Cheese blends for process cheese are heated to at least 65.5°C, but more commonly to about 85°C.[407] Hydrolysis of pyro- and polyphosphates occurs during melting and afterwards and the extent of degradation varies with cheese type used. This degradation is speculated to be due to phosphatase.[406] Some characteristics of process cheese products along with the temperature of heat treatment are shown in Table 3.16.

3.19.5 pH and Microbiological Stability

The pH value of process cheese is important from the standpoint of protein configuration and solubility and microbiological stability.[400] In process cheese compositions, pH may vary from 5.0 to 6.5. At the lower pH, process cheese may become crumbly and at higher pH value, it may become soft. At higher pH value, cheese is more susceptible to microbiological spoilage.[399] Sodium salts are used in process cheese formulations to produce desired body, texture, flavor, and degree of product safety. The sodium salt emulsifiers, usually phosphates, or citrates together with NaCl already in cheese or added when process cheese is made contribute to the total electrolyte level in the cheese formulation. The pH, moisture, and total electrolyte level play a critical role in product safety, preventing growth and toxin production by *C. botulinum* in shelf-stable pasteurized process cheese spreads.[414]

It has been established that pasteurized process cheese with relatively high pH (5.6 to 6.2) and a moisture of about 50% has an excellent record of safety against *Clostridium botulinum*.[415]

Table 3.16 SOME CHARACTERISTICS OF PROCESS CHEESE, PROCESS CHEESE FOOD, AND PROCESS CHEESE SPREADS

Type of Product	Ingredients	Cooking Temperature	Composition	pH	Author
Process cheese	Natural cheese, emulsifiers, NaCl, coloring	71–80°C	Moisture and fat[a] contents correspond to the legal limits for natural cheese	5.6–5.8	Kosikowski
		74–85°C	45% Moisture		Thomas
Process cheese food	Same as above plus optional ingredients such as milk, skim milk, whey, cream, albumin, skim milk cheese; organic acids	79–85°C	No more than 44% moisture, no less than 23% fat	5.2–5.6	Kosikowski
Process cheese spread	Same as process cheese food plus gums for water retention	88–91°C	No less than 44% and no more than 60% moisture	<5.2	Kosikowski
		90–95°C	55% Moisture		Thomas

Source: Ref. 399. (Kosikowski, ref. 43; Thomas, ref. 397).

[a] 1% higher for Cheddar cheese.

In pasteurized process cheese, and particularly in shelf-stable products which are not commercially sterile, salt plays a critical role in concert with other factors such as pH, moisture, and water activity in preventing growth of *Clostridium botulinum.*[416] In 1988, addition of nisin (250 ppm) to specific process cheese spread compositions was approved by the FDA as a safety factor against *C. botulinum.**

3.20 References

1. Ensrud, B. 1981. *The Pocket Guide to Cheese.* Frederick Muller, Limited, London.
2. Carlson, A., G. C. Hill, and N. F. Olson. 1987. Kinetics of milk coagulation: 1. The kinetics of kappa casein hydrolysis in the presence of enzyme deactivation. *Biotechnol. Bioengin.* **29**:582–589.
3. Fox, P. F. 1987. *Cheese: Chemistry, Physics and Microbiology,* Vol. 1. Elsevier Applied Science Publishers, London.
4. Anonymous. 1979. *Newer Knowledge of Cheese and Other Cheese Products.* National Dairy Council, Rosemont, IL.
5. U.S. Department of Commerce, Bureau of the Census, International Trade Administrations. 1989. Reprinted in *U.S. Industrial Outlook 1989—Food, Beverages, and Tobacco,* U.S. Government Printing Office, Washington, D.C., Jan.

---

* The author gratefully acknowledges the help of Barbara J. Kostak for critically reading through the manuscript.

6. Food and Drug Administration. 1978. *Cheese and Related Products.* Code of Federal Regulations. Food and Drugs, Title 21, Part 133, Washington, D.C. Office of the Federal Register, National Archives of the United States.

7. Bassette, R., and J. S. Acosta. 1988. Composition of Milk Products. *In Fundamentals of Dairy Chemistry.* N. P. Wong, R. Jenness, M. Keeny, and E. H. Marth (eds.), Van Nostrand Reinhold, New York, pp. 39–79.

8. Whitehead, H. R., and G. A. Cox. 1935. The occurrence of bacteriophage in cultures of lactic streptococci. *N. Z. J. Sci. Technol.* **16**:319–320.

9. Davies, F. L., and B. A. Law. 1984. *Advances in the Microbiology and Biochemistry of Cheese and Fermented Milk.* Elsevier Applied Science Publishers, London.

10. Second symposium on lactic acid bacteria—genetics, metabolism and applications. 1987. *FEMS Microbiol. Rev.* **46**:199–379, Abstr. pp. 1–99.

11. Third symposium on lactic acid bacteria—genetics, metabolism and applications. 1990. *FEMS Microbiol. Rev.* **87**:1–188, Abstr. pp. 1–177.

12. Bramley, A. J., and C. H. McKinnon. 1990. The microbiology of raw milk. *In* R. K. Robinson (ed.), *The Microbiology of Milk,* Vol. 1, 2nd edit.: *Dairy Microbiology.* Elsevier Applied Science Publishers, London.

13. Johnson, E. A., J. H. Nelson, and M. Johnson. 1990. Microbiological safety of cheese made from heat-treated milk, Part 1. Executive summary, introduction and history. *J. Food Prot.* **53**:441–452.

14. Lau, K. Y., D. M. Barbano, and R. R. Rasmussen. 1990. Influence of pasteurization on fat and nitrogen recoveries and Cheddar cheese yield. *J. Dairy Sci.* **73**:561–570.

15. Johnson, E. A., J. H. Nelson, and M. Johnson. 1990. Microbiological safety of cheese made from heat-treated milk, Part III. Technology, discussion, recommendations, bibliography. *J. Food Prot.* **53**:610–623

16. Olson, N. F. 1991. Does saving money mean sacrificing yield? *Dairy Ice Cream Field* **174**:28–29.

17. Paluch, L. J., M. E. Johnson, B. Riesterer, and N. F. Olson. 1990. Cheddar cheese manufactured from milk HTST pasteurized at 73.3C, 75.6C and 77.8C. *J. Dairy Sci.* **73**(Suppl. 1):115.

18. Lau, K. Y., D. M. Barbano, and R. R. Rasmussen. 1991. Influence of pasteurization of milk on protein breakdown in Cheddar cheese during aging. *J. Dairy Sci.* **74**:727–740.

19. Kristoffersen, T. 1963. Measuring thiamine disulfide reducing substances in Cheddar cheese. *J. Dairy Sci.* **48**:1135–1136.

20. Kristoffersen, T., I. Gould, and G. A. Purvis. 1964a. Cheddar cheese flavor. III. Active sulfhydryl group production during ripening. *J. Dairy Sci.* **47**:599–603.

21. Kristoffersen, T. 1978. Recognizing the proper compounds to get the most out of cheese flavor. *Dairy Ice Cream Field* **161**:80E–80F.

22. D'Aoust, J.-Y., D. B. Emmons, R. McKeller, G. E. Timbers, E. C. D. Todd, A. M. Sewell, and D. W. Warburton. 1987. Thermal inactivation of *Sallmonella* species in fluid milk. *J. Food Prot.* **50**:494–501.

23. Farber, J. M., G. W Sanders, J. I. Speirs, J. Y. D'Aoust, D. B. Emmons, and R. McKeller. 1988. Thermal resistance of *Listeria monocytogenes* in inoculated and naturally contaminated raw milk. *Int. J. Food Microbiol.* **7**:277–286.

24. D'Aoust, J.-Y., C. E. Park, R. A. Szabo, E. C. D. Todd, D. B. Emmons, and R. McKeller. 1988. Thermal inactivation of *Campylobacter* species, *Yersinia enterocolitica,* and hemorrhagic *Escherichia coli* 0157:H7 in fluid milk. *J. Dairy Sci.* **71**:3230–3236.

25. Johnson, E. A., J. H. Nelson, and M. Johnson. 1990. Microbiological safety of cheese made from heat-treated milk, part II. Microbiology. *J. Food Prot.* **53**:519–540.

26. Lawrence, R. C., H. A. Heap, and J. Gilles. 1984. A controlled approach to cheese technology. *J. Dairy Sci.* **67**:1632–1645.

27. Chapman, H. R., and M. E. Sharpe. 1990. Microbiology of Cheese. *In* R. K. Robinson (ed.), *The Microbiology of Dairy Products,* Vol. 2, 2nd edit: *Dairy Microbiology.* Elsevier Applied Science Publishers, London.

28. Martley, F. G., and R. C. Lawrence. 1972. Cheddar cheese flavor. II. Characteristics of single strain starters associated with good or poor flavor development. *N. Z. J. Dairy Sci. Technol.* **7**:38–44.

29. Garvie, E. I. 1984. Taxonomy and identification of bacteria important in cheese and fermented dairy products. *In* F. L. Davies and B. A. Law (eds.), *Advances in the Microbiology and Biochemistry of Cheese and Fermented Milk.* Elsevier Applied Science Publishers, London.

30. Cogan, T. M., and C. Daly. 1987. Cheese starter cultures. *In* P. F. Fox (ed.), *Cheese: Chemistry, Physics and Microbiology* Vol. 1. Elsevier Applied Science Publishers, London.

31. Holt, J. G. 1986. *Bergey's Manual of Systematic Bacteriology,* Vol. 2. Williams & Wilkins, Baltimore and London.

32. Hurst, A. 1981. Nisin. *Adv. Appl. Microbiol.* **27**:25–123.

33. Gasson, M. J. 1984. Transfer of sucrose fermenting ability, nisin resistance and nisin production in *Streptococcus lactis* 712. *FEMS Microbiol. Lett.* **21**:7–10.

34. Gonzalez, C. F., and B. S. Kunka. 1985. Transfer of sucrose fermenting ability and nisin production phenotype among lactic streptococci. *Apl. Environ. Microbiol.* **49**:627–633.

35. Steele, J. L., and L. McKay. 1986. Partial characterization of the genetic basis for sucrose metabolism and nisin production in *Streptococcus lactis. Appl. Environ. Microbiol.* **51**:57–64.

36. Jackson, H. W., and M. E. Morgan. 1954. Identity and origin of the malty aroma substance from milk cultures of *Streptoccous lactis* var. *maltigenes. J. Dairy Sci.* **37**:1316–1324.

37. Hirsch, A. 1952. The evolution of the lactic streptococci. *J. Dairy Res.* **19**:290–293.

38. Davey, G. P., and B. C. Richardson. 1981. Purification and some properties of diplococcin from *Streptococcus cremoris* 346. *Appl. Environ. Microbiol.* **41**:84–89.

39. Davey, G. P. 1981. Mode of action of diplococcin, a bacteriocin from *Streptococcus cremoris* 346. *N. Z. J. Dairy Sci. Technol.* **16**:187–190.

40. Cogan, T. M. 1976. The utilization of citrate by lactic acid bacteria in milk and cheese. *Dairy Indust. Int.* **41**:12–16.

41. Heiberg, R. 1985. Factors affecting the properties of bulk starters. *Scand. J. Dairy Technol. Know-How* **NM2**:65–71.

42. Cogan, T. M. 1983. Some aspects of the metabolism of dairy starter cultures. *Irish J. Food. Sci. Technol.* **7**:1–13.

43. Kosikowski, F. V. 1977. *Cheese and Fermented Milk Foods.* Edwards Bros., Ann Arbor, MI.

44. Rajagopal, S. N., and W. E. Sandine. 1990. Associative growth and proteolysis of *Streptococcus thermophilus* and *Lactobacillus bulgaricus* in skim milk. *J. Dairy Sci.* **73**:894–899.

45. Premi, L., W. E. Sandine, and P. R. Elliker. 1972. Lactose-hydrolysing enzymes of *Lactobacillus* species. *Appl. Microbiol.* **24**:51–57.

46. Turner, K. W., and F. G. Martley. 1983. Galactose fermentation and classification of thermophilic lactobacilli. *Appl. Environ. Microbiol.* **45**:1932–1934.

47. Radke-Mitchell, L., and W. E. Sandine. 1984. Associative growth and differential enumeration of *Streptococcus thermophilus* and *Lactobacillus bulgaricus:* a review. *J. Food Prot.* **47**:245–248.

48. Labrapoulos, A. E., W. F. Collins, and W. K. Stone. 1982. Starter culture effects on yogurt fermentation. *Cult. Dairy Prod. J.* **17**:15–17.

49. Moon, N. J., and Reinhold. 1974. Selection of active and compatible starters for yogurts. *Cult. Dairy Prod. J.* **9**:10–12.

50. Bautista, E. S., R. S. Dahiya, and M. L. Speck. 1966. Identificaton of compounds causing symbiotic growth of *Streptococcus thermophilus* and *Lactobacillus bulgaricus* in milk. *J. Dairy Res.* **33**:299–307.

51. Pettie, J. W., and H. Lolkema. 1950. Yogurt. II. Growth stimulating factors for *S. thermophilus. Netherlands Milk Dairy J.* **4**:209–224.

52. Driessen, R. M., J. Ubbels, and J. Stadhouders. 1982. Evidence that *L. bulgaricus* in yogurt is stimulated by carbon dioxide produced by *S. thermophilus. Netherlands Milk Dairy J.* **36**:135–144.

53. Galesloot, Th. E., F. Hassing, and H. A. Veringa. 1968. Symbiosis in yogurt. 1. Stimulation of *Lactobacillus bulgaricus* by a factor produced by *Streptococcus thermophilus. Netherlands Milk Dairy J.* **22**:50–63.

54. Pulusani, S. R., and D. R. Rao, and G. R. Sunki. 1979. Antimicrobial activity of lactic cultures: partial purification and characterization of antimicrobial compounds produced by *Streptococcus thermophilus. J. Food Sci.* **44**:575–578.

55. Rasic, J. L., and J. A. Kurmann. 1978. *Yogurt. Scientific Grounds, Technology, Manufacturing and Preparations.* Technical Dairy Publishing House, Copenhagen, Denmark.

56. Pulusani, S. R., and D. R. Rao. 1984. Stimulation by formate of antimicrobial activity of *Lactobacillus bulgaricus* in milk. *J. Food Sci.* **49**:652–653.

57. Naylor, J., and M. E. Sharpe. 1958. Lactobacilli in Cheddar cheese 1. The use of selective media for isolation and serological typing for identification. *J. Dairy Res.* **25**:92–103.

58. Laleye, L. C., R. E. Simard, B. H. Lee, R. A. Holley, and R. N. Giroux. 1987. Involvement of heterofermentative lactobacilli in development of open texture. *J. Food Prot.* **50**:1009–1012.

59. Lee, B. H., L. C. Laleye, R. E. Simard, M. H. Munsch, and R. A. Holley. 1990. Influence of homofermentative lactobacilli on the microflora and soluble nitrogen components in Cheddar cheese. *J. Food Sci.* **55**:391–397.

60. Dacre, J. C. 1958. Characteristics of a presumptive *Pediococcus* occurring in New Zealand Cheddar cheese. *J. Dairy Res.* **25**:409–413.

61. Dacre, J. C. 1958. A note on the Pediococci in New Zealand Cheddar cheese. *J. Dairy Res.* **25**:414–417.

62. Franklin, J. G., and M. E. Sharp. 1963. The incidence of bacteria in cheese milk and Cheddar cheese and their association with flavor. *J. Dairy Res.* **30**:87–99.

63. Fryer, T. F., and M. E. Sharpe. 1966. Pediococci in Cheddar cheese. *J. Dairy Res.* **33**:325–331.

64. Turner, K. W., and T. D. Thomas. 1980. Lactose fermentation in Cheddar cheese and the effect of salt. *N. Z. J. Dairy Sci. Technol.* **15**:265–276.

65. Elliott, J. A., and H. T. Muligan. 1968. Pediococci in Canadian Cheddar cheese. *Can. Inst. Food Technol. J.* **1**:61–63.

66. Bhunia, A. K., M. C. Johnson, and B. Ray. 1988. Purification, characterization and antimicrobial spectrum of bacteriocin produced by *Pediococcus acidilactici. J. Appl. Bacteriol.* **65**:261–268.

67. Tzanetakis, N., and E. L. Tzanetakis. 1989. Biochemical activities of *Pediococcus pentosaceus* isolates of dairy origin. *J. Dairy Sci.* **72**:859–863.

68. Bhowmik, T., and E. H. Marth. 1990. Role of *Micrococcus* and *Pediococcus* species in cheese ripening: a review. *J. Dairy Sci.* **73**:859–866.

69. Bhowmik, T., and E. H. Marth. 1990. β-Galactosidase of *Pediococcus* species: induction, purification and partial characterization. *Appl. Microbiol. Biotechnol.* **33**:317–323.

70. Tetlow, A. L., and D. G. Hoover. 1988. Fermentation products from carbohydrate metabolism in *Pediococcus pentosaceus* PC 39. *J. Food Prot.* **51**:804–806.

71. Godinho, M., and P. F. Fox. 1981. Effect of NaCl on the germination and growth of *Penicillium roqueforti*. *Milchwissenschaft* **36**:205–208.

72. Moreau, C. 1980. *Penicillium roqueforti:* morphology, physiology, importance in the cheesemaking, mycotoxins. *Le Lait* **60**:254–271.

73. Samson, R. A., C. Eckardt, and R. Orth. 1977. The taxonomy of *Penicillium* species from fermented cheeses. *Antonie van Leeuwenhoek* **43**:341–350.

74. Gripon, J. C. 1987. Mold-ripened cheeses. *In* P. F. Fox (ed.), *Cheese: Chemistry, Physics and microbiology.* Vol. 2. Elsevier Applied Science Publishers, London.

75. Kundrat, W. 1952. Effect of pH on the development of different mold strains. *Dairy Sci. Abstr.* **14**:618.

76. Modler, H. W., J. R. Brunner, and C. M. Stine. 1974. Extracellular protease of *Penicillium roqueforti* II. Characterization of a purified enzyme preparation. *J. Dairy Sci.* **57**:528–534.

77. Auclair, J. E., and A. Hirsch. 1953. The inhibition of microorganisms by raw milk. 1. The occurrence of inhibitory and stimulatory phenomena. Methods of estimation. *J. Dairy Res.* **20**:45–59.

78. Reiter, B., and J. D. Oram. 1962. Nutritional studies on cheese starters 1. Vitamins and amino acid requirements of single strain starters. *J. Dairy Res.* **29**:63–77.

79. Valerie, M., E. Marshall, and B. A. Law. 1984. The physiology and growth of dairy lactic acid bacteria. *In* F. L. Davies and B. A. Law (eds.), *Advances in Microbiology and Biochemistry of Cheese and Fermented Milk,* Elsevier Applied Science Publishers, London.

80. Kempler, G. M., and L. L. McKay. 1979. Genetic evidence for plasmid-linked lactose metabolism in *Streptococcus lactis* subsp. *diacetylactis*. *Appl. Environ. Microbiol.* **37**:1041–1043.

81. Klaenhammer, T. R., L. L. McKay and K. A. Baldwin. 1978. Improved lysis of group N streptococci for isolation and rapid characterization of plasmid deoxyribonucleic acid. *Appl. Environ. Microbiol.* **35**:592–600.

82. Pearce, L. E., N. F. Sage, and J. Hastings. 1973. Non-acid milk. The reappearance of an old problem. *N. Z. J. Dairy Sci. Technol.* **8**:165–166.

83. Mattick, A. T. R., and A. Hirsch. 1947. Further observations on an inhibitory substance (nisin) from lactic streptococci. *Lancet* **11**:5–12.

84. Oxford, A. E. 1944. Diplococcin, an antibacterial protein elaborated by certain milk streptococci. *Biochem. J.* **38**:178–182.

85. Costilow, R. N., and M. L. Speck. 1951. Inhibition of *Streptococcus lactis* in milk by fatty acids. *J. Dairy Sci.* **34**:1104–1110.

86. Costilow, R. N., and M. L. Speck. 1951. Inhibitory effect of rancid milk on certain bacteria. *J. Dairy Sci.* **34**:1119–1127.

87. Maxcy, R. B. 1964. Influence of surface active agents on some lactic streptococci. *J. Dairy Sci.* **47**:1285–1290.

88. Maxcy, R. B., and C. W. Dill. 1967. Adsorption of free fatty acids on cells of certain microorganisms. *J. Dairy Sci.* **50**:472–476.

89. Anders, R. F., and G. R. Jago. 1964. The effect of fatty acids on the metabolism of lactic acid streptococci. 1. Inhibition of bacterial growth and proteolysis. *J. Dairy Res.* **31**:81–89.

90. Anders, R. F., and G. R. Jago. 1970. The effect of fatty acids on the metabolism of pyruvate in lactic acid streptococci. *J. Dairy Res.* **37**:445–456.

91. Anders, R. F., D. M. Hogg, and G. R. Jago. 1970. Formation of hydrogen peroxide by group N streptococci and its effect on their growth and metabolism. *Appl. Microbiol.* **19**:608–612.

92. Subramaniam, C. S., and N. F. Olson. 1968. Effect of hydrogen peroxide on activity of lactic starter cultures in milk. *J. Dairy Sci.* **51**:517–519.

93. Gilliland, S. E., and M. L. Speck. 1969. Biological response of lactic streptococci and lactobacilli to catalase. *Appl. Microbiol.* **17**:797–800.

94. Nath, K. R., and B. J. Wagner. 1973. Stimulation of lactic acid bacteria by a *Micrococcus* isolate: evidence for multiple effects. *Appl. Microbiol.* **26**:49–55.

95. Reiter, B. 1985. The biological significance of the non-immunoglobulin protective proteins in milk: lysozyme, lactoferrins, lactoperoxidase. *In* P. F. Fox (ed.), *Develoments in Dairy Chemistry*-3. Elsevier Applied Science Publishers, London.

96. Wright, R. C., and J. Tramer. 1958. Factors influencing the activity of cheese starters. The role of milk peroxidase. *J. Dairy Res.* **25**:104–118.

97. Reiter, B. 1973. Some thoughts on cheese starters. *J. Soc. Dairy Technol.* **26**:3–15.

98. Stadhouders, J., and H. A. Veringa. Some experiments related to the inhibitory action of milk peroxidase on lactic streptococci. *Netherlands Milk Dairy J.* **16**:96–116.

99. Elliot, K. A. C. 1932. Milk peroxidase. Its preparation, properties, and action with H_2O_2 on metabolites. With a method for determining small amounts of H_2O_2 in complex mixtures. *Biochem. J.* **26**:10–24.

100. Roginski, H., M. C. Broome, and M. W. Hickey. 1984. Non-phage inhibition of group N streptococci in milk. 1. The incidence of inhibition in bulk milk. *Aust. J. Dairy Technol.* **39**:23–27.

101. Roginski, H., M. C. Broome, D. Hungerford, and M. W. Hickey. 1984. Non-phage inhibition of group N streptococci in milk. 2. The effects of some inhibitory compounds. *Aust. J. Dairy Technol.* **39**:28–32.

102. Guirguis, N., and M. W. Hickey. 1987. Factors affecting the performance of thermophilic starters. 2. Sensitivity to the lactoperoxidase system. *Aust. J. Dairy Technol.* **42**:14–16, 26.

103. Gilliland, S. E., and M. L. Speck. 1968. D-Leucine as an auto-inhibitor of lactic streptococci. *J. Dairy Sci.* **51**:1573–1578.

104. Olson, H. C., and S. E. Gilliland. 1970. The influence of heat treatments on the rates of acid production by lactic cultures. *Cult. Dairy Prod. J.* **5**:2–7.

105. Speck, M. L. 1962. Starter culture growth and action in milk. *J. Dairy Sci.* **45**:1281–1286.

106. Shankar, P. A., and F. L. Davies. 1977. Associative bacterial growth in yogurt starter; initial observations on stimulatory factors. *J. Soc. Dairy Technol.* **30**:31–32.

107. Veringa, H. A., Th. E. Galesloot, and H. S. Davelaar. 1968. Symbiosis in yogurt (II). Isolation and identification of a growth factor for *Lactobacillus bulgaricus* produced by *Streptococcus thermophilus*. *Netherlands Milk Dairy J.* **22**:114–120.

108. Wright, R. C., and J. Tramer. 1957. The influence of cream rising upon the activity of bacteria in heat-treated milk. *J. Dairy Res.* **24**:174–183.

109. Stadhouders, J., and F. Hassing. 1974. Enhancement of the acid production of some lactic streptococci by milk peroxidase. *19th Int. Dairy Congr. New Delhi* **1E**:369–370.

110. Emmons, D. B., and J. A. Elliott. 1967. Effect of homogenization of skim milk on rate of acid development, sediment formation and quality of cottage cheese made with agglutinating cultures. *J. Dairy Sci.* **50**:957.

111. Heap, H. A. 1982. Sensitivity of starter cultures to penicillin and streptomycin in bulk-starter milk. *N. Z. J. Dairy Sci. Technol.* **17**:81–86.

112. Kosikowski, F. V., and G. Mocquot. 1958. *Advances in Cheese Technology. FAO Agriculture studies. No. 38.* Food and Agriculture Organization of the United Nations.

113. Pearce, L. E., S. A. Brice, and A. M. Crawford. 1973. Survival and activity of lactic streptococci following storage at 22°C and 4°C. *N. Z. J. Dairy Sci. Technol.* **8**:41–45.

114. Olson, H. C. 1959. Preservation of lactic cultures. *J. Dairy Sci.* **42**:388.

115. Harvey, R. J. 1965. Damage to *Streptococcus lactis* resulting from growth at low pH. *J. Bacteriol.* **90**:1330–1336.

116. Marquis, R. E., N. Porterfield, and P. Matsumura. 1973. Acid–base titration of streptococci and the physical states of intracellular ions. *J. Bacteriol.* **114**:491–498.

117. Whitehead, H. R., and G. A. Cox. 1934. Observations on sudden changes in the rate of acid formation by milk cultures of lactic streptococci. *J. Dairy Res.* **5**:197–207.

118. Whitehead, H. R., and G. A. Cox. 1935. The occurrence of bacteriophage in cultures of lactic streptococci. *N. Z. J. Sci. Technol.* **16**:319–329.

119. Whitehead, H. R., and G. A. Cox. 1936. Bacteriophage phenomenon in cultures of lactic streptococci. *J. Dairy Res.* **7**:55–62.

120. Whitehead, H. R. 1953. Bacteriophage in cheese manufacture. *Bacteriol. Rev.* **17**:109–123.

121. Lawrence, R. C., and L. E. Pearce. 1968. The case against the unpaired single startr strains. *N. Z. J. Dairy Sci. Technol.* **3**:137–139.

122. Hynd, J. 1976. The use of concentrated single strain cheese starters in Scotland. *J. Soc. Dairy Technol.* **29**:39–45.

123. Heap, H. A., and R. C. Lawrence. 1976. The selection of starter strains for cheesemaking. *N. Z. J. Dairy Sci. Technol.* **11**:16–20.

124. Limsowtin, G. K. Y., H. A. Heap, and R. C. Lawrence. 1977. A multiple starter concept for cheesemaking. *N. Z. J. Dairy Sci. Technol.* **12**:101–106.

125. Lawrence, R. C., H. A. Heap, G. Limsowtin, and A. W. Jarvis. 1978. Symposium: Research and development trends in natural cheese manufacturing and ripening Cheddar cheese starters: current knowledge and practices of phage characteristics and strain selection. *J. Dairy Sci.* **61**:1181–1191.

126. Lawrence, R. C., and H. A. Heap. 1986. *The New Zealand Starter System.* International Dairy Federation Bulletin 199.

127. Thunell, R. K., F. W. Bodyfelt, and W. E. Sandine. 1984. Economic comparisons of Cheddar cheese manufactured with defined-strain and commercial mixed strain cultures. *J. Dairy Sci.* **67**:1061–1068.

128. Cox, W. A. 1977. One-Day Symposium: Characteristics and use of starter cultures in the manufacture of hard pressed cheese. *J. Soc. Dairy Technol.* **30**:5–15.

129. Timmons, P., M. Hurley, F. Drinan, C. Daly, and T. M. Cogan. 1988. Development and use of a defined strain starter system for Cheddar cheese. *J. Soc. Dairy Technol.* **41**:49–53.

130. Stadhouders, J., and G. J. M. Leenders. 1984. Spontaneously developed mixed-strain cheese starters. Their behavior towards phages and their use in the Dutch cheese industry. *Netherlands Milk Dairy J.* **38**:157–181.

131. Stadhouders, J. 1986. The control of cheese starter activity. *Netherlands Milk Dairy J.* **40**:155–173.

132. Peebles, M. M., S. E. Gilliland, and M. L. Speck. 1969. Preparation of concentrated lactic streptococcus starters. *Appl. Microbiol.* **17**:805–810.

133. Stanley, G. 1977. One-day symposium: the manufacture of starters by batch fermentation and centrifugation to produce concentrates. *J. Soc. Dairy Technol.* **30**:36–39.

134. Chavarri, F. J., M. O. Paz, and M. Nunez. 1988. Cryoprotective agents for frozen concentrated starters from non-bitter *Streptococcus lactis* strains. *Biotechnol. Lett.* **10**:11–16.

135. Rasmussen, H. 1977. ''Revolutionary'' cheese starter cultures live up to the name. *Dairy Ice Cream Field* **160**:70H–70N4, 88.

136. Wigley, R. C. 197. The use of commercially available concentrated starters. *J. Soc. Dairy Technol.* **30**:45–47.

137. Lloyd, G. T., and E. G. Pont. 1973. The production of concentrated starters by batch culture. *Aust. J. Dairy Technol.* **28**:104–108.

138. Richardson, G. H., C. T. Cheng, and R. Young. 1976. Lactic bulk culture system utilizing a whey-based bacteriophage inhibitory medium and pH control. 1. Applicability to American style cheese. *J. Dairy Sci.* **60**:378–386.

139. Richardson, G. H. 1985. Increasing cultured product quality and yield through careful strain selection and propagation. *Cult. Dairy Prod. J.* **20**:20–27.

140. Tofte Jespersen, N. J. 1976. Concentration of cultures. A look at some old and new methods. *Dairy Ice Cream Field* **159**:58A–58G.

141. Huggins, A. R. 1984. Progress in dairy starter culture technology. *Food Technol.* **38**:41–50.

142. Bergere J. L. 1968. Production massive de cellules de Streptocoque lactiques. I. Methods generales d'etude et facteurs de la crossance de *Streptococcus lactis* souche C10. *Le Lait* **48**:1–11.

143. Gilliland, S. E., and M. L. Speck. 1974. Frozen concentrated cultures of lactic starter bacteria. A review. *J. Milk Food Technol.* **37**:107–111.

144. Bergere, J. L., and I. Hermier. 1968. La production massive du cellules de Streptocoques lactiques. II. Croissance de *Streptococcus lactis* dans un milieu A pH constant. *Le Lait* **48**:13–30.

145. Cogan, T. M., D. J. Buckley, and S. Condon. 1971. Optimum growth parameters of lactic streptococci used for the production of concentrated cheese starter cultures. *J. Appl. Bacteriol.* **34**:403–409.

146. Lloyd, G. T., and E. G. Pont. 1973. Some properties of frozen concentrated starters produced by continuous culture. *J. Dairy Res.* **40**:157–167.

147. Pont, E. G., and G. L. Holloway. 1968. A new approach to the production of cheese starter. Some preliminary investigations. *Aust. J. Dairy Technol.* **23**:22–29.

148. Pettersson, H. E. 1975. Growth of a mixed species lactic starter in a continuous ''pH-stat'' fermentor. *Appl. Microbiol.* **29**:437–443.

149. Turner, K. W., G. P. Davey, G. H. Richardson, and L. E. Pearce. 1979. The development of a starter handling system to replace traditional mother cultures. *N. Z. J. Dairy Sci. Technol.* **14**:16–22.

150. Lewis, J. E. 1986. *Cheese Starters: Development and Application of the Lewis System.* Elsevier Applied Sciences Publishers, London.

151. Heap, H. A., and R. C. Lawrence. 1988. *In* R. K. Robinson (ed.), *Developments in Food Microbiology,* Vol. 4. Elsevier Applied Science Publishers, London.

152. Tamime, A. Y. 1990. *In* R. K. Robinson (ed.), *Dairy Microbiology. The Microbiology of Milk Products.* Vol. 2, 2nd edit. Elsevier Applied Science Publishers, London.

153. Bolle, A. C., G. J. M. Leenders, and J. Stadhouders. 1985. A study of the efficiency of capturing bacteriophages from air by the Pall Emflon filter, Type AB1FR7PV. *Nizo Rapporten.* R121.

154. Ultradepth filters. Ultrafilter international, 555 Oakbrook Parkway (660). Norcross, Georgia, 30093.

155. Anonymous 1990. The white room. *Dairy Indust. Int.* **54**:39.

156. Collins, E. B., F. E. Nelson, and C. E. Parmelee. 1950. The relation of calcium and other constituents of a defined medium to proliferation of lactic streptococcus bacteriophage. *J. Bacteriol.* **60**:533–542.

157. Reiter, B. 1956. Inhibition of lactic streptococcus bacteriophage. *Dairy Indust.* **21**:877–879.

158. Hargrove, R. E. 1959. A simple method for eliminating and controlling bacteriophage in lactic starters. *J. Dairy Sci.* **42**:906.

159. Zottola, E. A., and E. H. Marth. 1966. Dry-blended phosphate treated milk media for inhibition of bacteriophages active against lactic streptococci. *J. Dairy Sci.* **49**:1343–1349.

160. Sozzi, T. J. 1972. A study of the calcium requirement of lactic starter phages. *Milchwissenschaft* **27**:503–507.

161. Gulstrom, T. J., L. E. Pearce, W. E. Sandine, and P. R. Elliker. 1979. Evaluation of commercial phage inhibitory media. *J. Dairy Sci.* **62**:208–221.

162. Ledford, R. A., and M. L. Speck. 1979. Injury of lactic streptococci by culturing in media-containing high phosphates. *J. Dairy Sci.* **62**:781–784.

163. Galesloot, Th. E., and F. Hassing. 1962. Enhele waarnemingen betreffende het gedrag van zuursels in entbalkte melk en in melk voorzien van calciumbindende zouten. *Netherlands Milk Dairy J.* **16**:117–130.

164. Henning, D. R., W. E. Sandine, P. R. Elliker, and H. A. Hays. 1965. Studies with bacteriophage inhibitory medium. 1. Growth of single strain lactic streptococci and *Leuconostoc. J. Milk Food Technol.* **28**:273–277.

165. LaGrange, W. S., and G. W. Reinbold. 1968. Starter culture costs in Iowa Cheddar cheese plants. *J. Dairy Sci.* **51**:1985–1990.

166. LaGrange, W. S. 1987. Iowa's American Type cheese plant starter culture costs. *J. Dairy Sci.* **70**:367–372.

167. Nannen, N. L., and R. W. Hutkins. 1991. Intracellular pH effects in lactic acid bacteria. *J. Dairy Sci.* **74**:741–746.

168. Richardson, G. H., C. T. Cheng, and R. Young. 1977. Lactic bulk culture system utilizing a whey-based bacteriophage inhibitory medium and pH control. 1. Applicability to American style cheese. *J. Dairy Sci.* **60**:378–386.

169. Chen, Y. L., and G. H. Richardson. 1977. Lactic bulk culture system utilizing whey-based bacteriophage inhibitory medium and pH control. III. Applicability to cottage cheese manufacture. *J. Dairy Sci.* **60**:1252–1255.

170. Ausavanodom, N., R. S. White, G. Young, and G. H. Richardson. 1977. Lactic bulk culture system utilizing whey-based bacteriophage inhibitory medium and pH control. II. Reduction of phosphate requirements under pH control. *J. Dairy Sci.* **60**:1245–1251.

171. Wright, S. L., and G. H. Richardson. 1982. Optimization of whey-based or no fat dry milk-based media for production of pH controlled bulk lactic cultures. *J. Dairy Sci.* 65:1882–1889.

172. Mermelstein, N. H. 1982. Advanced bulk starter medium improves fermentaton processes. *Food Technol.* **36**:69–76.

173. Wilrett, D. L., W. E. Sandine, and J. W. Ayres. 1982. Evaluation of pH controlled starter media including a new product for Italian and Swiss-type cheeses. *Cult. Dairy Prod. J.* **17**:5–9.

174. Wigley, R. C. 1980. Advances in technology of bulk starter production and cheesemaking. *J. Soc. Dairy Technol.* **33**:24–30.

175. Fox, P. F. 1987. Cheese manufacture: chemical, biochemical and physical aspects. *Dairy Indust. Int.* **52**:11–13.

176. Wilson, H., and G. W. Reinbold. 1965. *American Cheese Varieties. Pfizer Cheese Monographs,* Vol. 2. Chas. Pfizer & Co., New York.

177. Wilster, G. H. 1980. *Practical Cheesemaking,* 13th edit. O. S. U. Book Stores, Corvalis, OR.

178. Van Slyke, L. L., and W. V. Price. 1979. *Cheese.* Ridgeview, CA.

179. Lawrence, R. C., J. Gilles, and L. K. Creamer. 1983. The relationship between cheese texture and flavor. *N. Z. J. Dairy Sci. Technol.* **18**:175–190.

180. Czulak, J., J. Conochie, B. J. Sutherland, and H. J. M. Van Leeuwen. 1969. Lactose, lactic acid and mineral equilibria in Cheddar cheese manufacture. *J. Dairy Res.* **36**:93–101.

181. Lawrence, R. C., and J. Gilles. 1987. Cheddar cheese and related dry-salted cheese varieties. *In* P. F. Fox (ed.), *Cheese: Chemistry, Physics and Microbiology,* Vol. 2. Elsevier Applied Science Publishers, London.

182. Robertson, P. S. 1966. Reviews of the progress of dairy science. Section B. Recent developments affecting the Cheddar cheese making process. *J. Dairy Res.* **33**:343–369.

183. Vedamuthu, E. R., and G. W. Reinbold. 1967. Starter cultures for Cheddar cheese. *J. Milk Food Technol.* **30**:247–252.

184. Reinbold, G. W. 1972. *Swiss Cheese Varieties. Pfizer Cheese Monographs,* Vol. V. Chas. Pfizer & Co., New York.

185. Reinbold, G. W. 1963. *Italian Cheese Varieties. Pfizer Cheese Monographs,* Vol. 1. Chas. Pfizer & Co., New York.

186. Oberg, C. J. 1991. Controlling body and texture of Mozzarella cheese: Microbiological and chemical methods. *Proceedings of Cheese Research & Technology Conference,* March 6–7, Madison, WI.

187. Breene, W. M., W. V. Price, and C. A. Ernstrom. 1964. Manufacture of Pizza cheese without starter. *J. Dairy Sci.* **47**:1173–1180.

188. Olson, N. F. 1969. *Ripened Semifort Cheeses. Pfizer Cheese Monographs,* Vol. IV. Chas. Pfizer & Co., New York.

189. Hickey, M. W. 1985. *Mold and Smear Cultures.* Australian Society of Dairy Technology, Technical Publication No. 29.

190. Morris, H. A. 1981. *Blue-Veined Cheeses. Pfizer Cheese Monographs,* Vol. VII. Chas. Pfizer & Co., New York.

191. El-Gazzar, F. E., and E. H. Marth. 1991. Ultrafiltration and reverse osmosis in dairy technology: a review. *J. Food Prot.* **54**:801–809.

192. Honer, C., and A. Horwich. 1983. Cheese and ultrafiltration: Where we are today. *Dairy Rec.* **84**:80–82.

193. Olson, N. F. 1983. Ultrafiltration and cheese manufacture. *Dairy Rec.* **84**:85–86.

194. Steen-Hansen, M. 1985. *The Alcurd for Continuous Cheesemaking Using UF-Retentate.* Australian Society of Dairy Technology, Technical Publication. No. 29.

195. Norman, L., and T. Garrett. 1987. The Sirocurd process for cheese manufacture. *J. Soc. Dairy Technol.* **40**:68–70.

196. Sood, V. K., and F. V. Kosikowski. 1979. Process Cheddar cheese from plain and enzyme treated retentate. *J. Dairy Sci.* **62**:1713–1718.

197. Kosikowski, F. V. 1980. Cheddar cheese from water reconstituted retentates. *J. Dairy Sci.* **63**:1975–1980.

198. DeBoer, R., and P. F. C. Nooy. 1980. Low-fat semi-hard cheese from ultrafiltered milk. *N. Eur. Dairy J.* **3**:52–61.

199. DeKoening, P. J., R. De Boer, P. Both, and P. F. C. Nooy. 1981. Comparison of proteolysis in a low-fat semi-hard type of cheese manufactured by standard and by ultrafiltration techniques. *Netherlands Milk Dairy J.* **35**:35–46.

200. Bush, C. S., C. A. Caroutte, C. H. Amundson, and N. F. Olson. 1982. Manufacture of Colby and Brick cheeses from ultrafiltered milk. *J. Dairy Sci.* **66**:415–421.

201. Anonymous. 1983. Production of Mozzarella cheese by ultrafiltration. *N. Eur. Dairy J.* **6**:165–169.

202. Fernandez, A., and F. V. Kosikowski. 1985. Low moisture Mozzarella cheese from whole milk retentates of ultrafiltration. *J. Dairy Sci.* **69**:2011–2017.

203. Sharma, S. K., L. K. Ferrier, and A. R. Hill. 1989. Effect of modified manufacturing parameters on the quality of Cheddar cheese made from ultrafiltered (UF) milk. *J. Food Sci.* **54**:573–577.

204. Green, M. L. 1990. The cheesemaking potential of milk concentrated up to four-fold by ultrafiltration and heated in the range 90–97°C. *J. Dairy Res.* **57**:549–557.

205. Green, M. L. 1990. Cheddar cheesemaking from whole milk concentrated by ultrafiltration and heated to 90°C. *J. Dairy Res.* **57**:559–569.

206. Fox, P. F. 1987. Significance of salt in cheese ripening. *Dairy Indust. Int.* **52**:19–22.

207. Geurts, T. J., P. Walstra, and H. Mulder. 1980. Transport of salt and water during salting of cheese. 2. Quantities of salt taken up and of moisture lost. *Netherlands Milk Dairy J.* **34**:229–254.

208. Marcas, A., M. Alcalá, F. León, J. Fernández-Salguero, and M. A. Estaban. 1981. Water activity and chemical composition of cheese. *J. Dairy Sci.* **64**:622–626.

209. Lowrie, R. S., and R. C. Lawrence. 1972. Cheddar cheese flavor IV. A new hypothesis to account for the development of bitterness. *N. Z. J. Dairy Sci. Technol.* **7**:51–53.

210. Lawrence, R. C., and J. Gilles. 1982. Factors that determine the pH of young Cheddar cheese. *N. Z. J. Dairy Sci. Technol.* **17**:1–14.

211. Lawrence, R. C., and J. Gilles. 1980. The assessment of the potential quality of young Cheddar cheese. *N. Z. J. Dairy Sci. Technol.* **15**:1–12.

212. Thomas, T. D., and K. N. Pearce. 1982. Influence of salt on lactose fermentation and proteolysis in Cheddar cheese. *N. Z. J. Dairy Sci. Technol.* **16**:253–259.

213. Walstra, P., and T. Van Vliet. 1986. The physical chemistry of curd making. *Netherlands Milk Dairy J.* **40**:241–259.

214. Webb, B. H., A. H. Jhonson, and J. R. Alford. 1974. *Fundamentals of Dairy Chemistry* 2nd edit. Avi, Westport, CT.

215. Nauth, K. R., J. T. Hynes, and R. D. Harris. 1991. Cheese. *In* Y. H. Hui (ed.), *Encyclopedia of Food Science & Technology,* Vol. 1. John Wiley & Sons, New York.

216. Schormuller, J. 1968. *The Chemistry and Biochemistry of Cheese Ripening. Advances in Food Research,* Vol. 16. Academic Press, New York and London.

217. Kilara, A., and K. M. Shahani. 1978. Lactic fermentations of dairy foods and their biological significance. *J. Dairy Sci.* **61**:1793–1800.

218. Clark, W. S. Jr., and G. W. Reinbold. 1967. The low temperature microflora of young Cheddar cheese. *J. Milk Food Technol.* **30**:54–58.

219. Stadhouders, J. 1959. Hydrolysis of protein during the ripening of Dutch cheese. *International Dairy Congress Proceedings* **2**:703–708.

220. O'Keefe, R. B., P. F. Fox, and C. Daly. 1976. Contribution of rennet and starter proteases to proteolysis in Cheddar cheese. *J. Dairy Res.* **43**:97–107.

221. Noomen, A. 1978. Activity of proteolytic enzymes in simulated soft cheeses (meshanger type). 1 Activity of milk protease. *Netherlands Milk Dairy J.* **32**:26–48.

222. Fox, P. F. 1989. Proteolysis during cheese manufacture and ripening. *J. Dairy Sci.* **72**:1379–1400.

223. Delfour, A., J. Jolles, C. Alais, and B. Jolles. 1965. Casein-glycopeptides: characterization of methionine residue and of the N-terminal sequence. *Biochem. Biophys. Res. commun.* **19**:452–455.

224. Ledford, R. A., J. H. Chen, and K. R. Nath. 1968. Degradation of casein fractions by rennet extract. *J. Dairy Sci.* **51**:792–794.

225. Hill, R. D., E. Lahav, and D. Givol. 1974. A rennin-sensitive bond in a β-casein. *J. Dairy Res.* **41**:147–153.

226. Visser, S. 1981. Proteolytic enzymes and their action on milk proteins. A review. *Netherlands Milk Dairy J.* **35**:65–88.

227. Creamer, L. K., and B. C. Richardson. 1974. Identification of the primary degradation product of α_{s1}-casein in Cheddar cheese. *N. Z. J. Dairy Sci. Technol.* **9**:9–13.

228. Pelissier, J. P., J. C. Mercier, and B. R. Dumas. 1974. Proteolysis of bovine α_{s1}- and β-casein by rennin. Proteolytic specificity of the enzyme and bitter peptides released. *Annis Biol. Anim. Biochem. Biophys.* **14**:343–362.

229. Mulvihill, D. M., and P. F. Fox. 1977. Proteolysis of α_{s1}-casein by chymosin: influence of pH and urea. *J. Dairy Res.* **44**:533–540.

230. Mulvihill, D. M., and P. F. Fox. 1979. Proteolytic specificity of chymosins and pepsins on β-caseins. *Milchwissenschaft* **34**:680–683.

231. Mulvihill, D. M., and P. F. Fox. 1980. Proteolysis of α_{s1}-casein by chymosin in dilute NaCl solutions and in Cheddar cheese. *Irish J. Food Sci. Technol.* **4**:13–23.

232. Mulvihill, D. M., and P. F. Fox. 1978. Proteolysis of β-casein by chymosin: influence of pH, urea and sodium chloride. *Irish J. Food Sci. Technol.* **2**:135–139.

233. Ledford, R. A., A. C. O'Sullivan, and K. R. Nath. 1966. Residual casein fractions in ripened cheese determined by polyacrylamide-gel electrophoresis. *J. Dairy Sci.* **49**:1098–1101.

234. O'Keefe, A. M., P. F. Fox, and C. Daly. 1978. Proteolysis in Cheddar cheese: role of coagulant and starter bacteria. *J. Dairy Res.* **45**:465–477.

235. Nath, K. R., and R. A. Ledford. 1973. Growth response of *Lactobacillus casei* variety *casei* to proteolysis products in cheese during ripening. *J. Dairy Sci.* **56**:710–715.

236. Green, M. L., and P. M. D. Foster. 1974. Comparison of the rates of proteolysis during ripening of Cheddar cheeses made with calf rennet and swine pepsin as coagulants. *J. Dairy Res.* **41**:269–282.

237. Law, B. A., and J. Kolstad. 1983. Proteolytic system in lactic acid bacteria. *Antonie Van Leeuwenhoek* **49**:225–245.

238. Thomas, T. D., and O. M. Mills. 1981. Proteolytic enzymes of starter bacteria. *Netherlands Milk Dairy J.* **35**:255–273.

239. Thomas, T. D., and G. G. Pritchard. 1987. Proteolytic enzymes of dairy starter cultures. *FEMS Microbiol. Rev.* **46**:245–268.

240. Kok, J. 1990. Genetics of the proteolytic system of lactic acid bacteria. *FEMS Microbiol. Rev.* **87**:15–42.

241. Khalid, N. M., and E. H. Marth. 1990. Lactobacilli—their enzymes and role in ripening and spoilage: a review. *J. Dairy Sci.* **73**:2669–2684.

242. Kamaly, K. M., and E. H. Marth. 1989. Enzyme activities of lactic streptococci and their role in maturation of cheese: a review. *J. Dairy Sci.* **72**:1945–1966.

243. El-Soda, M., M. Korayem, and N. Ezzat. 1986. The esterolytic and lipolytic activities of lactobacilli III. Detection and characterization of the lipase system. *Milchwissenschaft* **41**:353–355.

244. Yu, J. H. 1986. Studies on the extracellular and intracellular lipase of *Lactobacillus casei* 1. On the patterns of free fatty acids liberated from milk fat reacted with the lipases. *Korean J. Dairy Sci.* **8**:167–177.

245. Olson, N. F. 1990. The impact of lactic acid bacteria on cheese flavor. *FEMS Microbiol. Rev.* **87**:131–148.

246. Thomas, T. D. 1987. Cannibalism among bacteria found in cheese. *N. Z. J. Dairy Sci. Technol.* **22**:215–219.

247. Ohmiya, K., and Y. Sato. 1969. Studies on the proteolytic action of dairy lactic acid bacteria. Part IX. Autolysis and proteolytic action of *Streptococcus cremoris* and *Lactobacillus helveticus*. *Agric. Biol. Chem.* **23**:1628–1635.

248. Law, B. A., M. E. Sharpe, and B. Reiter. 1974. The release of intracellular dipeptidase from starter streptococci during Cheddar cheese ripening. *J. Dairy Res.* **41**:137–146.

249. Ohmiya, K., and Y. Sato. 1970. Studies on the proteolytic action of dairy lactic acid bacteria. Part X. Autolysis of lactic acid bacterial cells in aseptic rennet curd. *Agric. Biol. Chem.* **34**:457–463.

250. Bie, R., and G. Sjostrom. 1975. Autolytic properties of some lactic acid bacteria used in cheese production. Part II. Experiments with fluid substrates and cheese. *Milchwissenschaft* **30**:739–747.

251. Langsrud, T., A. Landaas, and H. B. Castberg. 1987. Autolytic properties of different strains of group N. streptococci. *Milchwissenschaft* **42**:556–560.

252. Feirtag, J. M., and L. L. McKay. 1987. Isolation of *Streptococcus lactis* C_2 mutants selected for temperature sensitivity and potential use in cheese manufacture. *J. Dairy Sci.* **70**:1773–1778.

253. Amantea, G. F., B. J. Skura, and S. Nakai. 1986. Culture effect on ripening characteristics and rheological behavior of Cheddar cheese. *J. Food Sci.* **51**:912–918.

254. Nasr, M. M., and N. A. Younis. 1986. Effect of adding alanine, phenylalanine and proline on the properties of Romi cheese. *Egypt. J. Food Sci.* **14**:385–390.

255. Marsili, R. 1985. Monitoring chemical changes in cheddar cheese during aging by high performance liquid chromatography and gas chromatography techniques. *J. Dairy Sci.* **68**:3155–3161.

256. Kaminogawa, S., T. R. Yan, N. Azuma, and K. Yamauchi. 1986. Identification of low molecular weight peptides in Gouda-type-cheese and evidence for the formation of these peptides from 23N-terminal residues of α_{s1}-casein by proteinases of *Streptococcus cremoris* H61. *J. Food Sci.* **51**:1253–1256, 1264.

257. Petterson, H. E., and G. Sjostrom. 1975. Accelerated cheese ripening: a method for increasing the number of lactic starter bacteria in cheese without detrimental effect to the cheese makeup process, and its effect on cheese ripening. *J. Dairy Res.* **42**:313–326.

258. Law, B. A., and M. E. Sharpe. 1977. The influence of the microflora of the Cheddar cheese on flavor development. *Dairy Indust. Int.* **42**:10–14.

259. Mulder, H. 1952. Taste and flavor forming substances in cheese. *Netherlands Milk Dairy J.* **6**:157–168.

260. McGugan, W. A. 1975. Cheddar cheese flavor. A review of current progress. *J. Agric. Food Chem.* **23**:1047–1050.

261. Reiter, B., and M. E. Sharpe. 1971. Relationship of the microflora to the flavor of Cheddar cheese. *J. Appl. Bacteriol.* **34**:63–80.

262. Ohren, J. A., and S. L. Tuckey. 1969. Relation of flavor development in Cheddar cheese to chemical changes in the fat of the cheese. *J. Dairy Sci.* **52**:598–607.

263. Kristofferson, T. 1973. Biogenesis of cheese flavor. *J. Agric. Food Chem.* **21**:573–575.

264. Manning, D. J. 1974. Sulfur compounds in relation to Cheddar cheese flavor. *J. Dairy Res.* **41**:81–87.

265. Harvey, R. J. and J. R. L. Walker. 1960. Some volatile compounds in New Zealand Cheddar cheese and their possible significance in flavor formation. *J. Dairy Res.* **27**:335–340.

266. Maning, D. J. 1979. Cheddar cheese flavor studies. *J. Dairy Res.* **46**:523–529.

267. McGugan, W. A., D. B. Emmons, and E. Lammond. 1979. Influence of volatile and non-volatile fractions on intensity of Cheddar flavor. *J. Dairy Sci.* **62**:398–402.

268. Aston, J. W., and L. K. Creamer. 1986. Contribution of the components of the water-soluble fraction to the flavor of Cheddar cheese. *N. Z. J. Dairy Sci. Technol.* **21**:229–248.

269. Aston, J. W., and J. R. Dulley. 1982. Cheddar cheese flavor. *Aust. J. Dairy Technol.* **37**:59–64.

270. Aston, J. W., and K. Douglas. 1983. The production of volatile sulfur compounds in Cheddar cheese during accelerated ripening. *Aust. J. Dairy Technol.* **38**:66–70.

271. Lloyd, G. T., and E. H. Ramshaw. 1985. Objective assessment of flavor development during maturation of cheese. Australian Society of Dairy Technology Technical Publication No. 29.

272. Urbach, G. 1982. The effect of different feeds on the lactone and methyl ketone precursors of milk fat. *Lebensm.-Wiss. U. Technol.* **15**:62–67.

273. Urbach, G. 1990. Headspace volatiles from cold-stored raw milk. *Aust. J. Dairy Technol.* **45**:80–85.

274. Liardon, R., J. O. Bosset, and B. Blanc. 1982. The aroma composition of Swiss Gruyère cheese. 1. The alkaline volatile components. *Lebensm.-Wiss. U. Technol.* **15**:143–147.

275. Bosset, J. O., and R. Liardon, 1984. The aroma composition of Swiss Gruyère cheese II. The neutral volatile components. *Lebensm.-Wiss. U. Technol.* **17**:359–362.

276. Bossett, J. O., and R. Liardon. 1985. The aroma composition of Swiss Gruyère cheese III. Relative changes in the content of alkaline and neutral volatile components during ripening. *Lebensm.-Wiss. U. Technol.* **18**:175–185.

277. Vandeweghe, P., and G. A. Reineccius. 1990. Comparison of flavor isolation techniques applied to Cheddar cheese. *J. Agric. Food Chem.* **38**:1549–1552.

278. Barlow, J., G. T. Lloyd, E. H. Ramshaw, A. J. Miller, G. P. McCabe, and L. McCabe. 1989. Corrections and changes in flavor and chemical parameters of Cheddar cheeses during maturation. *Aust. J. Dairy Technol.* **44**:7–18.

279. Foster, E. M., F. E. Nelson, R. N. Doetsch, and J. C. Olson Jr. 1957. *Dairy Microbiology.* Prentice-Hall, Englewood Cliffs, NJ.

280. Harvey, C. D., R. Jenness, and H. A. Morris. 1981. Gas chromatographic quantitation of sugars and nonvolatile water-soluble organic acids in commercial Cheddar cheese. *J. Dairy Sci.* **64**:1648–1654.

281. Huffman, L. M., and T. Kristofferson. 1984. Role of lactose in Cheddar cheese manufacturing and ripening. *N. Z. J. Dairy Sci. Technol.* **19**:151–162.

282. Pearce, K. N. 1982. The effect of salt on the rate of proteolysis of casein in Cheddar cheese. *Proc. 21st Int. Dairy Congr.* Moscow, Vol. 1, Book 1, p. 519.

283. Fox, P. F., and B. F. Walley. 1971. Influence of sodium chloride on the proteolysis of casein by rennet and by pepsin. *J. Dairy Res.* **38**:165–174.

284. Richardson, B. C., and K. N. Pearce. 1982. The determination of plasmin in dairy products. *N. Z. J. Dairy Sci. Technol.* **16**:209–220.

285. Thomas, T. D., and K. N. Pearce. 1981. Influence of salt on lactose fermentation and proteolysis in Cheddar cheese. *N. Z. J. Dairy Sci. Technol.* **16**:253–259.

286. Creamer, L. K., and B. C. Richardson. 1974. Identification of the primary degradation product of α_{s1}-casein in Cheddar cheese. *N. Z. J. Dairy Sci. Technol.* **9**:9–13.

287. Phelan, J. A., T. Guinee, and P. F. Fox. 1973. Proteolysis of β-casein in Cheddar cheese. *J. Dairy Res.* **40**:105–112.

288. Marcos, A., M. A. Esteban, F. Leon, and J. Fernandez-Salguero. 1979. Electrophoretic pattern of European cheese: comparison and quantitation. *J. Dairy Sci.* **62**:892–900.

289. Farkye, N. Y., and P. F. Fox. 1991. Preliminary study on the contribution of plasmin to proteolysis in Cheddar cheese. Cheese containing plasmin inhibitor, 6-aminohexanoic acid. *J. Agric. Food Chem.* **39**:786–788.

290. Creamer, L. K., and N. F. Olson. 1982. Rheological evaluation of maturing Cheddar cheese. *J. Food Sci.* **47**:632–636, 646.

291. Franklin, J. G., and M. E. Sharpe. 1963. The incidence of bacteria in cheese milk and Cheddar cheese and their association with flavor. *J. Dairy Res.* **30**:87–99.

292. Vegarud, G., H. B. Castberg, and T. Langsrud. 1983. Autolysis of group N. streptococci. Effect of media composition modifications and temperature. *J. Dairy Sci.* **66**:2294–2302.

293. Law, B. A., M. J. Castanon, and M. E. Sharpe. 1976. The contribution of starter streptococci to flavor development in Cheddar cheese. *J. Dairy Res.* **43**:301–311.

294. Peterson, S. D., and R. T. Marshall. 1989. Non-starter lactobacilli in Cheddar cheese: a review. *J. Dairy Sci.* **73**:1395–1410.

295. Allen, L. A., and N. R. Knowles. 1934. Studies in the ripening of Cheddar cheese. *J. Dairy Res.* **5**:185–196.

296. Mabbit, L. A., and M. Zielinska. 1955. The importance of lactobacilli in the production of Cheddar cheese flavor. 1. Growth of lactobacilli in cheese serum. *J. Dairy Res.* **22**:377–383.

297. Tokita, F., and T. Nakanishi. 1964. The isolation and identification of a peptide with a growth factor for *Lactobacillus casei* in Edam cheese. *Milchwissenschaft* **19**:521–524.

298. Hickey, M. W., A. J. Hillier, and G. R. Jago. 1983. Peptidase activities in lactobacilli. *Aust. J. Dairy Technol.* **38**:118–123.

299. Peters, V. J., J. M. Prescott, and E. E. Snell. 1953. Peptides and bacterial growth. IV. Histidine peptides as growth factors for *Lactobacillus delbrueckii*. *J. Biol. Chem.* **202**:521–532.

300. Mabbit, L. A., and M. Zielinska. 1956. The use of a selective medium for the enumeration of lactobacilli in Cheddar cheese. *J. Appl. Bacteriol.* **19**:95–101.

301. Thomas, T. D. 1986. Oxidative activity of bacteria from Cheddar cheese. *N. Z. J. Dairy Sci. Technol.* **21**:37–47.

302. Davis, J. G. 1935. Studies in Cheddar cheese. IV observations on the lactic flora of Cheddar cheese made from clean milk. *J. Dairy Res.* **6**:175–190.

303. Fryer, T. F. 1970. Utilization of citrate by lactobacilli isolated from dairy products. *J. Dairy Res.* **37**:9–15.

304. Cantoni, C., and M. R. Molnar. 1967. Investigations on the glycerol metabolism of lactobacilli. *J. Appl. Bacteriol.* **30**:197–205.

305. Branan, A. L., and T. W. Keenen. 1969. Growth stimulation of *Lactobacillus* species by lactic streptococci. *Appl. Microbiol.* **17**:280–285.

306. Puchades, R., L. Lemieux, and R. E. Simard. 1989. Evolution of free amino acids during the ripening of Cheddar cheese containing lactobacilli strains. *J. Food Sci.* **54**:885–886, 946.

307. Lee, B. H., L. C. Laleye, R. E. Simrad, R. A. Holley, D. B. Emmons, and R. N. Giroux. 1990. Influence of homofermentative lactobacilli on physiochemical and sensory properties of Cheddar cheese. *J. Food Sci.* **55**:386–390.

308. Lee, B. H., L. C. Laleye, R. E. Simard, M. H. Munsch, and R. A. Holley. 1990. Influence of homofermentative lactobacilli on the microflora and soluble nitrogen components in Cheddar cheese. *J. Food Sci.* **55**:391–397.

309. Broome, M. C., D. A. Krause, and M. W. Hickey. 1990. The use of non-starter lactobacilli in Cheddar cheese manufacture. *Aust. J. Dairy Technol.* **45**:67–73.

310. Laleye, L. C., R. E. Simard, C. Gosselin, B. H. Lee, and R. N. Giroux. 1987. Assessment of Cheddar cheese quality by chromatographic analysis of free amno acids and biogenic amines. *J. Food Sci.* **52**:303–307, 311.

311. Patton, S. 1963. Volatile acids and the aroma of Cheddar cheese. *J. Dairy Sci.* **46**:856–858.

312. Forss, D. A., and S. Patton. 1979. Review of the progress of dairy science: mechanism of formation of aroma compounds in milk and milk products. *J. Dairy Res.* **46**:691–705.

313. Forss, D. A., and S. Patton. 1966. Flavor of Cheddar cheese. *J. Dairy Sci.* **49**:89–91.

314. Law, B. A., M. J. Castanon, and M. E. Sharpe. 1976. The effect of non-starter bacteria on the chemical composition and the flavor of Cheddar cheese. *J. Dairy Res.* **43**:117–125.

315. Law, B. A. 1982. Flavor compounds in cheese. *Perfumer Flavorist* **7**:9–21.

316. Stadhouders, J., and H. A. Veringa. 1973. Fat hydrolysis by lactic acid bacteria in cheese. *Netherlands Milk Dairy J.* **27**:77–91.

317. Nakae, T., and Y. A. Elliot. 1965. Volatile fatty acids produced by some lactic acid bacteria. 1. Factors influencing production of volatile fatty acids from casein hydrolysate. *J. Dairy Sci.* **48**:287–292.

318. Dulley, J. R., and P. A. Grieve. 1974. Volatile fatty acid production in Cheddar cheese. *Aust. J. Dairy Technol.* **29**:120–123.

319. Singh, A., R. A. Srinivasan, and A. T. Dudani. 1973. Studies on exocellular and endocellular lipases of some of the lipolytic bacteria. *Milchwissenschaft* **28**:164–166.

320. Moskowitz, G. J., and S. S. Noeleck. 1987. Enzyme modified cheese technology. *J. Dairy Sci.* **70**:1761–1769.

321. Manning, D. J., and H. M. Robinson. 1973. The analysis of volatile substances associated with Cheddar cheese aroma. *J. Dairy Res.* **40**:63–75.

322. Manning, D. J., and H. E. Nursten. 1985. Flavor of milk and milk products. *In Developments in Dairy Chemistry,* P. F. Fox (ed.), Vol. 3. Elsevier Applied Science Publishers, London.

323. Keen, A. R., and N. J. Walker. 1974. Diacetyl, acetoin, 2,3-butyleneglycol, 2-butanone and 2-butanol concentrations in ripening Cheddar cheese. *J. Dairy Res.* **41**:65–71.

324. Farrer, K. T. H., and K. J. Weeks. 1970. Some aspects of chemistry of cheese. *Food Technol. Aust.* **22**:620–621, 623.

325. Horwood, J. F. 1989. Headspace analysis of cheese. *Aust. J. Dairy Technol.* **44**:91–96.

326. Hutkins, R. W., and H. A. Morris. 1987. Carbohydrate metabolism by *Streptococcus thermophilus:* a review. *J. Food Prot.* **50**:876–884.

327. Gilles, J., K. W. Turner, and F. G. Martley. 1983. Swiss-type cheese. 1. Manufacturing and sampling procedures. *N. Z. J. Dairy Sci. Technol.* **18**:109–115.

328. Auclair, J., and J. P. Accolas. 1983. Use of thermophilic lactic starters in the dairy industry. *Antonie Van Leeuwenhoek* **49**:313–316.

329. Nath, K. R., and B. J. Kostak. 1986. Etiology of white spot defect in Swiss cheese made from pasteurized milk. *J. Food Prot.* **49**:718–723.

330. Turner, K. W., H. A. Morris, and F. G. Martley. 1983. Swiss type cheese. II. The role of thermophilic lactobacilli in sugar fermentation. *N. Z. J. Dairy Sci. Technol.* **18**:117–123.

331. Langsrud, T., and G. W. Reinbold. 1973. Flavor development and microbiology of Swiss cheese. A review. III. Ripening and flavor production. *J. Milk Food Technol.* **36**:593–609.

332. Wood, H. G. 1981. Metabolic cycles in the fermentation by propionic acid bacteria. *Curr. Top. Cell. Regul.* **18**:255–287.

333. Seuvre, A. M., and M. Mathlouthi. 1982. Contribution to the study of gas release during the maturation of a French Emmental cheese. *Lebensm.-Wiss. U. Technol.* **15**:258–262.

334. Crow, V. L., and K. W. Turner. 1986. The effect of succinate production on other fermentation products in Swiss-type cheese. *N. Z. J. Dairy Sci. Technol.* **21**:217–227.

335. Kaneuchi, C., M. Seki, and K. Komagata. 1988. Production of succinic acid from citric acid and related acids by *Lactobacillus* strains. *Appl. Environ. Microbiol.* **54**:3053–3056.

336. Thomas, T. D. 1987. Acetate production from lactate and citrate by non-starter bacteria in Cheddar cheese. *N. Z. J. Dairy Sci. Technol.* **22**:25–38.

337. Crow, V. L. 1986. Metabolism of aspartate by *Propionibacterium freudenreichii* subsp. *shermanii*: effect on lactate fermentation. *Appl. Environ. Microbiol.* **52**:359–365.

338. Crow, V. L. 1987. Properties of alanine dehydrogenase and aspartase from *Propionibacterium freudenreichii* subsp. *shermanii. Appl. Environ. Microbiol.* **53**:1885–1892.

339. Brendehaug, J., and T. Langsrud. 1983. Amino acid metabolism in propionibacteria: resting cells experiments with four strains. *J. Dairy Sci.* **26**:281–289.

340. Fluckiger, E. 1980. Formation of CO_2 and eyes in Emmental cheese. *Schweiz. Milchzeit.* 106:473–474.

341. Fluckiger, E. 1980. Formation of CO_2 and eyes in Emmental cheese. *Schweiz. Milchzeit.* 106:479–480.

342 Rousseau, M., and C. Le Gallo. 1990. Scanning electron microscopic study of the structure of Emmental cheese during manufacture. *Le Lait.* **70**:55–66.

343. Ollikainen, P., and K. Nyberg. 1988. A study of plasmin activity during ripening of Swiss-type cheese. *Milchwissenschaft* **43**:497–499.

344. Langsrud, T., and G. W. Reinbold. 1973. Flavor development and microbiology of Swiss cheese—A review. 1. Milk quality and treatments. *J. Milk Food Technol.* **36**:487–490.

345. Mitchell, G. E. 1981. The production of selected compounds in a Swiss-type cheese and their contribution to cheese flavor. *Aust. J. Dairy Technol.* **36**:21–25.

346 Biede, S. L., and E. G. Hammond, 1979. Swiss cheese flavor: 1. Chemical analysis. *J. Dairy Sci.* **62**:227–237.

347. TenBrink, B., C. Damink, H. M. L. J. Joosten, and J.. H. J. Huis in't Veld. Occurrences and formation of biologically active amines in foods. *Int. J. Food Microbiol.* **11**:73–84.

348. Olson, N. F. 1969. *Ripened Semi-Soft Cheese. Pfizer Cheese Monographs*, Vol. IV. Chas. Pfizer & Co., New York.

349. Walstra, P., A. Noomen, and T. J. Guerts. 1987. Dutch type varieties. *In* P. F. Fox (ed.), *Cheese: Chemistry, Physics and Microbology*, Vol. 2. Elsevier Applied Science Publishers, London. England.

350. Langeveld, L. P. M., and T. E. Galesloot. 1971. Estimation of the oxidation–reduction potential as an aid in tracing the cause of excessive openness in cheese. *Netherlands Milk Dairy J.* **25**:15 – 23.

351. Creamer, L. K. 1970. Protein breakdown in Gouda cheese. *N. Z. J. Dairy Sci. Technol.* **5**:152–154.

352. Van den Berg, G., and P. J. De Konings. 1990. Gouda cheesemaking with purified calf chymosin and microbiologically produced chymosin. *Netherlands Milk Dairy J.* **44**:189–205.

353. Stadhoudors, J., and G. Hup. 1975. Factors affecting bitter flavor in Gouda cheese. *Netherlands Milk Dairy J.* **29**:335–353.

354. Visser, F. M. W. 1977. Contribution of enzymes from rennet, starter bacteria and milk to proteolysis and flavor development in Gouda cheese. 1. Description of cheese and aseptic cheesemaking techniques. *Netherlands Milk Dairy J.* **31**:120–133.

355. Kleter, G. 1977. The ripening of Gouda cheese made under strictly aseptic conditions. 2. The comparison of activity of different starters and the influence of certain *Lactobacillus* strains. *Netherlands Milk Dairy J.* **31**:177–187.

356. Badings, H. T. 1984. Flavors and off-flavors. *In* P. Walstra and R. Jenness (eds)., *Dairy Chemistry and Physics*, p. 336. John Wiley & Sons, New York.

357. Badings, H. T., and R. Neeter. 1980. Recent advances in the study of aroma compounds of milk and dairy products. *Netherlands Milk Dairy J.* **34**:9–30.

358. Sloot, D., and P. D. Harker. 1975. Volatile trace components in Gouda cheese. *J. Agric. Food Chem.* **67**:960–968.

359. Coghill, D. 1979. The ripening of blue vein cheese: a review. *Aust. J. Dairy Technol.* **34**:72–75.

360. Hartley, C. B., and J. J. Jezeski. 1954. The microflora of Blue cheese slime. *J. Dairy Sci.* **37**:436–445.

361. Kinsella, J. E., and D. Hwang. 1976. Biosyntesis of flavors by *Penicillium roqueforti. Biotechnol. Bioengin.* **18**:927–938.

362. Gripon, J. C. 1987. Mold-ripened cheeses. *In* P. F. Fox (ed.), *Cheese: Chemistry, Physics and Microbiology.* Vol. 2. Elsevier Applied Science Publishers, London.

363. Trieu-cuot, P., and J. C. Gripon. 1982. A study of proteolysis during Camembert cheese ripening using iso-electric focusing and two-dimensional electrophoresis. *J. Dairy Res.* **49**:501–510.

364. Gripon, J. C., and J. Bergere. 1972. Le systèm protéolytique de *Penicillium roqueforti.* 1. Conditions de production et nature du systemes protéolytique. *Le Lait* **52**:497–514.

365. Gueguen, M., and J. Lenoir. 1976. Caractères du système protéolytique de *Geotrichum candidum. Le Lait* **56**:439–448.

366. Friedman, M. E., W. O. Nelson, and W. A. Wood. 1953. Proteolytic enzymes from *Bacterium linens. J. Dairy Sci.* **36**:1124–1134.

367. King. R. D., and G. H. Clegg. 1979. The metabolism of fatty aids, methylketones and secondary alcohols by *Penicillium roqueforti* in Blue cheese slurries. *J. Sci. Food Agric.* **30**:197–202.

368. Yamaguchi, S., and T. Mase. 1991. Purification and characterization of mono- and diacylglycereol lipase isolated from *Penicillium camemberti* U-150. *Appl. Microbiol. Biotechnol.* **30**:720–725.

369. Hwang, D. H., Y. J. Lee, and J. E. Kinsella. 1976. β-ketoacyl decarboxylase activity in spores and mycelium of *Penicillium roqueforti. Int. J. Biochem.* **7**:165–171.

370. Franke, W., A. Platzeck, and G. Eichhorn. 1961. Zur Kenntnis des fettsaureabbaus durch schmmelpilze. *Arch. Microbiol.* **40**:73–93.

371. Lawrence, R. C., and J. C. Hawke. 1968. The oxidation of fatty acids by mycelium of *Penicillium roqueforti. J. Gen. Microbiol.* **51**:289–302.

372. Wong, N. P., R. Ellis, L. A. LaCroix, and J. A. Alford. 1973. Lactones in Cheddar cheese. *J. Dairy Sci.* **56**:636.

373. Jolly, R. C., and F. V. Kosikowski. 1975. Quantification of lactones in ripening pasteurized milk Blue cheese containing added microbial lipases. *J. Agric. Food Chem.* **23**:1175–1176.

374. Law, B. A. 1984. Microorganisms and their enzymes in the maturation of cheeses. *In* M. E. Bushell (ed.), *Progress in Industrial Microbiology*, Vo. 19. Elsevier, New York.

375. Choisy, C., M. Desmazeaud, J. C. Gripon, G. Lamberet, J. Lenoir, and C. Tourneur. 1986. Microbiological and biochemical aspects of ripening. *In* A. Eck (ed.), *Cheesemaking: Science and Technology.* Lavoisier, Paris.

376. Adda, J. 1986. Flavor formation. *In* A. Eck (ed.), *Cheesemaking: Science and Technology*, p. 336. Lavoisier, Paris.

377. Garey, J. C., E. M. Foster, and W. C. Frazier. 1941. The bacteriology of Brick cheese. 1. Growth and activity of starter bacteria during manufacture. *J. Dairy Sci.* **24**:1015–1025.

378. Langhus, W. L., W. Y. Price, H. H. Sommer, and W. C. Frazier. 1945. The "smear" of Brick cheese and its relation to flavor development". *J. Dairy Sci.* **28**:827–838.

379. Lubert, D. J., and W. C. Frazier. 1955. Microbiology of the surface ripening of Brick cheese. *J. Dairy Sci.* **38**:981–990.

380. Iya, K. K., and W. C. Frazier. 1949. The yeast in the surface smear of Brick cheese. *J. Dairy Sci.* **32**:475–476.

381. Szumski, S. A., and J. F. Cone. 1962. Possible role of yeast endoproteinases in ripening surface-ripened cheeses. *J. Dairy Sci.* **45**:349–353.

382. Purko, M., W. O. Nelson, and W. A. Wood. 1951. The associative action between certain yeasts and *Bacterium linens*. *J. Dairy Sci.* **34**:699–705.

383. Ades, G. L., and J. F. Cone. 1969. Proteolytic activity of *Brevibacterium linens* during ripening of Trappist-type cheese. *J. Dairy Sci.* **52**:957–961.

384. Mulder, E. G., A. D. Adamse, J. Antheunisse, M. H. Deinema, J. W. Woldendorf, and L. P. T. M. Zeventhuizen. 1966. The relationship between *Brevibacterium linens* and bacteria of the genus *Arthrobacter*. *J. Appl. Bacteriol.* **29**:44–71.

385. Foissy, H. 1978. Amino peptidase from *Brevibacterium linens*: production and purification. *Milchwissenschaft* **33**:221–223.

386. Sharpe, M. E., B. A. Law, B. A. Phillips, and D. G. Pilcher. 1977. Methanethiol production by coryneform bacteria: strains from dairy and human skin sources and *Brevibacterium linens*. *J. Gen. Microbiol.* **101**:345–349.

387. Parliment, T. H., M. G., Kolor, and D. J. Rizzo. 1982. Volatile components of Limburger cheese. *J. Agric. Food Chem.* **30**:1006–1008.

388. Cervantes, M. A., D. B. Lund, and N. F. Olson. 1983. Effects of salt concentration and freezing on Mozzarella cheese texture. *J. Dairy Sci.* **66**:204–213.

389. Masi, P. and F. Addeo. 1986. An examination of some mechanical properties of a group of Italian cheeses and their relation to structure and conditions of manufacture. *J. Food Engin.* **5**:217–229.

390. Thunell, R. K. 1989. Culture performance in Mozzarella cheesemaking. Twenty-sixth Marschall Italian Cheese Seminar, Madison, WI.

391. Creamer, L. K. 1976. Casein proteolysis in Mozzarella-type cheese. *N. Z. J. Dairy Sci. Technol.* **11**:130–135.

392. Woo, A. H., and R. C. Lindsay. 1984. Concentration of major free fatty acids and flavor development in Italian cheese varieties. *J. Dairy Sci.* **67**:960–968.

393. Marth, E. H., and T. L. Thompson. 1979. Bacterial changes in parmesan cheese during ripening. Marschall International Cheese Conference, pp. 21–33.

394. Fox, P. F., and T. P. Guinee. 1987. Italian cheeses. *In* P. F. Fox (ed.), *Cheese: Chemistry, Physics and Microbiology*, Vol. 2. Elsevier Applied Science Publishers, London.

395. Conner, T. 1988. Advances in accelerated ripening of cheese. *Cult. Dairy Prod. J.* **23**:21–25.

396. Code of Federal Regulations. Title 21. Office of Federal Register, National Archives and Records. 21 CFR 100.120.

397. Thomas, M. A. 1977. *The Processed Cheese Industry*. Dept. of Agriculture, Sydney, New South Wales, Australia.

398. Caric, M., and M. Kalab. 1987. Processed cheese products. *In* P. F. Fox (ed.), *Cheese: Chemistry, Physics and Microbiology*, Vol. 2. Elsevier Applied Science Publishers, London.

399. Caric, M., M. Gantor, and M. Kalab. 1985. Effects of emulsifying agents on the microstructure and other characteristics of process cheese—a review. *Food Microstruct.* **4**:297–312.

400. Shimp, L. A. 1985. Process cheese principles. *Food Technol.* **39**:63–70.

401. Gupta, S. K., C. Karahadian, and R. C. Lindsay. 1984. Effect of emulsifier salts on textural and flavor properties of processed cheese. *J. Dairy Sci.* **67**:764–778.

402. Lee, B. O., D. Paquet, and C. Alais. 1986. Biochemical study of cheese melting. IV. Effect of melting salts and proteins on peptization. Use of a model system. *Le Lait* **66**:257–267.

403. Molins, R. A. 1991. *Phosphates in Food*. CRC Press, Boca Raton, FL.

404. Templeton, H. L., and H. H. Sommers. 1936. Studies on emulsifying salts used in process cheese. *J. Dairy Sci.* **19**:561–572.

405. Palmer, H. J., and W. H. Sly. 1944. Cheese melting salt and their properties. *J. Soc. Chem. Indust.* **63**:363.

406. Price, W. V., and M. G. Bush. 1974. The process cheese industry in the United States: a review II. Research and Development. *J. Milk Food Technol.* **37**:179–198.

407. Rayan, A. A., M. Kalab, and C. A. Ernstrom. 1980. Microstructure and rheology of process cheese. *Scan. Electron Microsc.* **III**:635–643.

408. Taneya, S., T. Kimura, T. Izutsu, and W. Buchheim. 1980. The submicroscopic structure of processed cheese with different melting properties. *Milchwissenschaft* **35**:479–481.

409. Caric, M., L. Kulik, D. Gaverie, B. Pejie, M. Stipetic and I. Bebic. 1989. *Mljekarstvo* **39**:95. Cited in *Dairy Indust. Int.* 1990. **55**:12–13.

410. Mashali, R. I. 1987. Alexandria. *J. Agric. Res.* **32**:191. Cited in *Dairy Indust. Int.* 1990. **55**:12–13.

411. Tatsumi, K., T. Nishiya, H. Yamamoto, K. Ido, N. Hanawa, K. Ito and K. Tamaki. 1989. *Rep. Res. Lak. Snow Brand Milk Prod.* Co. No. 88 73. Cited in *Dairy Indust. Int.* 1990. **55**:12–13.

412. Blond, G., E. Haury and D. Lorient. 1988. *Sci. Aliments* **8**:325. *Dairy Indust. Int.* 1990 **55**:12–13.

413. Anonymous 1989. Process (Rennes) No. 104047. Cited in *Dairy Indust. Int.* 1990 **55**:12–13.

414. Tada, M., I. Shinoda, and H. Okai. 1984. L-Ornithyltaurine, a new salty peptide. *J. Agric. Food Chem.* **32**:992–996.

415. Tanaka, N. 1982. Challenge of pasteurized process cheese spreads with *Clostridium botulinum* using in-process and post-process inoculation. *J. Food Prot.* **45**:1044–1050.

416. Tanaka, N., E. Traisman, P. Plantinga, L. Finn, W. Flom, L. Meske, and J. Guggisberg. 1986. Evaluation of factors involved in antibotulinal properties of pasteurized process cheese spreads. *J Food Prot.* **49**:526–531.

CHAPTER

4

Concentrated and Dried Dairy Products

Marijana Carić

4.1 History and Definitions

One of the first preservation methods developed, milk drying is centuries old. Primitive cultures used the sun's energy to concentrate and dry milk. Records indicate that the Japanese manufactured concentrated milks as far back as the 7th century, while Marco Polo, in his wanderings in the 13th century, described a product considered to be milk powder.[1]

The real beginning of the concentrated and dried dairy industry began in the 19th century when Nicholas Appert, a French inventor, described his procedure for concentrating and drying milk. Concentrating milk consisted of evaporation to two thirds volume in an open kettle, filling bottles, and heating them in a water bath for 2 h.[2] These reports indicated a quality product after 18 months of storage. Further manipulation of the concentrated milk produced dried milk.

Malbec (in 1826) and Newton (in 1835) attempted to prolong shelf life of concentrated milk by adding sugar. In 1856 Borden produced condensed milk industrially by applying a partial vacuum evaporation process. This was the first widespread marketing of condensed milk. Meanwhile, the English (Grimwade in 1856) began commercial production of milk powder with a patent that used Na_2CO_3 (K_2CO_3) and sucrose.[1] The production of milk powder without additives started at the end of the 19th century (1898) after many previous attempts. Percy in 1872 was granted a patent in the United States in which he described the principle of spray drying and he is considered to be the inventor.[3] Stauff's patent in 1901 formed the basis of the first industrial spray drying equipment. His patent was purchased by the Merril Soul Company in 1905.[3] At about the same time, roller drying equipment was developed for industrial application.

Numerous investigations and inventions in the following years produced more sophisticated technology that greatly improved the qualities of concentrated and dried dairy products. One of the important innovations in technology and quality was instantization.

The instantization process, patented by Peebles in 1955,[1–3] has significantly improved the quality and economical aspects of drying technology. The process is characterized by a two-stage drying that causes agglomeration.

The availability of new dry dairy products with better quality and lower manufacture cost was realized in the 1970s and 1980s by several developments. These industrial applications included concentrating and fractionating by membrane processes, ultrafiltration, reverse osmosis, and electrodialysis or ion exchange. Further enhancement was achieved in 1983 when a three-stage drying procedure with integrated fluid bed was introduced.

Concentrated and dried dairy products are milk products with an extended shelf life. Concentrated milk products are obtained by partial water removal, while the water content of dried products is usually <4%. The concentrated products are sterilized or their osmotic pressure is increased so that no microorganisms survive.

Concentrated and dried milk products have several advantages, including:

1. *Storage:* Requires small space under regular storage conditions and retains high quality at the same time.
2. *Economy:* Because mass and volume are reduced, transportation costs are less.
3. *Balance:* Surplus milk can be reconstituted when fresh milk supplies are low.
4. *Use in emergencies:* Can be used under adverse conditions such as wars, epidemics, or earthquakes when fresh milk is unavailable.
5. *Formulations:* Suitable for tailored food products such as those designed for sportsmen, convalescents, or geriatric individuals.

4.2 Unsweetened Condensed Milk

4.2.1 Processing Chart and Preparing Raw Milk

The manufacture of unsweetened condensed milk is based on evaporation, that is, partial removal of water from milk followed by the addition of sugar. This process also extends the shelf life by suppressing the microorganisms present in the milk via plasmolysis.

The processing procedure[1,4,5] of unsweetened condensed (evaporated) milk is shown in Figures 4.1 and 4.2.

It is necessary to choose the raw material carefully; milk quality for unsweetened condensed milk production has to fulfill even more rigorous criteria than milk used in most other technological processes. This is necessary because unsweetened condensed milk or evaporated milk contains concentrated milk solids and is planned for long storage.

After raw milk is clarified by centrifugal separators, it is cooled to 4°C by plate heat exchangers and stored in tanks at the same temperature.

Standards for milk and dairy products regulate the ratio of milkfat to the nonfat solids in unsweetened condensed milk. In the United States, federal standards for evaporated milk prescribe not less than 7.5% by weight of milkfat and not less than 25% by weight of total milk solids in the final product.[6] In Great Britain the ratio of fat to nonfat solids is 10.0:20.0; in West Germany it is 7.5:17.5.[4] The standardization of ratios is most often done by separator-standardizers.

4.2.2 Preheating and Evaporation

The basic reasons for preheating are to increase the concentrated milk stability during sterilization and to modify the viscosity of the final product. A practical consequence of preheating is that milk enters the evaporator already hot, and stays for a shorter

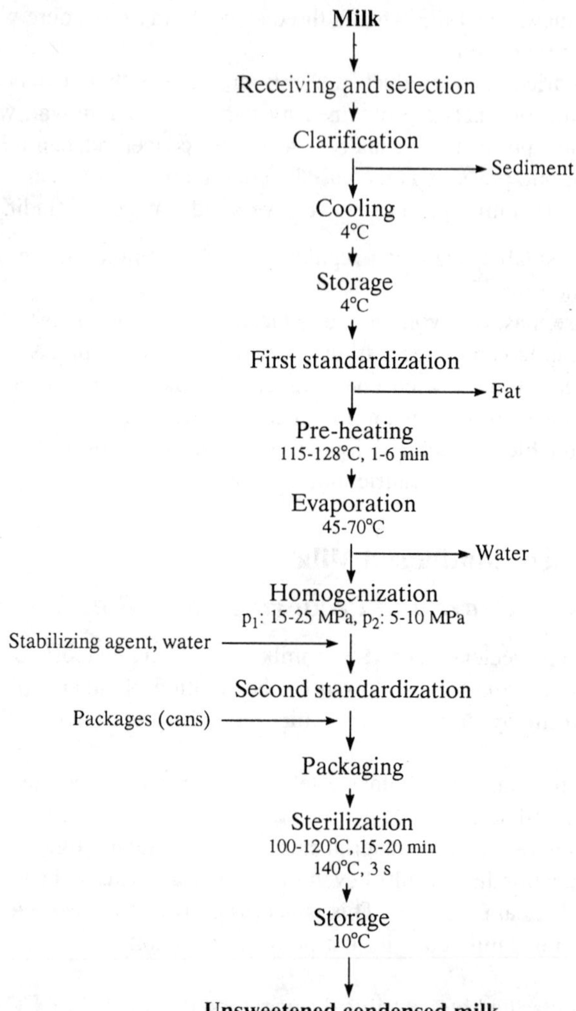

Figure 4.1 Flow chart of unsweetened condensed milk production.

time. The effect of preheating on thermal stability can be explained as follows. In milk, an equilibrium exists between acid and alkaline equivalents. The sum of the acid equivalents—P_2O_5, Cl, SO_3, CO_2, citrates, casein, albumin, and globulin—is approximately equal to the sum of the alkaline ones: CaO, MgO, K_2O, and Na_2O. The acid equivalents show a distinctly stabilizing effect on the protein system in milk, whereas the alkaline, especially Ca^{2+} and Mg^{2+}, perform in the opposite manner, leading to the aggregation of the casein micelles and their destabilization and precipitation.[7] The amounts of soluble calcium and phosphorus in milk decrease during heat treatment. Since the decrease of soluble calcium is more significant than

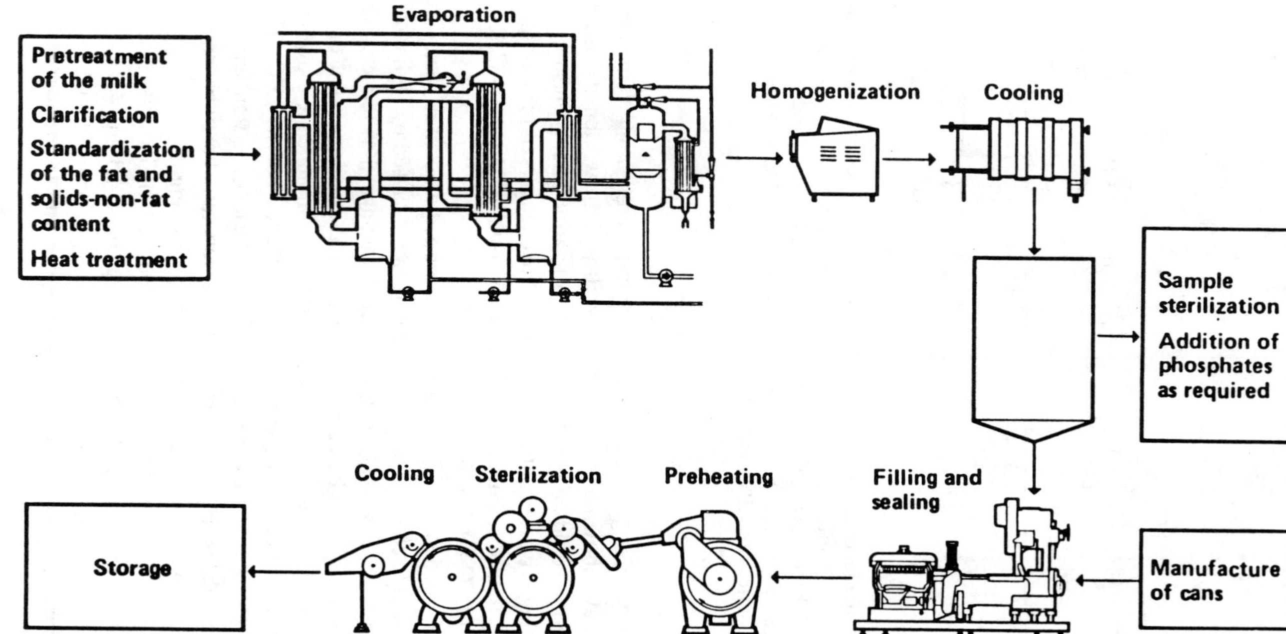

Figure 4.2 Process line for unsweetened condensed milk. (Courtesy of α-Laval.)

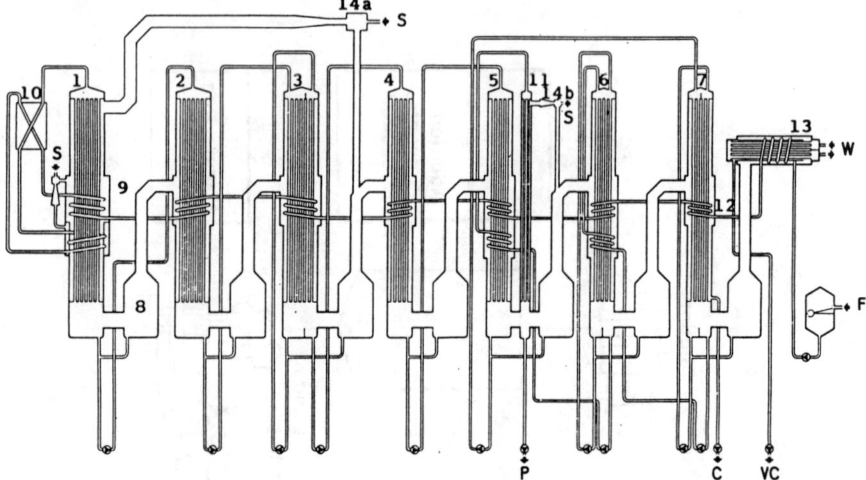

Figure 4.3 Falling film evaporator with TVR: 1. First effect; 2. second effect; 3. third effect; 4. fourth effect; 5. fifth effect; 6. sixth effect; 7. seventh effect; 8. vapor separator; 9. pasteurizing unit; 10. heat exchanger; 11. finisher; 12. preheater; 13. condenser, 14a and 14b. thermocompressor. F = feed, S = steam, C = condensate, VC = vacuum, W = water, P = product. (Courtesy of APV Anhydro.)

that of phosphorus,[7,8] a greater system stability after preheating is obtained. Recent investigations have shown that the thermal stability of unsweetened condensed milk during sterilization could be increased by centrifugation after preheating.

The preheating is carried out in continual heat exchangers of plate or tubular type. The time–temperature regimen of preheating is usually 93 to 100°C for 10 to 25 min or 115 to 128°C for 1 to 6 min.

Concentration is done by evaporating a determined amount of water from milk. To avoid undesirable changes of the milk components caused by high temperatures, the evaporation is always conducted in a partial vacuum, thus reducing the evaporating temperature. This is based on the fact that the boiling point of a liquid is lowered to a pressure below atmospheric. In order to eliminate the growth of staphylococci, the evaporation temperatures used are never below 45°C. The heating medium is usually a low-pressure steam with the heat being transferred indirectly through tubes or plates. Both tubular and plate evaporators may be single-effect or multiple-effect of two, three, four, or more units up to eight.

The falling film tubular evaporator, first introduced in Germany in 1953,[2,9,10] is the leading evaporator used in the dairy industry (Figs. 4.3 and 4.4). The liquid is introduced at the top of the evaporator and is evenly distributed on the inner surface of tubes. The tubes are about 3 to 5 cm in diameter, and 15 m long. They are fixed together in a corpus, called a calandria.

The interspace between the tubes is heated by steam. This type of evaporator operates at a much lower temperature and has a number of advantages such as a

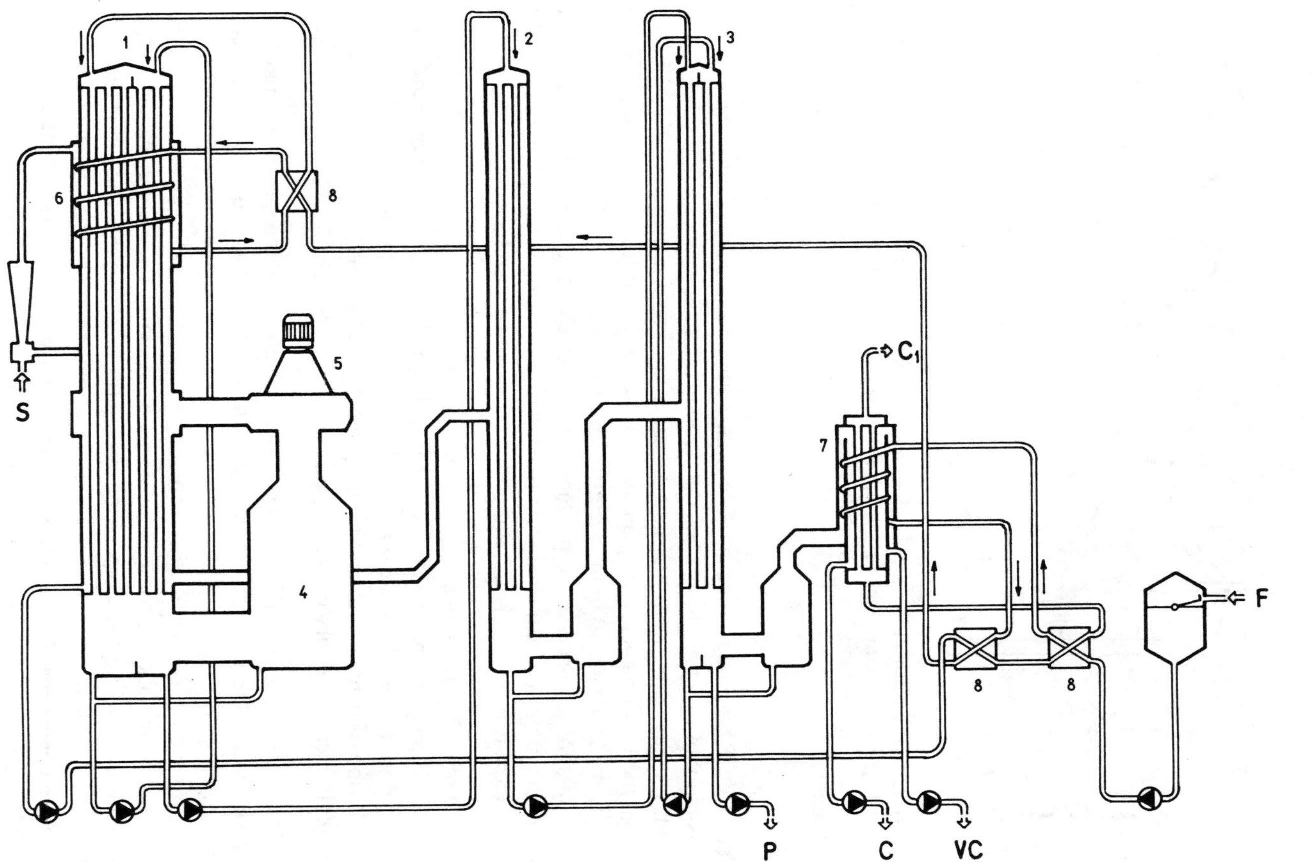

Figure 4.4 Falling film evaporator with MVR: 1. First effect; 2. second effect; 3. third effect; 4. vapor separator; 5. mechanical compressor/high-pressure fan; 6. pasteurizing unit; 7. condenser; 8. preheater. F = feed, S = steam, C = primary condensate outlet, C_1 = secondary condensate outlet, VC = vacuum, P = product. (Courtesy of APV Anhydro A/S.)

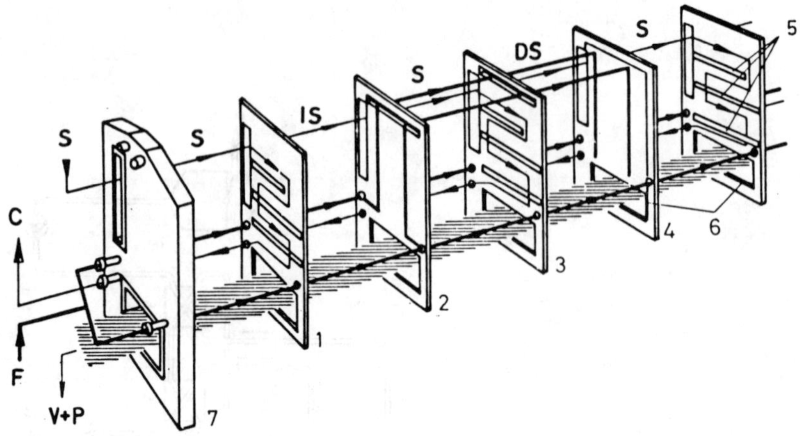

Figure 4.5 Arrangement of plates for one complete feed pass in plate evaporator. 1–4. Plates; 5. steam spacers; 6. joint rubbers; 7. head. F = feed, S = steam section, IS = inlet section, DS = discharge section, C = condensate, P+V = product + vapor. (Courtesy of APV Anhydro.)

decrease of heat-induced changes in milk components, low energy consumption, possibility of using multiple evaporators, and easy maintenance.

Plate evaporators were first introduced in the dairy industry in 1957.[10] In this design the heating bodies are plate heat exchangers used for heat transfer from the heating medium (steam) to milk (Fig. 4.5).

A mixture of concentrated product and its vapor is discharged into a separator, where the product is extracted from the vapor. This vapor is used in the next cycle. There are advantages in both falling film and plate heat exchangers:

1. The operation is simple.
2. Establishing and maintaining a stable regimen of operation, control, and adjustment is easy and accurate. Once adjusted, further regulation is automatic.
3. The job of maintenance and cleaning are simple. An automatic cleaning system (CIP, cleaning in place) may be used.

Plate evaporators are used in some other industries more frequently than in dairy manufacture. One reason may be the increased capacities required in the dairy industry.

In order to make use of the secondary vapor and thus improve the economic aspect of evaporation, two or more stages of the same type are connected in line, forming a multiple-effect evaporation system (Figs. 4.2 to 4.5). The vapor generated in the first cycle or effect during milk evaporation serves as the heating medium in the subsequent cycle. In this way, it is possible to reutilize the thermal energy brought into the system by the live steam. For the whole system to be effective, sufficient thermal energy brought by vapor from the previous stage must be available to initiate evaporation in the subsequent stage. A higher vacuum and corresponding lower pressure must be applied. Because of the pressure difference, the vapor moves to the next effect. The temperature difference between two neighboring effects is usually ≥5°C. This allows maximum boiling temperature to be as low as 70°C, corresponding to an absolute pressure of 230 mm Hg.[2]

Every evaporation assembly, single or multiple-effect, consists of an evaporator, condenser, equipment for vacuum creation, separator for separating vapor from the concentrated product, and a steam recompression system.

The energy crisis in the 1970s made it necessary to develop improved evaporation techniques in order to minimize the total energy consumption. As a result, there are two systems for vapor recompression in operation at present: (1) evaporation with thermal vapor recompression (TVR) and (2) evaporation with mechanical vapor recompression (MVR).[10] At TVR (see Fig. 4.3), the heating medium in the first effect is the product vapor from one of the next calandria, which is compressed to a higher temperature by a steam injection. The vapor generated in one calandria is used as the heating medium in the next one.

The MVR evaporator (Fig. 4.4) is a newer system that is superior to the conventional ones in areas where electrical energy is cheap or where natural gas is available. The heating medium in the first effect is vapor generated in one of the associated effects, or in the same effect, compressed by a turbo-compressor or high-pressure fan to a higher pressure, corresponding to the rise of the condensation temperature. Like TVR, the heating medium in each effect is vapor from the previous calandria. Vapor from the final effect is transferred to the suction side of the turbo-compressor or the fan and condensate is used for preheating the feed.

In addition to evaporation, only reverse osmosis is widely used in the industrial procedures for water removal, for example, dairy processing. However the concentration is only 20 to 25% total solids, as better heat economy is achieved at this level.[4,12]

4.2.3 Homogenization and Second Standardization

Homogenization is carried out in order to improve the stability of milkfat emulsion by decreasing the average diameter of milkfat globules. At the same time, the globules attain uniform diameter, forming a polydispersive system of milkfat with markedly narrower distribution. The diameter of milkfat globules in nonhomogenized milk varies in the range of 0.1 to 15 μm, the average diameter being 3 to 5 μm, which results in a wide dispersion. After homogenization, about 85% of the fat globules are smaller than 2 μm in diameter (0.1 to 2 μm), and all are under 3 μm, resulting in a fine dispersion.[13] Homogenization is carried out at high pressure (Fig. 4.1).

The absence of a cream layer in homogenized milk is not only the consequence of the average diameter of fat globules and Stokes' law. This effect is also attributed to the physicochemical changes observed during homogenization. Fat globules do not cluster together in homogenized milk, as they do in nonhomogenized milk. Some experiments show the existence of protein changes caused by homogenization, which are similar to denaturation.

Other changes of a physical nature, caused by homogenization, are: (1) more intensive white color as the consequence of an increased number of fat globules which have reflecting and light-scattering effects; (2) increased viscosity as the consequence of adsorption of proteins from solution on newly formed fat globule surface; (3) increased surface tension from removal of surface-active material from the skim milk phase; (4) decreased coagulation capability, because casein is partly absorbed as an ingredient in the newly formed fat globule membranes.

Among chemical changes, the most important are: (1) increased lipolytic rancidity, which is the consequence of the relatively greater total fat globule surface and better contact with lipase; (2) decreased oxidative changes, which are attributed to phospholipid migration from the surface layer into the skim milk phase, with formation of sulfhydryl compounds that have antioxidative properties; (3) decreased protein stability, similar to heat-induced denaturation with shifts in salt equilibrium. This last observation is not yet completely understood.

During the stage of second standardization, the ratio of milkfat to nonfat milk solids is adjusted (if the first standardization was not conducted) or the total dry matter is standardized (if the first standardization has been carried out).

The ability of milk to withstand intensive heat treatment (sterilization) is very important in this processing and depends to a great extent on its salt balance. During repeated standardization, or even during first standardization, stabilizing salts are added to milk in order to increase its heat stability. For this purpose, calcium, potassium, or sodium carbonates and bicarbonates, potassium or sodium citrates, phosphates, and other salts are used. In common practice, ready-made commercial mixtures are added.

The essential effect of these salts can be explained by the great affinity that the anions have toward calcium. Because it is bound by added anions, calcium from milk cannot adversely affect the stability of the protein system which is the basic condition for the stability of the whole system.

Test sterilization of evaporated milk is usually first carried out in the laboratory to determine the appropriate concentration of stabilizing agents.

4.2.4 Packaging, Sterilization, and Storage

Unsweetened condensed milk is usually packaged in cans of various sizes depending on use and then sterilized. Continuous flow sterilization of evaporated milk is also common, followed by packaging under aseptic conditions.

The sterilization of filled and sealed cans is carried out in continuous sterilizers at 100 to 200°C for 15 to 20 min. Flow sterilization of concentrated milk before packaging is a short-time sterilization by direct or indirect high-temperature heating (HTST) at 130 to 140°C in an ultrahigh temperature (UHT) plant. Sterilization is followed by filling into cans, closing under aseptic conditions, and labeling.

Evaporated milk can successfully be stored up to a year without any significant quality change at temperatures of 6 to 8°C. Therefore, it is suggested that this product not be kept at ≥10°C although it can withstand room temperature (20°C).[14]

4.3 Sweetened Condensed Milk

4.3.1 Processing Chart, Raw Milk, and First Standardization

Sweetened condensed milk is manufactured by removing part of water from fresh milk (usually by evaporation) and adding sugar to the concentrated milk in order to extend its shelf life. The procedure is based on osmoanabiosis, that is the prevention of the growth of microorganisms by increasing the osmotic pressure of the medium.

The technological process of sweetened condensed milk production[1,4,5,9] showing operations and equipment is shown in Figs. 4.6 and 4.7.

The comparison of this scheme with that of unsweetened condensed milk production shows that unsweetened condensed milk will last for long periods due to sterilization, whereas sweetened condensed milk is long lasting because of the increased osmotic pressure. Most of the differences in their processing originate from this basic observation.

The production of sweetened condensed milk using hydrolyzed lactose is also possible. Milk is cooled to 5 to 10°C after pasteurization and lactose is hydrolyzed by β-galactosidase, obtained from *Saccharomyces fragilis*. If the hydrolysis is carried out at 37°C for 3 h or 8°C for 24 h with 1% enzyme added to the initial milk, lactose is hydrolyzed by 95 to 99%. The sweetness of the final product is approximately the same as that of sweetened condensed milk obtained by conventional procedure with sucrose addition.[4]

In order to prevent lactose crystallization in sweetened condensed milk, acid-hydrolyzed sugar syrup may be added. The sugar syrup is separately hydrolyzed using HCl at 80 to 90°C for 20 to 30 min at pH 6.5 to 6.7, with subsequent operations being similar to those in traditional sweetened condensed milk production.

Application of ultrafiltration in sweetened condensed milk production has recently been investigated. It shows certain advantages when compared to the traditional procedure.

Milk selected for sweetened condensed milk manufacture must satisfy the same rigorous quality criteria as the manufacture of unsweetened condensed milk. Clarifying, cooling, and storage are carried out in the same way as described in the processing of unsweetened condensed milk.

In this stage of processing, the milkfat to solids-not-fat ratio is adjusted so that the composition quality requirements for the final product are met.

4.3.2 Heat Treatment, Evaporation, Sugar Addition, and Second Standardization

Heat treatment has a special importance in sweetened condensed milk production, because it is the most intensive thermal treatment in the procedure. There is no sterilization during sweetened condensed milk production because evaporation temperatures are low in multiple-effect vacuum evaporators ($t < 70°C$). Therefore the main goal of heat treatment is the total destruction of osmophilic and thermophilic microorganisms and inactivation of enzymes, particularly lipase and proteases. In

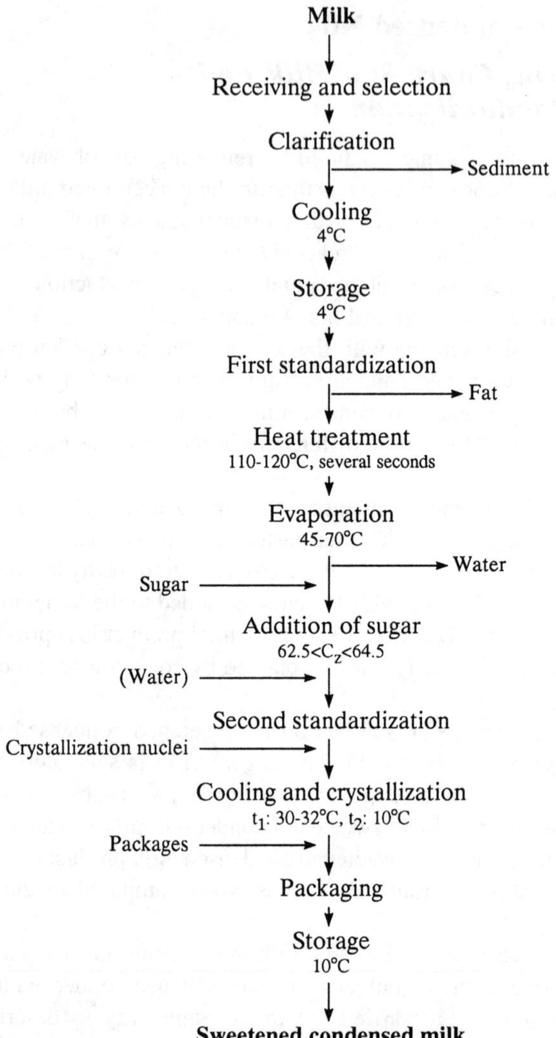

Figure 4.6 Flow chart of sweetened condensed milk production.

addition, heat treatment decreases fat separation and inhibits oxidative changes. The milk is also warmed prior to evaporation, and this has positive economical and technological effects.

Heat treatment in sweetened condensed milk processing affects the final product viscosity (i.e., the tendency to viscosity increase during storage). This defect is called "age thickening" and is usually not caused by heat-resistant microbial proteases, but is rather the consequence of physicochemical changes in casein. The most frequently applied temperatures in this technological process are 100 to 120°C in continual heat exchangers of plate or tubular type.[9]

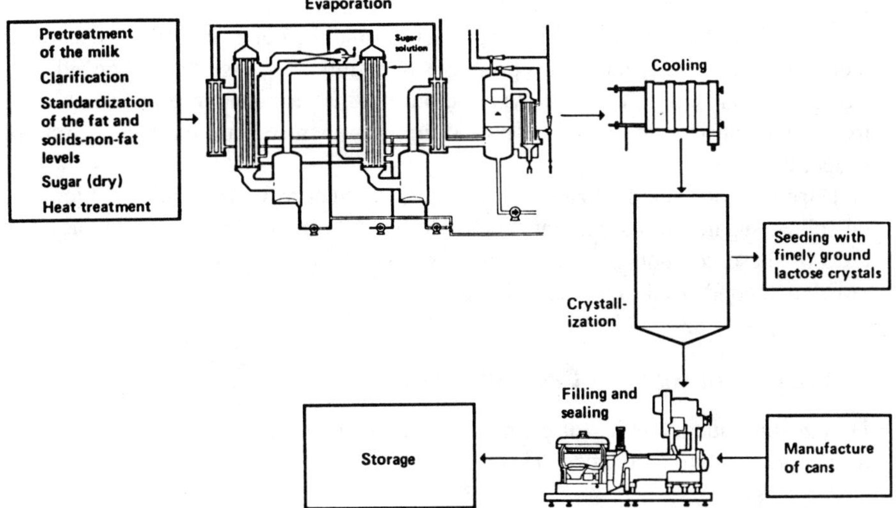

Figure 4.7 Process line for sweetened condensed milk. (Courtesy of α-Laval.)

Evaporation, described in detail in Section 4.2.2, is carried out in sweetened condensed milk production using the same equipment and process parameters. The degree of condensation depends on standards (mandatory or voluntary) for final product composition and is usually a 2:1 ratio or slightly above.

Sugar addition is the way to prolong the shelf life of this product, because heat sterilization has not been performed. Usually the selected sugar is sucrose, although glucose, dextrose, or others could also be applied, particularly when the product has special application. The main advantages of sucrose over other sugars are good solubility, low susceptibility to fermentation, and the preference of consumers. The sucrose must be microbiologically safe, with no acids or invert sugar present.

During sweetened condensed milk production for direct consumption, sucrose is added either in crystal form or as a solution. In the latter case, the sugar is dissolved in water at 95°C; the solution is heated before addition into the milk by a high pasteurization regimen. The amount of added sugar has to be such that its concentration in the aqueous phase of the final product ranges from 62.5 to 64.5%.[1,4,9] This parameter, called "sugar number" or "sugar index," is calculated in the following way:

$$C_z = \frac{S}{S + W} \times 100 \tag{4.1}$$

where $62.5 < C_z < 64.5$.

S = sucrose content in sweetened condensed milk (%)

W = water content in sweetened condensed milk (%).

When the sugar number or C_z is <62.5, there may be bacteria-induced changes in the final product. Higher sugar index than the permissible maximum (C_z >64.5) could, at lower temperatures, cause lactose crystallization. Sugar addition before heat treatment increases thermoresistance of bacteria and their enzymes during heat treatment and significantly intensifies product susceptibility to "age thickening" during storage. Sugar addition before evaporation also has a negative effect on viscosity changes during storage. The optimal time for sucrose addition is at the end of evaporation.

During second standardization (restandardization or repeated standardization), total solids, sugar, and fat contents are controlled. On the basis of data obtained, the ratio of these components to total milk solids, as well as the total dry matter of sweetened condensed milk, are adjusted.

4.3.3 Cooling with Crystallization

During subsequent cooling of the product after evaporation and sugar addition, lactose crystallization is induced. This is caused by:

- Temperature decrease
- High lactose concentration (>10%)
- Presence of high concentrations of added sugar (about 40%)
- Relatively small amount of water.

Lactose has the capability to create supersaturated, metastable solutions in which mass crystallization occurs. However, if crystals larger than 15 μm develop, the product has a texture defect, known as "sandiness," which affects the mouthfeel when the product is consumed. To avoid this fault, inoculation with powdered lactose crystals (0.5 kg/1000 kg of milk) is used and the process completed with rapid cooling and simultaneous agitation. In this way, more than 4.10^{11} crystals per m^3 are formed, the size of which does not exceed 10 μm. Inoculation could also be done by adding 0.5% skim milk powder, formerly centrifuged 1% sweetened condensed milk, or 0.2 to 0.3% whey powder.[1,4,15]

Cooling with crystallization is accomplished by using double-wall tank-crystallizers, fluid flow continual coolers, or vacuum crystallizers.

Sweetened condensed milk designed for the retail market is usually packaged in cans (similar to unsweetened condensed milk), tubes, plastic forms, and others. For use by institutional consumers or industries, the milk is packed in metal barrels, metal cylindric drums, or similar large containers. The storage of sweetened condensed milk is the same as for unsweetened.

4.4 Other Concentrated Dairy Products

Recently a number of other concentrated dairy products, some traditional, some new, have been developed since the advent of membrane technology. Some of these products, such as whey protein concentrates (WPC), are usually further processed by drying and used in powder form. They are then considered dried dairy products or dried dairy ingredients. Some of the traditional and well-established concentrated dairy products deserve discussion.

Condensed skim milk is obtained by a simple concentration of skim milk by vacuum evaporation or reverse osmosis. Condensed skim milk is cheaper and, for some purposes, a product of better quality than skim milk powder. It serves as a source of milk solids in various food products. When ultrafiltration is used, a concentration of 30% dry matter is possible and it may reach a protein content of 70 to 75%. However, in general, the protein level may vary from 50 to 80%. The concentrate obtained by ultrafiltration from either skim milk or whole milk consists of undenatured high-quality milk proteins, which find wide application in the infant food industry, dietetic products manufacture, dairy processing, and others.

Unsweetened condensed milk may be flavored with coffee, cocoa, or other ingredients. The production procedure is similar to that for sweetened condensed milk, with the addition of a flavoring agent prior to sterilization. Sweetened condensed skim milk is produced by concentrating skim milk and subsequently adding sugar.

''Block milk'' is a product derived from concentrated milk with sugar addition. It has a high concentration of total solids (84 to 90%) and can be cut with a knife. Increased density gives the product all the advantages of concentrated or dry products, and the presence of 16% water facilitates product dissolution.

Caramelized condensed milk, originally called ''Dulce de leche'' by Spanish natives, is a traditional, regional product, but is produced industrially as well. It is marketed mostly as a paste, but may also be in powder or tablet form. A kind of caramelized condensed milk in paste form is produced by concentrating and caramelizing milk with 18 to 20% of sucrose or glucose, with or without a flavor supplement of dried products.

4.5 Dried Dairy Products

4.5.1 Milk Powder

4.5.1.1 Processing Chart, Raw Milk, and Standardization

Milk powders are dairy products from which the water has been removed to the greatest extent possible, thus preventing the growth of microorganisms.

A milk powder production line[1,4,5,9–11,16,17] is shown in the flow chart in Fig. 4.8.

Skim milk powder (SMP) processing is similar to milk powder processing except for two differences. The skim milk fat content is decreased to 0.05 to 0.10% from higher values (standardized) common in whole milk powder manufacture and the skim milk may be heated before evaporation. SMP heat treatment depends on the kind of powder produced. SMP produced by a ''low heat method'' is simply pasteurized, while heat treatment by a ''high heat method'' requires heating at 85 to 88°C for 15 to 30 min in addition to pasteurization.[4,10,11] Intensive heat treatment of skim milk during powder processing is applied in the production of powder to be used in the bakery industry, where a high degree of milk protein denaturation is

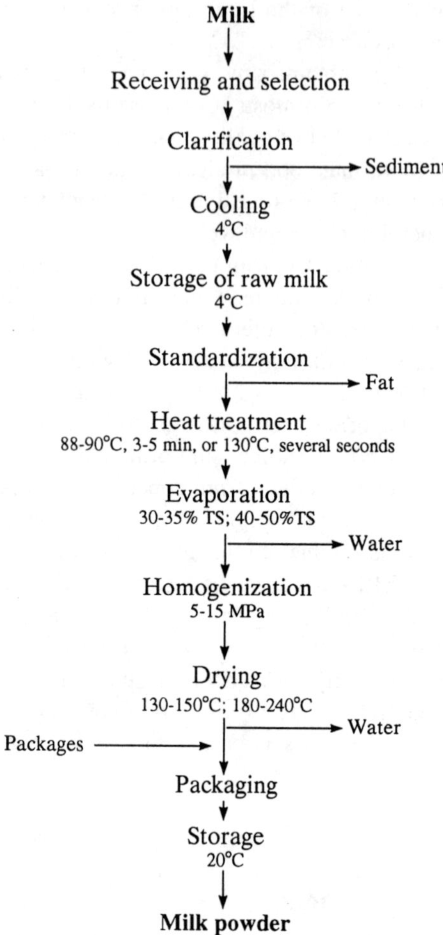

Figure 4.8 Flow chart of milk powder production.

required. There is no need to homogenize skim milk destined for powder production in view of the low fat content.

Milk used in the production of powder must be of high chemical, sensory, and bacteriological quality. The same rigorous criteria apply as in the production of sweetened and unsweetened condensed milk. They are regulated by law and differ in various countries.

Further processing includes clarification by centrifugal separators or filtration, cooling in plate heat exchangers to 4°C, and storage in tanks at the same temperature. This is followed by standardization which is to adjust the ratio of milkfat to total solids as required in the final product.

4.5.1.2 *Heat Treatment, Evaporation, Homogenization, and Drying*

Heat treatment is usually carried out at temperatures higher than those required for pasteurization. The aim is to destroy all pathogenic and most of the saprophytic microorganisms; to inactivate enzymes, especially lipase, which could cause lipolysis during storage; and to activate the SH groups of β-lactoglobulin, thus increasing resistance of the powder to oxidative changes during storage. In order to avoid the possibility of total solids variation or milk recontamination by steam injection, heat treatment is commonly performed in an indirect way, via tubular or plate heat exchangers.

Evaporation is a mandatory operation in powder processing for several reasons: milk powder produced from evaporated milk consists of larger powder particles containing less occluded air and has longer shelf life. Viscosity of the milk increased due to higher concentrations of total solids results in larger powder particles. Omitting evaporation of the milk prior to its drying would not be economically feasible, because the demand for energy would be severely increased. Energy consumption in modern evaporators of multiple effect with steam recompression is about 10 times lower than in spray drying. In addition, omission of evaporation would result in an inferior quality of powder. For roller drying, the concentration during evaporation is raised to 33 to 35% total solids, whereas for spray drying it is up to 40 to 50%. This difference in concentrations during evaporation is caused by the drying technique. Higher concentration rates during roller drying would form a thick layer on the rollers, followed by slower drying and irreversible changes of proteins, lactose, and fat. Concentrating the milk destined for spray drying beyond the 50% total solids limit would further increase viscosity and cause difficulties during atomization.

Homogenization is not an obligatory operation in milk powder processing, but is usually applied in order to decrease the free fat content. Fat globules depleted of protective membranes reduce milk powder solubility and increase their susceptibility to oxidative rancidity. During homogenization (pressure = 5 to 15 MPa), free fats are transformed into fat globules; membranes are regenerated on their surfaces because of adsorption of proteins.

On an industrial scale, milk is most commonly dried by roller drying or spray drying in a stream of hot air. Various modifications of both systems include:

1. Roller drying
   - Roller drying at atmospheric pressure
   - Vacuum roller drying
2. Spray drying
   - Centrifugal atomization
   - Pressure atomization
   - Foam spray drying
   - Steam swept wheel atomization
   - Venturi spraying

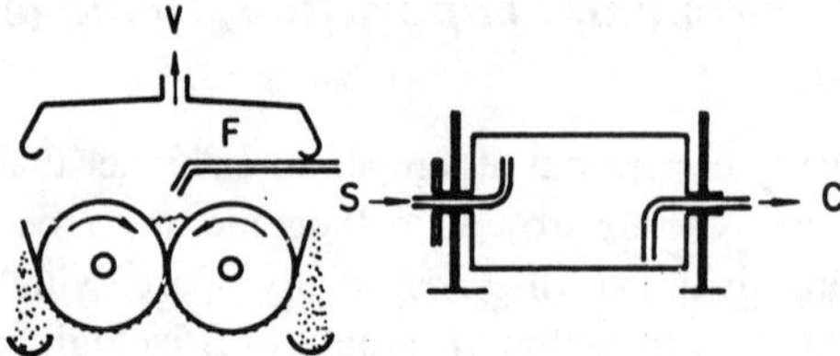

Figure 4.9 Steam and condensate flow in roller dryer. F = feed, S = steam, V = vapor, C = condensate (1).

Two-stage spray drying (a system producing nonagglomerated powders, 1970)
Three-stage spray drying (a system producing either agglomerated or nonagglomerated powders, 1980).

In addition to these systems, interest has been generated in the following drying methods:

1. Foam mat drying: product in foam form.
2. Drying in vacuum chambers: product in paste form.
3. Freeze drying: product in powder form.

There have also been attempts to develop completely new drying methods. However, all such attempts have failed for different reasons, leaving only spray drying and roller drying for industrial application in dairy technology. Because the product quality and process economy are superior and constantly improving with spray drying, this method is of the greatest value today and in the foreseeable future. Trends of the past 10 years indicate that intensive efforts have been made to improve and modify the spray drying system. The two major objectives are to improve product quality while decreasing energy consumption.

4.5.1.3 Roller Drying

This method is commonly used in the production of skim milk powder as well as whole milk powder, which find applications in other industries (confectionery, feed blends, etc.) because of low product solubility. Direct contact of a layer of concentrated milk with the hot surface of rotating rollers adversely affects the milk components, and causes irreversible changes in most of them. Examples include lactose caramelization, lactose degradation with high energy of activation, Maillard's reactions between certain amino acids and lactose, protein denaturation, etc. Products of Maillard-type reactions may cause a scorched flavor, while protein denaturation results in poor solubility. Vacuum roller drying, at 91 to 98 KPa, operates at temperatures below 100°C and eliminates an oxygen effect, providing better powder characteristics than by roller drying under atmospheric pressure. Drying equipment is either constructed of one or two rotating rollers[1,4] (Fig. 4.9).

The construction used most frequently in the dairy industry is a double drum dryer that operates at atmospheric pressure. Dry, saturated steam at a temperature of

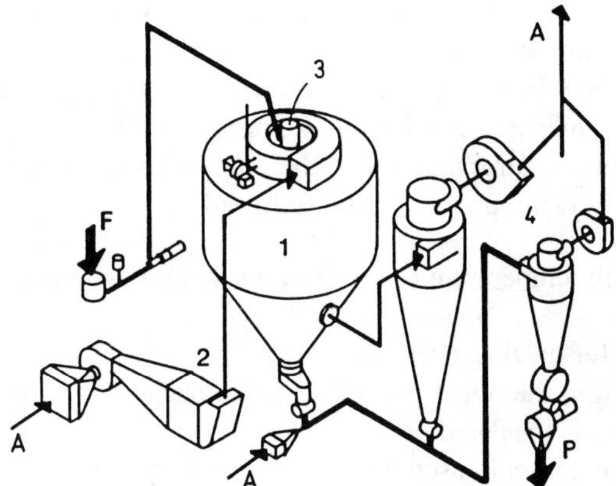

Figure 4.10 Process line for spray drying (one-stage drying, centrifugal atomization). 1. Spray drying chamber; 2. air heater; 3. atomizer; 4. cyclone system; 5. control board: F = feed; A = air; P = product (powder). (Courtesy of NIRO ATOMIZER.)

up to 150°C and a pressure of up to 621 MPa is used for heating the roller when introduced into its axis. The steam condensate is continuously removed by a pump located at the other end of the roller's axis. Milk temperatures reach approximately the same value as the steam during drying. Dry film, scraped off by knives, falls on spiral conveyer belts located along each roller, where it is finely crushed ("diced") and transported to a hammer mill, where it is pulverized.

4.5.1.4 Spray Drying

As mentioned earlier, spray drying is mainly used for drying milk and milk products. Evaporated milk is atomized into fine droplets and exposed to a hot air flow in a spray drying chamber (Fig. 4.10) which may be in horizontal or vertical positions. Although horizontal drying chambers are common in the United States, vertical drying chambers with flat or conical bottoms are used more often.[3,4,10] The ambient air is filtered, heated by steam or a liquid phase heating system (oil/gas) up to 150 to 300°C, and introduced into the drying chamber at a velocity up to 50 m/s.[1] Air is usually filtered before being heated. However, air can be also heated by mixing it with combustion gases in a direct gas-fired heater, where burning products of gas or oil and hot air enter the chamber. Because there is a direct contact between milk and combustion gas, this method of air heating is not commonly used in the dairy industry, although it has a 100% heat efficiency and low investment and maintenance costs. One obvious reason is contamination of the milk powder with nitrogen oxide present in combustion gases. Carcinogenic substances, such as nitrosamines, may result from reactions of nitrogen oxides, amines, and other ingredients present in milk.[10]

Air heating by indirect methods involves heating by steam in tubular or plate heat exchangers, liquid phase heating, or indirect oil or gas heating.

In relation to the flow of milk, the air stream moves through the spray dryer in a concurrent flow (same direction), countercurrent flow (opposite direction), or in a mixed flow (angular). In spite of the inferior heat economy, concurrent flow is preferred in the dairy industry, as it improves product quality.

In order to decrease air quantity and thermal losses in the spray dryer, it is important that the following procedures be followed.

1. Maintain a high temperature of the inlet air and a low temperature of the outlet air.
2. Use outlet air for heating inlet air.
3. Use inlet air from the upper part of the plant because it is the warmest and the drying chamber is insulated.
4. The two most commonly used devices for recovery of heat and mass from spray dryer exhaust include the sanitary spray scrubber and sanitary venturi scrubber.

A milk atomizer's basic function in spray drying is to provide a high surface-to-mass ratio, thus enabling quick heat transfer with a high evaporation rate. The two atomizing designs most commonly used are the centrifugal (rotary) atomizer and the pressure (nozzle) atomizer.

A centrifugal atomizer is advantageous in drying viscous materials and suspensions. However, to achieve versatility in production, most dryers are now constructed for both atomizing possibilities. By operating with a centrifugal atomizer, the same high-pressure feed pump can be used without pressure. Milk is dispersed in the centrifugal atomizer, at rotating speeds of 10,000 to 20,000 rpm, or by a pressure of 17.2 to 24.5 MPa in the pressure nozzles. In this way, fine particles with large specific surface areas are obtained. As the milk dispersion rate increases, the specific surface area is increased as well. This provides a rapid and intensive heat transfer from air to milk, and mass transfer from milk to air. By dispersing 1 L of milk into droplets of 50 μm in diameter, a total surface area of 120 m^2 is gained. Due to increased surface area and high latent heat of water evaporation (2.26 MJ/kg sprayed particles), moisture is quickly lost and the temperature of the incoming air drops immediately. When the inlet air temperature is up to 215°C, the temperature in the chamber drops almost instantly to that of outlet air. This is approximately a 95°C drop in one-stage drying.

Regardless of the kind of the atomizer used, the powder particles during drying gain spherical shapes (because of surface tension) with trapped air, thus gaining a low bulk density. Each spray dried particle is spherical and has a diameter between 10 and 250 μm. These particles contain evenly distributed vacuoles of occluded air in their interior.[18] Although most vacuoles are average in size, some are small (Fig. 4.11).

The surface of spray-dried particles is usually smooth, but may become wrinkled. The tendency to form wrinkles is increased by higher inlet air temperatures and larger temperature differences between the hot air and powder particles. The presence of particles of different morphology in the same sample is ascribed to the different drying conditions to which the individual particles were exposed.

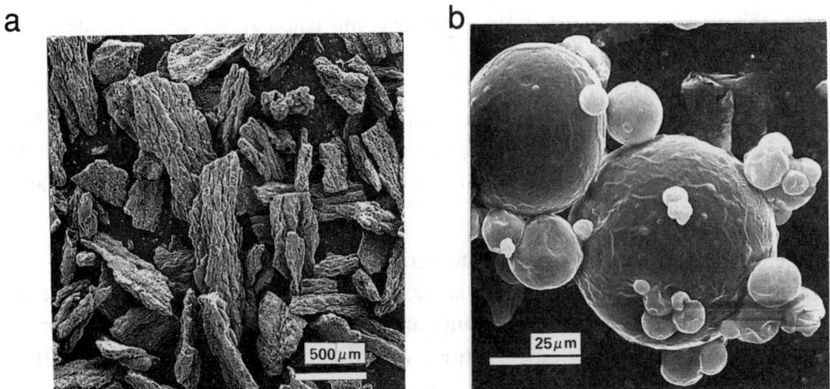

Figure 4.11 Microstructure of milk powder: (a) Roller dried; (b) Spray dried.[18]

Atomization methods, either centrifugal or nozzle, have no special effect on particle structure. The bulk density of spray-dried powders varies (0.50 to 0.70 g/cm^3).[1,18] Bulk density could be improved by introducing steam into the atomizer (''steam swept wheel''), using special atomizer constructions and adjusting drying parameters. Atomization parameters affect some important properties of the final product: bulk density, size and size distribution of powder particles, incorporated air content, moisture content, and others.

The product is removed immediately after drying to stop further contact between powder and hot air. Long contact with hot air could result in penetration of fat at the particle surface, which causes adhesions and overheating of the powder. It is necessary to provide adequate velocity for carrying particles of certain diameters. This velocity is calculated on the basis of Stokes' law and is called terminal velocity. With the enlargement of the particle diameter, the terminal velocity increases. Cyclone separators are used for powder recovery. With this system, 90% of the powder having particle size larger than 10 μm, 98% of the powder with particle sizes larger than 20 μm, and 99% of dry particles with diameters larger than 30 μm can be recovered. The efficiency of cyclones is calculated on the basis of material balance, that is, ratio of total solids of raw material entering the process versus total solids of powder at the outlet. The formula is:

$$E_f = \frac{\mathrm{SM_o}}{\mathrm{SM_e}} \times 100\ (\%) \tag{4.2}$$

where

SM_e = dry matter entering the process (kg)

SM_o = dry matter outlet of the process (kg).

Today, there is a system of several cyclones with large diameters combined with one cyclone of smaller diameter, which provides simultaneous cooling of the powder.

Spray drying has numerous important advantages compared to the other drying techniques:

1. The whole process is rapid: drying is accomplished at low temperatures, giving the product excellent properties.
2. There is no noticeable oxidation, vitamin loss, protein denaturation, lactose transformation, or other adverse effect from heat. Spray drying is also used for drying many pharmaceutical, biological, and thermolabile materials. Products obtained by spray drying are of quality similar to that obtained by freeze drying and do not require further processing.

Because spray drying is fully automatic, even a high-capacity operation with a high productivity requires minimal labor. Because the product comes into contact with the wall of the closed chamber only in powder form, there is neither the problem of equipment maintenance or corrosion, nor microbiological quality in the finished product. Spray devices can be used for drying all kinds of products that can be pumped, even if they are adhesive or very viscous (i.e., casein, caseinates, cream, blends, etc.). Spray-dried products also have a fine structure. There is not a large quantity of the product in the chamber simultaneously, which will be advantageous during the eventual breakdown in processing.

4.5.1.5 Packaging and Storage

Powder should be packed in suitable containers that protect it from moisture, air, light, etc. The following wrappings are generally used: paper, multilayer boxes or bags with a polyethylene layer inside, metal barrels with polyethylene bags inside, or tins covered with aluminum foil at the contact surface. When planning the quantity of wrapping material, it is necessary to consider the bulk density of the product, because it is highly affected by processing parameters and techniques. When the product is intended for long storage, it is packaged in an atmosphere of inert gas, mostly nitrogen, or in a partial vacuum (4.0 to 5.3 KPa) to avoid oxidative changes of fat and other components.

Properly produced and packed milk power, with low oxygen content, is stored at ambient temperature.

4.5.2 Instant Milk Powder

Instant milk powder has better reconstitution properties than other milk powders. The instantization process patented by Peebles in 1995[19–21] significantly improved the quality and economical aspects of the drying technology. The properties of dried products improved positively by the instantizing process include wettability, penetrability, sinkability, dispersibility, and dissolvability. The process permits a better equilibrium among these variables.

Instant characteristics are attained by agglomeration, which causes an increase of the amount in air incorporated between powder particles. During reconstitution the air is replaced by water. Incorporated air enables a larger quantity of water to come into contact immediately with the powder particles during reconstitution.

In a noninstantized product, a viscous layer forms around clustered powder particles and hinders further water absorption. This slows the dissolving process.

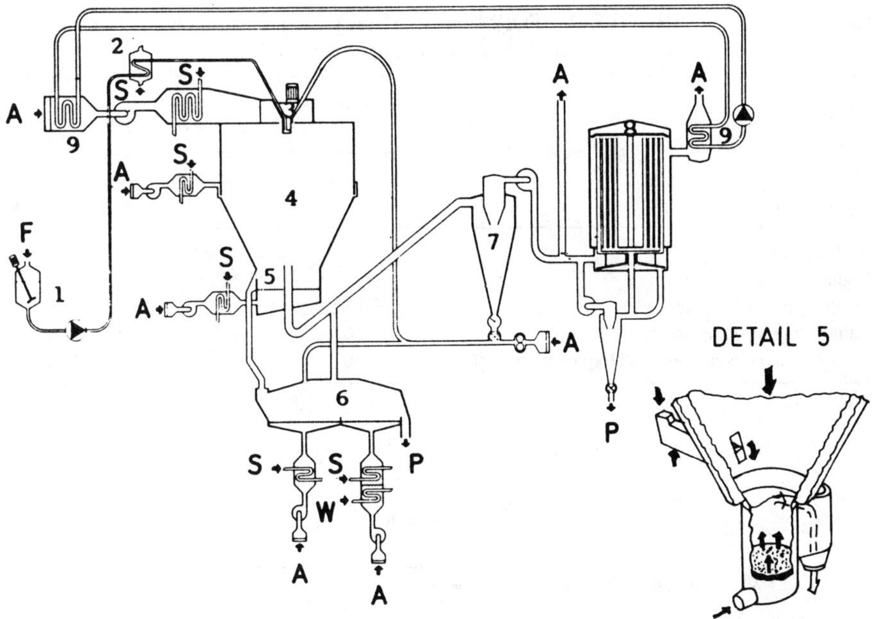

Figure 4.12 Three-stage drying. 1. feed tank; 2. concentrate preheater; 3. atomizer; 4. spray drying chamber; 5. integrated fluid-bed; 6. external fluid bed; 7. cyclone; 8. bag filter; 9. liquid coupled heat exchanger. F = feed, A = air, S = steam, W = water, P = product. Detail: 5. Integrated fluid bed[3] (Courtesy of APV Anhydro).

The main advantages of a two-stage drying system are[3,4,16,17]:

1. Increased heat utilization (specific heat consumption 15 to 20% lower than that for a single-stage dryer)
2. Improved product quality (solubility, bulk density, occluded air, etc.)
3. Higher capacity.

The development of the three-stage drying procedure has made possible greater energy savings than the two-stage dryer (Fig. 4.12).

Three-stage drying involves a spray dryer as the first stage, a static fluid bed integrated in the base of the drying chamber as the second stage, and an external vibrating fluid bed as the third drying stage. By moving the second drying stage into a drying chamber, it is possible to achieve even higher moisture removal at the end of the first drying stage than by two-stage drying.[3,4,10,16,17] Lower temperatures are applied which results in a powder of better quality and thermal efficiency (Table 4.1).

There are two basic types of instantizing (Fig. 4.13): The ''rewet'' process, where the instantization is carried out after the powder is obtained in dry form, and the ''straight through'' process, where instantization is accomplished during drying.

Table 4.1 COMPARATIVE PERFORMANCE DATA FOR THREE-STAGE DRYING VERSUS TWO-STAGE DRYING. BASIS: SKIM MILK

	Three-Stage Integrated Fluid Bed Spray Dryer		
Drying system	With Nozzle Atomizer	With Rotary Atomizer	Two-Stage
Feed solids (%)	48	50	50
Product rate (kg/h)	2140	1720	
Residual moisture content (%)	3.5	3.5	3.5
Main drying air temperature (°C)	280	215	220
Specific heat consumption (kcal/kg evap. water)	850	866	972

From ref. 16.

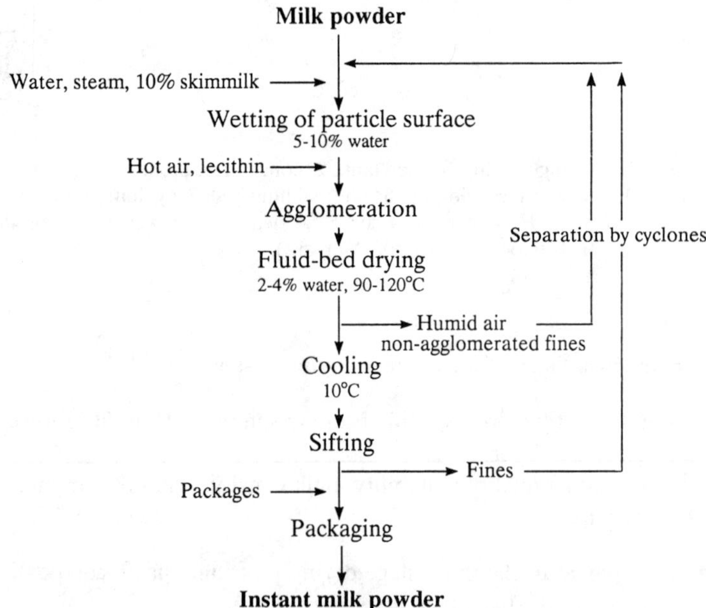

Figure 4.13a Flow chart of instantized milk powder production. (a) Rewet procedure.

Dry milk powder is the starting material for the rewetting procedure. The powder is dispersed in the wetting chamber and gains a water content of 5 to 10%, causing formation of powder particle agglomerates. The agglomerated product is transferred to the vibrating fluid bed dryer and cooler (Figs. 4.12 and 4.13), where it is redried in a hot air stream at 90 to 120°C and immediately cooled to approximately 10°C.

The powder layer in the fluid bed dryer is about 10 cm high, with a residence time of 10 to 12 min.[10] Because two-stage drying results in a powder of lower

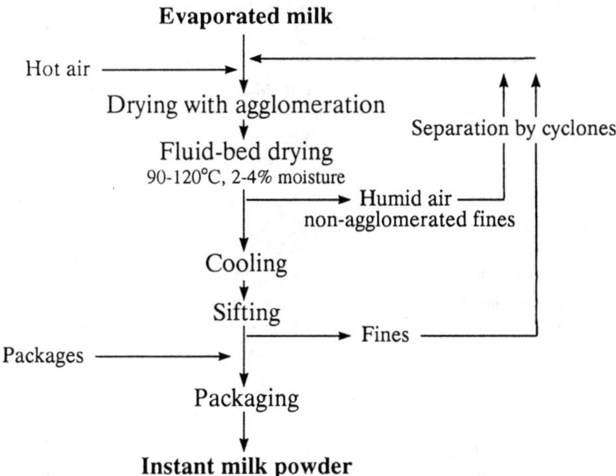

Figure 4.13b Flow chart of instantized milk powder production. (b) Straight through procedure.

temperature at the last drying stage than that from a single-stage drying, it is preferred.

The final product from the fluid bed dryer has a water content <4% (2 to 4%). Hot air, blown upwards through the fine perforated plate of the fluid bed dryer, ''fluidizes'' the powder and carries the smallest powder particles or fines to the cyclone, where separation takes place. The air is discharged into the atmosphere after heat and solids have been recovered and the fines are returned to the start of the process.

This instantization process is more complicated when the treated product contains fat, for example, whole milk or cream. Free fat forms a hydrophobic layer on the particle surfaces, impregnating them and decreasing their water-binding capability. In order to prevent this and improve the recombination properties, powder particles are coated during instantization with a surface-active agent, usually lecithin (0.2% lecithin in the powder),[23] during instantization.[22]

During ''straight through'' instantization (Fig. 4.13), the agglomeration process is carried out in wet powder, immediately after powder particles have been formed. It differs from the two-stage process, where dry powder enters the process.

Because of low outlet air temperature and other drying parameters, the discharged powder contains moisture. The powder is subsequently transferred through two vibrating dryers, where excess water is removed and hot air stream carries the fines into the cyclone system. After separation, the fines are fed back into the atomization zone to be agglomerated with the wet powder.

The newer type of three-stage spray drying chamber construction contains a Filtermat dryer that has a main drying chamber and three smaller additional chambers for crystallization, final drying, and coating. This spraying design has a lower height, high productivity, and production versatility. In addition, the ''instant'' properties

Table 4.2 COMPOSITION OF HUMAN AND COW MILK

Component (%)	Human Milk	Cow Milk
Water	87.43	86.61
Fat	3.75	4.14
Protein	1.63	3.58
Lactose	6.98	4.96
Ash	0.21	0.71
Nonfat solids	8.82	9.25
Total solids	12.57	13.39

From ref. 25.

of the product are excellent. The most recent development is a multistage spray dryer developed by Storck.[23] The drying chamber is directly connected to the external fluid bed through the well mix section. This reduces the transportation time for the moist powder.

4.5.3 Infant Formulas

Infant formulas were designed as a substitute for human milk. Some mothers cannot or do not wish to breast-feed. Such formulas are usually derived from cow's milk that has been modified to simulate breast milk as much as possible. The use of infant formulas started at the beginning of the 20th century and resulted in a product with increasing resemblance to breast milk. Science and technology in the past decade have significantly improved the formulations.

Instead of using modified cow's milk as a base, modern infant formulas may contain other forms of basic ingredients: milk and whey products (proteins, lactose); soybean proteins (soybean protein isolate); and protein hydrolysates. Formulas without cow's milk as a component are designed for infants with milk intolerance, milk allergies, or special needs.

According to Table 4.2, human and cow's milk differ in the relative content and chemical composition of macronutrients (lactose, proteins, minerals, and fat).[25,26] The quantity of protein in cow's milk is 3.5 times higher than that in breast milk and contains 80% casein and 20% whey proteins. Human milk protein is composed of 20% casein and 80% whey proteins. β-lactoglobulin, which represents the largest amount of whey protein in cow's milk, is not found in human milk at all.

Therefore, to simulate breast milk, cow's milk must be modified to:

1. Reduce protein and mineral content, especially sodium.
2. Change milk protein ratio in favor of whey proteins.
3. Increase the Ca/P ratio from 1.2 to 2.0.
4. Increase the carbohydrate content and add vitamins, but less complicated.
5. Modify fat (this step presents special problems because of its stability).

Table 4.3 NUTRIENT LEVELS OF INFANT FORMULAS (PER 100 KCAL)[a]

Nutrient	FDA 1971 Regulations Minimum	1976 Recommendations[b] Minimum	Maximum
Protein (g)	1.8	1.8	4.5
Fat			
(g)	1.7	3.3	6.0
(% cal)	15.0	30.0	54.0
Essential fatty acids (linoleate)			
(% cal)	2.0	3.0	—
(mg)	222.0	300.0	—
Vitamins			
A (IU)	250.0	250.0 (75 μg)[c]	750.0 (225 μg)[c]
D (IU)	40.0	40.0	100.0
K (μg)	—	4.0	—
E (IU)	0.3	0.3 (with 0.7 IU/g linoleic acid)	—
C (ascorbic acid) (mg)	7.8	8.0	—
B (thiamine) (μg)	25.0	40.0	—
B_2 (riboflavin) (μg)	60.0	60.0	—
B_6 (pyridoxine) (μg)	35.0	35.0 (with 15 μg/g of protein in formula)	—
B_{12} (μg)	0.15	0.15	—
Niacin			
(μg)	—	250.0	—
(μg equiv)	800.0	—	—
Folic acid (μg)	4.0	4.0	—
Pantothenic acid (μg)	300.0	300.0	—
Biotin (μg)	—	1.5	—
Choline (mg)	—	7.0	—
Inositol (mg)	—	4.0	—
Minerals			
Calcium (mg)	50.0[d]	40.0[d]	—
Phosphorus (mg)	25.0[d]	25.0[d]	—
Magnesium (mg)	6.0	6.0	—
Iron (mg)	1.0	0.15	—
Iodine (μg)	5.0	5.0	—
Zinc (mg)	—	0.5	—
Copper (μg)	60.0	60.0	—
Manganese (μg)	—	5.0	—
Sodium (mg)	—	20.0 (mEq)[e]	60.0 (17 mEq)[e]
Potassium (mg)	—	80.0 (14 mEq)[e]	200.0 (34 mEq)[e]
Chloride (mg)	—	55.0 (11 mEq)[e]	150.0 (29 mEq)[e]

[a] Adapted from ref. 26.
[b] Modified from Committee on Nutrition (American Academy of Pediatrics, 1976).
[c] Retinol equivalents.
[d] Calcium-to-phosphorus ratio must be no less than 1.1:1.0 nor more than 2.0:1.0.
[e] Milliequivalents for 670 kcal/L of formula.

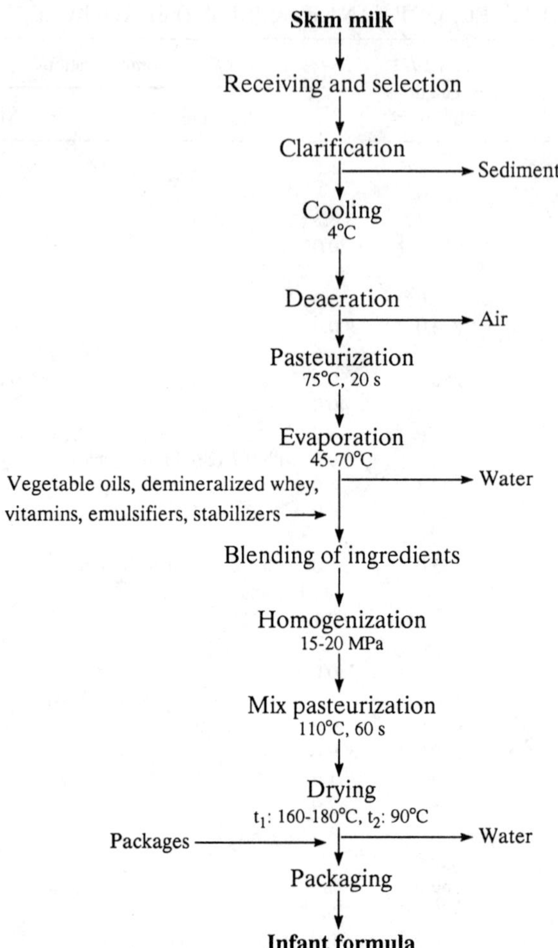

Figure 4.14 Flow chart of infant formula production.

Recommendations for the composition of infant formulas are presented in Table 4.3.[26]

The most difficult modification is to add immuno factors to cow's milk, as these substances are normally present in breast milk, but not in cow's milk. The well-known deficit has prompted many health professionals to recommend breast feeding whenever possible.

The manufacture of infant formulas requires different processes. One is the "dry procedure," where all ingredients are blended in dry form; the other is the "wet procedure" where mixing is done in the wet state prior to drying. Frequently, these methods are combined. In the dry process, the goal is to produce an even blend.

As evident from Fig. 4.14, the spray-drying regimens in the production of infant formulas differ from those for milk powder production because of the high content of lactose and fat (in infant formulas). In addition to lower inlet air temperature, the total solids content of the feed (for infant formulas) is lower than that of milk powder and the drying chamber must be specially constructed to provide cooling. Drying of infant formulas is usually accomplished in the two- or three-stage drying process.

When combining the two procedures, water-soluble components are added to the milk before drying, whereas less soluble components are added in a dry form to the blend after drying. The wet procedure provides the best mixing, resulting in sterile products, whereas the dry procedure is cheaper in capital outlay and operation. The combined method has some advantages for both and is the most used.

4.5.4 Other Products

Apart from those dried dairy products discussed in the early sections, there are three others that are briefly mentioned here.

4.5.4.1 Reconstituted Milk Powder

Milk powder may be reconstituted with potable water to form beverages or may be processed into various other dairy products such as pasteurized milk, sterilized milk, fermented dairy product, and cheeses. Reconstituted milk processing plants have been well established in the dairy industry for over 20 years. Skim milk powder is the most commonly used for reconstitution. It has several advantages over whole milk powder when used for reconstitution:

1. Longer shelf life.
2. Easy adjustment of the ratio of fat to nonfat dry solids.
3. Easier substitution with vegetable fat.
4. Easier recombination.

4.5.4.2 Modified Milk Powder

Modified milk powder has one or more components substituted with ingredients of other origin. This process enables food processors to utilize milk components and by-products in combination with nutritive ingredients of other origin, thus lowering production costs and designing nutritive characteristics for specific purposes. Each of the three macro components in milk may be replaced with ingredients of other origin: lactose with sucrose; milk proteins with vegetable proteins, and milkfat with vegetable fats. Some substitutions may be made simultaneously as needed for a specific product.

4.5.4.3 Imitation Milk Powder

Imitation milks are similar to milk, but unlike modified milks contain no milk components. Indirectly, sodium caseinate produced from casein is sometimes used as the protein ingredient in imitation milks. The other ingredients in the product are mostly the same as those used in modified milk production. Imitation milk powder production is used in a manner similar to modified milk powder and has multiple advantages:

1. It has low productions costs (as the price of vegetable fat and protein is much lower than for the corresponding milk components).
2. It serves well those parts of the world where there are no cattle and no milk production.
3. It has a longer shelf life compared to milk powder.
4. It has a wide variation in composition, depending on the availability of ingredients.

Other dry dairy products include anhydrous milkfat, dried dairy beverages, dietetic dry products, coffee whiteners, dry fermented milk products, dry cream, dry cheese products, dry ice cream mix, dry buttermilk, and single cell protein.

4.6 Dried Dairy Ingredients

4.6.1 Whey Powder

Whey, a by-product from cheese and casein manufacture, was traditionally returned to the farmers as animal feed or as a fertilizer for spreading in the fields. Today, large cheese factories are common and world cheese production continues to rise. It is not economical to use whey in the traditional manner. Industrial processors have been using heat concentrating and drying to make whey a more profitable entity. In addition to the traditional dry whey products, there are other dry products derived from whey as shown in Fig. 4.15.

The basic advantage of processing whey into powder is that there is no residue, whereas the drawback is the need for expensive equipment and a large energy consumption. Converting whey into powder requires a large processing capacity but the price of the final product is low in comparison with other dried or concentrated products (for example, whey protein concentrates).

Whey can be transformed into powder by different techniques and the quality of the product varies with the technology applied (Fig. 4.16).[3,4,11]

For example, different processing procedures affect caking tendency (0 to 100%), lactose crystallization rate (0 to 95%), free water content (1 to 4%), and so on. Caking tendency is affected by the degree of lactose crystallization, as well as the number and size distribution of the crystals.

Procedure a (flow chart in Fig. 4.16), in addition to resulting in a highly hygroscopic product, also uses a great amount of energy because whey can only be concentrated up to 45% of total solids in the evaporator.

By introducing lactose crystallization between evaporation and drying (Fig. 4.16, Procedure b), powder quality and process economy are improved. Crystallization starts in flash coolers or specially designed vacuum coolers and continues in crystallization tanks for 4 to 24 h, with constant agitation during filling and emptying of

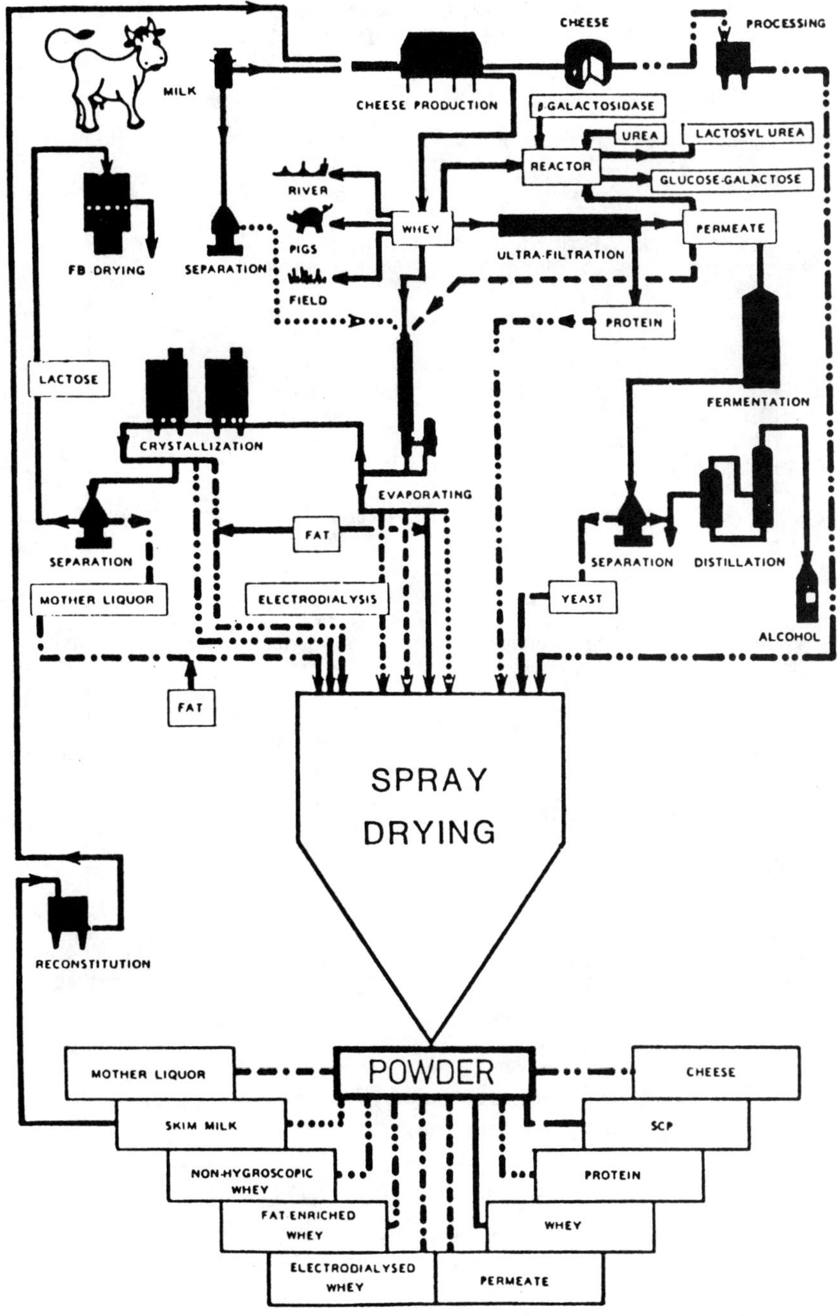

Figure 4.15 Dry dairy products derived from whey. (Courtesy of A/S NIRO Atomizer.) SCP = Single cell protein

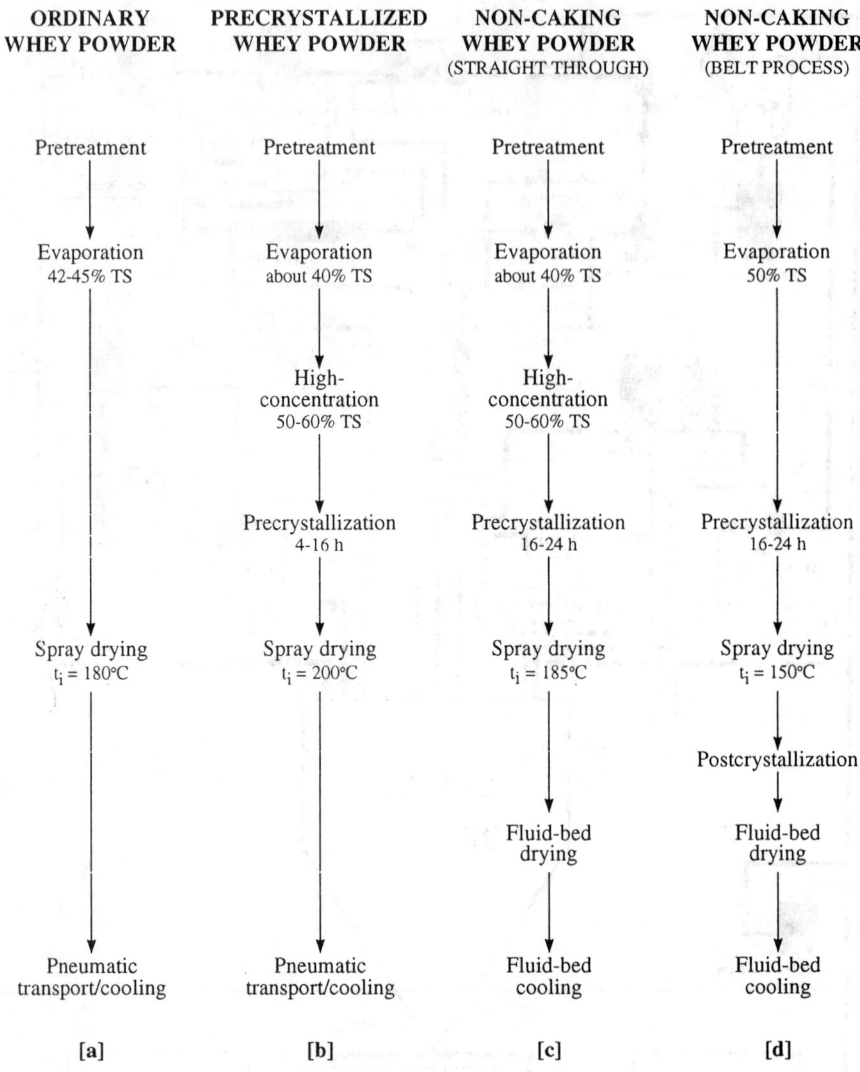

Figure 4.16 Four different procedures of spray drying whey.

the tanks. For crystallization nuclei, pulverized α-lactose monohydrate (0.1%) or crystallized whey powder (8.2%) is used. Quick cooling in flash coolers is accomplished at temperatures up to 30°C which transforms β-lactose into the α-form. The mass is further cooled in the crystallization tank to 10°C at a rate of 3°C/h. During procedures b, c, and d (Fig. 4.16), 50 to 75%, 75 to 85%, and 85 to 95% of the lactose crystallize, respectively.

Whey powder is composed of large agglomerated particles in Procedures c (100 to 500 μm) and e (up to 3000 μm). It has excellent free-flowing characteristics

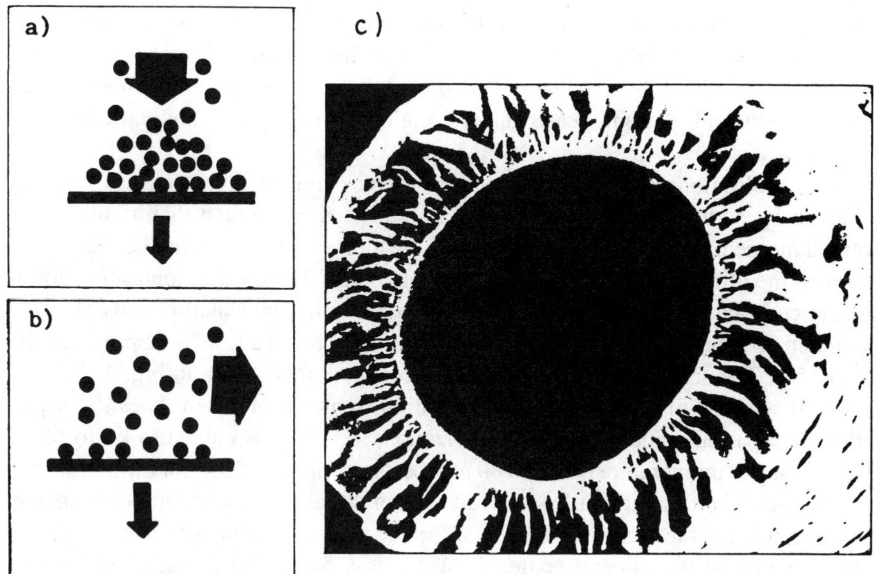

Figure 4.17 Dead-end (a) versus cross-flow (b) ultrafiltration. (c) Cross section of asymmetric membrane of hollow fiber type.

and is not hygroscopic, with no caking tendencies. It is used extensively in food processing.

In all four procedures, reverse osmosis may be used for partial whey concentration (up to 25% total solids), prior to evaporation. This is an energy saving measure. It must be emphasized that the two concentrating plants may be located in different places.

4.6.2 Whey Protein Concentrates

There are several industrial methods suitable for the production of various whey protein concentrates (WPC). The interest in whey processing is a result of two factors. One is a worldwide shortage of high-quality animal proteins that whey proteins may alleviate, and the other is the problem with the disposal of whey. The high biological oxygen demand (BOD) of whey makes this cheese by-product a pollutant so that it is more desirable to process it than to dump it.

In addition to traditional methods such as evaporation and drying, modern methods used in industrial whey processing include ultrafiltration, microfiltration, reverse osmosis (hyperfiltration), and demineralization (electrodialysis, ion exchange). The most commonly used membrane method in dairying is ultrafiltration. Its industrial application was aided by the introduction of cross flow instead of dead-end filtration and the invention of asymmetric membranes[27] (Fig. 4.17).

During the ultrafiltration of whey, low molecular weight compounds such as lactose, minerals, nonprotein nitrogen, and vitamins are separated in the permeate, whereas proteins are concentrated in the retentate. This permits a WPC with 20 to 60% protein in total solids and low quantities of lactose and mineral matter to be obtained. Permeate, a by-product of this processing, is used for producing lactose, alcohol, single cell protein, yeast, galactose, glucose, cattle feed, and various pharmaceuticals.

As ultrafiltration proceeds, an increased protein content of up to 98% may be achieved by adding water to the feed.[28] This proceure is called diafiltration. The optimal moment to start diafiltration is when the total solids content has been reached at which the ultrafiltration flux is still relatively high. That level of total solids must be kept constant during diafiltration in order to minimize the water quantity needed. To obtain 80% protein in total solids, the latter should reach a level of approximately 22 to 25%. The scheme of continuous WPC production is shown in Fig. 4.18.[28]

Sweet whey is first subjected to clarification (removal of casein fines, fat separation, and pasteurization). After pasteurization, the whey is cooled to 60 to 65°C and held at this temperature for 30 to 60 min before cooling to 50°C for ultrafiltration. This heat-and-hold treatment has the function of stabilizing the calcium phosphate complex, and thus reduces the fouling of the membranes during ultrafiltration. Further reduction of the mineral content in WPC is achieved by adjusting pH of the whey to pH 5.7 to 6.0 with HCl. In this way, the solubility of calcium is increased, followed by its greater portion in the permeate. After ultrafiltration, the retentate is pasteurized, evaporated, and dried. Although in Fig. 4.18 evaporation is included in the process, a better solution is to directly dry the product. Depending on the protein content, total solids may be increased from 22 to 25% up to 44% during ultrafiltration, and WPC may be dried directly as obtained from the ultrafiltration plant. This provides a better quality of high protein product. To reduce or avoid protein denaturation, lower temperatures than those for drying milk are used: 160 to 180°C for the inlet temperature and less than 80°C for the outlet air temperature (Fig. 4.18).

4.6.3 Casein Products

4.6.3.1 Casein

Casein is the major milk protein. In addition to the protein moiety, it also contains phosphorus, calcium, and citrate in the structure of its micelles.[29–31]

As the initial pH value of milk is decreased from 6.5, casein starts losing its colloidal dispersibility and stability and begins to precipitate at pH 5.3. Maximum precipitation takes place at pH 4.6, which is the isoelectric point of casein. Casein may also be precipitated by proteolytic enzymes. Depending on the reagent used, the following kinds of casein are produced.[32–35]

1. Acid casein is obtained by precipitating milk with an acid such as hydrochloric, sulfuric, or lactic acid.
2. Sweet casein results from the action of chymosin.
3. Low-viscosity casein is produced by treating milk simultaneously with proteolytic enzymes and an acid.

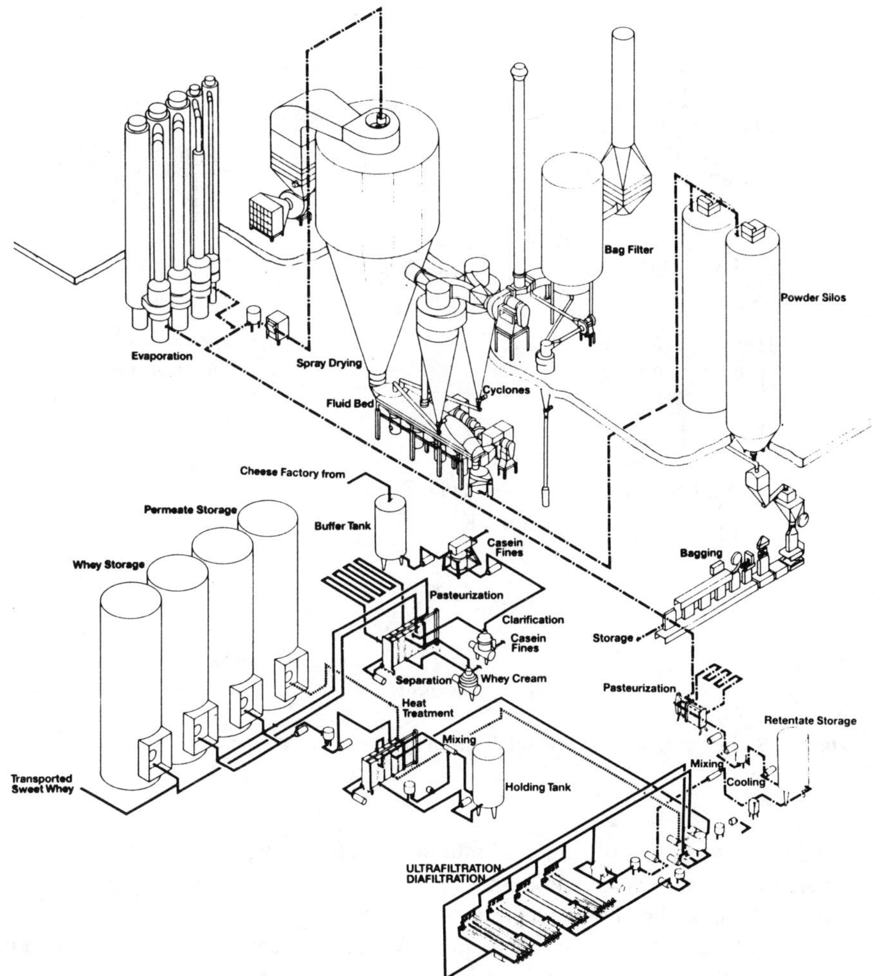

Figure 4.18 Processing plant for production of WPC from sweet whey.

The basic operations in the production of casein are the same irrespective of the type of casein produced. The flow chart of acid casein production, together with sodium caseinate production, is shown in Fig. 4.19.

The precipitation of casein in skim milk is initiated by changing the pH value of the milk using hydrochloric, sulfuric, or lactic acid. The nature of the coagulum (curd) obtained by direct precipitation of skim milk depends on the temperature of precipitation, the intensity of agitation, and the final pH value of the precipitate. The best results are obtained by atomizing a diluted acid solution such as 1.3 to 1.4 *N* HCl in a countercurrent direction to the flow of the milk maintained at 30 to 35°C.

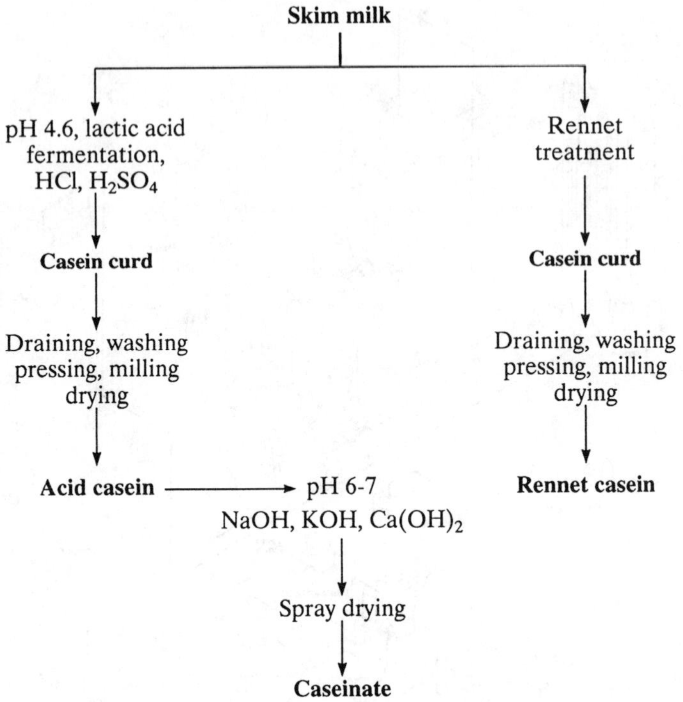

Figure 4.19 Production of commercial casein and caseinate products.

In the next step, steam is injected into the mixture in order to rapidly increase its temperature to cause coagulation, that is, 40 to 45°C. The mixture is subsequently directed into an inclined tube where it coagulates.

Skim milk may also be coagulated in a two-section plate heat exchanger. Acid is injected into the skim milk after it passed through the first section of the heat exchanger, where it was heated to 30°C by heat recuperated from whey processing. The acidified skim milk is then heated to 45°C by hot water in another section of the heat exchanger. The yield of casein may be as high as 99%.

The procedure is the same regardless of the type of the acid used. Hydrochloric and sulfuric acids are most commonly used. The selection of a particular acid depends on economic factors. Preference has been given to hydrochloric acid because it is usually available at a lower cost than sulfuric acid.

An economical, high-capacity production of casein is based on the use of lactic acid as a precipitating agent. Lactic acid can be produced inexpensively by the fermentation of lactose. In New Zealand, almost all acid casein is produced in this way, using cultures of *Streptococcus lactis* and/or *Streptococcus cremoris*.

Initially, this process, as well as all subsequent wet operations, were carried out in cheese vats. The skim milk was inoculated at 25 to 27°C with 0.5 to 1.5% of a mixed lactic acid bacteria starter culture. The coagulation of the skim milk was completed within 16 to 18 h. The temperature of the coagulum was then increased to 50 to 60°C by steam injection. The coagulum was cut with cheese knives and the curd was agitated to facilitate syneresis until the final temperature was reached. The whey was then drained and the curd was washed with water.

In 1963, Muller and Hayes[36] designed a process for the manufacture of low viscosity casein to be used in the paper industry. Such casein can be produced by enzymatic coagulation of milk. Viscosity of a comparable regular acid casein solution is 2 Pa·s whereas a 15% solution of enzymatically produced casein has viscosity of 0.3 to 0.4 Pa·s. In a continuous manufacturing procedure, approximately 40% of the volume of the skim milk to be processed is treated with pepsin and then blended with the remaining skim milk. Curd is formed following acid injection into the blend.

After the coagulation of the curd is completed, it is important to separate the whey from it as soon as possible. This can be accomplished by draining the whey from the holding tank through a decanter or an inclined dewheying screen.

The freshly precipitated casein, from which whey has been separated, is washed in order to remove residual acids, salts, whey proteins, and lactose. The curd should be washed at least three times, with each washing lasting 15 to 20 min in order to ensure that the lactose content in the final product is reduced to a minimum. In the countercurrent flow arrangement, the volume of the washing water is approximately one half of the volume used in the parallel flow washing. The dry matter content of the washed curd is approximately 45%.

In a continuous washing process, the curd is moving through a set of several tanks. To separate the curd from the washing water, the top of each tank is equipped with a 90-mesh draining screen, inclined 60° from the vertical line which separates the curd from the washing water.[33]

In order the preserve the desired curd characteristics during washing, it is important to maintain the pH value of the washing water at 4.6, which is the isoelectric point of casein. If water pH is lower than 4.6, a gelatinous layer may form on the curd particle surface and obstruct the washing. Continuous casein pressing may be accomplished by using a centrifuge, a screw press equipped with a pair of rotating screws pressing and moving the curd, or a mechanically driven roller press equipped with a pair of stainless steel rollers.

The curd is usually milled before drying in order to obtain particles of a uniform size. These will dry evenly through the entire casein mass, thus avoiding incomplete drying of a part of them and scorching of others. Vibrating dryers (fluid-bed dryers) of the type used to dry other milk products are used most frequently to dry casein.

Recently, a new drying procedure called "attrition" drying has been designed. The dryer consists of a rotor and a stator. The curd is ground during this procedure, exposing a large surface to hot air circulating in the dryer and making the drying proceed very rapidly. The resulting powder particles have irregular shapes with a large number of cavities and readily disperse in water.

The objective of tempering is to cool the casein and to evenly distribute the moisture in it. Hot casein, which has an uneven moisture distribution, is plastic and very difficult to grind.

Table 4.4 APPROXIMATE PERCENTAGE COMPOSITION OF COMMERCIAL CASEIN AND CASEINATE PRODUCTS

Components	Sodium Caseinate	Calcium Caseinate	Acid Casein	Rennet Casein	Coprecipitate
Protein, N × 6.38 (min)	94.0	93.5	95.0	89.0	89–94
Ash (max)	4.0	4.5	2.2	7.5	4.5
Sodium	1.3	0.05	0.1	0.02	—
Calcium	0.1	1.5	0.08	3.0	—
Phosphorus	0.8	0.8	0.9	1.5	—
Lactose (max)	0.2	0.2	0.2	—	1.5
Fat (max)	1.5	1.5	1.5	1.5	1.5
Moisture (max)	4.0	4.0	10.0	12.0	5.0
pH	6.6	6.8	—	7.0	6.8

Grinding produces uniform dimensions of the casein particles. They range from 300 to 600 μm in diameter. Particles obtained by attrition drying are considerably smaller, that is, to 150 μm in diameter.[33]

Ground casein is classified according to particle dimensions. It is sifted through a series of gradually increasing mesh number sieves. Classified casein is packaged in bags that are of the same kind as those used for milk powder packaging.

The approximate composition of commercial casein and casein products is presented in Table 4.4.

Casein is used in many industries such as the paper industry, the manufacture of water-based paints, the production of adhesives, the food industry, the manufacture of plastics, the production of casein fibres, the tanning industry, and the manufacture of animal feeds and pet foods.

4.6.3.2 Sodium Caseinate

Casein consists of electrically charged proteins. The charges form polar regions along the polypeptide chain. This makes casein an ampholyte that is capable of reacting either with hydroxides or with acids depending on the pH value of the medium. Casein reacts with various metal ions and forms caeinates such as sodium caseinate, calcium caseinate, and others.

Sodium caseinate is commonly manufactured by a continuous process[32–35] in which thoroughly washed acid casein is used as the starting material. In addition to raw casein, dry acid casein is also suitable as the starting material in the production of sodium caseinate. Irrespective of the starting material used, the manufacture of sodium caseinate consists of the formation of a casein suspension, solubilization of casein using sodium hydroxide, and drying the sodium caseinate produced (Fig. 4.19). Raw acid casein is milled in a continuous mill and subsequently suspended in a hot water tank.

The casein suspension is pumped from the holding tank into another tank while the sodium hydroxide solution is simultaneously injected through a mixer. Water is also added in order to maintain the total solids content of the caseinate solution below the 20 to 22% level. The total solids content of the solution destined for spray drying is 25 to 31% lower than that of milk, which is usually in the 45 to 55% range. The low dry matter content, dictated by the requirement to maintain a low viscosity of the sodium caseinate solution, increases the production costs. The viscosity of sodium caseinate solutions is a logarithmic function of the total solids concentration. In order to increase the solids concentration to a maximum, a relatively high solubilization temperature of 90 to 95°C is applied. The viscosity is lowest in the pH range of 6.6 to 7.0. The raw acid casein must be completely free of lactose; otherwise conditions favorable to the induction of Maillard reactions leading to the discoloration of the product would develop.

The homogeneous sodium caseinate solution obtained in the preceding operation is usually spray dried in a stream of hot air. Only rarely is sodium caseinate dried by roller drying. The total solids content of the solution destined for spray drying ranges from 20 and 22% and may be exceptionally as high as 25%. The highest permissible caseinate concentration is determined experimentally for every individual spray dryer.

All sodium caseinate produced commercially is used in the food industry. The following foods are examples of products containing sodium caseinate: various kinds of sausages, meat-based instant breakfast and milk-based instant breakfast, modified milk, whipped cream, coffee whiteners, ice cream, desserts, sauces, soups, casein bread, doughs, crackers, biscuits, dietetic products, and various protein-enriched products. The two main reasons for using sodium caseinate as an ingredient in foods are its functional properties and nutritive value.

4.6.3.3 Coprecipitates

In coprecipitate processing, high-temperature treatment of skim milk leads to the interaction of the β-lactoglobulin fraction of the whey proteins with κ-casein. The heat-induced κ-casein—β-lactoglobulin complex is then coprecipitated with casein by an acid, or another chemical agent such as $CaCl_2$, or a mix of the two.[32–35] Other milk proteins are coprecipitated together with the casein–lactoglobulin complex. Coprecipitates were patented in the 1950s and became more popular in the 1970s. Their advantage over casein and its compounds is that they also consist of whey proteins that contain relatively high concentrations of sulfur-containing amino acids. This factor contributes to the biological value of coprecipitates. In addition, the coprecipitate procedure increases the recovery of milk proteins.

In order to produce coprecipitates, skim milk is preheated and the final heating of up to 90°C in the second stage is obtained by steam injection into the milk. $CaCl_2$ or acid is also injected through spray countercurrent to the direction of milk flow to provide full mixing. The mixture is transformed into curd in a holding tube (20 to 25s). The curd is separated from the whey and the coprecipitate is washed, pressed, and dried. At optimal process conditions it is possible to recover 95 to 97% of the milk proteins. There are three basic varieties of coprecipitates, each having different amounts of calcium[33]: low-calcium coprecipitate (LCC, 0.1 to 0.5% Ca), medium-calcium coprecipitate (MCC, 1.0 to 1.5% Ca), and high-calcium coprecipitate (HCC, 2.5 to 3.0% Ca). The calcium concentration in coprecipitates can be changed by changing basic parameters in the production process. A higher pH value at precipitation results in a higher calcium concentration in the product, whereas longer retention time at high temperature decreases calcium concentration.

Coprecipitates with different concentrations of calcium and polyphosphate and different ratios of serum protein and casein have various uses in the food industry. They each serve the same purpose as caseinates. The production process of coprecipitates has been developed in order to recover not only casein, which is about 80% of all milk protein, but other proteins as well. This increases the recovered protein to nearly 96%.

4.6.4 Lactose

Lactose is a disaccharide consisting of D-glucose and D-galactose. In the chemical nomenclature, lactose is called 4-O-β-D-galactopyranosyl-D-glucopyranose. It is the major component of total milk solids and can be isolated on a commercial scale from whole whey or from deproteinized whey.[37–39] More recently, as the use of membrane methods for the concentration and fractionation (ultrafiltration, hyperfiltration, etc.) of milk in the dairy industry is being expanded, the permeate obtained by the ultrafiltration of whey is being used as the starting material in the production of lactose.

Technological processes used to produce lactose may be divided into two basic groups:

1. Crystallization of lactose from whey in the presence of whey proteins.
2. Crystallization of lactose from deproteinized whey after the removal of whey proteins. Crude or refined lactose can be produced by either of these processes.

Lactose manufacture is shown in Fig. 4.20.[37]

The raw material for lactose production is evaporated in multistage vacuum evaporators or may be subjected to preliminary concentration by reverse osmosis, as well. The final concentration of lactose depends on whether proteins are present in the syrup. If lactose is produced from protein-containing whey, the syrup is evaporated to increase its dry matter content to 60 to 65%. In the production of lactose from deproteinized whey, the dry matter content of the syrup may be increased as high as 70%.

Lactose crystallization is initiated in the hot syrup that had been concentrated to oversaturation. The crystallization is initiated either spontaneously in oversaturated syrups that are in an unstable crystallization state, or following the introduction of seed crystals into syrups that are in the metastable crystallization state. The objective of crystallization is to produce a large number of similar sized crystals (0.2 mm average diameter) which would be easy to separate from the molasses.

A crystallizer is a double-walled closed tank having a conical bottom. It is equipped with slow-motion agitators and scrapers which prevent the formed lactose crystals from sticking to each other and from sedimenting.

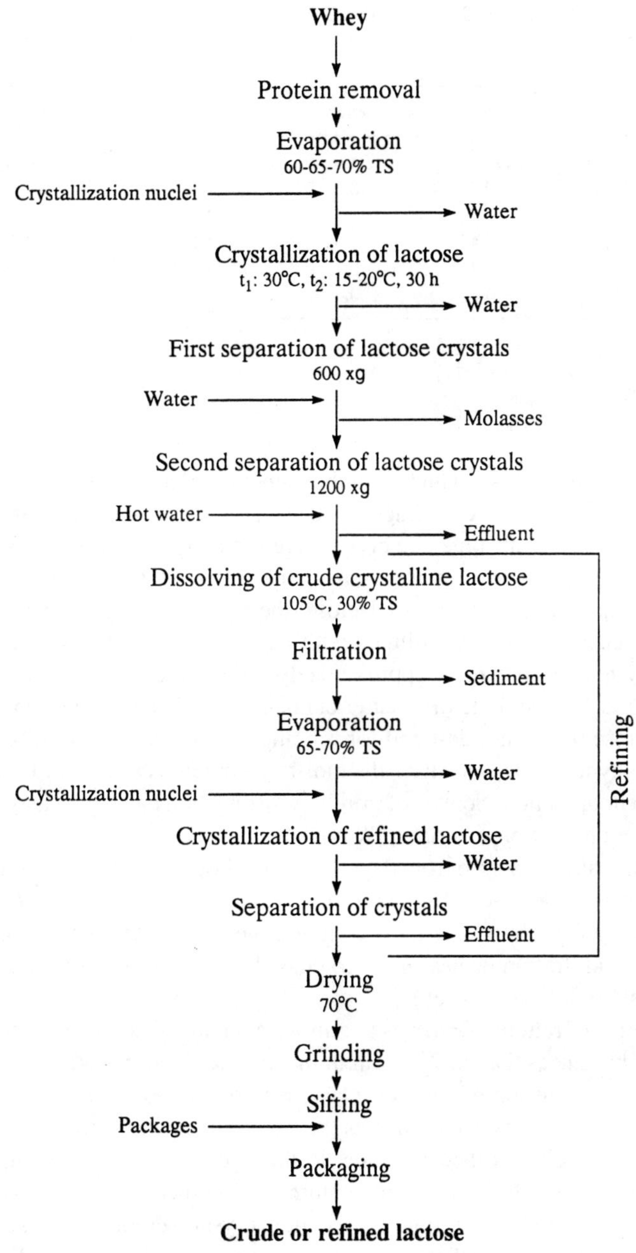

Figure 4.20 Flow chart of the production of crude or refined lactose.

Table 4.5 COMPOSITION OF COMMERCIAL LACTOSE PRODUCTS

	Lactose			
Component (%)	Technical	Row	Edible Grade	Pharmaceutical Grade
Lactose	98.0	94.0	99.0	99.4–99.85
Moisture	0.35	0.3	0.5	0.1– 0.5
Protein (N × 6.38)	1.0	0.8	0.1	0.01– 0.05
Ash	0.45	0.4	0.2	0.03– 0.09
Fat	0.2	0.1	0.1	0.001– 0.01
Acid (as lactic acid)	0.4	0.4	0.06	0.04– 0.03

From ref. 4.

Crude crystalline lactose, which is in the α-monohydrate form, is separated from the molasses in continuous centrifuges or decanters. Two centrifuges are used in a sequence. In the first centrifuge, the crystals are separated from the molasses, and in the other centrifuge, the crystals are washed with water. Molasses, which contain 38 to 48% of dry matter, including 30% lactose (the rest consists of proteins and salts), may also be recycled. They are diluted with fresh whey or with the wash water to contain a dry matter content of approximately 15%. Crude lactose has a moisture content of 10 to 14% and the dry matter contains approximately 99% lactose.

Crude lactose that is not destined for refining is dried at approx. 70°C in one of the numerous types of dryers where the moisture content is reduced to 0.1 to 0.5%. The subsequent operations consist of grinding, sifting, and packaging and are similar to those in the production of skim milk powder.

The manufacture of lactose from deproteinized whey differs from the manufacture of lactose using whole whey. The major difference is the removal of proteins at the beginning of the operation. The most common method for the removal of proteins is based on ultrafiltration or heat-induced coagulation by steam injected into whey acidified to pH 6.2 (Centri Whey).[40]

The objective of refining lactose is to remove contaminants such as proteins, salts, and colored substances that may remain in the mix. Refined lactose is almost chemically pure. It contains a minimum of 99.6% lactose and no protein.

The production process is the same as that for crude lactose except the separation of the lactose crystals and their washing. Refining consists of dissolving the crude crystalline lactose in water at high temperature, adding specific chemicals (e.g., charcoal and/or filtration aids), filtration, evaporation, crystallization of lactose, and separation of the crystals. The subsequent operations such as drying, grinding, sifting, and packaging are the same as those for crude lactose production.

Agglomerated lactose powder is produced using the same procedures as those used in the production of instant milk powder. This form of lactose is used in the pharmaceutical industry.

The average composition of commercial forms of lactose is presented in Table 4.5.

4.7 References

1. Hall, C. W., and T. I. Hedrick. 1975. *Drying of Milk and Milk Products.* AVI, Westport, CT. 338 pp.

2. Wiegand B. 1985. Evaporation. *In* R. Hansen (ed.), *Evaporation, Membrane Filtration and Spray Drying in Milk Powder and Cheese Production.* North European Dairy Journal, Vanløse, Denmark, pp. 91–178.

3. Masters, K. 1984. *Spray Drying Handbook*, 4th edit. George Godwin, London. 696 pp.

4. Carić, M. 1990. *Technology of Concentrated and Dried Dairy Products*, 3rd edit. Naučna Knjiga, Beograd, Yugoslavia, 293 pp. (in Serbian).

5. Kiermeier, F., and E. Lechner. 1973. *Milch und Milcherzeugnisse.* Paul Parey, Berlin, Germany, 443 pp.

6. Food and Drug Administration. 1978. *Standards, Food and Drugs: Evaporated Milk*, 131.130, 153 pp.

7. Swaisgood, H. E. 1986. Chemistry of milk protein. *In* P. F. Fox (ed.), *Developments in Dairy Chemistry-1. Proteins*, pp. 1–60. Elsevier Applied Science, London.

8. Holt, C. 1985. The milk salts: their secretion, concentrations and physical chemistry. *In* P. F. Fox (ed.), *Developments in Dairy Chemistry-3. Lactose and Minor Constituents*, pp. 143–182. Elsevier Applied Science, London.

9. Kessler, H. G. 1988. *Lebensmittel- und Bioverfahrenstechnik. Molkereitechnologie,* 3rd edit. A. Kessler, Freising, Germany, 582 pp.

10. Knipschildt, M. E. 1986. Drying of milk and milk products. *In* R. K. Robinson (ed.), *Modern Dairy Technology Advances in Milk Processing,* Vol. 1, pp. 131–234. Elsevier Applied Science, London.

11. Westergaard, V. 1983. *Milk Powder Technology, Evaporation and Spray Drying,* 3rd edit. Niro Atomizer, Copenhagen, Denmark, 147 pp.

12. Hallstrøm, B. 1985. Evaporation versus hyperfiltration. *In* R. Hansen (ed.), *Evaporation, Membrane Filtration and Spray Drying in Milk Powder and Cheese Production*, pp. 289–298. North European Dairy Journal, Vanløse, Denmark.

13. Walstra, P. 1983. Physical chemistry of milk fat globules. *In* P. F. Fox (ed.), *Developments in Dairy Chemistry-2. Lipids*, pp. 119–158. Elsevier Applied Science, London.

14. Carić, M. 1988. Nonenzymic browning of dairy products. *In Reactions of Nonenzymic Browning of Food Products*, pp. 105–141. Naučna Knjiga, Beograd, Yugoslavia (in Serbian).

15. Morrissey, P. A. 1985. Lactose: chemical and physiochemical properties. *In* P. F. Fox (ed.), *Developments in Dairy Chemistry-3. Lactose and Minor Constituents*, pp. 1–34. Elsevier Applied Science, London.

16. Masters, K. 1985. Spray drying. *In* R. Hansen (ed.), *Evaporation, Membrane Filtration and Spray Drying in Milk Powder and Cheese Production*, pp. 299–346. North European Dairy Journal, Vanløse, Denmark.

17. Pisecky, J. 1983. New generation of spray dryers for milk products, *Dairy Indust. Int* **48:** 21–24.

18. Carić, M., and M. Kaláb, 1987. Effects of drying techniques on milk powders quality and microstructure: a review. *Food Microstructure* **6:** 171–180.

19. Peebles, D. D. 1936. U.S. Patent 2,054,441.

20. Peebles, D. D., and D. D. Clary, Jr. 1955. U.S. Patent 2,710,808.

21. Peebles, D. D. 1958. U.S. Patent 2,835,586.

22. Pisecký, J. 1985. Technological advances in the spray dried milk. *J. Soc. Dairy Technol.* **38:** 60–64.

23. Carić, M. 1991. Drying technologies developments in dairying, Marschall Rhône-Poulenc International Dairy Science Award Lecture, ADSA Annual Meeting, Logan, Utah, USA.

24. Carić, M. 1993. *Concentrated and Dried Dairy Products.* VCH Publishers, New York.

25. Corbin, E. A., and E. O. Whittier. 1972. The composition of milk. *In Fundamentals of Dairy Chemistry*, pp. 1–36. AVI, Westport, CT.

26. Packard, V. S. 1982. *Human Milk and Infant Formula.* Academic Press, New York, 269 pp.

27. Loeb, S., and S. Sourirajan. 1964. U.S. Patent 3,133,132.

28. Ottosen, N. 1991. Membrane filtration for whey protein concentrate. Marketing Bulletin, APV Pasilac AS, Copenhagen, Denmark, pp. 3–22.

29. Schmidt, D. G. 1986. Association of caseins and casein micelle structure. *In Developments in Dairy Chemistry-1. Proteins,* pp. 61–86. Elsevier Applied Science, London.

30. Belitz, H. D., and W. Grosch. 1986. *Food Chemistry*, 2nd edit. Springer-Verlag, Berlin, Germany, 774 pp.

31. Kaláb, M., E. Phipps-Todd, and P. Allan-Wojtas. 1982. Milk gel structure XIII. Rotary shadowing of casein micelles for electron microscopy. *Milchwissenshaft* **37:** 513–518.

32. Morr, C. V. 1986. Functional properties of milk proteins and their use as food ingredients. *In* P. F Fox (ed.), *Developments in Dairy Chemistry-1. Proteins*, pp. 375–400. Elsevier Applied Science, London.

33. Muller, L. L. 1986. Manufacture of casein, caseinates and co-precipitates. *In* P. F. Fox (ed.), *Developments in Dairy Chemistry-1. Proteins*, pp. 315–338. Elsevier Applied Science, London.

34. Mulvihill, D. M. 1989. Caseins and caseinates: manufacture. *In* P. F. Fox (ed.), *Developments in Dairy Chemistry-4. Functional Milk Proteins*, pp. 97–130. Elsevier Applied Science, London.

35. Southward, C. R. 1986. Utilization of milk components: casein. *In* R. K. Robinson (ed.), *Modern Dairy Technology,* Vol. 1, pp. 317–368. Elsevier Applied Science, London.

36. Muller, L. L. and E. J. Hayes. 1963. The manufacture of low-viscosity casein. *Austr. J. Dairy Technol.* **18:** 184–188.

37. Østergaard, B. 1988. Lactose from ultrafiltration permeate. Marketing Bulletin, APV Pasilac AS, Copenhagen, Denmark, pp. 3–14.

38. Modler, H. W. 1984. Functional properties of nonfat dairy ingredients: a review. *J. Dairy Sci.* **68:** 2206–2214.

39. Brinkmann, G. E. 1976. New ideas for the utilization of lactose: principles of lactose manufacture. *Soc. Dairy Technol.* **29:** 101–107.

40. Anonymous. 1988. *Dairy Hand Book*, Alfa-Laval, AB, Lund, 219 pp.

CHAPTER

5

Dairy Microbiology and Safety

Purnendu C. Vasavada and Maribeth A. Cousin

5.1 Introduction

An understanding of the microbiological aspects of milk is essential to those who successfully pursue production, processing, and manufacturing of quality milk and dairy foods. The significance of microbes occurring in milk is multifold: microorganisms can be either beneficial or harmful depending on the circumstances of their presence and activities in milk and dairy products. The adverse publicity resulting from the foodborne illness outbreaks or widespread recalls of cheese, ice cream, and other dairy foods found to be contaminated with foodborne pathogens could be devastating to the economy of the dairy industry. More importantly, the consumer anxiety associated with safety and wholesomeness of milk and dairy foods can lead to lack of consumer confidence and further reduce consumption of milk and dairy foods.

The microbiological quality of raw milk is critical for production of superior quality dairy foods with reasonable and predictable shelf life. Milk and other ingredients used in manufacturing must be free of offensive odors and flavors and undesirable microorganisms capable of causing spoilage. The psychrotrophs—microorganisms capable of growth at 7°C (45°F) regardless of their optimum growth temperature—are the predominant spoilage organisms in milk and dairy foods. Although the majority of psychrotrophs are relatively heat sensitive and are readily inactivated by conventional pasteurization, the thermoduric psychrotrophs and psychrotrophic spore-formers can be important in causing flavor and texture defects and spoilage of long-shelf-life dairy products. Moreover, psychrotrophic bacteria, particularly *Pseudomonas* spp., are capable of producing heat-stable enzymes[1,2] during growth in milk prior to pasteurization. These enzymes have an adverse effect on the quality and yield of products prepared from such milk, for example, Adams et al.[3] and Patel et al.[4] have reported on studies of heat-stable proteases resulting from the growth of psychrotrophs in milk. The psychrotrophic bacteria and their role in quality and shelf life of milk and dairy products have been studied extensively. Several

Table 5.1 REQUIREMENTS FOR GRADE A MILK

	SPC	<100,000/ml
	Coliform	—
Raw milk	Temperature	<40°F
Pasteurized milk	SPC	<200,000/ml
	Coliform	<10/ml
	Phosphatase	Negative

excellent reviews on psychrotrophic bacteria and their importance in milk and dairy product quality have appeared in the literature.[5–11]

Recent findings of the psychrotrophic nature of the so-called emerging pathogens, that is, *Listeria monocytogenes*, Yersinia enterocolitica, and others have provided additional significance to the problem of psychrotrophic contamination in milk.

The beneficial aspects of microorganisms in milk mainly apply to the cultured or fermented dairy foods. Microorganisms capable of degrading lactose to lactic acid and other compounds such as acetic acid, propionic acid, acetaldehyde, and diacetyl are used as starter cultures in the dairy industry. Proper selection, propagation, and behavior of starter cultures is crucial in manufacturing a variety of cheeses, yogurts, and other cultured milk products. Recent developments in starter culture genetics have increased the potential for development of specific bacterial strains to improve flavor and nutritional quality, control pathogenic bacteria, and resist infection by bacteriophage.[12]

The significance of microorganisms, particularly pathogens, was clearly evident in the earlier days of the dairy industry as milk and milk products were the major vehicle for dissemination of pathogens. The diseases once commonly spread by milk include typhoid fever, septic sore throat, tuberculosis, diphtheria, scarlet fever, and undulant fever.[13,14] The advent of tuberculin testing, brucellosis eradication programs, and mandatory pasteurization led to a dramatic decline in the incidence of milkborne disease outbreaks.

The importance of microorganisms in milk is recognized in the fact that the microbiological limits for total bacterial count and the coliform count are the crucial parts of the milk quality grades and compliance with the regulations in the Pasteurized Milk Ordinance (PMO) (Table 5.1).

Because milk provides nutrients, near neutral pH, and a high water activity (A_w) preferred for reproduction of microorganisms, it can serve as a growth medium for a wide variety of microorganisms. The main objective of this chapter is to provide a broad overview of the microbiological aspects of milk and milk products.

5.2 General Dairy Microbiology

The microbiological analysis of milk and dairy products may reveal many diverse types of microorganisms. The magnitude and diversity of microbial populations vary considerably depending on the specific production, processing, and postprocessing storage and distribution conditions associated with a particular batch of milk. The routine microbiological analysis of milk and dairy products generally involves enumeration of spoilage and indicator organisms, although detection and enumeration of pathogenic organisms such as *Salmonella* may be involved in certain products. An understanding of general dairy microbiology is essential for appreciation of the significance of proper procurement, processing, handling, and storage of milk and dairy products.

5.2.1 Morphological Features

Microorganisms in milk and dairy products can be categorized in various groups based on their morphological features: the shape; size; presence of specific cellular structures, that is, flagella, spores, and capsule; Gram reaction; and the organization of a group of cells as in a pair, chain, cluster, etc. Yeasts and molds are among the largest microorganisms, being several times larger than bacteria. Viruses and rickettsia are smaller than bacteria. Bacteria are the most important microorganisms in milk and milk products, although bacteriophages causing problems in cultured dairy products manufacturing by attacking the starter culture are also of considerable concern. Various morphological features of microorganisms are shown in Figure 5.1.

5.2.2 Microorganisms Associated with Milk

5.2.2.1 Bacteria

The latest classification of milkborne microorganisms is given in *Bergey's Manual of Systematic Bacteriology*, Volumes I and II.[15,16] In *Bergey's Manual*, the bacteria are grouped according to the Gram reaction, morphology, and relation to growth with or without oxygen. Summary information about these groups is given below. Detailed information on microorganisms associated with milk and milk products may be found in several reference sources.[17–19]

The Spirochetes

The only organism of importance in dairy microbiology classified in this section is *Leptospira interrogans.* Organisms in the genus *Leptospira* are flexible, helicoidal rods (0.1 $\times$ 6 to 12 μm), Gram negative, motile, and obligately aerobic. They inhabit the kidney of the animal or human and are shed in the urine. Clinical effects of leptospira vary from an influenza type illness to a severe icteric form. The optimum growth temperature for the organism is 28 to 30°C.

Gram-Negative, Aerobic/Microaerophilic, Motile, Helical (Vibroid) Bacteria

In this group only one organism, *Campylobacter jejuni*, is important in milk and dairy foods. The organisms in genus *Campylobacter* are slender, curved rods (0.2 to 5.0 μm $\times$ 0.5 to 5 μm). They are microaerophilic in nature and require 3 to 15%

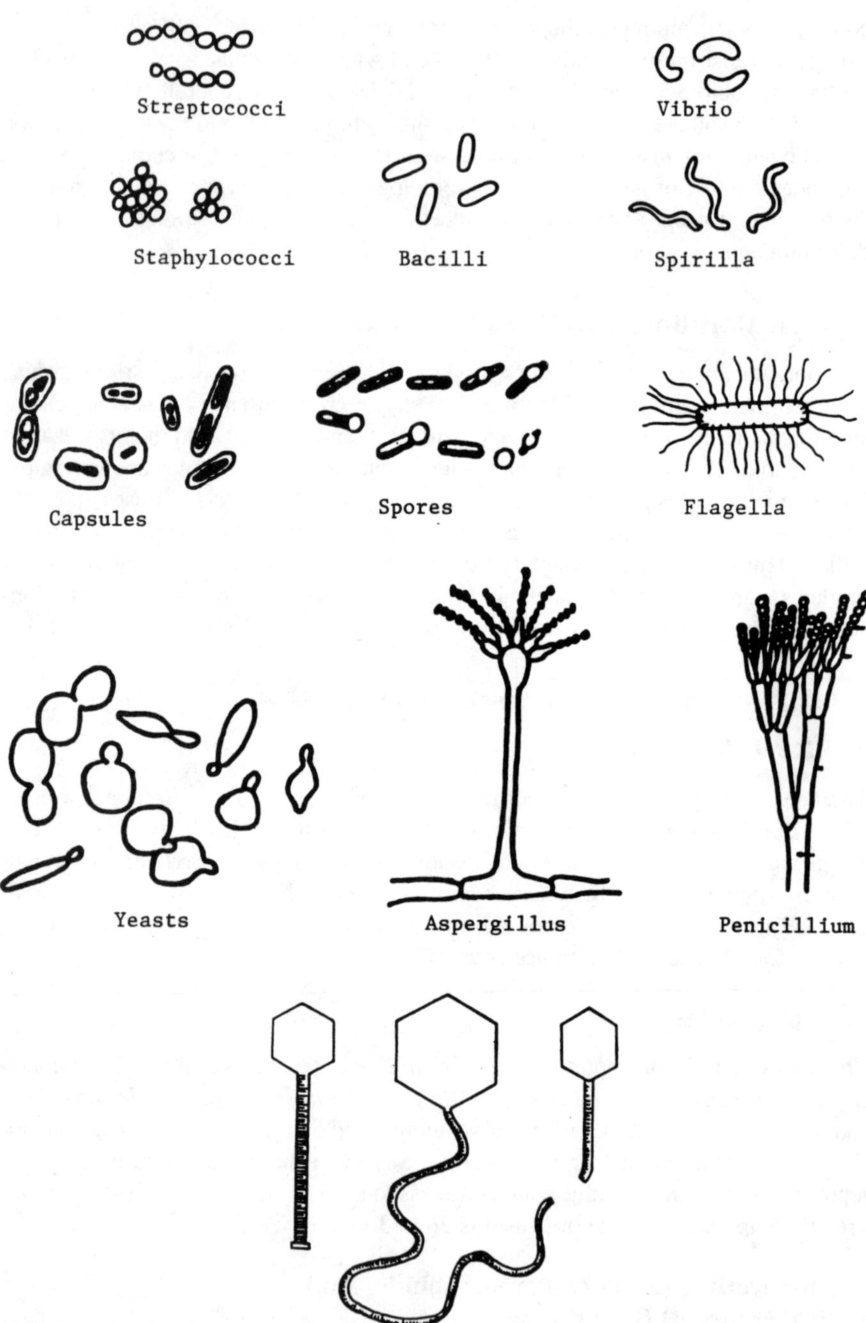

Figure 5.1 Morphological features of microorganisms.

O_2 and 3 to 5% CO_2 in their atmosphere for growth. The optimum growth temperature is 42°C, although *C. jejuni* is known to survive in milk stored at low temperatures.[20–22] *Campylobacter* is known to cause abortion in cows. In humans, *C. jejuni* has been the cause of several milkborne illness outbreaks associated with the consumption of raw or unpasteurized milk.[23–26]

C. jejuni can cause mastitis in cows.[27,28] Since 1980, the incidence of campylobacteriosis has increased drastically and the organism has surpassed *Salmonella* as the main causative agent of foodborne illness.[29,30]

Gram-Negative, Aerobic Rods and Cocci

The organisms important in dairy microbiology in this group include two families: the *Pseudomonadaceae* and the *Neissiriaceae* and four genera of uncertain status, *Alcaligenes, Alteromonas, Flavobacterium,* and *Brucella.*

Pseudomonas. The pseudomonads are Gram-negative, straight to curved (0.4 to 1.5 μm $\times$ 0.7 to 5.0 μm), motile (polar flagella) rods. The pseudomonads are capable of producing heat-stable proteases and lipases that can lead to flavor and texture defects in milk and milk products. *Pseudomonas fluorescens* and *Pseudomonas fragi* are the two organisms recognized among the predominant psychrotrophic organisms causing spoilage in milk and dairy products during storage at refrigeration temperatures. Prior growth of *P. fluorescens* in milk can increase percent insolubility, percent foam volume, and average dispersibility of freeze-dried, nonfat dried milk when numbers in raw milk exceed 1×10^6 cfu/m1.[18] *P. fragi* is rarely caseolytic but it is known to cause lipolytic and "fruity and fermented" defects characteristically associated with psychrotrophic spoilage of milk and dairy products.

P. fluorescens often produces diffusible fluorescent pigment (pyoverdin). *P. fragi* rarely produces fluorescent pigment, but some strains may produce a brown, diffusible pigment. Most pseudomonads are obligately aerobic although some strains can use nitrate as a terminal acceptor. The optimum growth temperature of pseudomonads is 25 to 30°C, but some strains can grow at 4°C.

Xanthomonas. This genus contains plant pathogenic organisms that are straight, motile rods, catalase positive, oxidase negative, and produce yellow pigment. It has been suggested that the organism *Pseudomonas maltiphilia* be transferred in this genus because it produces yellow pigment, is oxidase negative, and gives a negative nitrate reduction reaction.[18] The optimum growth temperature of the organism is ≈35°C. It does not grow at 4°C but may grow at 41°C. It has been isolated from water, milk, and frozen foods.

Alcaligenes. This genus contains organisms that are motile, aerobic rods (0.5 to 1.0 μm to 0.5 to 2.6 μm). The optimum growth temperature is 20 to 37°C although they are part of the typical psychrotrophic contaminants found in raw milk. *Alcaligenes viscolactis* is associated with ropiness in milk.

Flavobacterium. They are part of the psychrotrophic microflora in raw milk. The organisms are aerobic, nonmotile rods (0.5 μm $\times$ 1.0 to 3.0 μm), oxidase negative and phosphtase positive. Many strains produce yellow to orange pigment although some strains do not produce pigment. *Flavobacterium* species can hydrolyze casein and cause psychrotrophic spoilage of milk and dairy products.

Brucella. This genus contains coccobacillary or short rods (0.5 to 0.7 μm to 0.6 to 1.5 μm) that are catalase positive, oxidase positive, and grow optimally at 37°C. Many strains are pathogenic to cattle (*B. abortus*), pigs (*B. suis*), and goats (*B. melitensis*). *B. abortus* causes abortion in cattle and undulant fever in man. Annually about 100 to 200 cases of brucellosis occur in the United States. The organism primarily affects slaughterhouse workers, veterinarians, livestock producers, and others. *Brucella* are readily destroyed by pasteurization of milk. The organisms are difficult to work with in the laboratory and specific safety precautions should be taken when working with *Brucella*.[31]

It has been suggested that only *B. melitensis* should be recognized as a separate species with others being recognized as biovars.[18]

Alteromonas. This genus contains *Alteromonas putrefaciens*, a Gram-negative, facultative, rod-shaped (0.5 to 1.0 μm × 1.1 to 4.0 μm) organism, previously known as *Pseudomonas putrefaciens*. It grows optimally at 20 to 25°C and produces a nondiffusible reddish-brown or pink pigment. The organism is associated with spoilage of milk and dairy products, particularly surface taint of butter. Clinical isolates of the organism may grow at temperatures as high as 42°C. The organism may be an opportunistic pathogen in immunocompromised individuals.

Acinetobacter *and* Moraxella-*Like Organisms*. *Acinetobacter* are Gram-negative, aerobic, plump short rods. They are found in a variety of raw and prepared foods, including milk in which they may be the cause of ropiness or enzymatic defects. *Moraxella*-like organisms are oxidase positive psychrotrophs and may be associated with spoilage of milk and milk products during storage at refrigeration temperatures.

Gram-Negative, Facultative Anaerobic Rods

Families *Enterobacteriaceae* and *Vibrionaceae* listed in this group include genera whose members are often found to be associated with milk and dairy products. The latest edition of *Bergey's Manual* lists several genera—*Obesumbacterium*, *Xenorhabdus*, *Rhanella*, *Cedecea*, and *Tatumella*—that only recently have been recognized as part of this group and whose significance in dairy microbiology is not yet known.

The organisms in this group are either aerobic or facultatively anaerobic rods (0.3 to 1.0 μm × 1.0 to 6.0 μm), oxidase negative and mostly catalase positive, capable of producing acid and gas from carbohydrate sources. Many of the organisms inhabit intestinal tracts of man and animals and are often important in dairy and food microbiology as "indicator" organisms. The coliform test, one of the common tests used in dairy microbiology for evaluating milk quality, particularly postprocessing contamination and poor sanitary practices in dairy plants, is designed to enumerate coliform organisms belonging to this group. The coliforms are defined as aerobic and anaerobic, Gram-negative, non-spore-forming rods, able to ferment lactose with production of acid and gas at 32°C within 48 h. Typically, the coliform groups include members of the genera *Escherichia*, *Enterobacter*, *Klebsiella*, and *Citrobacter*.

Escherichia. This genus contains *E. coli* which is described as motile or nonmotile rods that ferment lactose, glucose, and other carbohydrates to form acid and gas. The organism is oxidase, urease, and H_2S negative and gives + + − − results with the IMVIC (indole, methyl red, Voges–Proskauer, and citrate) tests. Strains capable of producing gas from lactose within 24 h at 44°C are known as fecal coliforms and usually indicate potential for pathogenic contamination. Recently enterohemorrhagic *E. coli*, *E. coli* 0157:H7 which is capable of causing foodborne illness characterized by the hemolytic uremic syndrom (HUS), hemorrhagic colitis, and thrombotic thrombocytopenic purpura (TPP), has been recognized as an important emerging pathogen.[32] Although primarily associated with undercooked and raw ground beef, pork, and poultry, the organism has been isolated from unpasteurized milk from the bulk tank of a farm from which raw milk was suspected as the vehicle of *E. coli* 0157:H7 in a case of HUS in an infant.[32,33] Two typical characteristics of *E. coli* 0157:H7 bear mention here. Unlike most *E. coli* isolates of human origin, *E. coli* 0157:H7 does not ferment sorbitol and it lacks glucuronidase activity, the latter being responsible for negative results with a rapid fluorogenic assay for detecting *E. coli* based on the hydrolysis of 4-methyl umbelliferyl β-D-glucuronide (MUG) and subsequent formation of a fluorogenic product.[32]

Enterobacter. The genus *Enterobacter* contains eight species, although only five are included in the latest edition of *Bergey's Manual*. Of these, only *Enterobacter aerogenes* is of importance in dairy microbiology. *Enterobacter* vary in their ability to ferment lactose, for example, *E. cloacae* generally ferments lactose, *E. aerogenes* may take 3 to 7 days for lactose fermentation, and *E. agglomerans* gives variable results of lactose fermentation. Therefore, the extent to which the *Enterobacter* contribute to the ''coliform'' group is quite variable. *E. aerogenes* is one of the five of the eight *Enterobacter* spp. having clinical significance.[34]

Klebsiella. This genus contains *Klebsiella pneumoniae* and *Klebsiella oxytoca*, among others that may be found to be associated with milk and dairy environment. The organisms are nonmotile, encapsulated rods. *Klebsiella* spp. occur in a variety of sources: fresh vegetables and fruit, soil, dust, milk and dairy products, air, and water. A contaminated milk shaker mixer is reported to be the cause of *Klebsiella* contamination in high-calorie milk shakes.[35] *Klebsiella* spp. have also been known to be associated with mastitic infections in cows.[36] The major source of the contamination is bedding materia.[36,37]

Salmonella. Several classification schemes are currently used to classify various species and strains of *Salmonella*. The Edwards and Ewing scheme lists three species of salmonellae: *S. typhi*, *S. cholerasuis*, and *S. enteridis*. However, the principle scheme used for salmonellae is the Kauffmann–White scheme, based on the somatic (O); Capsular (Vi), or flagellar (H) antigenic profile of various *Salmonella* strains.

Salmonellae are facultatively anaerobic, Gram-negative, motile (peritrichous flagella) rods (0.3 to 1.0 μm × 1.0 to 6.0 μm) that produce gas from glucose and use citrate as carbon source. They are oxidase negative, catalase positive, produce H_2S, decarboxylate lysine and ornithine but are urease negative, and do not produce indole. They generally do not ferment lactose, although lactose-positive strains have been noted.

Foodborne outbreaks of salmonellosis have been caused by a variety of foods, primarily poultry, eggs, and meats. However, salmonellosis from consumption of milk and dairy products has been reported[13,14,38,39] in which raw or improperly pasteurized fluid milk, ice cream, and cheese were implicated as the vehicles of the organism.[13,14,29,40]

The salmonellae are ubiquitous, being found worldwide in a wide variety of sources including milk, meat, poultry, eggs, soil, water, sewage, pets and other animals, humans, feed processing environments, etc.

The optimum growth temperature for salmonellae varies from 35 to 37°C, although many strains are capable of growing at 5 to 7°C. The organism is heat sensitive and is readily inactivated by conventional pasteurization. However, *Salmonella seftenberg* is generally recognized as more heat-resistant than most salmonella strains. Thermal inactivation of salmonella depends on time–temperature of the heat treatment, pH, and moisture content (A_w) of the food.

Yersinia. Previously classified as pasteurella, the genus *Yersinia* includes *Yersinia pestis, Y. pseudotuberculosis, Y. frederikensenii, Y. kristensenii, Y. intermedia, Y. enterocolitica*, and *Y. ruckeri.*[41]

Y. enterocolitica is a Gram-negative, short, rod-shaped organism that is motile at <30°C but not at 37°C. The organism is a psychrotroph capable of growing, albeit slowly, in milk and dairy products stored at refrigeration temperatures. The optimal growth temperature for *Y. enterocolitica* is 32 to 34°C.[42–44] It is a poor competitor with common spoilage bacteria in milk at 4°C.[22] Yersiniae are remarkably tolerant to bile salt and can survive better under alkaline conditions.

Outbreaks of yersiniosis implicating milk and milk products have been reported.[14,42,45] *Y. enterocolitica* is widespread in nature, having been isolated from water, sewage, soil, and a wide variety of animals, particularly pigs.[43] It should be noted that only certain strains of *Y. enterocolitica* are considered pathogenic for man, with most strains being environmental strains. Pathogenic strains of Yersinia may be distinguished from nonpathogenic strains based on esculin hydrolysis and salicin fermentation.[42,43]

Aeromonas. The genus *Aeromonas* belongs to the family *Vibrionaceae. Aeromonas* are facultatively anaerobic, Gram-negative cocci or rods with rounded ends (0.3 to 1.0 μm × 1.0 to 3.5 μm). They are generally motile, with polar flagella, and produce oxidase and catalase. The optimum and maximum growth temperatures for *Aeromonas* are 28°C and 42°C, respectively. However, strains capable of growth at 5°C have been reported.[20,46–49] *Aeromonas* spp. can grow in nutrient broth containing 5% salt at 28°C.[46,49]

A. hydrophila is considered an opportunistic pathogen. The organism has been isolated from a variety of aquatic sources including the Great Salt Lake and the Chesapeake Bay, as well as from the feces of healthy farm animals, including cows, pigs, sheep, and horses. It was found significantly more often in the feces of cows than in any other species. *A. hydrophila* is a relatively heat sensitive organism that is readily inactivated by pasteurization.

The *Citrobacter* species in this genus utilize citrate as a sole source of carbon and may ferment lactose, albeit slowly. Recognized species of *Citrobacter* include *C. freundii*, *C. diversus*, and *C. amalonaticus*. *C. freundii* can grow on media designed for *Salmonella* and is often confused with salmonella. Citrobacter occur in the intestine and have been isolated from feces, water sewage, and foods of animal origin.

Serratia. This genus includes *S. marcescens* which may be found in the environment and foods. It is an aerobic, Gram negative, motile (peritrichous flagella), rod-shaped organism, capable of producing a characteristic red pigment, prodigiosin. *Serratia* may be important as a potential spoilage organism in some foods.

Hafnia. *Hafnia alvei*, formerly known as *Enterobacter hafnia* or *Enterobacter alvei*, has been implicated as the cause of mild gastroenteritis in hospitals and in community outbreaks associated with milk,[34] although most *H. alvei* strains are not considered to be pathogenic for humans. The organism resembles *Salmonella* spp. and can be isolated on media designed for salmonella. It does not ferment lactose. *Hafnia* spp. are found in sewage, soil, water, and feces of man and animals. It may be important in spoilage of milk and milk products.

Chromobacterium. This genus contains two species, *C. violaceum* and *C. lividum*. The Chromobacteria are facultatively anaerobic, oxidase-positive, Gram-negative, often slightly curved rods (0.6 to 0.9 μm $\times$ 1.5 to 3.5 μm) capable of producing violet or dark blue pigment. The violet pigment, violacein, produced by *C. violaceum* has antibiotic properties. The organisms are found in water, soil, and foods and may occasionally cause infections in animals.

Rickettsia *and* Chlamydia. The genus *Coxiella* of the family *Rickettsiaceae* is the only member of this group important in dairy microbiology. *C. burnetii* is the causative agent of Q fever. *C. burnetii* are Gram-negative or -positive, short, rod-shaped organisms that may occasionally appear as diplobacilli or cocci. They are obligate parasites that grow in the vacuoles rather than in the cytoplasm or nucleus of host cells, especially ticks that transmit Q fever to cattle, sheep, goats, and other animals. The organism is shed in the milk of the infected animal or during parturition. Consumption of raw milk contaminated with *C. burnetii* may lead to Q fever in humans. The organism can withstand drying and elevated temperatures, but is readily inactivated by proper pasteurization of milk. The time–temperature for commercial high temperature–short time (HTST) pasteurization of milk is designed to destroy *C. burnetii* in milk.

Gram-Positive Cocci

This group includes aerobic or facultatively anaerobic Gram-positive, usually nonmotile spherical (0.5 to 1.5 μm diameter) shaped organisms that are important in the dairy industry as foodborne pathogens, causative agents of mastitis, thermoduric spoilage organisms, and lactic starter cultures used in the manufacture of fermented dairy foods.

Micrococcus. These are strictly aerobic, catalase-positive organisms that are found in soil, water, dust, and on skins of human and animals. Micrococci occur in a variety of foods, including milk and dairy products. Optimum growth temperatures of micrococci range from 25 to 37°C, although most strains can grow at 10°C but not at 45°C. Micrococci can ferment glucose aerobically but not anaerobically and grow in the presence of 5% salt.

Staphylococcus. This genus contains *S. aureus*, *S. hyicus*, *S. epidermidis*, and two lesser known species, *S. chromogenes* and *S. caprae*.

Staphylococcus aureus is a well-known pathogen that can produce a heat-stable enterotoxin implicated in several outbreaks of foodborne illness. It is coagulase positive and produces a variety of hemolysins and a thermostable nuclease. Some strains of *S. aureus* produce an antibiotic-like substance, staphylococcin, that can inhibit other staphylococcal strains. *Staphylococcus aureus* contamination in milk and dairy products indicates contaminations from human sources. In contrast, *S. epidermidis* is found on human skins and is coagulase negative. The organism produces hemolysin and thermostable nuclease; however activity of these two compounds is weak compared to that produced by *S. aureus*.

Staphylococcus hyicus and *S. chromogenes* have been isolated from skins of pigs and cows and from the milk of cows suffering from mastitis. Most strains of *S. hyicus* do not show coagulase activity, although enterotoxigenic strains of *S. hyicus* have been reported.[50] *S. chromogenes* was considered a subspecies of *S. hyicus* until 1986 when it was proposed as a separate species by Hajek et al.[18] It is coagulase negative and shows a negative or weakly positive thermostable nuclease activity.

S. caprae is a facultatively anaerobic organism isolated from goat's milk. It is coagulase negative and has characteristic hemolysis and fermentation reactions useful in characterization and differentiation from other staphylococci.[18]

Streptococcus. This genus contains several organisms known to be associated with milk. The lactic streptococci and enterococci belonged to the genus *Streptococcus* until recently but now they have been classified in the genera *Betacoccus* and *Enterococcus*, respectively.

The organisms classified in the genus *Streptococcus* are Gram-positive cocci, occurring in pairs or chains, mostly nonmotile and facultatively anaerobic with some strains being strictly anaerobic. *Streptococcus pyogenes* is a pathogen associated with scarlet fever. Optimum growth temperature for growth of this organism is 37°C with no growth occurring at 10 or 45°C. The genus *Streptococcus* also includes three species important as the causative agent of bovine mastitis—*S. agalactiae*, *S. dysgalactiae*, and *S. uberis*. *S. agalactiae* infections are readily controllable through mastitis prevention programs instituted in the United States and herds may be certified *Strep. ag.* free following eradication of the organism in a herd through successful mastitis control programs. *S. pyogenes* is classified in the Lancerfield group A whereas *S. agalactiae* and *S. dysgalactiae* are classified in Lancefield group B and C, respectively. These organisms have complex growth requirements and are characterized on the basis of hemolysis on blood agar and hydrolysis of hippurate and esculin.

Two other important species of the genus *Streptococcus* are *Streptococcus zooepidemicus* and *Streptococcus salivarius* subsp. *thermophilus.* The latter is a thermophilic bacterium used as a part of a mixed strain starter culture used in the manufacture of yogurt and Italian cheeses. *S. thermophilus*, as it was known earlier, grows at 35 to 37°C. It can grow at 45°C, but not at 10°C.

S. zooepidemicus is primarily an animal pathogen causing septicemia in cows. It has been implicated as the cause of a food poisoning outbreak associated with the consumption of raw milk cheese.[51] The milk was obtained from cows with mastitis caused by this organism.

Lactococcus. This genus contains a group of organisms formerly known as mesophilic lactic streptococci—*S. lactis*, *S. cremoris*, and *S. diacetylactis.* These organisms are classified in Lancefield group N and have complex growth requirements. Several strains are capable of producing nisin and bacteriocins which can inhibit foodborne pathogens. The lactococci can produce lactic acid and other compounds responsible for the characteristic flavor and aroma of fermented milk products such as cheeses, cultured buttermilk, and sour cream. Some strains of *L. lactis* subsp. *lactis* are also known to cause malty off-flavor in milk and dairy products. *L. lactis* var. *diacetylactis* can metabolize citric acid to produce CO_2 and diacetyl; the latter is responsible for the characteristic "nutty," or "buttery" aroma of cultured butter, buttermilk, and sour cream. Some strains can also produce H_2O_2 and acetic acid and inhibit pseudomonads, coliforms, and other contaminating organisms, including salmonella. The characteristics, functions, and plasmid-mediated properties of *Lactococcus* have been reviewed elsewhere.[52–56]

Leuconostoc. This genus contains Gram-positive, spherical to lenticular shaped organisms that occur in either pairs or chains. *Leuconostocs* occur in milk and dairy products, plant materials, fruits, and vegetables. They are heterofermentative, producing lactic acid, ethanol, and CO_2 from glucose. Some leuconostocs, for example, *L. mesenteroides*, produce extracellular polysaccharides leading to slime formation in sugar solutions and other products. *L. mesenteroides* subsp. *cremoris* and *L. mesenteroides* subsp. *dextranicum* are mesophilic organisms used in combination with lactococci in the production of cream cheese, cottage cheese, cultured buttermilk, and quarg. The so-called dairy strains of *Leuconostocs* generally do not produce slime, although dextran-producing strains of *L. mesenteroides* may be used to impact body and texture of products such as ice cream. Some leuconostocs produce acetic acid from citrate and may be used to control psychrotrophic spoilage (e.g., slime production) in fermented milk products. Growth temperatures for leuconostocs range from 10 to 37°C, although most prefer 18 to 25°C. *L. lactic* may grow at temperatures up to 40°C. It can also survive at 60°C for 30 min.

Endospore-Forming Rods and Cocci

This group includes the genera *Bacillus*, *Clostridium*, *Sporolactobacillus*, and *Desulfotomaculum* with only *Bacillus* and *Clostridium* being the genera of significance in dairy microbiology.[18]

Bacillus. These are Gram-positive, aerobic or facultatively anaerobic rod-shaped organisms that are generally motile, and produce catalase and acid but not gas from glucose. They occur in soil, air, water, dust, feed, and other sources, including milk and dairy products. Several bacilli possess proteolytic and lipolytic activity and can cause a variety of quality defects in milk and dairy products, for example, *B. cereus* can cause "bitty" cream defect.[6,10,57]

B. stearothermophilus can cause proteolytic defects in milk and cheese as well as sweet curdling defect due to its renninlike enzyme activity. *B. coagulans* and *B. licheniformis* are important spoilage organisms in ultrahigh temperature (UHT) and evaporated and condensed milk products.

Under aerobic conditions, bacilli form endospores which are an inactive or dormant state of the organism. The spores are heat resistant, allowing the organism to survive various heat treatments, including pasteurization. There are psychrotrophic, mesophilic, and thermophilic strains of the genus *Bacillus*. These organisms can grow at temperatures from -5°C to about 45°C, although *B. stearothermophilus* strains can grow at 55 to 75°C. The optimum growth temperature for most bacilli is 20 to 40°C.

B. stearothermophilus is important as the test organism for confirming antibiotic residue contamination in milk by the disc-assay procedure.[58]

B. cereus is recognized as a significant cause of foodborne intoxication. A typical illness is characterized as the so-called "diarrheal" or "emetic" syndrome associated with production of separate enterotoxins.[59–61]

Clostridium. These organisms are Gram-positive, facultative or strictly anaerobic spore-forming rods that are catalase negative and gelatinase positive. Clostridia are found in soil and sediment as well as in the intestinal tracts of man and animals. Important species in this group include *Clostridium botulinum*, *C. perfringens*, *C. sporogenes*, *C. butyricum*, and *C. tyrobutyricum*.

C. botulinum is recognized as the causative agent of botulism worldwide. There are seven types of *C. botulinum*, types A to G based on the serological specificity of the neurotoxin(s). Of these, only types A, B, and E have been involved in human illness. *C. botulinum* has been isolated from soils, sediments, water, thermally processed milk, and cured foods, particularly meat, fish, and honey. Botulism outbreaks implicating process cheese contaminated with *C. botulinum* have been reported.[62]

C. perfringens is recognized as a food-poisoning organism worldwide. The strains of *C. perfringens* have been classified into five types, A to E, based on the production of four extracellular toxins—α, β, ϵ, and ι. *C. perfringens* occur in a wide variety of raw and processed foods including meat, poultry, and fish. *C. perfringens* is also found in soil, sediments, and intestinal tracts of animals. It may cause mastitis in cows.[63]

C. butyricum and *C. tyrobutyricum* may be responsible for delayed gas production in cheese linked to the "late blowing" defect and rancidity in Emmenthal and Gruyère cheeses. *C. butyricum* may also be responsible for excessive gas production from glucose, that is, stormy fermentation. Both of these organisms produce acetic acid and butyric acid and may cause rancidity in certain cheeses. Optimum growth occurs at 30 to 37°C, although most strains may grow at 25°C and some at 10°C. Contaminated silage and dust are the two primary sources of spores of these organisms. Many dairy plants enumerate *Clostridium* spores in the incoming milk as a means of controlling the seasonal problem of late gas production and rancidity in cheeses. Another species, *C. sporogenes*, may also be responsible for "late blowing" of cheese. *C. sporogenes* also produces butyric acid as well as ammonia and H_2S and may be responsible for "sulfide" defects in cheeses.

Regular, Non-Spore-Forming Gram-Positive Rods

Important genera in this group include *Lactobacillus*, *Listeria*, and *Kurthia*.

Lactobacillus. These organisms are facultative or microaerophilic, Gram-positive, nonmotile rods of varying morphology ranging from coryneform coccobacillary or short rods to long and slender rods (0.5 to 1.6 μm $\times$ 1.5 to 11.0 μm), capable of homo- or heterofermentative metabolism. *L. delbrueckii* subsp. *lactis* and *L. delbrueckii* subsp. *bulgaricus*, previously known as *L. lactis* and *L. bulgaricus*, and *L. helveticus* are thermophilic starter cultures used in the production of yogurts and Swiss and Italian cheeses. Both of these organisms require vitamins and amino acids as growth factors. The optimum growth temperature for these organisms is $\approx$40°C.

Among other lactobacilli important as starter cultures in the dairy industry are *L. acidophilus*, *L. casei*, and *L. brevis*. The differential characteristics of the lactobacilli and other dairy starter culture organisms have been discussed recently by Tamine[64]

Listeria. This genus contains *L. monocytogenes* which is perhaps the most important pathogen involved in several outbreaks of listeriosis and widespread recalls of dairy products during the 1980s.[14,65,66] The organism is ubiquitous in nature, having been isolated from a wide variety of sources including water, sewage, soil, vegetation and plant materials, and milk.[67,68] Listeria may be found in improperly fermented silage and dairy barn environments. They may be involved in mastitis in dairy cows.[69] *L. monocytogenes* may exist intracellularly in phagocytes, and at one time were thought to be able to survive pasteurization.[70] However, subsequent research has proved that commercial HTST pasteurization treatment is adequate for inactivation of *L. monocytogenes*.[71]

The genus *Listeria* includes eight species. They are small, Gram-positive rods (0.4 to 0.5 μm $\times$ 0.5 to 2.0 μm) with rounded ends. Often, they may be seen as

Table 5.2 DIFFERENTIATION OF *LISTERIA* SPECIES[a]

Biochemical Test	*mono-cytogenes*	*ivanovii*	*innocua*	*welshi-merei*	*seeligeri*	*grayi*	*murrayi*
Dextrose	+	+	+	+	+	+	+
Esculin	+	+	+	+	+	+	+
Maltose	+	+	+	+	+	+	+
Rhamnose	+	−	V[b]	V	−	−	V
Xylose	−	+	−	+	+	−	−
Mannitol	−	−	−	−	−	+	+
Hippurate hydrolysis	+	+	+	+	+	−	−
Voges-Proskauer	+	+	+	+	+	+	+
Methyl red	+	+	+	+	+	+	+
β-hemolysis	+	+	−	−	+	−	−
Urea hydrolysis	−	−	−	−	−	−	−
Nitrate reduction	−	−	−	−	−	−	+
Catalase	+	+	+	+	+	+	+
H_2S on TSI	−	−	−	−	−	−	−
H_2S by lead acetate strip							

[a] From Lovett.[72]
[b] V = variable.

short chains, lying parallel or in a ''V'' shape. Listeria exhibit a characteristic tumbling motility and a typical ''umbrella'' pattern in an appropriate motility medium when grown at 20°C. On a solid medium, listeria produce typical blue-gray colonies when viewed by 45°C incident transmitted light (Henry's illumination). Biochemically, Listeria resemble members of the genera *Brochothrix, Erysipelothrix, Lactobacillus,* and *Kurthia,* but can be differentiated from them based on motility, catalase reaction, and glucose fermentation. A detailed differentiation of *Listeria* species is given in Table 5.2.

Kurthia. These are Gram-positive, strictly aerobic, usually motile, often occurring as unbranched or coccoid rods (0.8 to 1.2 μm × 2.0 to 4.0 μm). They are found in meats and meat products, meat processing plants, intestinal contents, and in milk. They are oxidase negative and grow optimally at 25 to 30°C. The presence of *Kurthia* may indicate improper handling of the product or contamination with animal feces.

Irregular, Non-Spore-Forming, Gram-Positive Rods

This group contains several diverse bacteria, including the so-called coryneform group and genera important in dairy microbiology—*Corynebacterium, Arthrobacter, Brevibacterium, Caseobacter, Microbacterium, Aureobacterium, Propionibacterium,* and *Actinomyces.*

The genus *Corynebacterium* contains facultatively anaerobic or some aerobic, straight or curved rods with tapered ends that are nonmotile and that form metachromatic granules. The *Corynebacterium* spp. may be human, animal, or plant pathogens. *C. bovis* and *C. striatum* may be associated with mastitis and *C. renale* can cause urinogenital infections in cows. Pathogenic corynebacteria grow optimally at 37°C and some produce exotoxins. Not all species of *Corynebacterium* are pathogenic.

The genus *Arthrobacter* contains short irregular rods, cocci, or pleomorphic bacteria arranged in V-forms depending on their growth conditions and phase. They are usually strictly aerobic, nonmotile organisms that may occur in soil and the dairy farm environment and may form a part of ''coryneforms'' bacteria in milk. Optimum growth temperatures for these organisms are 25 to 30°C.

The organisms in the genera *Brevibacterium, Caseobacter*, and *Aureobacter* are nonmotile, obligately aerobic or anaerobic irregular rods, cocci, or pleomorphic forms similar to *Arthrobacter* spp. *Brevibacterium linens* is used in ripening of certain cheeses, for example, Limburger cheese where proteolytic action and methanethiol production by the organism are important in developing the characteristic flavor, aroma, and texture of the cheese. These organisms grow optimally at 20 to 30°C, although some strain of *Brevibacterium* may also grow at 37°C. Several species of these genera produce characteristic pigments, for example, *B. linens* produces yellow to deep orange carotenoid pigment whereas *A. liquifaciens* produces a bright yellow pigment. These organisms are found in milk, cheese, dairy products, and dairy equipment and some have an important function in cheese ripening.

Propionibacteria are the members of the genus *Propionibacterium* that contain nonmotile, anaerobic or aerotolerant pleomorphic rods, diphtheroids, or club-shaped organisms with one end rounded and the other tapered. Often, propionibacteria exhibit V or Y shapes and ''Chinese character''-like cellular arrangements. The propionibacteria are responsible for the ''eye'' formation and development of flavor and aroma characteristically found in Swiss and Emmenthal cheeses. Besides formation of CO_2 and proline, the propionibacteria also produce propionic and acetic acid by fermentation of carbohydrates. The *Propionibacterium* spp. may be pigmented and some can produce slime. The optimum growth temperature for propionibacteria is 30 to 32°C.

The genus *Actiomycetes* contains two species important in dairy microbiology—*A. bovis*, the causative agent of lumpy jaw in cattle, and *A. pyogenes*, potentially causing summer mastitis in dairy cows. The salient characteristics of the two *Actinomyces* spp. are given in Table 5.3.

Table 5.3 SALIENT CHARACTERISTICS OF *ACTINOMYCES*

Characteristic	*A. bovis*	*A. pyogenes*
Hemolysis	β	
Casein hydrolysis	−	+
Nagase production	+	—
Aerobic growth	+[a]	—
Growth temperature	36°C	30°C

[a] With added CO_2.

Mycobacteria

The genus *Mycobacterium* includes two species important in dairy microbiology: *M. tuberculosis*, the causative agent of tuberculosis in humans, and *M. bovis*, which causes tuberculosis in cattle, dogs, cats, primates, and man. The *Mycobacterium* spp. are generally aerobic or rarely facultatively anaerobic, slightly curved or straight, generally Gram-positive rods. These organisms can withstand acid/alcohol decolorization and hence are termed ''acid-fast'' organisms. They may be isolated and characterized using procedures described by Jenkins et al.[73] Mycobacteria have been isolated from raw milk.[74] They do not grow at 25°C or 45°C, but grow optimally at 37°C. The advent of pasteurization of milk was primarily designed to inactivate *M. tuberculosis*, and control tuberculosis.

5.2.2.2 Yeast and Molds

Yeasts and molds are unicellular or multicellular members of a higher group of microorganisms called fungi. They are ubiquitous in nature and are found to occur in soil, air, water, decaying organic matter, and a variety of foods including milk and dairy products.

Yeasts are microscopic, ovoid, elongate, or elliptical or spherical organisms that are several times larger than the common bacteria (Fig. 5.1). They are very active biochemically and can grow over a wide range of pH, temperature, and alcohol concentrations. The limiting water activity (A_w) of most spoilage yeasts is 0.88, although osomophilic yeasts may grow at an A_w value of 0.60.[17,19]

The true yeasts (ascosporogenous) reproduce by sexual reproduction as well as by asexual spores and chlamydospores. In contrast, the false yeasts (asporogenous) or wild yeasts do not show sexual reproduction. They reproduce asexually by fragmentation of mycelium into blastospores or by budding. Important genera of yeasts in milk and milk products include *Saccharomyces*, *Kluyveromyces*, *Candida*, *Debaryomyces*, *Rhodotorula*, and *Torulopsis*. Table 5.4 lists some characteristics and significance of yeasts important in dairy microbiology.

Molds are multicellular organisms that grow in the form of a tangled mass of mycelium that is composed of filamentous structures called hyphae. Some molds characteristically form cross-walls or septae in their hyphae, whereas others do not. The septate or nonseptate mycelium is an important morphological feature in differentiating molds. Unlike true bacteria and most yeasts, molds reproduce sexually by ascospores, oospores, or zygospores. Asexual reproduction in molds is by sporangiospores, conidiospores, arthrospores, and chlamydospores; the latter two may be somewhat difficult to inactivate by heat and may cause problems in the food industry, particularly in the canning industry. Some characteristics and significance of molds important in dairy microbiology are given in Table 5.5.

5.2.2.3 Viruses

Viruses are ultramicroscopic, obligate parasites consisting of a nuclear material (DNA or RNA) surrounded by a protein coat. They require biological host cells for

Table 5.4 IMPORTANT GENERA OF YEASTS IN MILK AND DAIRY PRODUCTS

Yeast	Characteristics	Significance
Saccharomyces	Oval, ellipsoidal or cylindrical cells, multilateral budding, generally white, creamy colonies on agar, lactose fermentation and nitrate assimilation negative	*S. cerevisiae* usually used in the baking and brewing industry. *S. cerevisiae* isolated from Stracchino cheese and kefir. Contaminant in raw milk may be associated with mastitis
Candida	Yeastlike organisms—fungi imperfecti, short-ovoid or longer cells, lactose fermentation and nitrate assimilation may be positive or negative	*C. utilis* is fodder yeast, *C. kefyr* found in Kefir, buttermilk and cheese, *C. lacticondensi* in condensed milk
Kluyveromyces	Subglobose, ellipsoidal or cylindrical cells, true yeast, asexual reproduction by budding. May form pseudomycelium, sugar fermentation, including lactose positive; nitrate assimilation negative	*K. marxians* var *lactis* associated with yogurt, gassy cheese, milk, Italian cheese, buttermilk and cream. Produce β-galactosidase. *K. marxians* var *marxians (K fragilis)* found in Kefir and Koumiss
Debaryomyces	Spherical—short oval cells, ascosporogenous, may form pseudomycelium, lactose fermentation negative, nitrate assimilation positive	Found on surfaces of spoiled foods. *D. Hansenii* found in cheese

their growth and replication. They are classified into various groups according to their morphology, host range, physicochemical characteristics, serological properties, and ability to lyse the host cell during replication. Viruses infect humans, animals, and plants and cause disease in susceptible hosts. Bacterial viruses or bacteriophages are important in the cheese industry. Because phages are host specific, they may be used for typing or characterizing bacterial species or strains. In addition to bacteriophages, viruses of importance in the dairy industry include those causing poliomyelitis, cowpox, central European tickborne fever, and hepatitis.

Cowpox virus is a causative agent of lesions, vesicles, or pustules on teats of the cow. It may be transmitted to the milkers, producing lesions on the back of the hands or forearms and face. The virus is oval in shape, consisting of a multilayered covering around a double-stranded DNA core.

Poliomyelitis virus is an icosohedral particle containing a single-stranded RNA core, but no envelope. It can infect the central nervous system of the subject and cause paralysis. The polio virus may be transmitted through raw and pasteurized milk. It is inactivated by heat treatment of 74 to 76°C, unless occurring in a concentration of $>5 \times 10^{11}$/ml.[18]

Central European tickborne fever is a virus-borne disease that may be transmitted through raw goat's milk. The virus is spherical shaped, consisting of a single-stranded RNA core enclosed within an envelope. It is readily inactivated by heat treatment of 60°C for 10 min.

Table 5.5 SOME CHARACTERISTICS AND SIGNIFICANCE OF MOLD SPECIES IMPORTANT IN DAIRY MICROBIOLOGY

Organism	Characteristics	Significance
Aspergillus	Septate mycelium, globose conidia of varying color including yellow-green black to brown	Contamination in cheese, butter. *A. flavus* produces aflatoxin, some spp. important as commercial source of protease and industrial fermentations.
Penicillium	Septate mycelium, brushlike conidiophore bearing blue-green conidia	*P. roqueforti* used in the manufacture of Roquefort, Stilton, Gorgonzola and other blue-veined cheeses. *P. camemberti* important in the manufacture of Camembert, Brie and other cheeses. *P. casei* similar to *P. roqueforti*, associated with Swiss cheese.
Geotrichum	Yeastlike fungi which is usually white, septate mycelium, arthrospores are cylindrical with rounded ends.	*G. candidum* important as ''dairy mold'' or ''machinery mold.'' Found in several cheeses and as contamination on plant machinery.
Scopulariopsis	Produce characteristic truncated, spherical conidia with a thickened basal ring around truncation.	Found in cheeses; *S. brevicaulis* causes ammonia odor in some mold-ripened cheeses.
Sporendonema	Conidiospore formed within conidiophore, produces discrete colonies.	*S. sebi* important as ''mold buttons'' in sweetened condensed milk
Mucor	Aseptate mycelia, bear columella and a sporangium which contains smooth, round conidiospores.	Found in large number of foods, including cheeses. Some strains e.g., *M. miehei* and *M. pusillus*, are important sources of ''renninlike'' enzymes used as cheese coagulants.
Rhizopus	Aseptate mycelia, sporangiophores arise at nodes bearing thick tufts of ''rhizoids,'' mycelia bear columella and sporangia which contain dark sporangiospores.	Widely distributed in nature. Frequently found in cheese and other foods. *R. stolonifer*, known as the ''bread mold.'' Some strains are important for industrial fermentations.

In the cheese industry, bacteriophages are the primary cause of slow or dead vat problems due to the failure of a lactic starter culture.[75,76] The replication process of a phage in a susceptible bacterial host follows four distinct steps: (1) adsorption and attachment, (2) injection of the genetic material into the host cell, (3) production of the phage particles within the host cell, and (4) lysis of the host cell. The newly formed phages released in the environment on lysis of the host cell attach to new host cells to continue the cycle. Occasionally, the genetic material from the phage may become integrated into the host chromosome or it may be maintained in the host cytoplasm as an extrachromosomal nuclear material, the plasmid. In either case, the phage replicates along with the host cell without causing lysis of the host cell. This is known as lysogeny, which is rather a stable event and may continue indefinitely until such a time when the phage is activated and the cell produces new phage particles that are released on the lysis of the cell.

Bacteriophages of *Lactobacillus* spp. typically have isometric or prolate heads and tails varying in length from 80 to 200 nm (Figure 5.1) Some may possess distinct collars and base plates. They are classified in Bradley's group B.[77] Phages of thermophilic streptococci typically have isometric heads and 200- to 300-nm-long tails. They do not possess collars, only a small base plate, often with a central fiber and are classified in Bradley's group B. Phages of lactobacilli and leuconostocs are morphologically diverse. They are grouped in Bradley's group A or B.

For more information on bacteriophages of lactic acid bacteria refer to reviews by Davies and Gasson,[76] Klaenhammer,[78,79] Sanders,[80] and Sechaud, et al.[81]

5.3 Growth of Dairy Microbes in Milk and Dairy Products

Although J. Forster observed the growth of microorganisms at 0°C in 1887, it was not until 1902 that the term ''psychrophile'' was applied to this group by Schmidt-Nielsen.[82] Psychrophile comes from the Greek *psychros*, meaning cold, and *philos*, meaning loving. Hence, this word implies that these microorganisms grow optimally at low temperatures. Over the years psychrophiles have been defined in several ways based on growth at low temperature, optimum temperature of growth, temperature of enumeration, and other criteria not related to temperature.[11] Because thermophiles are defined by their optimum growth temperatures, microorganisms that grow at low temperatures should be similarly defined. This led Mossel and Zwart[83] and Eddy[8] to propose that microorganisms that grow at low temperatures but have higher temperature optima be defined as psychrotrophs. Morita[84] called mesophilic microorganisms that grow at 0°C psychrotolerant or psychrotrophic because psychrophilic microorganisms have temperature optima of 15°C, maxima of 20°C, and minima of 0°C or below. Psychrotrophs are microorganisms that can grow at refrigerated temperatures but that have temperature optima above 20°C. Psychrotrophs are those microorganisms that can produce visible growth at 7 ± 1°C within 7 to 10 days, regardless of their optimal growth temperatures.

5.3.1 Relative Growth Rates of Psychrotrophs

The Arrhenius equation is used to express the relationship between growth rate and temperature: $\log k = E/2.303\,RT + C$ where k = growth rate; E = activation energy or μ; R = gas constant; T = absolute temperature; and C = constant.[85–87] When the $\log k$ versus $1/T$ is plotted, then the linear slope $= -\mu/2.302R$. With this plot the temperature profile for a microorganism can be determined because a psychrophile has a linear slope to 0°C, but psychrotrophs show a nonlinear slope around 5°C and nonpsychrotrophic mesophiles become nonlinear at higher temperatures.

The curve is not completely linear because μ changes with temperature. Phillips and Griffiths[87] showed that the Arrhenius equation did not reflect the temperature profiles of several psychrotrophs grown in dairy products because the μ values depended upon the bacterium and its growth medium. However, a square root plot: $r = b(T - T_0)$ where r = growth rate constant, b = slope of the regression line, T = temperature (K), T_0 = temperature below which the microorganisms cannot grow, predicted the effects of temperature on the growth of psychrotrophs in dairy products. This confirmed the research of Reichardt and Morita[88] when they studied 16 psychrotrophic and psychrophilic bacteria, but could not establish a relationship between μ and the optimum growth rate that was valid to classify microorganisms as psychrophiles, psychrotrophs, nonpsychrotrophic mesophiles, and thermophiles. Stannard et al.[89] concluded that the square root plot can be used to establish the lowest growth temperature that can serve as a classification tool for psychrotrophs, nonpsychrotrophic mesophiles, and thermophiles.

Microorganisms that grow at low temperatures must have substrate uptake, cell permeability, enzyme systems, and synthetic pathways that function at low temperatures. Some theories about the growth of psychrotrophs involve the generation of low μ-values, presence of unsaturated fatty acids in the cell membranes, conformational changes in the ribosomal proteins and regulatory enzymes, substrate uptake, and cell permeability.[6,9,84–86]

The generation times of Gram-negative psychrotrophs range from 3.5 to 17 h at 5 to 7°C.[6,90] Spohr and Schütz[91] reported that *P. fluorescens* had generation times of 6 and 4.5 h at 4 and 8°C, respectively, with no lag period. *Pseudomonas* species generally have the fastest generation at these temperatures. Gram-positive bacteria have generation times ranging from 6 to 36 h at 5 to 7°C. *Micrococcus* species have generation times over 20 h compared to the *Bacillus* species.[90] Griffiths and Phillips[92] reported that *B. circulans* had generation times of 19 to 36 h at 2°C in whole milk. All strains of *Bacillus* studied had generation times of 7 to 23 h at 6°C in whole milk. For *Pseudomonas* species, the presence of air can shorten the generation time, especially at low temperatures. Hence, this genus generally becomes dominant when raw milk is stored for several days. Bloquel and Veillet-Poncet[93] reported that raw milk initially had 41% Gram-negative bacteria (mainly *Pseudomonas* and *Achromobacter* species), but after 96 h at 4°C, 88% were Gram-negative bacteria comprised of about 73% fluorescent pseudomonads and only 12% were Gram-positive bacteria (mainly *Micrococcus* species). Shelley et al.[94] reported that >90% of raw milk samples in an Australian study were pseudomonads, particularly *P. fluorescens* and *P. fragi*, which produced heat-stable lipases. Kroll et al.[95] found that the proteolytic microflora of raw milk consisted of 83% *P. fluorescens.*

The major Gram-negative and Gram-positive bacteria are listed in Tables 5.6 and 5.7, respectively. Psychrotrophic fungi can be associated with refrigerated milk and dairy products, and among the yeast genera are *Candida, Cryptococcus, Debaryomyces, Kluyveromyces, Pichia, Rhodotorula, Saccharomyces, Torulopsis,* and *Trichosporon.*[6,99] Mold genera that have psychrotrophic strains include *Alternaria, Aspergillus, Cladosporium, Fusarium, Geotrichum, Mucor, Penicillium,* and *Rhizopus.*[98,100] Fungi become important in refrigerated dairy product spoilage when

Table 5.6 GRAM-NEGATIVE BACTERIA ISOLATED FROM MILK AND DAIRY PRODUCTS[a]

Genus	Cell Shape	Relative to Oxygen	Proteinase	Lipase
Achromobacter	Rods	Aerobe	+	+
Acinetobacter	Rods to cocci	Aerobe	+	+
Aeromonas	Rods to cocci	Facultative anaerobe	+	+
Alcaligenes	Rods to cocci	Aerobe	+	+
Alteromonas	Rods	Aerobe	+	+
Chromobacterium	Rods	Aerobe or facultative anaerobe		+
Citrobacter	Rods	Facultative anaerobe		+
Cytophaga	Rods	Aerobe	+	
Enterobacter	Rods	Facultative anaerobe	+	
Escherichia	Rods	Facultative anaerobe	+	+
Flavobacterium	Rods	Aerobe	+	+
Klebsiella	Rods	Facultative anaerobe		
Moraxella	Rods to cocci	Aerobe	+	+
Proteus	Rods	Facultative anaerobe	+	
Pseudomonas	Rods	Aerobe	+	+
Serratia	Rods	Facultative anaerobe	+	+

[a] After Bloquel and Veillet-Poncet,[92] Cogan,[95] Cousin,[6] Mottar,[97] Suhren,[90] and Walker.[98]

Table 5.7 GRAM-POSITIVE BACTERIA ISOLATED FROM MILK AND DAIRY PRODUCTS[a]

Genus[b]	Cell Shape	Relative to Oxygen	Proteinase	Lipase
Arthrobacter	Rods to cocci	Aerobe	+	
Bacillus	Rods	Aerobe or facultative anaerobe	+	+
Clostridium	Rods	Anaerobe	+	+
Corynebacterium	Rods	Aerobe or facultative anaerobe		
Lactobacillus	Rods	Facultative anaerobe		
Microbacterium	Diphtheroid rods	Aerobe		
Micrococcus	Cocci	Aerobe	+	+
Staphylococcus	Cocci	Facultative anaerobe	+	+
Streptococcus	Cocci	Facultative anaerobe		

[a] Bloquel and Veillet-Poncet,[92] Cogan,[95] Cousin,[6] Suhren,[90] and Walker.[98]
[b] Most of these genera have thermoduric psychrotrophic strains.

water activity (A_w), acidity, and processing method become more favorable for them than for bacteria in cheese, yogurt, and other fermented dairy products.

5.3.2 Sources of Psychrotrophs in Milk

Psychrotrophs can get into milk in many ways. Generally water, soil, vegetation, air, bedding materials, cow udders, dairy equipment, and tanker trucks are the major sources of psychrotrophs. The incidence of psychrotrophs in milk depends on the type of microorganisms and the numbers present, the conditions of production, the temperature and length of storage, the season of the year, and other such factors.[6,90]

Griffiths and Phillips[92] found that psychrotrophic *Bacillus* species were in 58% of the milks in bulk tank milk collected between May and June in Scotland. Counts of spores were between 30 to 920/L (average 460/L). Unclean bulk tanks contributed most to the presence of the spores of *Bacillus cereus*, *B. circulans*, *B. mycoides*, and other *Bacillus* species in milk. Coghill and Juffs[101] isolated similar *Bacillus* species from milk and cream in Australia. McKinnon and Pettipher[102] found that heat-resistant spore-forming bacteria in milk came from the teat of the cow and to a lesser extent from improperly cleaned milking equipment. Other heat-resistant, psychrotrophic bacteria that have been isolated from milk include species of *Aerococcus*, *Arthrobacter*, *Corynebacterium*, *Microbacterium*, *Micrococcus*, and *Streptococcus*.[103–105] Heat-resistant ascospores of *Byssochlamys nivea* can also be isolated from raw milk and may be present in cheese and other fermented dairy products.[106] *D* values at 92°C ranged from 1.6 to 1.9 S for cream.

The growth of microorganisms in milk and dairy products will be a function of storage temperature, time, and generation time (growth rate) of contaminants. Griffiths and Phillips[107] developed a relationship between storage temperature and microbial growth using linear relationships between temperature and the square root of the specific growth rate of psychrotrophs over 2 to 22°C and between temperature and the square root of the reciprocal of lag time. There were highly significant relationships between these factors of specific growth rate and lag time and temperature. Generally *Pseudomonas* spp. were isolated most frequently at 2°C but as the temperature increased to 21°C, they made up only 10% of the population. Other Gram-negative bacteria (*Acinetobacter*, *Alcaligenes*, *Flavobacterium*, *Moraxella*, and *Aeromonas*) remained constant regardless of temperature. The *Enterobacteriaceae*, (mainly species of *Enterobacter*, *Escherichia*, *Citrobacter*, *Klebsiella*, and *Serratia*) remained constant at 3 to 10°C, but became more dominant as the temperatures increased to 21°C. Similarly, Gram-positive cocci (*Micrococcus* and *Streptococcus* species) increased as temperatures increased and they predominated above 16°C. Hence, milk stored at below 5°C will most likely be populated by *Pseudomonas* species.

5.3.3 Significance of the Presence and Growth of Psychrotrophs

Psychrotrophic microorganisms can grow in refrigerated dairy products resulting in spoilage due to degradation of carbohydrates, proteins, or lipids. Psychrotrophs carry out many biochemical reactions that are seen at higher temperatures, but reaction rates are slowed by low temperatures. Slight biochemical changes occur early in the growth phase of some psychrotrophs, but several weeks at refrigerated temperatures may be necessary for extreme changes to occur.

Little information is available on the carbohydrate metabolism of *Pseudomonas* species in milk because most research has focused on proteolytic and lipolytic degradation of milk. Spohr and Schütz[91] reported that pyruvate accumulated in cells of

P. fluorescens when 10^5 to 10^6 cfu/ml were reached because malate was decarboxylated to pyruvate and carbon dioxide. L-Lactate decreased when cell counts reached 10^7 to 10^8 cfu/ml and glucose-3-P increased. Bacterial lipases and proteases were noted when microbial numbers reached 10^7 to 10^8 cfu/ml. Citric acid cycle intermediates can stimulate the synthesis of proteases and lipases by psychrotrophs.[108] Glucose, lactate, pyruvate, acetate, and citric acid repressed the production of proteinases by psychrotrophs, especially *Pseudomonas* species. These psychrotrophs will normally use nonprotein and nonlipid carbon sources before using the more complex proteins and lipids.

Proteolytic and lipolytic enzymes can be produced when conditions are favorable for their production. McKellar[108] has reviewed the conditions that regulate enzyme synthesis. Generally, temperature, pH, oxygen, and nutrients exert the greatest effect on proteinase and lipase synthesis. Griffiths[109] reported that both proteinases and lipases were maximally synthesized during the late exponential to stationary growth phases when *P. fluorescens* strains were grown at 6 to 21°C but not at 2°C. McKellar[110] found that proteinase production was 55% greater at 20 than at 5°C. Similar results for maximal proteinase and lipase production during late logarithmic and stationary phases were reported by Stead.[111] More lipase was produced at low temperatures of 4 to 10°C than at 20°C.[112,113] *Flavobacterium* spp. also produced more proteinase in the late logarithmic and early stationary phases, but they could not grow well at 7°C,[114] The amount of enzymes synthesized depended on the strain and there was little difference between synthesis in whole and skim milk. Griffiths and Phillips[115] found that aeration of milk increased lipolysis (mainly due to native milk lipoprotein lipase) and decreased proteolysis due to catabolite repression by glucose. Bucky et al.[113] also reported that lipase production increased when milk was aerated and could be noted in the early logarithmic growth for *P. fluorescens.* When milk was flushed with nitrogen, proteolytic psychrotrophs grew slowly, but did not produce proteinases after 18 days of storage at 4°C, suggesting that oxygen is important for proteinase production.[116]

Much research has been done on the growth of psychrotrophs in milk and dairy products.[6,110,115,117] Degradation of proteins and lipids by psychrotrophs or their heat-stable enzymes has been the subject of several recent reviews.[99,117–119] The presence of psychrotrophs or their enzymes in milk and dairy products directly correlates with decreased shelf life, development of off-flavors and odors, decreased product yield, gel formation in liquid products, and defective manufactured products.[6,95,120–122] There are several published reports over the last 10 years on the decreased yield in cheese manufacture due to proteolysis that results in loss of casein protein with the whey.[123–129] The length of milk storage at low temperatures, counts of $>10^6$ psychrotrophs/ml of milk, temperature of storage, and types of psychrotrophs affect the amount of decreased yield for both nonripened and ripened cheeses.

The proteinases produced by psychrotrophs in milk selectively degrade the casein proteins, especially κ-, β-, and α-casein.[6,99] Patel et al.[4] found that extracellular heat-resistant proteases of *Pseudomonas* spp. degraded α-, κ-, β-, and γ-caseins to different degrees depending on the strain. α-Casein was a good substrate for most of these strains. Generally, κ-casein is degraded first and this has implications for cheese manufacture where κ-casein is important for rennet coagulation.[99] Also, the age gelation of UHT milk has been attributed to κ-casein. The size of the casein micelle decreased with increasing growth of psychrotrophic bacteria to populations $>10^8$ cfu/ml.[130] Hence, the degradation of casein during milk storage can have detrimental effects on final dairy product quality.

Most psychrotrophs are killed by normal pasteurization temperatures; however, some species and strains of *Arthrobacter*, *Bacillus*, *Clostridium*, *Corynebacterium*, *Lactobacillus*, *Microbacterium*, *Micrococcus*, and *Streptococcus* can survive pasteurization and cause problems in finished products.[6,103] Cromie et al.[104,105] have shown that aseptically packaged pasteurized milk changes the spoilage microflora to *Bacillus* species. Also, some of the lipase and proteinase activity will remain after pasteurization, even after UHT processing, because these enzymes are heat stable. Proteinases can have high heat resistances at UHT processing. Two *Pseudomonas* proteinases had D values of 4.8 and 6.2 min at 140°C.[122] Cogan[95] reviewed the heat resistance of lipases and proteinases from psychrotrophs that grew in milk and reported values from 0.2 to 54 min at 66 to 74°C for lipases and 54 to 950 min for proteinases at 71 to 74°C. Similar information is reviewed by Kroll[2] and Linden.[131] Low-temperature inactivation of these enzymes has been reported at temperatures from 50 to 60°C depending on the enzyme studied.[2,131,132] Leinmüller and Christophersen[133] reported that a proteinase from *P. fluorescens* was completely inactivated after 15 min at 50°C. Kumera et al.[134] recently presented data suggesting that the production of proteinases helped to stabilize lipases to heat. Therefore, the presence of enzymes produced by psychrotrophs growing in milk and dairy products can lead to both quality and economic losses for dairy processors. Ways to prevent psychrotrophic growth are very important for dairy product quality.

5.4 Inhibition and Control of Microorganisms in Milk and Dairy Products

From the time milk leaves the cow's udder until it is processed, packaged, and distributed, it can become contaminated with microorganisms. If these microorganisms are allowed to grow, they can eventually cause spoilage of the milk or milk products. There are many ways that microorganisms can be prevented from growing in milk. Use of natural antimicrobial systems, addition of antimicrobial agents, production of inhibitors by microorganisms, and use of physical methods to kill or remove microorganisms are the most common ways to prevent microorganisms from spoiling milk. These four areas will be briefly reviewed.

5.4.1 Natural Antimicrobial Systems

Milk contains several nonimmunological proteins that have antimicrobial properties.[135–139] The four most common proteins that have been studied are lactoperoxidase, lactoferrin, lysozyme, and xanthine oxidase. These proteins are involved in complex systems that cause microorganisms to become inactivated. Lactoperoxidase forms an antimicrobial system with hydrogen peroxide and thiocyanate. Lactoferrin is an iron-binding protein that binds both Fe^{3+} and the carbonate anion. Lysozyme is a protein that can have a direct or indirect enzymatic effect or a nonenzymatic effect on microorganisms. Xanthine oxidase is involved in the generation of hydrogen peroxide which can either be used for the lactoperoxidase system or as a direct antimicrobial agent. Each one of these proteins is briefly discussed in the following sections.

5.4.2 Lactoperoxidase

The lactoperoxidase system has been extensively studied. Reviews by Ekstrand,[135] Reiter,[136,137] and Reiter and Härnulv,[139] can be consulted for more detail on the history, background, and biological functions of this inhibitory enzyme. The lactoperoxidase enzyme catalyzes the reaction of $H_2O_2 + SCN^- \rightarrow OSCN^- + H_2O$; hence, both hydrogen peroxide and thiocyanate are essential to the antimicrobial activity. Lactoperoxidase is present in bovine milk in the whey proteins at concentrations from 10 to 30 μg/ml of milk depending on the cow and its breed.[136,137,140] Lactoperoxidase is a basic glycoprotein with a molecular weight of about 77,000 and iron (Fe^{3+}) heme group.[135] It has its highest activity at pH 4 to 7 which would be in the range for fresh milk. Hernandez et al.[141] isolated and further characterized lactoperoxidase from bovine milk.

There is little hydrogen peroxide in milk, but it can be produced by lactic acid bacteria that contaminate the milk. Also, if free oxygen is present in milk, hydrogen peroxide can be produced by reactions with xanthine oxidase, copper sulfhydryl oxidase, and ascorbic acid.[136,137,140] Because hydrogen peroxide is not very stable, it can be reduced by catalase or bound to enzymes, such as lactoperoxidase.

Thiocyanate is present in bovine milk in up to 15 ppm, especially in milk that has a high somatic cell count.[136,137,139,140] Thiocyanate is a common anion that is present in many animal tissues (mammary glands, salivary glands, stomach, kidneys, etc.) and secretions (cerebral fluid, saliva, lymph fluid, plasma, etc.). The type of feed, especially clover and feed containing glucosides, affects the concentration of thiocyanate. The health of the cow affects the thiocyanate level because cows with diseases such as mastitis contain more leucocytes and obtain the increased thiocyanate concentration from the blood.[136,137,139,140]

The mode of bacterial inhibition by the lactoperoxidase system involves a change in the cytoplasmic membrane because hypothiocyanate ($OSCN^-$) binds to the free SH–groups of key enzymes, causing the pH gradient to drop and potassium and amino acids to leak from the cell.[135–137,140,142,143] This prevents the uptake of carbohydrates, amino acids, and other nutrients because their transport mechanisms are inhibited. Further activities of the cell involved in protein, DNA, and RNA synthesis are disrupted. Gram-negative bacteria are more readily killed and lysed by the lactoperoxidase system than the Gram-positive bacteria. This is probably due to the differences in both cell wall composition and thickness. Some Gram-positive streptococci are resistant to the hypothiocyanate.

The lactoperoxidase system occurs naturally in several environments. In calves, the intestinal flora is colonized by lactobacilli that produce hydrogen peroxide which activates the lactoperoxidase system.[136,137,139,140] This can prevent undesirable bacteria, such as *E. coli*, from becoming established in the intestinal mucosa. The lactoperoxidase system is also active in the mouth of humans and this may help to prevent acid production in dental plaques which may reduce dental caries. The lactoperoxidase system inhibits many of the bacteria that cause mastitis in cows. Because the lactoperoxidase system is naturally active in mammalian environments, considerable research has been done to determine if this antimicrobial system has any toxic effects on the host,[139] This research has shown that there are no toxic effects on mammalian cells as well as HeLa cells and Chinese hamster ovary cells.

Because the lacoperoxidase system is considered a natural antimicrobial system in milk, various practical applications for its use have been proposed and researched. Among the most common ideas for dairy processing are to help preserve both refrigerated and nonrefrigerated milk to destroy bacterial pathogens in milk, and to extend the shelf life of refrigerated milk and cultured dairy products. The lactoperoxidase system has been successfully used to extend the shelf life of refrigerated raw milk. Reiter[144] showed that the *Pseudomonas fluorescens* growth can be slowed by about 200 h at 4°C and 20 h at 30°C by the activation of the lactoperoxidase system. Similar results were shown with a mixed population of common psychrotrophic bacteria. At 4°C it took longer than 6 days for the multiplication of this mixed flora once the lactoperoxidase system was activated by addition of hydrogen peroxide. At the dairy farm, a 3 log cycle lower count in lactoperoxidase-treated milk versus untreated milk was observed after 6 days of storage at 5°C. Zajac et al.[145,146] showed that the keeping quality of refrigerated (4°C) farm milk could be extended by the activation of the lactoperoxidase system using sodium thiocyanate (11.2 ppm) and sodium percarbonate (10 ppm H_2O_2) at regular intervals of 48 h. The count of both psychrotrophs and coliforms remained constant or decreased in the milks where the lactoperoxidase system was activated. Martinez et al.[147] activated the lactoperoxidase system every 48 h in both raw and pasteurized milk by maintaining concentrations of thiocyanate and hydrogen peroxide at 0.25 mM. This treatment effectively extended the shelf lives of both raw and pasteurized milk at 4, 8, and 16°C by 3 to 6 days depending on the storage conditions as measured by sensory analysis, titratable acidity, proteolysis, and lipolysis. Kamau et al.[148] activated the lactoperoxidase system of raw milk by adding 2.4 mM thiocyanate and 0.6 mM hydrogen peroxide. The milk was then pasteurized, cooled, and stored at 10°C with 150 rpm agitation for 22 days. The treated milk had an increased shelf life of 22 days compared to the control milk because counts were 10^3 and 10^7 cells/ml, respectively, Hernandez et al.[141] found that commercial pasteurization reduced the lactoperoxidase activity by 70 percent. Ekstrand et al.[149] heated milk to 80°C or higher and noted that the antibacterial effect of lactoperoxidase was decreased, possibly due to the exposure and oxidation of sulfhydryl groups. Generally, low-temperature pasteurization does not inactivate the lactoperoxidase system, whereas temperatures >80°C destroy activity. These studies show that activation of the lactoperoxidase system can extend the refrigerated keeping quality for both raw and pasteurized milk.

Because psychrotrophs grow in raw refrigerated milk and produce proteolytic and lipolytic enzymes, they can create problems for products made from this stored milk. Research has been done on the use of the lactoperoxidase system to improve the quality of cheese and other cultured dairy products. Reiter[144] and Reiter and Härnulv[138] reported that milk where lactoperoxidase system was activated resulted in cheese that was judged as normal in flavor after 4 months of storage, whereas the control cheese from untreated milk was labeled as rancid and had high free fatty acid profiles. Ahrné and Björck[150] reported that the lactoperoxidase system could inhibit lipoprotein lipase activity in milk, and lipolysis was decreased. The treated cheeses also gave higher yields because the proteolytic degradation by psychrotrophs was suppressed. Lara et al.[151] also noted a 1 to 2% (wet weight) increase in the lactoperoxidase-activated raw and pasteurized milk cheeses, respectively; however, acid production and microbial growth of the starters were reduced. Zall et al.[152,153] noted that acid production during cheddaring and weaker curds were seen for Cheddar cheese produced from milk with an activated lactoperoxidase system. These cheeses also did not develop the typical Cheddar flavor within 6 months as expected. Cottage cheese made from this milk was also judged by trained panelists as having a distinctly different flavor.[153] Yogurt and buttermilk made from milk that had an activated lactoperoxidase system took longer to make than controls because the culture grew slower.[152] The experimental buttermilk had an objectionable flavor, but the yogurt could not be differentiated from the control. Kamau and Kroger[154] also found that the rennet coagulation time and acid production by starter cultures were slower in the lactoperoxidase-activated systems than in control milks. Earnshaw et al.[155] added lactoperoxidase, potassium thiocyanate, glucose oxidase, glucose, and urea peroxide to cottage cheese to simulate the lactoperoxidase system. This system effectively reduced the populations of added *Pseudomonas* spp., *E. coli*, and *Salmonella thyphimurium*. The use of the lactoperoxidase system for controlling the growth of psychrotrophs in milk used for cultured product manufacture has both desirable and undesirable consequences. The treated milk has lower microbial counts and generally results in high product yields; however, the coagulation rate, acid production, and flavor are not produced in a time similar to that of control products.

The lactoperoxidase system inhibits *E. coli* and other Gram-negative bacteria in milk. Because milk and dairy products have been implicated in several foodborne disease outbreaks in recent years, there has been renewed interest in ways to prevent pathogens from growing to dangerous levels. Research has been done on the use of the lactoperoxidase system to inhibit some pathogenic bacteria that can grow in milk. Zajac et al.[156] found that the lactoperoxidase system decreased the vegetative cells of *Bacillus cereus*, but had no effect on the spores, because the plasma membrane is not accessible. *Campylobacter jejuni* rapidly decreased in raw or heated milk when the lactoperoxidase system was activated.[157] The lactoperoxidase system was also effective against strains of *Listeria monocytogenes* and *Listeria innocua* depending on the cell number, temperature, and medium.[158,159] Generally, low numbers (<100 cfu/ml) could be inactivated at 4 to 35°C and decreases in populations were noted for higher temperatures. Kamau et al.[160] reported that both *L. monocytogenes* and *S. aureus* were inactivated more rapidly when heated at 50 to 60°C after the lactoperoxidase system was activated. There were both decreased lag times and lower *D* values, showing that these bacteria were more sensitive to heat once the lactoperoxidase system was activated. The safety of milk in relation to foodborne pathogens can be increased by the use of the lactoperoxidase system in combination with heat and other preservation methods.

The lactoperoxidase system could also be beneficial in countries where cooling milk before transporting to dairy processing plants is not possible. The activation of the lactoperoxidase system with 10 ppm thiocyanate and sodium percarbonate to generate 8.5 ppm of hydrogen peroxide resulted in increased keeping quality of milk during transportation at 27 to 30°C.[138] Björck et al.[161] showed that the activation of the lactoperoxidase system with 5 ppm thiocyanate and 7.5 ppm hydrogen peroxide helped to preserve milk in Kenya. The reaction was inversely related to the temperature of storage. The bacteriostatic effect lasted for 7 to 8 h at 30°C, 11 to 12 at 25°C, 15 to 16 h at 20°C, and 24 to 26 h at 15°C during laboratory trials. In actual field conditions, the milk was treated at the collection station and then sent to the dairy plant which took 3 to 6 h at 27 to 30°C. After the activation of the lactoperoxidase system 88% of the samples had a resasurin reading of 6 after 10 min compared to 26% for the controls. Ridley and Shalo[162] studied the use of activation of the lactoperoxidase system and a combination of the lactoperoxidase system and evaporative cooling to extend the shelf life of milk in Kenya. The lactoperoxidase system reduced the total plate count by 1 log cycle and the combination of lactoperoxidase plus evaporative cooling reduced the count by 2 log cycles. In Sri Lanka both the bovine and buffalo milks were stabilized by the activation of the lactoperoxidase system once the milk reached collection centers 3 to 6 h after milking.[163] With temperatures ranging from 20 to 33°C, the milk could be kept for 4 to 9 h longer than when not treated. These results show that the use of the lactoperoxidase system in countries where milk cannot be refrigerated can help to extend the shelf life during transportation and storage before milk can be shipped to processing plants.

5.4.3 Lactoferrin

Lactoferrin is an iron-binding protein in milk that has antimicrobial activity.[135–138] Bovine milk contains 0.02 to 0.35 mg/ml of lactoferrin.[136,137] Lactoferrin is a glycoprotein with a molecular weight of 76,500 that has two metal binding sites that bind ferric ions and bicarbonic ions. The citrate concentration of milk is important because it can exchange the iron chelated by lactoferrin and this can cause loss of the bacteriostatic activity. Lactoferrin inhibits only bacteria with high iron requirements, such as coliforms but has no effect on bacteria that require a low amount of iron.[136,137] The bacteriostatic effect of lactoferrin is temporary because some Gram-negative bacteria can adapt to low iron and synthesize iron chelators. Ellison et al.[164] found that lactoferrin damaged the outer membrane of Gram-negative bacteria and caused permeability problems. Very little research has been done on the use of lactoferrin as an antimicrobial agent in milk.

5.4.4 Lysozyme

Lysozyme is a small basic protein that has a molecular weight of 15,000.[136,137] Bovine milk contains 13 μg of lysozyme/100 ml. Lysozyme has three functions: (1) a direct enzymatic effect that degrades the bacterial cell peptidoglycans and polysaccharides of Gram-positive bacteria; (2) an indirect enzymatic effect is seen when the peptidoglycan is cleaved to yield muramyldipeptide and an immunostimulating effect is produced; and (3) the positively charged lysozyme can neutralize the negatively charged groups on the bacterial cell membranes.[135] Lysozyme has found its greatest use in inactivating vegetative cells and germinating spores of *Bacillus* and *Clostridium.*[136,137] Wasserfall and Teuber[165] used 500 U/ml of egg white lysozyme to kill vegetative cells of *Clostridium tyrobutyricum*; however, spores were resistant. A 1-day delay in outgrowth of spores into vegetative cells could account for the "late gas" defect in Edam and Gouda cheeses. Countries such as Germany, Italy, Denmark, the Netherlands, France, Spain, as well as Australia, have experimented with (some have even approved) the use of lysozyme to prevent the "late gas" defect due to butyric acid fermentation by *Clostridium* species in semihard and hard cheeses including Gouda, Emmenthal, Provolone, Edam, and others.[166,167] Lysozyme hydrolyzes the peptidoglycan in clostridia and other Gram-postive bacteria. Lysozyme is added to the cheese milk and 99% stays with the casein and remains active during ripening. Bester and Lombard[168] found that 250 U/ml of lysozyme inhibited vegetative growth of *C. tyrobutyricum* but spores were not inhibited and germination was stimulated. *Lactobacillus* spp. were inhibited only if concentrations of lysozyme were >500 to 1000 U/ml. Starter cultures composed of *Lactococcus* and *Leuconostoc* species were not affected; hence, they could grow normally in the presence of lysozyme. El-Gendy et al.[169] also showed that 0.02% hydrogen peroxide could inhibit *Clostridium* species involved in "late gas" formation in cheese. Griffiths and Phillips[170] found that lysozyme did not inhibit growth of psychrotrophic *Bacillus* spp. in milk. Therefore, lysozyme can find specific uses to prevent gas formation in cheese by *Clostridium* species.

5.4.5 Xanthine Oxidase

Xanthine oxidase is an enzyme that is associated with the fat globule membrane in bovine milk. This enzyme contains iron and molybdenum and catabolizes purines producing uric acid, superoxide, and hydrogen peroxide; however, in milk there are few free purines and the xanthine oxidase reacts with acetaldehydes produced by lactic acid bacteria to produce the hydrogen peroxide.[136] Hydrogen peroxide is bactericidal by itself or can be used to activate the lactoperoxidase system. Roginski et al.[171] found that the xanthine oxidase–hypoxanthine system produced sufficient hydrogen peroxide to allow the lactoperoxidase system to stimulate growth and acid production by some *Streptococcus cremoris* and inhibited growth and acid production by *S. lactis* and some strains of *S. cremoris.* This led to the recommendation that starters used in cheesemaking should be resistant to the lactoperoxidase system and also low producers of hydrogen peroxide. Xanthine oxidase can act synergistically with lactoperoxidase and thiocyanate to complete the lactoperoxidase system. Interactions between lysozyme and lactoferrin, and xanthine oxidase and lactoperoxidase can further enhance the antimicrobial nature of these systems.[136,137,140]

More research needs to be done on the antimicrobial properties of nonimmunological proteins that occur naturally in milk. Dairy processors will need to use these systems more effectively to increase the shelf life of milk and dairy products. Also, the legislative hurdles against the use of activation of these natural systems in milk will have to be resolved before they can be effectively used.

5.4.6 Lactic Acid Bacteria and Bacteriocins

Lactic acid bacteria (*Lactobacillus*, *Lactococcus*, *Leuconostoc*, *Pediococcus*, and *Streptococcus* species) can preserve foods by producing compounds that inhibit other microorganisms. The traditional preservative activities have been the use of carbohydrates and the subsequent production of lactic and acetic acids that lower the pH of the food.[172,173] In addition to these activities several lactic acid bacteria can produce inhibitory compounds, such as hydrogen peroxide, diacetyl, bacteriocins, and other compounds.

Lactic acid bacteria have been used as inocula in milk to inhibit both spoilage and pathogenic microorganisms. Martin and Gilliland[174] found that lactobacilli isolated from yogurt inhibited psychrotrophic bacteria in autoclaved milk at 5.5°C; however, when a *L. bulgaricus* strain was added to raw milk at 5.5°C, there was no inhibition of the psychrotrophs. Champagne et al.[175] used two mesophilic *Lactococcus* strains to inhibit psychrotrophic bacteria in raw milk. Addition of more than 7×10^6 cells/ml was needed to reduce the level of psychrotrophs in raw milk at 7°C. Cell-free filtrates of *Lactococcus lactic* subsp. *lactis*, *Lactococcus lactis* subsp. *cremoris*, *Lactobacillus casei*, *Lactobacillus plantarum*, and *Leuconostoc mesenteroides* inhibited various pathogenic and spoilage bacteria, such as *Enterobacter aerogenes*, *Proteus vulgaris*, *Pseudomonas aeruginosa*, *Bacillus subtilis*, *E. coli*, *Salmonella typhimurium*, and *S. aureus*.[176] The antimicrobial activity was strongest against Gram-negative bacteria. Batish et al.[177] found that *Lactococcus lactis* subsp. *lactis* var. *diacetylactis* inhibited *Aspergillus fumigatus* from growing in milk and prevented *A. parasiticus* from producing aflatoxin B_1. The inhibitory compounds were not identified in these various experiments.

Lactic acid bacteria produce antimicrobial compounds called bacteriocins (Table 5.8). They are proteins or protein complexes that have activity against other bacteria, usually in the same or a closely related genus.[178] Some bacteriocins inhibit foodborne pathogenic bacteria.[179] One bacteriocin that has been widely studied and is now commercially used is nisin.[180–185] Nisin is a polypeptide produced by *L. lactis* subsp. *lactis* that is active against Gram-positive bacteria, including *Listeria monocytogenes* and sporeformers.[180–183] The outgrowth of bacterial spores is prevented when nisin is present. Nisin is stable to acid and shows greatest activity as the pH decreases.[180–182] High pH and high temperature generally degrade nisin. The cytoplasmic membrane is the target of nisin. Henning et al.[183] suggested that nisin interacted with the phospholipids in the cytoplasmic membrane and thus disrupted

Table 5.8 BACTERIOCINS PRODUCED BY LACTIC ACID BACTERIA THAT HAVE ANTIMICROBIAL POTENTIAL FOR USE IN FOOD PRODUCTS[a]

Bacterium	Bacteriocin	Antimicrobial Activity Against Strains of
Lactobacillus helveticus	Lactocin 27	*Lactobacillus acidophilus* and *L. helveticus*
	Helveticin J	*L. helveticus, L. bulgaricus, L. lactis*
Lactobacillus acidophilus	Lactacin B	*L. leichmannii, L. bulgaricus, L. helveticus, L. lactis*
	Lactacin F	*L. fermentum, S. faecalis,* Enterococci
Lactobacillus plantarum	Lactolin	Not given
	Plantaricin A	*L. plantarum, Pediococcus pentosaceus, L. paramesenteroides*
Lactococcus lactis subsp. *lactis*	Nisin	Gram (+) bacteria Prevents outgrowth of *Bacillus* and *Clostridium* spores
Lactococcus lactis subsp. *cremoris*	Diplococcin	Other dairy *Lactococcus* species
	Lactostrepcins	Group A, C, G Streptococci *L. helveticus, L. citrovorum, L. paracitrovorum*
Pediococcus pentosaceus	pediocin A	*Clostridium botulinum, Clostridium sporogenes, Staphylococcus aereus, Lactobacillus brevis, Lactococcus lactis* subsp. *lactis, Listeria, monocytogenes,* other pediococci

[a] Klaenhammer.[178]

membrane function. Sulfhydryl groups in the cytoplasmic membrane were inactivated by nisin, affecting both spore and the vegetative cell.[181,182] Nisin is thought to inhibit the swelling process for spore germination. Nisin (100 RU/ml) enhanced spore germination for some psychrotrophic strains of *Bacillus* in milk[186] and made them easier to inactivate by heat. Nisin is rapidly degraded in the stomach, does not result in sensitized human intestinal microflora, and is accepted for food use in 49 countries.[181] In the United States, nisin use is limited to pasteurized cheese and process cheese spreads.[180] The antibotulinal effectiveness of nisin has been shown.[187] FDA set the daily intake to 2.9 mg nisin/day/person. In other countries nisin is used for preserving processed cheese spreads, pasteurized dairy desserts, milk in countries without adequate refrigeration, and canned evaporated milks.[181,182,188] Additional success has been observed with pasteurized double cream[189] and prevention of butyric acid fermentation in cheese.[190]

Other bacteriocins have also been evaluated for their antimicrobial activity. Pediocin AH, produced by *Pediococcus acidilactici*, adsorbed to Gram-positive bacterial surfaces and caused loss of potassium and other cellular components.[191] Another pediocin, PA-1-bacteriocin, was bactericidal to *Listeria monocytogenes.*[192] *Lactobacillus acidophilus* produces lactacin B that is bactericidal to other *Lactobacillus* species as well as *Enterococcus faecalis.*[193] Pulusani et al.[194] partially purified antimicrobial compounds produced by *Streptococcus thermophilus* that were low molecular weight (700) amines; however, they were not classified as bacteriocins. These compounds inhibited Gram-positive and Gram-negative bacteria, including *Salmonella* and *Shigella* species. Although it is not commonly considered a lactic acid bacterium, *Bifidobacterium bifidum* produced antibacterial activity that inhibited *S. aureus*, *Bacillus cereus*, *E. coli*, *Pseudomonas fluorescens*, *Salmonella typhosa*, and *Shigella dysenteriae* in skim milk medium.[195] Several lactic acid bacteria produce antimicrobial compounds; however, not all of them have enough specificity to be of general use for preserving dairy products. The ones that are active against bacterial foodborne pathogens, such as *L. monocytogenes*, should undergo more research and product trials to determine the extent of their preservation potential.

Other preservation compounds have been evaluated for specific applications. Three of these are Micrograd, natamycin, and nitrate. Microgard is a preservative that is made by fermenting grade A skim milk with *Propionibacterium shermanii* followed by pasteurization.[172] This product is approved for food use by the FDA because it extends the shelf life of foods, especially refrigerated dairy products. Microgard is bacteriostatic to mainly Gram-negative bacteria and some molds and yeasts but not Gram-positive bacteria.[172] Weber and Broich[128] showed that Microgard at 0.4% in cottage cheese was bacteriostatic against Gram-negative bacteria and increased the keeping quality at 7°C by 91%. In yogurt and sour cream, 0.5% Microgard inhibited molds and yeasts. Salih et al.[196] showed that Microgard extended the shelf life of yogurt and cottage cheese. The effect was concentration dependent for inhibition of yeasts in yogurt. Molds and Gram-negative bacteria were inhibited in cottage cheese. Gram-positive pathogenic *Bacillus cereus*, *Listeria monocytogenes*, and *S. aureus* were not inhibited by Micrograd and some strains were even stimulated by this product.[197] Gram-negative pathogenic bacteria, such as *Salmonella typhimurium*, *S. paratyphi*, *Yersinia enterocolitica*, and *Aeromonas hydrophila* were sensitive to Microgard at pH 5.3 in an agar assay. Microgard is used at 1% in cottage cheese, yogurt, and dairy based salad dressing with the greatest use in cottage cheese.[197] Microgard contains propionic acid, diacetyl, acetic acid, and lactic acid in addition to the heat-stable proteinaceous components with a molecular weight of 700.[172] Natamycin or pimarcin is an antibiotic produced by *Streptomyces natalaensis* that inhibits molds and yeasts.[180] The FDA has approved the use of 200 to 300 μg/ml maximum concentrations of natamycin for inhibition of mold on the surface of cheese that has a standard of identity that allows use of mold inhibitors. Morris and Castbert[198] showed that natamycin at 1000 ppm prevented unacceptable mold and yeast growth on blue cheese during curing. Natamycin did not penetrate into cheese nor did it cause the cheese to have off-flavors like those treated with potassium sorbate.[199] Butyric acid fermentation is a problem in some European cheeses, such as Edam, Gouda, Emmenthal, Gruyère, and others. Nitrate from 1 to 15 g/100 L of milk helped to decrease the level of spores in cheese because the xanthine oxidase could reduce nitrate to nitrite and prevent growth of germinating spores.[200] The need to control specific groups of microorganisms will result in the use of inhibitors that are approved for limited use. This is demonstrated by the selective approval of chemical inhibitors for specific foods.

5.4.7 Potassium Sorbate

Potassium sorbate has been used by itself or in combination with other chemicals to control mold growth in dairy products. Potassium sorbate (10 to 20% solution) is used as a dip or spray to inhibit mold growth on cheese surfaces.[180] In addition potassium sorbate can be used on packaging material at a rate of 1 to 6 g/m^2. A maximum of 0.2 to 0.3% sorbic acid is allowed in various types of processed cheese, cheese food, and cheese spreads. Both potassium sorbate and sorbic acid are generally recognized as safe (GRAS) preservatives in the United States and are permitted in many countries worldwide. Although there have been conflicting reports about the ability of *Aspergillus* and *Penicillium* species to grow and produce mycotoxins in the presence of potassium sorbate, none of these were done with milk or dairy products.[201–205] At temperatures of 25 to 28°C, mycotoxins were produced; however, at 12°C potassium sorbate either inhibited or greatly reduced mycotoxin production. Liewen and Marth[204] reported that some molds isolated from Cheddar cheese treated with sorbic acid could grow in the presence of this preservative. Several *Penicillium* species grew in the presence of ≥3000 ppm of sorbic acid at either 4 or 25°C, but none of the aspergilli grew in levels >2000 ppm at 25°C. Tsai et al.[206] isolated several different penicillia from moldy surplus cheese, but could find no correlation between sorbate resistance and mycotoxin production. About 10% of the isolates could produce patulin, penicillic acid, or ochratoxin; however, toxins were not produced much when the isolates were inoculated into processed American and Cheddar cheeses. Liewen and Marth[207] have reviewed the inhibition and growth of molds in the presence of sorbic acid. Potassium sorbate was effective in preventing mold growth in Gouda cheese; but the rind was discolored, the flavor was not acceptable, and the preservative migrated 5 mm below the rind and some could be detected in the center of the cheese.[199] Ahmad and Branen[208] reported that a combination of 0.2% potassium sorbate and 150 ppm butylated hydroxyanisole (BHA) inhibited *A. flavus* growth in broth. BHA at 150 to 400 ppm inhibited *A. flavus* or *P. expansum* growth in processed cheese spread depending on the method of application. Potassium sorbate can be effective in preventing mold growth on cheese depending on the microbes present. Sorbate-resistant molds do not produce mycotoxins in the presence of 0.3% potassium sorbate.[209]

Potassium sorbate in combination with other chemicals has prevented growth of psychrotrophic bacteria in milk. Gilliland and Ewell[210] reported that the use of both *Lactobacillus lactis* and 0.1 to 0.2% potassium sorbate inhibited psychrotrophic bacteria. Strains of *L. lactis* that produced hydrogen peroxide were more effective in inhibiting the psychrotrophs. When >0.1% potassium sorbate was added to pasteurized milk, a sweet taste was noted; therefore, 0.075% potassium sorbate was combined with 0.005% hydrogen peroxide to prevent psychrotrophic growth in milk.[211] These treated milks lasted over 26 days at 6.8°C compared to 10 to 12 days for the controls. The use of potassium sorbate plus hydrogen peroxide may help to extend the shelf life of milk. To date no chemical preservatives have been allowed in fluid milk in the United States and many other countries.

5.4.8 Carbon Dioxide

Carbon dioxide (CO_2) has been used to control microbial growth in many foods. There has been some interest in using it to prevent psychrotrophic bacterial growth in raw milk. King and Mabbitt[212] found that CO_2 at 10 to 30 mM/L decreased the growth of psychrotrophs in milk held at 4 to 10°C. The decrease was greatest at 4°C and 30 m*M* CO_2/L. The presence of CO_2 causes a decrease in pH from 6.7 to 6.0.[212–214] If this decrease is too great due to more than 30 m*M* CO_2/L, then the casein in milk becomes unstable and bitterness is noted.[213] The presence of 30 m*M* CO_2/L increased the shelf life of poor quality raw milk (total count $>10^5$ cfu/ml) and good quality raw milk (total count $<10^4$ cfu/ml) by 1.2 and >3 days, respectively.[213] This increase in shelf life is important because milk is normally cold stored for a few days before processing. CO_2-treated raw milk was used to make cheese and yogurt with no adverse effect. The CO_2 does not need to be removed from milk before use in manufacturing, but it can be removed by warming under vacuum.[212,214] One problem with this method could be activation of *Bacillus* spores with CO_2 and an increase in their heat resistance.[215] However, this research was done with saturated CO_2 and not low levels. The use of CO_2 will depend on cost, ease of use, and legislation.

5.4.9 Removal of Microorganisms by Physical Methods

Two physical methods of removing microorganisms from milk have been researched and are currently used for raw milk, especially in Europe. These two methods are thermization and centrifugation.

Thermization is a prepasteurization heat treatment of milk once it arrives at the dairy to decrease the psychrotrophic population and increase the storage life of the milk before it is processed.[216,217] The bacterial population did not increase significantly for 4 days after thermization at 65°C for 15 s and storage at 4°C; however, that of untreated milk increased significantly.[216] Thermization did not affect the pH, the whey proteins, or the ability of milk to coagulate. Gilmour et al.[217] showed that thermization at 60 to 70°C for 10 to 15 s decreased the level of proteolytic and lipolytic microorganisms in milk. The higher the temperature, the more was the reduction in population. Humbert et al.[218] suggested that 65°C for 20 s was adequate for extending the shelf life of raw milk for 4 days before processing. Thermization has been studied for use with cheese manufacture. Milk that was thermized at 65°C for 15 s had a 3 log cycle reduction in psychrotrophic count and prevented growth of proteolytic and lipolytic bacteria at 50°C for 7 days.[219] The yield of Cheddar cheese was not affected by thermization.[219,220] Johnston et al.[220] reported that thermization at 65°C for 15 s decreased lipase activity. Thermization of this cheese milk also decreased all types of microorganisms—mesophiles, psychrotrophs, coliforms, and proteolytic and lipolytic psychrotrophs, except spores. Cheddar cheese from nonthermized milk that was stored for 3 days had lower sensory scores and higher fat breakdown than thermized milk cheese.[221] Thermization may show its greatest value in decreased lipolytic changes as the cheese ages. Thermized milk (65°C for

15 sec) was used to produce dried skim milk.[222] Thermization decreased the psychrotrophic count to <100 cfu/ml, but had no effect on spore or thermoduric counts. Thermization can be used to reduce some pathogens in raw milk. *Yersinia enterocolitica* was not recovered from thermized milk if levels were $<10^5$ cells/ml.[223] No results have been reported for other dairy pathogens.

Preheat treatments can also be used to decrease enzyme levels and activate spores so that pasteurization can then affect the other bacteria. Psychrotrophic bacteria produce proteases and lipases that are stable to heat treatments given to milk and dairy products. These enzymes can be irreversibly inactivated by temperatures close to those for enzyme activity.[2] This has been called "low temperature inactivation" and has been well documented.[2] Protease from *Pseudomonas fluorescens* was inactivated rapidly at 55°C in both pH 6.6 and 4.5 solutions of casein.[224] Self-digestion resulting in low molecular weight compounds was suggested. Guamis et al.[225] found that 45°C for 5 min or 85°C for 45 s completely inactivated proteases of *Flavobacterium* sp. and drastically decreased *Cytophaga* sp. protease, but were not effective toward *P. fluorescens* protease. Temperatures of 50 to 57°C were generally successful at inactivating proteases of *Pseudomonas* spp.; however, above 60°C there was little inactivation.[2] The mechanism was an autolytic process. Low temperature inactivation has also been shown for lipases. Senyk et al.[226] found that 43 to 100% inactivation of lipase activity could be noted after treatments of 57.2 to 82.2°C for 10 s. Variable results were reported by Fitz-Gerald et al.[227] for 20 lipases heated at 55 to 100°C. More research is needed on the inactivation of lipases by low temperatures.

Psychrotrophic *Bacillus* spp. need to be effectively controlled in milk. Because pasteurization does not eliminate *Bacillus* spores, temperatures that activate spores before pasteurization are needed. Griffiths and Phillips[228] found that 95°C for 5 to 15 s was sufficient to heat activate 13 *Bacillus* spp. in milk. After 24 h at 8°C, a pasteurization treatment of 74°C for 15 s was given to the milk. Premaratne and Cousin[229] found that a heat treatment of 80°C for 10 min followed by incubation at 32°C for 4 h and then a pasteurization of 72°C for 15 s eliminated a *Bacillus cereus* contaminant that concentrated during ultrafiltration. In products where *Bacillus* spores can be problematic, an activated heat treatment followed by a germination step will be necessary before the milk is pasteurized and further processed.

The removal of bacteria by centrifugation has been used in Europe for preprocessing of cheese milks. Sillen[230] reported that the use of centrifuges to remove bacteria (bactofugation) at 60°C resulted in a 98 to 99% reduction in anaerobic spores that cause late fermentation in cheese. During centrifugation, about 3% of the milk has the high bacterial count. This fraction can be heated at UHT temperatures of 135 to 140°C to kill the bacteria and then it is remixed with the rest of the milk. Waes and Van Heddeghem[231] outlined a method for removal of bacterial spores from milk to be used for the manufacture of Edam, Gouda, and Tilsiter cheese. Centrifugation plus addition of 2.5 g of KNO_3/100 L centrifuged milk may be necessary to prevent butyric acid fermentation of these cheeses. Bactofugation can also be used for the removal of bacteria from milk to be used for UHT processing of milk.[232] Centrifugation can be useful in removing bacteria from milk; however, it can also result in loss of 2.5 to 3.5% of the total milk with most loss being protein.[231] Centrifugation can be useful for preprocessing of some milk before it is used to make a dairy product.

5.5 Mastitis

Mastitis is defined as an inflammation of the mammary gland regardless of the cause.[36] Most mastitis occurs in a subclinical form where the characteristic signs and symptoms are not readily detectable by visual examination of milk using a strip cup or by manual palpation of the udder. The clinical form of mastitis is evident by inflammatory swelling, fibrosis, and the atrophy of mammary tissue. Acute inflammatory swelling is also accompanied by hurt and pain. The clinical mastitis is also evident by marked abnormality in secretion, such as blood clots or abnormal color in milk.

5.5.1 Effect on Milk Composition

Mastitis greatly affects the composition of milk.[233] Generally, the concentrations of fat, solids-not-fat, lactose, casein, β-lactoglobulin, α-lactalbumin, and potassium are lowered and those of blood serum albumin, immunoglobulin, and chloride are increased.[36,37,63,234] Mastitic milk also contains elevated somatic cell counts (SCC),[63,234,235] which is the measurement most commonly used as an indicator of mastitis. During mastitis, the ability to synthesize lactose is impaired resulting in decrease in lactose content. Also, blood salts and protein are passed into the milk, leading to elevated salt concentrations. Mastitis has been implicated in rancidity. A linear relationship between lipolysis and cell counts to 1,000,000 cells/cm^3 has been reported.[236] However, reports of higher or lower lipolytic activity in mastitis milk[237,238] and no effects on lipases activity by high cell counts in milk[239] have been published. The above reports notwithstanding, the levels of free fatty acids in milk indicative of rancidity are often used as an indicator of mastitis.[37] Other characteristic changes in milk composition caused by mastitis include decrease in casein and calcium concentrations, elevated catalase and N-acetyl-β-D-glucosaminidase (nagase) activity, and change in pH.[36,37,234] A comparison of composition of normal and abnormal (mastitic) milk is given in Table 5.9.

5.5.2 Economic Losses

Mastitis is considered to be one of the most important problems facing the dairy industry. Earlier estimates of the cost of mastitis to the dairy industry were approximately 225 to 500 million dollars per year or about $69 per cow annually.[63] Others have estimated the cost of mastitis $90 to $230 per cow. In the United Kingdom, the reported cost advantage of mastitis in low prevalence herds was £29 per cow per year compared to that in high prevalence herds.[36] Recent United States estimates

Table 5.9 COMPOSITION OF NORMAL AND ABNORMAL (MASTITIC) MILK

Constituents	Normal Milk (%)	Abnormal Milk (%)	Percent Change
Solids-Not-Fat	8.9	8.8	−1
Fat	3.5	3.2	−9
Lactose	4.9	4.4	−10
Total protein	3.61	3.56	−1
Casein	2.8	2.3	−18
Whey proteins	0.8	1.3	+62
Serum albumin	0.02	.07	+250
Sodium	.057	0.105	+84
Chloride	.091	.147	+61
Potassium	0.173	0.157	−9
Calcium	0.12	.04	−66

Table 5.10 ECONOMIC LOSSES ASSOCIATED WITH BOVINE MASTITIS[a]

	Percentage of Total
Subclinical	
Milk production loss	70
Clinical	
Death and premature culling of cows and reduced cows sale value	13
Discarded or downgraded milk	14
Treatment, labor, and veterinarian service	9

[a] Blood and Radostitis.[36]

indicate that mastitis costs dairy producers up to $2 billion per year or $180 to $200 per cow annually.[37]

Economic losses to the dairy industry are attributed to decreased milk production, discarding of abnormal milk, losses of milk from cows treated with antibiotics, culling of cows, veterinarian services, and treatment costs (Table 5.10).

The losses are compounded when mastitic milk is combined with milk for manufacturing purposes. Low cheese yields and lower quality grades of cheese are often related to mastitic milk.[240] Also, the potential for milk adulteration with residues of antibiotics and sulfa drugs used for treating mastitic cows poses a very serious problem for the dairy industry[37]

5.5.3 Common Mastitis Pathogens

Bacteria are the primary causes of mastitis. Other organisms, including yeast, mycoplasma, nocardia, and even some algae have been occasionally known to cause mastitis. The most common biological agents causing mastitis in dairy cows are *Streptococcus agalactiae* and *Staphylococcus aureus*. *E. coli* is also a significant pathogen in housed or confined cattle. Of secondary importance are *Streptococcus uberis*, *Streptococcus dysgalactiae*, coliforms including *Klebsiella* spp., and *Pseudomonas aeruginosa.*

The mastitis pathogens are commonly grouped in two categories: (1) contagious bacteria that are spread from infected quarters to other quarters and cows, and (2) environmental bacteria that are normally occurring in the cow's environment and infect the cow's udder on contact with the source.

Contagious bacteria include *Streptococcus agalactiae* and *Staphylococcus aureus. S. agalactiae* is an obligate parasite that normally inhabits sinuses and ducts of the mammary glands. Unlike *S. aureus*, it does not normally invade mammary tissue and is readily controlled by penicillin treatment. It is possible to eradicate *S. agalactiae* from a herd.

Staphylococcal mastitis is difficult to control because the biological agent, *S. aureus*, produces several toxins that cause injury to epithelial cells. It invades tissue, and may form abscess and spread to other portions of the mammary gland. *Staphylococcus aureus* produces coagulase, α and β toxins that are important traits in their identification. They also produce penicillinase and develop resistance to penicillin.

Many environmental pathogens are implicated as the causes of mastitis. They include nonagalactiae streptococci (*S. dysgalactiae* and *S. uberis*), coliforms (*Escherichia*, *Enterobacter*, *Klebsiella*), and *Pseudomonas* spp. These organisms occur in the cow's environment: feces, soil and plant material, bedding, stagnant water, mud, etc. and infect the udder surface and teat canals. Many of these pathogens are opportunistic and may cause serious outbreaks of mastitis. The environmental pathogens are difficult to control and cannot be eradicated from individual herds. Common pathogens causing mastitis, and their source, means of spread, and control measures are listed in Table 5.11.

Table 5.11 SOURCE, MEANS OF SPREAD, AND EFFECTIVE CONTROL MEASURES FOR COMMON MASTITIS PATHOGENS

Microorganism	Source	Means of Spread	Control Measures
Streptococcus agalactiae	Infected udders	Cow-to-cow at milking time	Teat dipping; dry cow treatment; lactation treatment; milking time hygiene
Staphylococcus aureus	Infected udders; teat sores	Cow-to-cow at milking time	Teat dipping; dry cow treatment; segregation; cull chronically infected cows
Environmental streptococci	Environment	Environment-to-cow	Improve barn, free stall and hold area sanitation; teat dipping; dry cow treatment
Coliforms	Environment	Environment-to-cow	Improve barn, free stall and holding area sanitation; Improve bedding management

5.5.4 Uncommon Mastitis Pathogens

Besides the common contagious and environmental bacteria, the mammary gland of the cow can be inhabited by many other microorganisms. The following are recorded but less frequent pathogens implicated as causes of mastitis[36]: *Pseudomonas aeruginosa, Streptococcus zooepidemicus, Streptococcus faecalis, Streptococcus pyogenes, Corynebacterium bovis, Corynebacterium ulcerans, Klebsiella* sp., *Enterobacter aerogenes, Mycobacterium bovis* and other *Mycobacterium* spp., *Serratia marcescens, Mycoplasma bovis, Nocardia* spp., *Bacillus cereus, Clostridium perfringens, Brucella abortus, Pasteurella multocida, Cryptococcus neoformans, Aspergillus fumigatus, A. nidulans, Candida* spp., and *Saccharomyces* spp.

Most organisms occurring in mammary glands are aerobic or facultatively anaerobic. However, a few anaerobic organisms have been isolated from udders.[36] These include *Peptococcus indolicus, Bacteroids melaiogenius, Clortoridium sporogenes,* and *Fusobacterium necrophorun.*[36] Algal agents causing mastitis include *Prototheca trispora* and *P. zopfii.*

A few so-called emerging pathogens have long been associated with mastitis in cows and other mammals. These include *Listeria monocytogenes,*[69] *Campylobacter jejuni,*[27,28] *Yersinia enterocolitica,* and *Leptospira pamona.*[63] These organisms when shed in raw milk increase the potential for spread of such diseases as listeriosis, campylobacteriosis, and yersiniosis to humans.

5.5.5 Factors Affecting the Incidence of Mastitis

The probability and frequency of occurrence of mastitis depends on the ability of the pathogens to set up infection which is affected by the characteristics of the pathogen, mechanism of transmission of the disease, and susceptibility of cows.

Pathogenic characteristics important in setting up infection include the ability of the organism to adhere to the mammary epithelium and to colonize the teat duct. Also, the ability of the organism to survive in the cow's immediate environment and its resistance to antibiotics are important in causing mastitis.

Transmission mechanisms of mastitis depend on the extent of infection in the environment including infected quarters, efficiency of milking machine, and cleaning and sanitation of milking equipment as well as hygiene in the milking parlor.

Cows vary in susceptibility to invading pathogens depending on their age, stage of lactation, inherited traits for disease resistance, structure of udder, and infections with other bacteria of low pathogenicity.[63]

Injury to teat or lesion on teat skin resulting from irritations and speed of milking also enhances probability of mastitis.

Proper understanding of these factors plays a major role in many of the control measures designed to minimize mastitic infections in cows.

5.5.6 Detection and Diagnosis

Customarily, detection and diagnosis of mastitis involves observations of the milk using strip cup and physical examination of udder (palpitations, etc.) for inflammation, teat injury, lesions, etc. The tests for detection of mastitis include such cowside tests as the California Mastitis Test (CMT) or electrical conductivity test (e.g., Mas-D-Tec). Other tests for detection of mastitis and abnormal milk include the Wisconsin Mastitis Test (WMT), the catalase test, the Nagase Test, the filter-DNA test, and the modified Whiteside Test. These are commonly used as screening tests. However, milk samples showing positive screening tests and therefore probability of mastitis in cows must be subjected to the confirmatory tests. The two confirmatory tests commonly used in the dairy industry are the Direct Microscopic Somatic Cell Counts (DMSCC) and the Electronic Somatic Cell Counts (EMSCC). Details of the diagnostic tests for mastitis are described elsewhere.[58,234,241]

Microbiological procedures for isolation and characterization of specific pathogens are often used in diagnosis of mastitis. Cultures of milk samples from individual quarters or of composite samples from all four quarters of individual cows are plated on blood agar for detecting mastitic pathogens. Differentiation and characterization of pathogens is done by colony morphology, hemolysis reaction, CAMP-esculin test, catalase and coagulation tests, and various other biochemical reactions.

Microbiological diagnostics of intramammary pathogens can be very useful in determining prevention and treatment of mastitis.

5.6 Pathogenic Bacteria in Milk and Dairy Products

Historically, the presence of pathogenic bacteria in milk and dairy products has been a matter of public health concern. Earlier, diseases such as tuberculosis, typhoid fever, diphtheria, and septic sore throat were commonly transmitted to humans through milk (Table 5.12). Hygienic milk production practices, improved udder health, proper cooling and careful handling and storage of raw milk, as well as the advent of tuberculin testing, brucella eradication, and mandatory pasteurization of milk minimized the threat of pathogenic bacteria in milk and dairy product. The once common milkborne diseases such as tuberculosis, brucellosis, and typhoid fever were virtually eliminated by the end of World War II.[13] However, the problem of pathogenic bacteria in milk and dairy products continued as evidenced by reports of

Table 5.12 OUTBREAKS OF MILKBORNE DISEASE, 1900–1980[a]

Years	Notable Diseases/Pathogens
1900–1920	Typhoid fever, streptococcal infections, diphtheria, salmonellosis, botulism
1920–1940	Typhoid fever, streptococcal infections, staphylococcal intoxication, salmonellosis, poliomyelitis, Haverhill fever, diphtheria
1940–1960	Staphylococcal intoxications, salmonellosis, shigellosis, brucellosis, Q fever
1960–1980	Salmonellosis, staphylococcal intoxication, brucellosis, *E. coli*, campylobacteriosis, yersiniosis, toxoplasmosis

[a] From Bryan[13] and Vasavada.[14]

Table 5.13 LARGE OUTBREAKS ASSOCIATED WITH MILK AND MILK PRODUCTS, 1981–1988[a]

Year	Product	Country	Pathogen	Number	
				Cases	Deaths
1981	Raw milk	Switzerland	*C. jejuni*	500	0
	Raw milk	Scotland	*S. typhimurium*	654	2
	Powdered milk	U.S.A.	*Y. enterocolitica*	239	0
1982	Pasteurized milk	U.S.A.	*Y. enterocolitica* 0:13	172	0
	Pasteurized milk	England and Wales	*C. jejuni*	400	0
	French brie/Camembert cheese	Scandinavia	*S. sonnei*	50	0
1983	Homemade Queso blanco	U.S.A.	*S. zooepidemicus*	16	2
	French brie/Camembert cheese		*E. coli* 0:27	169	0
	Pasteurized milk		*L. monocytogenes*	49	14
1984	Raw milk	England and Wales	*S. zooepidemicus*	12	8
	Cheddar cheese	Canada	*S. typhimurium* PT10	1,500	0
1985	Pasteurized milk	U.S.A.	*S. typhimurium*	18,284	7
	Pasteurized milk	U.S.A.	*S. aureus*	860	0
	Pasteurized milk	Sweden	*S. Saint pul*	153	0
	Powdered milk	U.K.	*S. ealing*	48	1
	Mexican style cheese	U.S.A.	*L. monocytogenes* 4b	181	65
	Vacherin cheese	Switzerland	*S. typhimurium*	22	0
1988	Raw milk	Canada	*E. coli* 0157:H7	30	0

[a] From D'Aoust.[40]

disease outbreaks caused by *Salmonella* spp., *S. aureus*, enteropathogenic *E. coli*, and *Bacillus cereus* in manufactured dairy products such as dry milk, ice cream, and a variety of cheeses made from raw or heated [but not pasteurized] milk[14,40] (Table 5.13). Research dealing with the manufacturing processes, behavior of pathogens during the manufacture and storage of dairy products, and role of starter culture activity in controlling pathogens in cheese milk resulted in industry-wide surveillance programs (e.g., salmonella in dry milk) that helped in minimizing the problem of pathogenic bacteria. However, well-publicized outbreaks of salmonellosis,[38,39,242] listeriosis,[243,244] yersiniosis,[45,245] and campylobacteriosis[20,23–26,246] occurred during the 1980s (Table 5.13). In addition to the familiar pathogens such as *Salmonella*, *S. aureus*, *E. coli*, and *B. cereus*, a new generation of foodborne pathogens such as *L. monocytogenes*, *Y. enterocolitica*, *C. jejuni*, *E. coli* 0157:H7, and *Streptococcus zooepidemicus* has emerged.[14,40] Recent surveys have identified a variety of pathogenic bacteria in raw milk (Table 5.14). Although most pathogenic bacteria, except some enterococci and sporeformers, are inactivated by commercial pasteurization, several incidences of product recalls and reports of disease outbreaks implicating

Table 5.14 INCIDENCE OF FOODBORNE PATHOGENS IN RAW MILK[a]

Pathogen	Country	Number of Samples Tested	Number of Samples Percent Positive
B. cereus	U.S. (1982)	100	9
C. jejuni	U.S. (1982)	108	0.9
	Netherlands (1981)	200	0
	U.S. (1982)	195	1.5
	England (1984–87)	1138	6.0
E. coli 0157:H7	U.S. (1986)	24	4.2
	Canada (1986)	1912	2.0
Listeria monocytogenes	Spain (1982–83)	85	45.0
	U.S. (1983)	121	12.0
	U.S. (1984)	650	4.1
	France (1986)	337	4.2
	Canada (1986)	445	1.3
Salmonella spp.	U.S. (1985)	678	4.7
	Canada (1985–86)	511	2.9
	England (1984–87)	1138	0.2
Yersinia enterocolitica	Canada (1977)	131	22.1
	France (1980)	56	83.9
	U.S. (1982)	100	12.0
	Northern Ireland (1985)	150	11.3

[a] From D'Aoust.[40]

milk, ice cream, cheese, etc. have occurred during the 1980s.[14] Inadequate pasteurization, poor manufacturing practices, and postprocessing contamination were the primary causes of pathogenic contamination in dairy products. The common refrigeration practices for controlling pathogenic bacteria in milk and dairy products may not always be adequate.[14,47] The listeriosis and salmonellosis outbreaks and well-publicized recalls of dairy products caused concern among the consumers and regulators regarding safety of the milk supply[14,247] and prompted the Dairy Products Safety Initiative by the U.S. Food and Drug Administration.[14]

The main characteristics and illnesses caused by the more common pathogens found in milk and dairy products are given in Table 5.15. The following is a brief discussion on the so-called emerging pathogens.

5.6.1 *Listeria Monocytogenes*

L. monocytogenes is a Gram-positive, non-spore-forming rod-shaped organism with coccoid or diphtheroid morphology. It is psychrotrophic and can grow at temperatures from 3 to 45°C, optimally at 30 to 37°C. The organism forms bluish-green colonies on trypticase soy agar (oblique illumination) and shows characteristic tum-

Table 5.15 GENERAL CHARACTERISTICS OF PATHOGENS IN MILK AND DAIRY PRODUCTS[a]

Pathogen	Gram Stain	Morphology	Temperature Range for Growth	Oxygen Requirement	Catalase Reaction	pH for Growth	Motility	Pathogenicity
L. monocytogenes	Positive	Small coccoid rods—no spores	2.5–42°C	Microaerophilic	Positive	5.6–9.8	Positive (20–25°C)	*β*. listeriolysin lipase
B. cereus	Positive	Large rods, spore forming	10–50°C[a]	Aerobic[b,c]	Positive	4.9–9.3	Positive petritrichous flagella	Heat-labile diarrheal toxin, enterotoxin and heat stable emetic enterotoxin
C. jejuni	Negative	Slender-curved ''vibrioid'' rods	30–45°C, optimum 42–45°C	Microaerophilic[c]	Positive	4.9–8.0	Motile single polar flagellum	Heat labile enterotoxin, cytotoxin, colonization, invasiveness
E. coli	Negative	Small coccobacilli	10–35°C[d]	Facultative anaerobic	Positive	5.6–6.8	Positive	Invasiveness, heat-labile and heat-stable enterotoxins, verotoxins
Salmonella spp.	Negative	Short rods	5–47°C	Aerobic	Positive	6.6–8.2	Positive peritrichous flagella	Invasiveness, heat-labile enterotoxin, heat-stable cytotoxin
S. aureus	Positive	Cocci in pairs or irregular clusters	10–45°C	Aerobic or anaerobic	Positive	4.5–9.3	Negative	Seven enterotoxins (A, B, C_1, C_2 C_3, D, and E), somewhat resistant to heat and proteolytic enzymes
Y. enterocolitica	Negative	Small rods	4–34°C[e]	Aerobic	Positive	6.8–9.0	Negative	Plasmid-mediated, (HT) enterotoxin virulence

[a] Psychrotrophic variants grow at 5°C.
[b] Vegative cells may grow anaerobically.
[c] Optimum growth in an atmosphere containing 5% O_2.
[d] Fecal coliforms and pathogenic *E. coli* except *E. coli* 0157:H7 grow at 44–45–5°C.
[e] Most strains grow best at 22–25°C.

bling motility when grown in trypticase soy broth at 25°C. *L. monocytogenes* is weakly β-hemolytic on media containing blood. It grows in a pH range of about 4.8 to 9.6 and is catalase positive.

L. monocytogenes is distributed widely in nature and has been isolated from a variety of sources including soil, manure, leafy vegetables, raw beef, and poultry.[72] It also has been isolated from mastitic milk, improperly fermented silage, and from unpasteurized raw milk.[14,65,67,248]

Listeriosis can manifest a variety of symptoms in humans, including meningitis, infectious abortion, perinatal septicemia, and encephalitis. Often, it is the cause of stillbirths or deaths of infants soon after birth. Surviving infants usually develop meningitis, which can be fatal or result in permanent mental retardation.[68,248]

L. monocytogenes is heat sensitive and is inactivated by pasteurization. Doyle et al.[70] reported that *L. monocytogenes* in the intracellular phase (in leucocytes) may survive pasteurization. However, further research[71] has indicated that conventional HTST pasteurization treatment is adequate to inactivate the organism.

5.6.2 *Yersinia Enterocolitica*

Y. enterocolitica is a Gram-negative, non-spore-forming, rod-shaped bacterium. It is psychrotrophic and will grow at temperatures from 0 to 45°C, optimally at 22 to 29°C. Because *Y. enterocolitica* tolerates alkaline conditions, this characteristic is used in its selection.[42,250–255] On a selective medium such as the Cefsulodin–Iragasan–Novobiocin (CIN) agar or the *Yersinia* Selective Agar (YSA), *Y. enterocolitica* forms characteristics ''bulls eye'' or ''target'' colonies.[42–44]

Y. enterocolitica is widely distributed in nature. It has been isolated from foods of animal origin, including milk and cheese, beef, pork, and lamb.[41] It also is known to occur in waters of lakes, wells, and streams.[41]

Yersiniosis is characterized by gastroenteritis, mesenteric lymphadenitis, and terminal ileitis. Often, yersiniosis symptoms mimic acute appendicitis. Such was the case in the well known outbreak of yersiniosis in Oneida County, NY, in which several children were subjected to unnecessary appendectomies after drinking chocolate milk contaminated with *Y. enterocolitica* serotype 0:8,[245] whereas *Y. enterocolitica* stereotype 0:3 is more prevalent in Europe and Canada.[42,43]

Although *Y. enterocolitica* and related bacteria have frequently been isolated from raw milk,[251,252,256] most isolates have been recognized as nonpathogenic, ''environmental'' strains. However, production of enterotoxin by *Yersinia* spp. isolated from milk has been recently reported by Walker and Gilmour.[257] The organism is heat labile and is readily inactivated by conventional pasteurization.[254,258]

5.6.3 *Campylobacter Jejuni*

C. jejuni is a Gram-negative non-spore-forming bacterium with a characteristic S, gull, or comma-shaped morphology. Under the phase-contrast microscope, *C. jejuni* exhibits a characteristic darting, ''cork-screw'' motility. The organism is microaerophilic in nature and can be readily grown in reduced oxygen atmosphere of 5% O_2,

10% CO_2, and 85% N_2.[20–22,259–261] The organism grows at temperatures from 30 to 47°C. *Campylobacter jejuni* is β-hemolytic on media containing blood and is catalase positive.[21,22]

C. jejuni has been isolated from feces of cattle, swine, sheep, goats, dogs, cats, rabbits, and rodents.[20,262] It causes mastitis in cows and has been isolated from raw milk.[260–262]

Campylobacter infections are more common than cases of salmonellosis and shigellosis combined.[21] Symptoms of campylobacteriosis include mild enteritis or sometimes severe enterocolitis. Often the patient experiences apparent recovery followed by relapse. Other symptoms include nausea, abdominal cramps, and bloody diarrhea.[22]

C. jejuni is sensitive to heat, drying, air (oxygen), and acidic pH. It is readily inactivated by normal pasteurization.[20,262]

5.6.4 *Escherichia Coli*

The presence of coliforms, particularly *E. coli*, in foods indicates the possibility of pathogenic contamination, polluted water supply, or a breakdown in sanitation. *E. coli* is a Gram-negative, non-spore-forming, rod-shaped organism. Four groups of *E. coli* have been recognized—enteropathogenic, enterotoxigenic, enteroinvasive, and colehemorrhagic.[32,264]

5.6.5 *Escherichia Coli* 0157:H7

E. coli 0157:H7 is recognized as an emerging pathogen. Several outbreaks involving *E. coli* 0157:H7 have been reported.[265,266] The organism causes hemorrhagic colitis or bloody diarrhea. This infection is usually characterized by severe abdominal cramps followed by watery and grossly bloody stools.[33,267,268] Vomiting is common, but fever is rare. The illness occasionally involves hemolytic uremic syndrome (HUS), which is characterized by serious kidney dysfunction with urea in the blood.[268]

Although *E. coli* 0157:H7 is often associated with ground beef, recently it was implicated as the cause of an outbreak in Ontario where several kindergarten children suffered from an illness after visiting a dairy farm where raw milk was served.[269] Recently, dairy cattle have been identified as a reservoir of *E. coli* 0157:H7. Some regulatory authorities warn that the increased slaughtering and processing of dairy cattle resulting from the dairy diversion program and culling for mastitis management may increase the potential of *E. coli* 0157:H7 contamination. The significance of *E. coli* 0157:H7 as a foodborne pathogen is not fully known.[32]

Enterotoxigenic *E. coli* 027:H20 is another strain of *E. coli* that caused gastroenteritis associated with eating imported Brie cheese.[264,266,267] Several similar outbreaks occurred in the United States and one in the Netherlands associated with consumption of cheeses from France.[267]

5.6.6 *Bacillus Cereus*

Bacillus spp., particularly *B. cereus*, *B. cirulans*, and *B. mycoides*, are the spore-forming psychrotrophic bacteria known to occur frequently in raw and pasteurized milks.[6,10] These Gram-positive motile, aerobic, spore-forming, rod-shaped organisms have been implicated as the cause of a variety of proteolytic defects, including bitterness and sweet-curdling in milk and cream.[6] Outbreaks of food poisoning caused by milk products containing *B. cereus* have been reported.[59–61]

On the mannitol–egg yolk–polymyxin (MYP) agar, *B. cereus* produces typical pink colonies surrounded by a precipitate zone, indicating lecithinase activity.[270] In addition to lecithinase production, *B. cereus* is characterized by growth and acid production from glucose anaerobically, reduction of nitrate to nitrite, production of acetyl methyl carbinol, decomposition of L-tyrosine, and growth in the presence of 0.001% lysozyme.[270] *B. cereus* is usually strongly hemolytic, producing a 2 to 4-mm zone of β hemolysis on a blood agar plate. Differentiation of *B. cereus* from other related organisms, for example, *B. cereus* var. *mycoides*, and *B. thuringiensis* is done by tests for motility, hemolysis, and protein toxin crystals.[61]

B. cereus strains can produce both emetic and diarrheogenic toxins[271] and have been known to cause two different forms of gastroenteritis.[59] The emetic toxin is responsible for symptoms of nausea and vomiting within a few (0.5 to 6) hours after consumption of food containing *B. cereus* toxin.[59] The diarrheogenic toxin is responsible for diarrhea, abdominal cramps, and tenesmus occurring 5 to 6 h after consumption of the contaminated food. The symptoms of diarrheogenic illness may include nausea but rarely vomiting.

The diarrheal toxin is produced during the late logarithmic phase of growth at temperatures and pH values of 18 to 43°C and 6 to 11, respectively. Production of emetic toxin by *B. cereus* occurs during the stationary phase of growth at temperatures and pH values ranging from 25 to 30°C and 2 to 11, respectively. The emetic toxin is extremely heat stable and can withstand heat treatment of 126°C for 90 min.[61]

Although starchy foods containing corn and corn starch, mashed potatoes, pudding, soups, and sauces are most frequently associated with diarrheal-type food poisoning outbreaks, fried and boiled rice dishes and macaroni and cheese have been implicated as vehicles of emetic-type illness.[59,60]

5.6.7 Economic Significance of Pathogens

According to Archer and Kvenberg,[272] 24 to 80 million episodes of acute foodborne disease occur in the United States annually. An examination of etiologic agents and food vehicles associated with 7458 outbreaks of foodborne illnesses (involving 237,545 cases) reported to the Center for Disease Control (CDC) between 1973 and 1987 revealed that a specific food vehicle was implicated in 3699 (50%) outbreaks. Dairy products were responsible for 4% of outbreaks (158) and 14% (29,667) cases. The massive outbreak of *Salmonella typhimurium* in Illinois was responsible for the large proportion of cases.[29] The outbreak was associated with 2% low-fat pasteurized milk produced by a dairy plant in Chicago. Of >150,000 persons who became ill

Table 5.16 ECONOMICS OF FOODBORNE DISEASE OUTBREAKS[a]

Food	Country	Etiological Agent	Number		Cost ($\times 10^3$)[b]		Cost/Case
			Ill	Death	Direct	Indirect	
Raw milk	Scotland (1981)	*S. typhimurium*	654	2	$ 153	$1,226	$ 2,108.00
	U.S. (1985)	*S. typhimurium*	16,284	7	?	?	?
Cheese	U.S. (1965)	*S. aureus*[c]	42	0	$ 490	—	$11,676.00
Cheddar/ Monterey	Canada (1977)	*S. aureus*[c]	15	0	$ 653	—	$ 43.00
Emmenthal Cheddar	U.S. (1976)	*S. heidelberg*	234	0	$ 251[d]	—	$ 1,073.00
Chocolate	Canada, U.S.A. (1973–74)	*S. eastbournre*	≥200	0	$62,063	—	$30,317.00

[a] From Todd,[275] and D'Aoust.[40]
[b] Cost estimates expressed in 1983 U.S. dollars.
[c] Contamination of starter cultures.
[d] Excludes cost to the manufacture.

there were >16,000 culture-confirmed cases; 2777 were hospitalized and 14 died.[30,273] In another noteworthy outbreak of *Listeria monocytogenes* infections due to Mexican-style soft cheese in California,[274] over 150 persons became ill. Over 50 deaths (fatality rate of 34%) were aborted fetuses or pregnant women and their newborn offspring were reported.[30] According to the CDC, dairy products were associated with 103 deaths and the death-to-case ratio was 5.0 per 1000.[29]

Economic losses associated with foodborne illness and recalls have been estimated by Todd.[275] These include direct costs attributed to expenses involved in epidemiological investigations of outbreaks, laboratory diagnosis, treatment, loss of income by patients, and financial losses to the food manufacturers as a result of product recalls and loss of sales. Indirect costs involve expenses related to litigation, settlement, and compensation for grief, pain and suffering, and loss of life.[275] Table 5.16 shows cost estimates of economic losses associated with disease outbreaks involving raw milk and cheese.

5.6.8 Mycotoxins and Amines

Besides the pathogenic bacteria and their toxins, the public health and food safety concerns associated with milk and dairy products deal with the presence of mycotoxins and amines in milk and cheese. Mycotoxins are toxic metabolites produced by certain molds during their growth on cereal grains such as corn, rice, sorghum and peanuts, and other oilseeds. Possible sources of mycotoxins in milk and cheese include consumption of contaminated feed by cow and subsequent passage of the ingested mycotoxins or metabolites into the cheese milk, growth of toxigenic mold on cheese, and organisms used in mold-ripened cheeses.[276]

Aflatoxins, produced by *Aspergillus flavus* and *A. parasitcus*, are of particular concern because they are potent liver carcinogens and cannot be inactivated by pasteurization and sterilization of milk. Aflatoxin B, present in contaminated feed, is converted into a carcinogenic derivative M, and secreted into milk.[277] Results of studies of cheese manufacturing using milk from cows fed aflatoxin B, or milk with M, added directly to it, have shown that 47% of the toxin present in the milk was recovered in Cheddar cheese, about 50% in Camembert cheese, and 45% in why.[278]

Other mycotoxins such as penicillic acid, patulin, cyclopiazonic acid, or PR toxins may also be found in cheeses, including Cheddar and Swiss cheese. Certain mold starter cultures used in the manufacture of mold-ripened cheeses such as Camembert and Roquefort cheese are also capable of producing mycotoxins in cheese.[276] Further information on the occurrence, synthesis, and control of aflatoxins and other mycotoxins is given below. Reviews by Applebaum et al.[279] Bullerman,[280] and Scott[277,281–283] may be consulted for additional information on the subject.

Biogenic amines, for example, histamine, tyramine, and tryptamine, found in cheese and other foods constitute a negligible risk to all but the rare individuals lacking monoamine oxidases (MAO).[284] However, the occurrence of these amines in food, particularly cheese, may be responsible for causing hypertensive response and even death from cerebral hemorrhage in persons on monoamine oxidase inhibitor (MAOI) therapy.[285,286]

Several outbreaks of apparent amine intoxication have occurred from consumption of Gouda, Swiss, and other cheeses containing $\geq$ 100 mg of histamine per 100 g of cheese.[284,287]

The toxic amines are produced in cheese by decarboxylation of the appropriate amino acids by certain bacteria, including strains of *Streptococcus faecium*, *Streptococcus mitis*, *Lactobacillus bulgaricus*, *Lactobacillus plantarum*, viridans streptococci, and *Clostridium perfringens*.[284,288] Voight and Eitenmiller[288] studied tyrosine and histidine decarboxylase activities in dairy-related bacteria and showed that the lactic starter bacteria (group N streptococci) were not likely to be producers of biogenic amines in cheese.

Certain diamines such as putrescine, cadeverine, and spermine enhance the toxic amount of histamine.[284] Therefore, conditions allowing the formation of diamines, particularly putrescine and cadeverine, should be monitored carefully. The production of biogenic amines in cheese depends on a number of factors including the presence of certain bacteria, enzymes, and cofactors necessary for amino acid decarboxylation; existence of the proper environment, that is, pH, temperature, and water activity during cheese ripening; and the presence of potentiating compounds (e.g., diamines). Proper control of the cheese manufacturing process, particularly regarding pH, salt, and moisture levels during ripening, is essential for minimizing the potential threat of biogenic amines.

5.7 Mycotoxins in Milk and Dairy Products

Many different genera of molds can be isolated from dairy products.[283] Table 5.17 lists the most common molds that have been isolated from these products. Species of mainly *Aspergillus*, *Fusarium*, and *Penicillium* can grow in milk and dairy prod-

Table 5.17 MOLDS FOUND IN MILK AND DAIRY PRODUCTS[a]

Product	Genera of Molds Identified[b]
Raw milk	*Alternaria, Aspergillus, Cladosporium, Fusarium, Geotrichum, Mucor, Penicillium, Rhizopus*
Pasteurized milk[c]	*Alternaria, Aspergillus, Aureobasidium, Chrysosporium, Cladosporium, Epicoccum, Geotrichum, Mucor, Paecilomyces, Penicillium, Phoma, Rhizopus, Scopulariopsis, Stemphylium, Trichosporon*
Dried milk	*Alternaria, Aspergillus, Cladosporium, Mucor, Penicillium*
Cream	*Aspergillus, Geotrichum, Penicillium, Phoma*
Butter	*Alternaria, Aspergillus, Cladosporium, Fusarium, Geotrichum, Mucor, Paecilomyces, Penicillium, Phoma, Rhizopus, Scopulariopsis, Verticillium*
Cheese	*Alternaria, Aspergillus, Cladosporium, Fusarium, Geotrichum, Mucor, Penicillium, Rhizopus*
Yogurt	*Cladosporium, Geotrichum, Monilia, Mucor, Penicillium*

[a] Scott.[283]
[b] Although toxigenic strains were isolated from some of these products, mycotoxins are rare.
[c] Vadillo et al.[289]

ucts and produce mycotoxins if the conditions are correct.[283,290–292] Mycotoxins are secondary metabolites that are produced by molds and their consumption can result in biological effects in animals and humans. The major biological effects of the mycotoxins have been classified as acute toxic, carcinogenic, emetic, estrogenic, hallucinogenic, mutagenic, and teratogenic.[292,293] The common mycotoxins that can be found in dairy products are listed in Table 5.18.

5.7.1 Presence of Mycotoxins in Milk and Dairy Products

Dairy products can become directly contaminated with mycotoxins by molds that grow on them and produce the toxins or indirectly by the carryover of mycotoxins into milk as a result of dairy cows consuming mycotoxin-contaminated feeds.[296,297] Aflatoxins and other mycotoxins can be produced during the growth of plants or during their subsequent storage. Stresses that occur during growth of crops can increase the chances of aflatoxin production, such as drought, reduced fertilization, and competition with weeds.

There have been several studies done on the carryover of mycotoxins from contaminated feed, either natural or artificial, into the milk of dairy cows. Schreeve et al.[298] showed that when 1 to 2 mg/kg of ochratoxin A or zearalenone was present in feeds, there was no significant carryover into the milk; however, aflatoxin B_1 at 20 μg/kg was converted to aflatoxin M_1 in milk at concentrations of 0.06 μg/kg. Patterson et al.[299] found that cows consuming 10 μg/kg of aflatoxin B_1 excreted about 0.2 μg/kg of aflatoxin M_1 in milk daily. Munksgaard et al.[300] fed four levels of aflatoxin B_1 from naturally contaminated cottonseed meal. At 57, 142, 226, and

Table 5.18 SOME MYCOTOXINS THAT CAN BE FOUND IN MILK AND DAIRY PRODUCTS[a]

Mycotoxin	Molds
Aflatoxins	*Aspergillus flavus* *Aspergillus parasiticus*
Citreoviridin	*Penicillium citreoviride* *Penicillium toxicarium*
Citrinin	*Penicillium*
Cyclopiazonic acid	*Aspergillus flavus* *Penicillium camemberti* *Penicillium cyclopium*
Deoxynivalenol	*Fusarium* species
Moniliformin	*Fusarium* species
Nivalenol	*Fusarium* species
Ochratoxin	*Aspergillus ochraceus* *Penicillium viridicatum*
Patulin	*Penicillium patulin*
Penicillic acid	*Aspergillus* species *Penicillium* series
Penitrem A	*Penicillium crustosum*
Sterigmatocystin	*Aspergillus nidulans* *Aspergillus versicolor*
T-2 Toxin	*Fusarium* species
Versicolorin A	*Aspergillus versicolor*
Zearalenone	Fusarium graminearum

[a] Scott,[283,294] van Egmond,[291,295] van Egmond and Paulsch.[292]

311 μg/day of aflatoxin B_1, the aflatoxin M_1 produced in milk ranged from 27 to 74, 38, to 128, 60 to 271, and 96 to 138 ng/kg, respectively. There was great variation from cow to cow on the amount of aflatoxin M_1 detected even if the same level of aflatoxin B_1 was fed. When Price et al.[301] fed 5 to 560 μg/kg levels of aflatoxin B_1-contaminated cottonseed to dairy cows as 15% of the total feed ration to 90 cows for 70 days, the 0.5 ppb aflatoxin M_1 action level was exceeded only when 280 μg/kg or more of aflatoxin B_1 was fed to the cows. When the level of aflatoxin B_1 was decreased, the level of aflatoxin M_1 also decreased and fell below the 0.5 ppb action level. Frobish et al.[302] noted that aflatoxin M_1 occurred in milk within 12 h of feeding Holstein cows with cottonseed meal containing 94 to 300 μg/kg of aflatoxin B_1. The level of aflatoxin M_1 fell to below 0.5 ppb within 24 h after cessation of feeding aflatoxin B_1 to the cows. Because 1.7% of the total aflatoxin B_1 was

converted to aflaxtoxin M_1, feeding cows 33 μg of aflatoxin B_1/kg in the diet would result in exceeding 0.5 ppb of aflatoxin M_1 in milk. Corbett et al.[303] studied the presence of aflatoxin M_1 in milk to estimate the level of aflatoxin B_1 in feed. Although aflatoxin B_1 levels were all below 20 ppb, aflatoxin M_1 was found in the milk of 40 cows in levels ranging from 0.001 to 0.273 ppb. When more aflatoxin M_1 was detected in the milk, the level of milk production was decreased for the herd. More research is needed to see the long-term effects of chronic ingestion of low levels of aflatoxin B_1 by dairy cows. Because all these previous studies have shown that consumption of aflatoxin-contaminated feeds resulted in aflatoxin M_1 in milk, Fremy et al.[304] analyzed milk for aflatoxin M_1 after cows consumed peanut cakes contaminated with aflatoxin B_1 that was treated or untreated with ammonia gas. In milk from cows that consumed treated peanut cakes no or only trace amounts of aflatoxin M_1 were detected, but >0.5 ppb aflatoxin M_1 was detected in milk. These and other research reports have shown that aflatoxin and other mycotoxins can be carried over from the feed into milk.

The second way that milk can be contaminated by mycotoxins is the growth of molds in or on dairy products. For maximum mycotoxin production, the proper conditions of nutrients, temperature, pH, aeration, competition, and time are all important. Many studies have been done on the proper conditions for mild growth and mycotoxin production in milk and dairy products. Some of the research done over the past decade on growth and mycotoxin production in dairy products will be briefly reviewed. The production of aflatoxins in dairy products has been researched often because these are the most potent mycotoxins known. Park and Bullerman[305,306] examined the effect of temperature on the production of aflatoxin in cheese and yogurt by *A. flavus* and *A. parasiticus*. Both species of *Aspergillus* grew best at 25°C in Cheddar cheese with growth being detected within 2.5 days.[305] As the temperature was decreased to 18, 15, and 5°C, the time to detect growth in Cheddar cheese took 4.6 and 5.2 days, 16 and 15 days, and nondetectable for *A. parasiticus* and *A. flavus*, respectively. Sporulation in Cheddar cheese took longer than growth. At 25°C *A. parasiticus* sporulated in Cheddar cheese in 5 days compared to 8.4 days for *A. flavus*. Sporulation at lower temperatures took considerably longer for both species and no sporulation was noted at 5°C. The effects of cycling temperatures from 5 to 25°C were used to see if changes in temperature affected the production of aflatoxin in Cheddar cheese.[305] More aflatoxin B_1 was produced by *A. flavus* at a constant temperature of 25°C than at the cycling temperatures of 5 to 25°C. *A. parasiticus* produced more aflatoxin G_1 than B_1 at 25°C than at the cycling temperatures. For both molds, much less aflatoxin was produced at 18 and 15°C and none was produced at 5°C.

Further research using other dairy products showed that *A. parasiticus* produced little to no aflatoxins on Cheddar cheese, cottage cheese, and yogurt.[306] *A. flavus* produced no aflatoxin in both Cheddar and cottage cheeses at 15°C, but did in yogurt. This is most likely due to the presence of more carbohydrate because aflatoxin is produced best on substrates with high carbohydrate instead of high protein. This was also shown by the high production of aflaxtoxin in rice. Similarly more aflatoxin was produced at 25°C than at 15°C. *A. flavus* was able to use small amounts of carbohydrate to produce aflatoxin in dairy products, but *A. parasiticus* could not.

The production of aflatoxin in the presence of lactic acid bacteria has been investigated, as these bacteria are important in cheese ripening. El-Gendy and Marth[307] found that when both *Lactobacillus casei* and *A. parasiticus* were grown together, there was both more mold growth initially but less aflatoxin production than when the mold was grown alone. The aflatoxin was also degraded more after 7 to 10 days of coincubation of *L. casei* and *A. parasiticus*. Mohran et al.[308] noted that the proteolytic activity of *Streptococcus thermophilus, Lactococcus lactis* subsp. *lactis* var. *diacetylactis, Lactobacillus casei*, and *Lactobacillus bulgaricus* was not altered with increasing levels of aflatoxin B_1, but decreased for *L. lactis* subsp. *lactis*. The presence of aflatoxin B_1 in milk can have an effect on the subsequent use of the milk to produce fermented dairy products; however, this depends on the species and aflatoxin concentration.

Most of the research that has been done on the production of aflatoxins in dairy products shows that aflatoxins are not produced unless there is sufficient carbohydrate; therefore, cheese is not a good substrate. Also, the storage of dairy products at temperatures below 10°C effectively prevents the toxigenic species of *Aspergillus* from growing. Other molds will generally out-compete the aflatoxin-producing aspergilli in dairy products.

Aspergillus versicolor is frequently found growing on cheese.[292] *A. versicolor* can produce a toxin called sterigmatocystin, which has a chemical structure similar to that of aflatoxin B_1. For *A. flavus* and *A. parasiticus* sterigmatocystin is a precursor to aflatoxin biosynthesis.[309] Sterigmatocystin is toxic, mutagenic, and carcinogenic and has an LD_{50} in rats of 120 to 166 mg/kg of body weight when given orally.[309] Sterigmatocystin has been found in hard cheeses, such as Edam and Gouda.[292,309,310] Northolt et al.[310] noted that *A. versicolor* was frequently isolated from hard cheeses stored in warehouses, especially aged cheese. *A. versicolor* could grow in the lower water of aged cheeses and even penetrate the plastic coating in the cheese. When cheeses were chemically analyzed, they had sterigmatocystin in the upper 1 cm of the cheese. The concentrations of sterigmatocystin in the upper 1 cm layer ranged from 5 to 600 μg/kg. Veringa et al.[309] found that lactose, fat, and glycerol all stimulated *A. versicolor*'s production of sterigmatocystin on cheese. Frequent turning of cheese promoted growth of and toxin production by *A. versicolor*. If several layers of plastic were used to coat the cheese, then the fatlike compounds, which are stimulatory to sterigmatocystin production, cannot diffuse through for the mold to grow. Once sterigmatocystin is produced, it is stable in the refrigerator, freezer, and warehouses for several weeks.[292]

In addition to *Aspergillus* species, several toxin-producing *Penicillium* species can be isolated from dairy products. Northolt et al.[310] showed that *P. verrucosum* var. *cyclopium* could be isolated from cheeses that were refrigerated in shops, homes, and warehouses. This species and several *Penicillium* and *Aspergillus* species[311] produce penicillic acid. The oral toxicity of penicillic acid is low. Four strains of *P. cyclopium* did not produce penicillic acid in either Gouda or Tilsiter cheeses at 16°C for up to 42 days. The water activity of the cheeses was 0.97. Penicillic acid is not produced very well in substrates low in carbohydrates and at water activities below 0.97, which may occur in cheese. Also *P. brevicompactum*, producer of mycophenolic acid, and *P. verrucosum* var. *verrucosum*, which produces citrinin, ochratoxin, viridicatin, and viridicatic acid, were isolated from cheeses stored in warehouses.[310] Ochratoxins are produced by species of *Penicillium* and *Aspergillus*.[312] Ochratoxin can cause kidney and liver problems in laboratory animals. In some Balkan countries human endemic nephropathy may be due to ochratoxin A. On Edam cheese at a water activity of 0.95, ochratoxin A was produced by *P. cyclopium* at temperatures from 20 to 24°C. The toxicity of these mycotoxins is much lower than that for aflatoxins. Also, these mycotoxins do not occur very frequently in cheeses.

Mold-ripened cheeses are made from strains of two *Penicillium* species, *P. camemberti* for Camembert and Brie cheeses and *P. roqueforti* for Roquefort and Blue cheeses.[294,313] Toxic metabolites can be produced by these species. The major toxic metabolites that can be produced by *P. roqueforti* are patulin, penicillic acid, citrinin, alkaloids (roquefortines A to D, festuclavine, marcfortine), PR toxin, mycophenolic acid, siderophores (ferrichrome, coprogen), and betaines (ergothioneine and hercynine). These mycotoxins have either not been detected or detected only in very low levels. Penicillic acid and PR toxin are not stable in cheese. Engel et al.[314] found that only Roquefort cheese from one factory had mycophenolic acid present. Strains of *P. roqueforti* produced 50 to 100 times lower levels of mycophenolic acid in Blue cheese compared to synthetic media. Because blue cheese is eaten in low quantities, there should be no toxicological effects observed in humans. *P. camemberti* produces cyclopiazonic acid that shows toxicity in rats. Cyclopiazonic acid was found in the crusts, but not in the interior, of some Camembert and Brie cheeses. Also, production was higher at 25°C than at 4 to 13°C.[315] In an effort to develop cyclopiazonic acid negative strains of *P. camemberti*, Geisen et al.[316] isolated mutants that either produced no detectable cyclopiazonic acid or only about 2% that of the parent strain. The latter mutant produced a new metabolite within 21 days at 25°C. Therefore, it may be possible to produce strains for cheese manufacture that have low or no detectable levels of cyclopiazonic acid. Care must be taken in the production of these strains to ensure that no new toxic compounds are produced. Generally, mycotoxins produced by these mold starter cultures pose no health hazards because the levels of consumption of these cheeses are low.

5.7.2 Fate of Aflatoxin M_1 in Dairy Product Manufacture and Storage

Because aflatoxin M_1 can be present in milk as a result of carryover from the feed consumed by cows, it is important to determine how stable it is during dairy product manufacture. Wiseman et al.[317] reported that aflatoxin M_1 was stable in milk and cream pasteurized at 64°C for 30 min. Aflatoxin M_1 was also stable in milk heated up to 100°C for 2 h.[318] Likewise, the aflatoxin was stable to pH from 4 to 6.6 for the 4 days of the trial. Several studies have been done on the manufacture, ripening, and storage of different varieties of cheese and other dairy products. Brackett and Marth[319] showed that aflatoxin M_1 concentrated in the curd with a 4.3-fold increase over that of the milk. The level of aflatoxin M_1 did not decrease in either Cheddar cheese or process cheese spread that was aged for over a year at 7°C. In fact, the initial and final levels were very similar. For Brick cheese, aflatoxin M_1 concentrated by 1.7-fold because the washing step removed some of the toxin;[320] however, the level of aflatoxin M_1 never dropped below the initial concentration for the 22 weeks of aging at 10°C. In the surface-ripened Limburger-like cheese, the level of aflatoxin M_1 after 22 weeks at 10°C was the same as the initial concentration, indicating that the aging did not degrade the toxin. In Mozzarella cheese, there was an 8.1-fold increase in aflatoxin M_1 and the levels remained constant for 19 weeks storage at 7°C.[321] For Parmesan cheese, the level of aflatoxin M_1 concentrated 5.8-fold over that of milk, but the level decreased in the cheese over 22 weeks of ripening at 10°C and then a slow increase was seen until 40 weeks of ripening.[321] It was postulated that the addition of lipase could allow more efficient recovery of aflatoxin M_1 initially because similar increases in concentrations of aflatoxin M_1 in Cheddar cheese ripening were noted when the lipolytic and proteolytic enzymes would be most active. Wiseman and Marth[322] showed that aflatoxin M_1 was stable for 2 months during both refrigerated and frozen storage of Baker's and Queso Blanco cheeses. Aflatoxin M_1 was also stable during ripening and frozen storage of Manchego-type cheese.[323] For products that are not ripened such as cottage cheese, yogurt, and buttermilk, the level of aflatoxin M_1 remained stable during storage at 7°C.[297,324] Aflatoxin M_1 content decreased in Kefir; however, this could have been a result of the analysis or the binding of casein to aflatoxin M_1.[322] Munksgaard et al.[300] reported an apparent increase in aflatoxin M_1 in yogurt stored at 5°C for 2 weeks; but the level in Ymer remained constant. Aflatoxin M_1 was also stable during skim and whole milk, nonfat dried milk, and buttermilk manufacture.[300,325] Lower amounts of aflatoxin M_1 were found in a butterlike spread, as the toxin concentrates with casein and not fat.[317,325] All of this research has shown that aflatoxin M_1 is stable during the manufacture and storage of dairy products. Also, the level of aflatoxin remains stable during both refrigerated and frozen storage.

Only a limited amount of research has been done on the fate of aflatoxins B_1, B_2, G_1, and G_2 in dairy products. Megalla and Mohran[326] studied the fate of aflatoxin B_1 in milk fermented by *Lactococcus lactis* subsp. *lactis* and found that aflatoxin B_1 was converted to nontoxic and less toxic components, namely B_{2a} and aflatoxicol, respectively. Aflatoxins B_1, B_2, G_1, and G_2 distributed more in curd than whey on a per weight basis in Manchego-type cheese manufacture. During manufacture, aflatoxins B_1 and B_2 were lost up to 10% compared to 31% for G_1 and G_2. During the 60-day ripening, there was no loss of aflatoxins B_1 and B_2 and aflatoxins G_1 and G_2 increased by 133%. Although there were variations in samples during both refrigerated storage for 60 days and frozen storage for 90 days, the presence of aflatoxins B_1, B_2, G_1, and G_2 appeared to be stable. These results plus those published in earlier reports indicate that aflatoxins B_1, B_2, G_1, and G_2 will remain during manufacture, ripening, and storage of cheese and other dairy products.

5.7.3 Elimination of Mycotoxins

Because aflatoxins are not destroyed during the manufacture, ripening, and storage of dairy products, research has been done to see if these and other mycotoxins can be degraded or inactivated by chemical, physical, or biological means. Aflatoxin M_1 was decreased by 45% when 0.4% potassium bisulfite was used at 25°C for 5 h.[327] The bisulfites may cause the oxidation to a bisulfite free radical that reacts with the dihydrofuran double bond of aflatoxin to give sulfonic acid products. Combinations of hydrogen peroxide, riboflavin, heat, and lactoperoxidase were used to see if aflatoxin M_1 could be inactivated in milk.[296] The best procedure resulted in 98% inactivation of aflatoxin M_1 after use of 1% H_2O_2 plus 0.5 m*M* riboflavin followed by heating at 63°C for 20 min. When milk was treated with 0.1% H_2O_2 plus 5 U of lactoperoxidase and held at 4°C for 72 h, 85% of aflatoxin M_1 was inactivated. These authors postulated that either singlet oxygen or hypochlorous acid were involved in the destruction of the aflatoxin. Some physical methods have been experimented with to determine if they are viable options for detoxifying milk. Bentonite was added to milk in 0.1 to 0.4 g/20 ml for 1 h at 25°C. It absorbed 65 to 79% of aflatoxin M_1;[328] however, the removal of bentonite from milk could cause some problems. Yousef and Marth[329,330] reported that 0.5 ppb of aflatoxin M_1 could be degraded by 100% in milk after a 60-min exposure to UV at a wavelength of 365 nm at room temperature. The temperature increased by 15°C during the 60-min treatment. When 1% hydrogen peroxide was added to the milk and it was irradiated for 10 min, total destruction of the 0.5 ppb aflatoxin M_1 was noted.[330] Degradation of aflatoxin M_1 by UV energy followed first-order reaction kinetics and was not affected by enzymes present in the milk.

In addition to physical and chemical methods, mycotoxins can be degraded by other microorganisms. *Flavobacterium aurantiacum* in a concentration of 7×10^{10} cells/ml completely degraded 9.9 μg of aflatoxin M_1/ml during 4 h at 30°C.[328] The mechanism by which this bacterium degrades aflatoxin is not known. Some microorganisms, such as *L. lactis* subsp. *lactis*, can convert aflatoxin B_1 into aflatoxicol and other metabolites that are either nontoxic or less toxic than B_1.[326,328] Degradation in other foods by other microorganisms is reported by Doyle et al.[328]

Feed can also be detoxified before it is fed to dairy cows. General reviews on methods to detoxify feeds have been published.[328,331] One example was reported by Price et al.[332] who ammoniated cottonseed meal to reduce the amount of aflatoxin B_1 fed to cows. When ammoniated feed was consumed, aflatoxin M_1 was below the limits of detectability; however, when untreated feed was consumed, the level of aflatoxin M_1 increased to about 1 μg/L in 7 days. When the contaminated feed was removed from the diet and treated feed consumed, the level of aflatoxin M_1 became nondetectable again. The Food and Drug Administration authorizes ammoniation of feeds in Arizona, California, Georgia, and North Carolina, but it has not declared this treatment as being safe for all states to use.[331] If measures to prevent the growth of mold and aflatoxin formation in feed commodities fail, then detoxification with ammonia can reduce aflatoxin by 97 to 98%. This ammoniation detoxification process is already used in different countries.

Mold growth and subsequent mycotoxin production can be prevented by use of antifungal agents, such as sorbates, propionates, and benzoates. Ray and Bullerman[333] have reviewed the agents that prevent both mold growth and mycotoxin production.

5.7.4 Regulation of Mycotoxins in Foods

The presence of mycotoxins, especially aflatoxins, in foods and feeds can cause potential harm to humans and animals; therefore, many countries have developed regulations to control the amount of mycotoxins that can be in foods, or feeds. Under the United States Federal Food, Drug, and Cosmetic Act, aflatoxins are considered poisonous or deleterious substances.[334,335] This falls under Section 402(a)(1) of the act. The Food and Drug Administration (FDA) established a guideline in 1965 that included an action level of 30 ppb aflatoxin in foods and feeds.[334,335] This action level was lowered to 20 ppb by 1969. In 1977 and 1978, aflatoxin M_1 was detected in market milk in the southeastern United States and in Arizona; hence, an action level of 0.5 ppb aflatoxin M_1 was then set for fluid milk.[335] Over 50 countries now have legislation for the presence of aflatoxins in foods and feeds.[290] Tolerances range from 5 to 20 ppb depending on the country and may be for either aflatoxin B_1 or the total amount of aflatoxins B_1, B_2, G_1, and Gl_2. Several countries also have set tolerances for aflatoxin M_1 in milk and dairy products ranging from 0 to 0.5 ppb.[290] Van Egmond[336] summarized data from 66 countries on the planned, proposed or existing legislation for aflatoxins B_1, B_2, G_1, G_2, and M_1 in foods, feeds, and milk and dairy products. Other mycotoxins, namely chetomin, deoxynivalenol, ochratoxin A, phomopsin, T-2 toxin, stachyobotriotoxin, and zearalenone are regulated in some countries.[290,336] The acceptable tolerance levels depend on the country and the food or feed.

Several surveys have been done to determine whether toxigenic molds or mycotoxins are present in milk and dairy products. The results of some of these surveys will be summarized. Bullerman[337] examined both domestic and imported cheese for mycotoxin-producing molds. *Penicillium* species were isolated from 86.4 and 79.8% of the domestic and imported cheeses, respectively. *Aspergillus* species were isolated only from 2.3 and 5.4% of the domestic and imported cheeses, respectively. *Cladosporium*, *Fusarium*, and other genera made up the rest of the molds isolated from these cheeses. Toxigenic species—*P. cyclopium*, *P. viridicatum*, *A. flavus*, and *A ochraceus*—were found in only 4.4% of domestic and 4% of imported cheese. When 118 imported cheeses from 13 countries were analyzed, 8 had aflatoxin M_1 in levels of 0.1 to 1 ppb.[338] Kivanc[339] found that 65% of molds isolated from Van hereby and pickled white Turkish cheeses were *Penicillium* species, and fewer than 4% were *Aspergillus* species. The rest of the molds were species of *Mucor*, *Geotrichum*, *Candidum*, and *Trichoderma*. No aflatoxin was detected in any of the cheeses. Blanco et al.[340] analyzed commercial UHT-treated milk over 1 year in Spain and found that 30% of the samples contained 0.02 to 0.1 ppb aflatoxin M_1. Most contaminated samples were detected in summer and autumn. Wood[341] examined 182 samples of milk and dairy products in the United States and found no measurable aflatoxin in them. From these studies, it appears that the presence of aflatoxins in milk and dairy products is very low and most samples meet the tolerance or action levels established for them.

The presence of molds and mycotoxins in dairy products and animal feeds will continue to be a concern until the health effects in humans and animals are better understood. The control of mold growth in foods and feeds will be important to prevent mycotoxin production. New and improved analytical methods will help to monitor the level of mycotoxins in foods and feeds.

5.8 Microbiology of Starter Cultures

Starter cultures are those microorganisms (bacteria, yeasts, and molds or their combinations) that initiate and carry out the desired fermentation essential in manufacturing cheese and fermented dairy products such as yogurt, sour cream, kefir, koumiss, etc. In cheesemaking, starters are selected strains of microorganisms that are intentionally added to milk or cream or a mixture of both, during the manufacturing process and that by growing in milk and curd cause specific changes in the appearance, body, flavor, and texture desired in the final end product.

Progress in dairy starter culture technology and advances in the scientific knowledge regarding the nature, metabolic activity, and behavior of starter cultures in milk, whey, and other media have provided new and improved starter cultures for the dairy industry. Research dealing with plasmid-mediated functions of starter cultures and mechanism of genetic exchange has led to utilization of recombinant DNA and other technologies for improvement of dairy starter cultures, particularly regarding development of bacteriophage-resistant strains. In this section, general information about starter bacteria is given. Several excellent reviews[64,75,342–345] have been published and may be consulted for further details regarding starter bacteria.

5.8.1 Terminology

The fermentation of lactose to lactic acid and other products is the main reaction in the manufacture of most cheese and fermented dairy products. Consequently, dairy starter cultures are also referred to as lactic cultures or lactic starters. In the dairy industry, single or multiple strains of cultures of one or more microorganism are used as starter cultures.

The taxonomy and scientific nomenclature of the lactic acid bacteria have been recently modified, for example, lactic streptococci, *S. cremoris, S. lactis,* and *S. diacetylactis* are now classified in the genus *Lactococcus* and referred to as *Lactococcus lactis* subsp. *cremoris, L. delbrueckii* subsp. *lactis*, and *L. lactis* subsp. *lactis* biovar *diacetylactis*, respectively. However, for the sake of convenience, the older names will be retained here. The nomenclature and some distinguishing characteristics of dairy starter cultures are listed in Table 5.19.

There are two main types of lactic starters: the mesophilic (optimum growth temperature of about 30°C) and the thermophilic (optimum growth temperature of about 45°C). Mesophilic cultures usually contain *S. cremoris* and *S. lactis* as acid producers and *S. diacetylactis* and *Leuconostocs* as aroma and CO_2 producers. Thermophilic starters include strains of *S. thermophilus*, and, depending on the product, *Lactobacillus bulgaricus, L. helveticus,* or *L. lactis*. Often, a mixture of thermophilic and mesophilic strains is used as a starter culture for manufacturing Italian pasta-

Table 5.19 NOMENCLATURE AND SOME DISTINGUISHING CHARACTERISTICS OF DAIRY STARTER CULTURES[a]

Organism	Current Nomenclature	Morphology	Growth		Type	Lactic Isomer	Percent Lactic Acid Produced in Milk	Citrate Metabolism	Fermentations					NH_3 from Arginine
			10°C	45°C					Glucose	Galactose	Lactose	Maltose	Sucrose	
Streptococcus lactis	*Lactosoccus lactis* subsp. *lactis*	GM + cocci	+	−	Mesophilic	L (+)	0.8	−	+	+	+	+	(d)	+
Streptococcus cremosis	*L. lactis* subsp. *cremoris*	GM + cocci	+	−	Mesophilic	L (+)	0.8	−	+	+	+	−	−	+
Streptococcus diacetylactis	*L. lactis* subsp. *lactis* var. *diaceytilactis*	GM + cocci	+	−	Mesophilic	L (+)	0.8	+	+	+	+	+	−	±
Leuconostoc cremoris	*L. mesenteroides* subsp. *cremoris*	GM + cocci	+	−	Mesophilic	D (−)	0.2	+	+	+	+	(d)	−	−
Streptococcus thermophilus	*S. salivarius* subsp. *thermophilus*	GM + cocci	+	−	Thermophilic	L (+)	0.6	−	+	−	+	−	+	−
Lactobacillus bulgaricus	*L. delbrueckii* subsp. *bulgaricus*	GM + rods	−	+	Thermophilic	D (−)	1.8	−	+	−	+	−	−	−
Lactobacillus helveticus	*L. helveticus*	GM + rods	−	+	Thermophilic	DL	2.0	−	+	+	+	(d)	−	−

[a] After Tamine,[64] Cogan and Accolas.[75]
\+ = positive reaction by > 90% strains
− = negative reaction by > 90% strains
(d) = delayed reaction

Table 5.20 LACTIC STARTER CULTURES, ASSOCIATED MICROORGANISMS, AND THEIR APPLICATIONS IN THE DAIRY INDUSTRY

Lactic Acid Bacteria	Associated Microorganisms	Products
Mesophilic *Streptococcus lactis,* *Streptococcus cremosis,* *S. lactis* var. *diacetylactis,* *Leuconostoc cremosis*	*S. lactis* var. *diacetylactis,* *Penicillium camemberti,* *P. roqueforti, P. caseicolum,* *Brevibacterium linens*	Cheddar, Colby Cottage cheese, Cream cheese, Neufachatel, Camembert, Brie, Roquefort, Blue, Gorgonzola, Limburger
Thermophilic *Streptococcus thermophilus,* *Lactobacillus bulgaricus,* *L. lactis, L. casei,* *L. helveticus,* *L. plantarum,* *Enterococcus faecium*	*Candida kefyr, Torulopsis,* spp., *L. brevis,* *Bifidobacterium bifidum,* *Propionibacterium fureudenreichii, P. shermanii*	Parmesan, Romano, Grana Kefir, Koumiss yogurt, Yakult, Therapeutic cultured milks, Swiss, Emmenthal, Gruyére
Mixed starters *S. lactis, S. thermophilus,* *E. faecium, L. helveticus,* *L. bulgaricus*		Modified Cheddar, Italian, Mozzarella, Pasta Filata, Pizza cheese

filata type cheese. Some thermophilic starters, such as those used in Beaufort and Grana cheese, contain only lactobacilli,[75] whereas some fermented milks made with thermophilic starters also contain *Lactobacillus acidophilus, L. bulgaricus,* and bifidobacteria for their healthful and therapeutic properties.[346] Table 5.20 lists the common starter cultures and their applications in cheese and fermented dairy products.

The lactic starter cultures are also subdivided into two groups: defined cultures and mixed cultures. Defined cultures constitute starters in which the number of strains is known. The concept of defined starter culture, mainly pure cultures of *Streptococcus cremoris,* was developed in New Zealand to minimize the problem of open textures in cheese thought to be caused by CO_2 produced by flavor-producing strains in mixed cultures. The application of defined cultures did control the open texture problem, however, and they were prone to slow acid production due to their susceptibility to bacteriophage.[75,347] The use of pairs of phage-unrelated strains and culture rotation to prevent buildup of phage in the cheese factory were practiced to minimize the potential for phage problems.[75,347] Eventually, the use of multiple strain starter and factory-derived phage-resistant strains was made to control the phage problem.[345,347,348]

Lactic starter cultures are also categorized based on flavor or gas production characteristics[64,75] for example, B or L cultures (for *Betacoccus* or *Leuconostoc*) contain flavor and aroma producing organisms, for example, *Leuconostoc* spp. D cultures contain *Streptococcus diacetylactis*; BD or DL cultures contain mixtures of both *Leoconostoc* and *S. diacetylactis* strains and O cultures do not contain any flavor/aroma producers but contain *S. lactis* and *S. cremoris* strain. This nomenclature is commonly used in the Netherlands.[349]

Often, the lactic starters routinely used in dairy plants without rotation are called P (practice) cultures as opposed to L (laboratory) cultures which have been subcultured in the laboratory. The P cultures are not usually affected by their own phages, and unlike L cultures, they can recover following the attack of so-called ''disturbing'' phage.

5.8.2 Function of Starter Cultures

5.8.2.1 Production of Lactic Acid

The primary function of lactic starter culture is the production of lactic acid from lactose. The lactic acid is essential for curdling of milk and characteristic curd taste of cultured dairy products. The manufacturing procedures for cheese and other fermented dairy products are designed to promote growth and acid production by lactic organisms. The production of lactic acid is also essential for development of desirable flavor, body, and texture of cheese and cultured dairy products. The rate of lactic acid production during the cheesemaking is affected by the temperature, calcium and phosphorus content of milk, the type and amount of starter culture used, etc. Lactic acid production also results in a decrease of lactose in cheese and whey. The presence of excessive lactose in the cheese is undesirable because it can be metabolized by nonstarter bacteria during ripening and lead to flavor and body defects in cheese.

The mechanisms of lactose metabolism differ considerably in different lactic acid bacteria,[350] *Streptococcus lactis* employs the phosphoenol pyruvate phosphotransferase system (PES/PTS) to transport lactose which is hydrolyzed to glucose and galactose and metabolized by the glycolysis and tagatose pathways, respectively. *Leuconostoc* spp. and thermophilic lactobacilli, on the other hand, transport lactose by a permease system. It is hydrolyzed to glucose and galactose by β-galactosidase and further metabolized. Lactose metabolism by different starter cultures is reviewed elsewhere.[52,54,75,343,351–353]

5.8.2.2 Flavor and Aroma and Alcohol Production

In addition to production of lactic acid, starter cultures also produce volatile compounds, for example, diacetyl, acetaldehyde, and ketones responsible for the characteristic flavor and aroma of cultured dairy products. Flavor-producing starter cultures metabolize citric acid to produce CO_2 which is necessary for ''eye'' formation in some cheeses. Some starter cultures, mainly yeast, produce alcohol, which is essential for the manufacturing of kefir and koumiss.

5.8.2.3. Proteolytic and Lipolytic Activities

The starter cultures produce proteases and lipases which are important during the ripening of some cheese. Protein degradation by proteinases is necessary for active growth of starter cultures as most lactic acid bacteria require amino acids or peptides for their growth. Proteinase negative ($Prot^-$) strains of lactic starters depends on $Prot^+$ strains in a multiple strain culture for growth in milk.

5.8.2.4 Inhibition of Undesirable Organisms

The production of lactic acid lowers the pH of the milk and inhibits many spoilage organisms as well as pathogens. A number of metabolites produced by lactic cultures can limit the growth of undesirable organisms, for example, Ibrahim[354] reported that lowering the pH with lactic acid in a simulated Cheddar cheese making resulted in the inhibition of *S. aureus*. Rapid growth and acid development by lactic acid bacteria suppress growth of many spoilage and pathogenic bacteria.

Besides lactic acid, production of H_2O_2 and acetic acid by some starter cultures, particularly those containing *Leuconostoc* or *S. diacetylactis*, can also inhibit pathogenic bacteria.[354] The amount of H_2O_2 produced by lactic acid bacteria may not be adequate in itself to control undesirable organisms in milk. However, it can allow the enzyme lactoperoxidase (LPS) to react with thiocyanate (SNC^-) and produce hypothiocyanate ($OSCN^-$), which can inhibit various pathogens including *S. aureus*, *E. coli* and *Campylobacter jejuni*.[355]

Certain strains of *S. lactis* produce nisin, which is inhibitory to various organisms including species and strains of the genera *Bacillus, Clostridium, Listeria*, etc. However, the application of the nisin-producing strains as cheese starters is limited because of their slow acid production and susceptibility to bacteriophages. There is considerable interest in developing nisin-producing cultures that may be suitable for use in the dairy industry.

Several lactic acid bacteria, particularly streptococci, are capable of producing bacteriocins that inhibit Gram-positive pathogens such as *Clostridium* or *Listeria*. However, the application of these strains as cheese starters may be limited because they inhibit other closely related strains in a cheese starter.

5.8.3 Growth and Propagation

Lactic starter cultures are generally available from commercial manufacturers in spray-dried, freeze-dried (lyophilized), or frozen form. Spray-dried and lyophilized cultures need to be inoculated into milk or other suitable medium and propagated to the bulk volumes required for inoculating a cheese vat as follows:

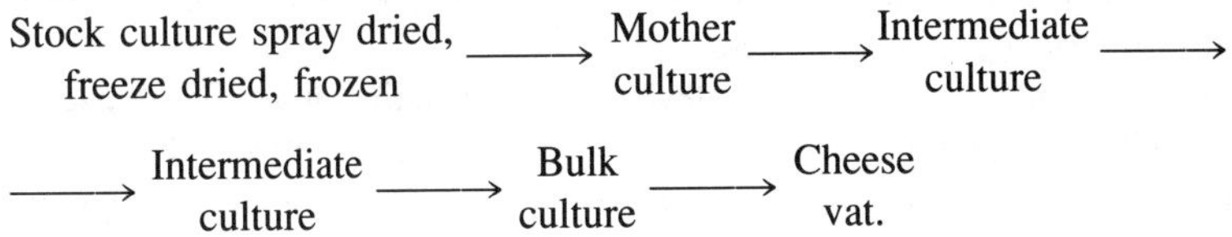

Many larger dairy plants develop their own cultures. However, preparing and maintaining bulk cultures requires specialized facilities and equipment. Much research and development in the starter culture technology has been aimed at designing specialized growth media for starters, protecting the starter cultures from sublethal stress and injury during freezing, and minimizing the theat of bacteriophage during starter culture preparations.

The specialized systems for starter culture propagation include the Lewis system, the Jones system, the Alfa-Laval system, etc.[64] The Lewis system[356] utilizes reusable polyethene bottles fitted with Astell rubber seals and two-way needles. The growth medium (10 to 12% reconstituted, antibiotic-free skim milk) is sterilized in the mother culture bottle. The stock culture is incubated through a two-way needle by squeezing the stock culture bottle. The bulk starter tank used in the Lewis system is pressurized to allow heating of the growth medium in the sealed vessel. The top of the tank is flooded with 100 ppm sodium hypochlorite solution to prevent any contamination during the inoculation of bulk starter.

The Jones system uses a specially designed bulk starter tank.[64] Unlike the Lewis system, this tank is not pressurized. The bulk starter tank is inoculated by providing the intermediate starter through a special narrow opening and a ring of flame or steam is used to prevent any contamination during the inoculation of bulk starter.

Recently, a combination of the Lewis/Jones system has been developed in the United Kingdom that improves on the Lewis technique of aseptic culture transfer and economizes by using cheaper, nonsealed tanks as in the Jones system. The details of the combined Lewis/Jones system have been described ty Tamime.[64]

The Alfa-Laval system uses filtered-sterilized air uner pressure, for transferring the culture. The mother and intermediate cultures are prepared in a special unit called a ''viscubator'' and transferred to the bulk starter tank using compressed air.[64,357]

5.8.3.1 pH Control Systems

There are two main reasons for using pH control systems in propagating bulk starter cultures: (1) to minimize daily fluctuations in acid development and thereby prevent ''over-ripening'' of the starter, and (2) to prevent the cellular injury that may occur to some starters when the pH of the medium drops below 5.0.

In the pH control systems, the acid produced by the starter culture is neutralized to maintain the pH at around 6.0.

The external pH control system, developed by Richardson et al.,[358,359] uses whey-based medium fortified with phosphates and yeast extract. The pH is maintained at around 6.0, by intermittent injection of anhydrous or aqueous ammonia, or sodium hydroxide. This system has been used successfully in the United States for production of most American-style cheeses.

The internal pH control system, developed by Sandine et al.,[360–363] uses a whey-based medium containing encapsulated citrate-phosphate buffers that maintain the pH at around 5.2. Unlike in the external pH control system, no addition of ammonia or NaOH is necessary. The internal pH control system is available as the phase 4 (Rhône–Poulenc—Marschall Products Division) and In-Sure (Chr. Hansen's Laboratory, Inc.) and is used in the United States and Europe for a variety of cheeses and fermented products such as buttermilk.[64]

5.8.3.2 Phage Inhibitory and Phage-Resistant Medium (PIM/PRM)

The PIM/PRM were developed following observations of Reiter[64] that bacteriophage of lactic streptococci were inhibited in a milk medium lacking in calcium. Hargrove[364] reported on the use of phosphates to sequester free calcium ions in milk or bulk-starter medium for inhibition of bacteriophage. The effectiveness of phosphates in the formation of PIM/PRM for phage control was confirmed by Christensen.[365–467] The PIM/PRM consisting mainly of milk solids, sugar, buffering agents such as phosphates and citrates and yeast extract have been widely used in the United States, Canada, and Europe for about 20 years.[345] However, the effectiveness of the PIM/PRM in inhibiting bacteriophage and stimulating growth of the starter culture media is somewhat limited.[64] Despite the absence of calcium, some phages can infect the the starter culture at its optimum growth temperature. Also, phosphates in the PIM/PRM can cause metabolic injury to some starter cultures.

The preparation of active bulk starter culture free of phage contamination is essential for cheese manufacturing. However, poor practices promoting phage contamination still exist in many commercial operations.[345,368] Factors important in bulk starter preparation and ways of minimizing bacteriophage problems in cheese factories have been reviewed by Huggins[345] and by Richardson.[368]

5.8.4 Inhibition of Starter Cultures

The inhibition or reduction in activity of lactic starter culture results in consequences ranging from "dead vat" or slow vat to production of poor quality cultured products. Also, sluggish starter culture produces acid at a slow rate and fails to control spoilage and potentially pathogenic bacteria. The primary cause of inhibition of starter cultures is the bacteriophage. Control of the bacteriophage problem depends on understanding of critical factors affecting phage infection and growth in lactic starter cultures,[369] factors dealing with bulk starter culture production, factory design, sanitation, and whey processing.[345]

Lactic starter cultures are very sensitive to antibiotic residues in milk,[171,370–372] for example, 0.01 IU/ml of penicillin may inhibit a mesophilic lactic starter and a yogurt culture.[64] The sensitivity of starter culture to a specific antibiotic residue depends on the species or strains of the starter culture, the antibiotic preparations, and the test for determining antibiotic concentrations. The problem of antibiotic residue is primarily associated with their use in mastitis therapy in the dairy cow and failure to withhold the milk from cows treated with antibiotics. This problem is currently receiving much attention in the United States dairy industry.

Residues of detergents and sanitizers used in the dairy industry for cleaning and sanitation may also inhibit starter culture growth and activity. The effects of commonly used cleaning compounds such as chloride, quaternary ammonium compounds, and alkaline detergents on the activity of various dairy starter cultures have been studied in detail.[373,374] Proper cleaning and sanitation, particularly adequate rinsing, is important in minimizing the inhibition of starter culture growth and activity by residues of cleaners and sanitizers.

Occasionally, inhibition of the growth of starter culture may be caused by naturally occurring antibacterial compounds present in milk. For example, lactin and the lactoperoxidase system (LPS) have been reported to cause inhibition of certain lactic cultures.[357,375,376]

5.8.5 Genetic Engineering for Improving Starter Cultures

Recent advances in the genetics of lactic acid bacteria, particularly progress in our understanding of the basic processes relating to transport, metabolism, and genetic regulation of sugar utilization, bacteriocin production, and phage resistance have created many opportunities for applying genetic engineering techniques for improving dairy starter cultures.[12]

In the past, fast-acid-producing and bacteriophage-insensitive strains were obtained through natural selection and mutation processes. However, many of these strains were unstable due to spontaneous loss of properties, apparently due to the loss of plasmid(s). The understanding of the functional properties of plasmids and of the mechanisms of genetic exchange and gene expression in lactic streptococci will allow the cloning of desirable traits into dairy starter cultures.

It is now well established that mesophilic lactic starter cultures harbor plasmids of diverse sizes and that some of these plasmids code for several major functions of lactic streptococci (Table 5.21). The knowledge of plasmid-mediated functions and plasmid transfer systems may be used to develop specific starter cultures that may:

Table 5.21 PLASMID-LINKED METABOLIC FUNCTION OF MESOPHILIC STREPTOCOCCI[a]

Function	Reference
Sugar utilization	LeBlank et al.[377] Gasson and Davies[54] McKay[55] Gonzalez and Kunka[373]
Proteinase activity	McKay[55] Kok et al.[379]
Citrate utilization	Kempler and McKay[380]
Bacteriocin production	Scherwitz et al.[381] Scherwitz and McKay[382]
Nisin production	McKay and Baldwin[383]
Bacteriophage resistance	McKay and Baldwin[383] Sanders and Klaenhammer[384] Chopin et al.[385] Sanders[80]

[a] Adapted from McKay.[55,56]

(1) produce desirable flavor compounds; (2) lower requirements for added sweeteners (sugar) in dairy fermentations; (3) produce enzymes necessary for cheese flavor, body, and texture development; and (4) resist bacteriophage attack during cheesemaking.[54–56,79,80,386,387]

Research by Klaenhammer and others has indicated that several mechanisms for bacteriophage resistance may exist in lactic streptococci.[78–80,384,387] These include prevention of phage absorption, restriction/modification controlled by the host and abortive infection via lysogenic immunity, or other mechanisms. Bacteriophage-resistant dairy streptococci have been obtained following conjugal transfer of a 30-megadalton plasmid, pTR 2030, from a lactose-negative *S. lactis* to a fast-acid producing *S. lactis* and *S. cremoris* strains.[80] The development and industrial utilization of phage-resistant strains containing the pTR 2030 have been reported.[79,80,178]

There exists a potential for application of genetic engineering for improvement of dairy starter. Laboratories in the United States, Australia, and Europe are actively engaged in research dealing with genetics of lactic acid bacteria. The use of genetically engineered lactic bacteria for dairy fermentation is limited although the genetic approach for developing improved strains for dairy industry appears promising.[12,388]

5.9 Methods for Microbiological Analysis of Milk and Dairy Products

Microbiological analysis of milk and dairy products is critical in evaluating quality, shelf life, and regulatory compliance of raw milk, ingredients, and finished products as well as in assessing the efficiency of manufacturing processes and cleaning and sanitation practices. Although there is much progress made in analytical methodology used for chemical analysis of milk components, cheese, whey, and other dairy products, the focus of microbiological testing in the dairy industry still remains on conventional plating methods and isolation and biochemical characterization of the microorganisms of interest. Unlike the chemical analysis of milk, where more traditional methods are used only for standardization of instrumental methods used for routine analysis, microbiological testing of milk and milk products is largely done by traditional plate count methods, most probable numbers (MPN) estimations, and empirical tests such as the methylene blue and resasurin tests. These slow and retrospective methods are often not suitable for perishable, relatively short shelf-life milk and milk products. During the past two decades, considerable interest in finding suitable alternatives to these time-, material-, and labor-intensive methods has led to development of several rapid and automated methods for routine microbiological testing of milk and dairy products.[91,389–399]

5.9.1 Conventional Methods

Routine microbiological testing of milk and dairy products involves plating procedures for detecting and enumerating microbial contamination in milk, dairy products, dairy equipment, and the dairy plant environment.

Table 5.22 METHODS FOR MICROBIOLOGICAL ANALYSIS OF MILK AND DAIRY PRODUCTS

Conventional Methods	Rapid and Automated Methods
Direct microscope count (DMC)	Bactoscan
Breed clump count	Biofoss
	Spiral plater
	Direct epifluorescent
	Filter technique (DEFT)
Standard plate count (SPC)	Plate-loop count
Pour plate	Petrifilm
Surface plate	ATP measurement
Drop plate	Bioluminescence
	Limulus test
Most probable numbers (MPN)	Electrical conductance
Three-tube method	Electrical impedance
Five-tube method	Electrical capacitance
Membrane filter	HGMF–Isogrid
	Direct epifluorescent
	Filter technique (DEFT)
Dye reductions	Microcalorimetry
Methylene blue	Flow cytometry
Resasurin	
RODAC plate	
Rinse-filter method	

Several procedures can be used to estimate a microbial population (Table 5.22). The four general methods commonly used for ''total'' numbers are Direct Microscope Counts (DMC), Standard Plate Counts (SPC), the Most Probable Numbers (MPN) methods, and the dye reduction tests. The following is a brief description of these methods:

Direct Microscopic Count (DMC) involves preparation of a smear on an outlined area of a microscopic slide, staining the slide with appropriate dye preparations, and microscopic examination of stained smears using the oil immersion lens. Usually a small amount (0.01 ml) of the sample or appropriate dilution of the sample is spread over a 1-cm^2 area. Microbial cells (individual or clumps) are counted in a given numbers of microscopic fields, and the total number of organisms per gram are determined by multiplying the average number of organisms per field by the microscopic factor (usually ≥500,000). The DMC method is widely used for determination of total microbial numbers in dry milks. The diret microscope somatic cell count (DMSCC), which employs essentially the same procedure, is used to confirm mastitis in cows or quality of bulk milk at the dairy farm. Further details of the direct microscopic count methods may be found in the Standard Methods for the Examination of Dairy Products (SMEDP)[58] and the IDF Document 168.[241]

Table 5.23 MODIFICATIONS OF THE STANDARD PLATE COUNT METHOD AND THEIR APPLICATIONS

Modification	Application
Preheat sample at 63°C for 30 min.	Thermoduric bacterial count (TBC) in milk and pasteurized products.
Incubate SPC plates at 7°C for 10 days.	Psychrotrophic bacterial count (PBC) for milk and dairy products.
Use SPC containing 10% sterile milk, incubation at 23–25°C for 48 h and flooding of plates w/ 10% acetic and/or 1% HCl.	Enumeration of proteolytic organisms
Preheat milk at >80°C for 10 min, incubation at 30°C for 77 and/or 55° for 24–48 h.	Enumeration of mesophilic/ thermophilic bacteria, and spores
Surface plating on spirit blue agar, incubation at 30°C for 6 days	Enumeration of lipolytic organisms
Acidified potato dextrose agar (PDA), incubation at 22–25°C for 5 days	Enumeration of yeasts and molds
Use violet-red bile agar, incubation at 35°C for 24 h	Enumeration of coliforms

Standard Plate Count (SPC) involves preparing a 10-fold serial dilution of the sample to be tested. A 1.0- or 0.1-ml sample of the dilution is placed in a sterile petri dish followed by pouring of the liquified sterile agar medium (SPC agar). The sample is mixed with the agar medium and agar is allowed to solidify. The petri dishes are incubated at 32°C for 48 h (or any other specified conditions). Following the incubation, the plates with 25 to 250 colonies are counted and the total number of microorganisms is determined by multiplying the average number of colonies by the dilution factor. The details of the sampling, diluting, plating, and incubating procedures and proper counting and reporting of the bacterial numbers in a sample of milk and milk products are described in the SMEDP.[58]

Various modifications of the SPC have been used to determine the numbers of psychrotrophic, thermoduric, proteolytic and lipolytic bacteria; coliforms; and yeast and molds in milk and dairy products (Table 5.23); for example, the psychrotrophic bacterial count (PMC) procedure involves the same method as the SPC, except that the plates are incubated at 7°C for 10 days.[58] Also, various methods designed for assessing the hygiene and keeping quality of milk are also based on the SPC method.[58,400]

Most Probable Numbers (MPN) involves the use of three sets of three or five tubes each containing a sterile medium. These tubes are inoculated from each of three consecutive 10-fold dilutions (10^0, 10^{-1}, 10^{-2} or 10^{-1}, 10^{-2}, 10^{-3}). The tubes are incubated and growth of the organisms is detected as turbidity or evidence of gas formation. Numbers of organisms in the original samples are determined by using standard MPN tables. The MPN is statistical in nature and the results are generally higher than SPC.[19]

Dye reduction involves the use of redox dyes such as methylene blue, resasuring, or 2,3,5-triphenyltetrazolium chloride (TTC). The method depends on the ability of microorganisms to reduce and hence change color or decolorize the dye. The time required for reduction of the dye is generally correlated with the metabolic activity and is universally proportional to the initial bacterial load of the sample. The dye reduction method is simple and economical. However, they are unsuitable for analysis of milk having low bacterial numbers[401] and are poorly correlated with the bacterial counts in refrigerated milk.[402] The dye reduction tests and their limitations are discussed in detail by Edmonson et al.[401]

5.9.2 Rapid Methods and Automation in Dairy Microbiology

In the past 20 years, interest in the field of rapid methods and automation in microbiology has been growing steadily. Several international symposia have been held on the subject since 1973.[403] The Sixth International Congress on Rapid Methods and Automation in Microbiology was held in June, 1990 in Helsinki, Finland. Developments in rapid methods and automation are discussed in detail in recent books such as *Rapid Methods and Automation in Microbiology*,[404] *Rapid Methods in Microbiology and Immunology*,[405] *Foodborne Microorganisms and Their Toxins: Developing Methdology*,[406] *Rapid Methods in Food Microbiology*,[407] *Impedance Microbiology*,[391] *The Direct Epifluorescent Filter Technique for the Rapid Enumeration of Microorganism*,[408] and *Instrumental Methods for Quality Assurance in Foods*.[403] Early developments in rapid methods and automations dealt with rapid identification and characterization of pathogenic microorganisms in a clinical setting. However, many of the procedures and instrumentations developed for the clinical laboratory have been successfully applied to microbiological analysis of milk and dairy products. Also rapid methods and automation for detection and enumeration of microorganisms suitable for use in the dairy industry have been developed in recent years.

The major areas of microbiological analysis of milk and dairy products include sample preparation, total viable cell counts, somatic cell counts, monitoring of microbial growth and activity and detection, and isolation and characterization of pathogenic organisms and toxins. All of these areas have been subjects of research and development to improve microbiological methods for milk and dairy product analysis.

5.9.2.1 Improvements in Sampling and Sample Preparation

Sampling of milk and milk products is critical in obtaining meaningful, reliable results. Different methods of sampling various products, care and handling of samples, storage and transportation, etc. are described in detail in reference sources such as the *Standard Methods for Examination of Dairy Products*[58] and the IDF.[409] Two noteworthy developments in instrumentations for sample preparation include the Stomacher (Tekmar, Cincinnati, OH) and the Gravimetric Diluter (Spiral Systems, Inc., Bethesda, MD).

The Stomacher uses two reciporcating paddles to crush the sample and diluent held in a polythene bag. Unlike the lab blender commonly used for sample preparation, there is no direct contact between the sample and the machine. Therefore, there is no need for cleaning and sterilization between use; also, the Stomacher minimizes the problem of aerosol formation. The Stomacher uses disposable sterile bags, thus eliminating the need for large numbers of glass or metal jars to be cleaned and resterilized. It is very easy to operate. Several reports on the comparison of total bacterial counts obtained using the Stomacher and the laboratory blender have indicated that satisfactory results can be obtained by using the Stomacher.

The Gravimetric Diluter eliminates the need for accurately weighing the sample (e.g., 10 g or 450 g) prior to adding the requisite amount of diluent to obtain a 1:10 or 1:100 dilution. The dilution operation is automated in that after weighing an amount of the sample, the machine delivers a specific volume of the diluent required to obtain the dilution. The Gravimetric Diluter is easy to operate and saves considerable time in routine microbiological analysis of milk products.

5.9.2.2 Modifications and Mechanization of Conventional Methods

Several labor and material saving methods developed for determining colony counts in milk and dairy products involve modifications and mechanization of conventional plate count procedure. These are not truly "rapid" methods as they require the same incubation period as the conventional methods. However, ease of operation, economizing of material and labor, and ability of handling large numbers of samples possible have popularized the use of modified methods in dairy industry.[399]

Agar Droplet Techniques

These are developed as a modification of the Miles-Misra method.[410] A variety of delivery systems (calibrated pipettes, etc.) are used to deliver 0.1-ml droplets of sample dilutions made in molten agar in a petri dish. After incubation at 30°C for 24 h, the microcolonies are counted under magnification. A diluter/dispenser and a projection viewer have been developed to aid rapid preparations of dilutions and dispensing of the agar droplet and facilitating counting of microcolonies.[407] Although data obtained with the droplet technique and conventional pour or spread plate methods show no significant difference with most samples, significantly higher counts with the droplet technique have been reported.[411] Despite this and other minor limitations, the agar droplet techniques are suitable for routine bacteriological testing of milk and dairy products.[410,411]

The Plate Loop Count (PLC)

This method involves the use of a calibrated loop, capable of delivering 0.001 ml of a sample. The loop is attached (preferably welded) to a Luer-Lock hypodermic needle, which in turn is attached to a continuous pipetting outfit adjusted to deliver 1.0 ml. A 0.001-ml sample is placed in the petri dish by delivering 1.0 ml of sterile diluent which eliminates the need for preparing serial 10-fold dilutions of the sample. The rest of the procedure for pouring, incubating, and counting plates is the same as the conventional SPC method. The PLC method is quite satisfactory for use with routine bacteriological testing of raw milk, except manufacturing grade raw milk, when counts exceed 200,000/ml.[412]

Noteworthy among the products on the market designed to facilitate conventional plate count methods are the Isogrid system, the Petrifilm plates, and the Spiral system with CASBA (computer-assisted spiral bioassay) data processor.

The Isogrid system[393,413,414] is a filtration method that uses a Hypobaric Grid Membrane Filter (HGMF) consisting of 1600 growth cells. The diluted sample is first filtered through a prefilter (5 μm) to remove large food particles and then through the HGMF. The HGMF is placed on a selective agar and incubated under specified conditions to allow the growth of microorganisms present in the food. The HGMF method is officially recognized by the AOAC and FDA and is used for detecting and enumerating *Salmonella* and coliforms, as well as for detecting aerobic plate counts.[415]

Petrifilm plates are dual-layer film systems coated with nutrients and a cold water soluble gelling agent. The diluted sample is inoculated on the Petrifilm surface, similar to the regular surface plating method, and the resulting petri plate is incubated under specified conditions to allow growth of the microorganisms. The standard plate count and coliform counts may be determined by the Petrifilm SM and Petrifilm VRB, respectively. Petrifilm plates have been evaluated extensively through collaborative studies[416–419] and are recognized as an official method for microbiological analysis of milk and dairy products.

The Spiral System[420,421] involves precise delivery of a continuously decreasing volume of a liquid sample onto the surface of an agar plate. Use of a hand or laser counter and a CASBA data handling system can facilitate throughput. The Spiral System greatly reduces media and dilution requirements. It is widely used for determination of aerobic plate counts of milk and dairy products.[58]

The Preliminary Incubation (PI) Count

Among the methods developed in recent years, various preincubation procedures for estimating psychrotrophic bacteria in milk products have received much attention. The 21°C/25 h incubation of milk followed by a conventional standard plate count procedure gives a good and reliable estimate of psychrotrophic bacteria.[422,423] Since Gram-negative psychrotrophs are the primary cause of spoilage in milk and dairy products, the preliminary incubation procedures are widely used to assess the potential shelf-life of pasteurized milk and cream.[398,424] Preincubations with selective inhibitors such as benzalkonium chloride, bile salts, crystal violet, penicillin, and nisin have also been used to determine spoilage potential and to predict shelf life.[390,425–427]

The Redigel system consists of sterile nutrients with a pectin gel in a tube and special petri dish previously coated with gelation material. A 1.0-ml sample (or appropriate dilution) is pipetted into the tube, mixed, and poured in the petri dish.

A pectin gel, resembling conventional agar medium, is formed in the petri dish, which is incubated and the colony count determined as in the conventional SPC procedure. Recently, the use of the Redigel system for determining microbial counts in milk and dairy products has been reported.[428–431] A high degree of statistical correlation was obtained when counts determined with the Redigel system were compared with that with the conventional method[428]

A comprehensive analysis of the Redigel, Petrifilm, Isogrid, and Spiral System using seven different foods, including raw milk, conducted by Chain and Fung[428] indicated that these systems compared favorably with conventional methods and a high degree of accuracy and agreement of the results were possible using alternative methods. A comparison of cost per viable cell counts was: SPC ($13.62), Petrifilm and Redigel ($8.22), Isogrid ($3.33), and Spiral System ($2.77).[394] The Isogrid and Spiral System require the initial purchase of specialized equipment. However, they require only one plate per sample compared to four to six plates required for the conventional SPC method.

Other applications of the Isogrid, Petrifilm, and Spiral System include enumeration of coliforms, *S. aureus*, and yeast and mold counts; detection of specific organisms such as *Salmonella*, *E. coli* 0157:H7, *Yersinia enterocolitica*, etc.; and determination of inhibitory properties and minimum inhibitory concentrations (MIC) of antibacterial compounds.

5.9.2.3 Methods Based on Microbial Growth and Metabolism

Several rapid and automated methods for microbiological analysis rely on parameters of microbial growth and metabolism such as adenosine triphosphate (ATP) levels, detection of electrical impedance or conductance, generation of heat or radioactive CO_2, presence of bacterial exopolysaccharides or enzyme activity, etc. These methods are based on the assumption that increase in bacterial numbers is correlated with the increase in various parameters of microbial growth and metabolism. A standard curve correlating various parameters with the colony counts is developed for comparison of unknown samples. Although theoretically it is possible to detect as low as one viable cell in a sample using these methods, populations of 10^6 to 10^7 organisms per milliliter are necessary for rapid (4 to 6 h) detection.

ATP levels in a sample are easily determined in terms of the bioluminescence resulting from the reaction between the ATP and the luciferin/luciferase enzyme system obtained from fireflies. The amount of light generated is proportional to the levels of ATP and hence levels of bacterial contamination. It is measured as relative light units (RLU) using instruments such as Lumac and Luminometer. The ATP levels measurement as an indication of microbial load is widely used in Europe for detecting postpasteurization contamination in milk and cream.[391,432–434] Because somatic cells in milk constitute a nonmicrobial source of ATP, treatment of samples to hydrolyze somatic cell ATP is necessary prior to determining ATP from bacterial cells. The ATP method may be readily automated to allow handling of large numbers of samples. It can also be used to monitor hygiene in dairy plants. A rapid (5-min) test for judging bacteriological quality of raw milk at receiving in dairy plant has been developed in Europe.[432] The principles and applications of ATP measurement tests have been reviewed recently by LaRocco et al.[435] and Stannard.[436]

The growth of microorganisms results in unique and significant changes in electrical conductivity and resistance in growth medium. The changes in electrical impedance, capacitance, or conductance are measured using specialized instruments such as the Bactometer and Malthus system.[392,437] The Bactometer is an instrument designed to measure impedance changes resulting from microbial metabolism and growth.[392] The impedance detection time (IDT), or simply detection time, is the time (h) when the electrical parameter being measured changes significantly from the starting value. The IDTs are inversely proportional to the initial levels of microorganisms present in the sample and are generally indicative of the time required to reach population of approximately 10^6/ml. Impedance changes are affected by the composition of growth medium, temperature of incubation, and specific growth kinetics of bacteria. Impedimetric methods have been used for a variety of dairy microbiology applications including detection of abnormal milk,[438] estimation of bacteria in raw or pasteurized milk[439–441] and dairy products,[389,390,425,442] detection of antibiotics,[443] measurements of starter culture activity,[444,445] and determining levels of bacteriophage.[446,447]

The Malthus system is similar to the Bactometer in that both systems involve continuous monitoring of changes in electrical parameters to obtain detection times. However, they differ in the electrical component measured, the frequency at which the measurements are made, and the specific design of electrode, measurement and the instrument.[437] The conductance curve generated by the Malthus system is similar to the impedance curve obtained by the Bactometer. In both systems, screen displays of green, yellow, and red colors indicating ''accept,'' ''caution,'' and ''reject'' or ''pass,'' ''caution,'' and ''fail'' levels of microbial population are available for use in routine monitoring of microbiological quality of samples being tested.[392,437]

The Malthus system has been used for detection of postpasteurization contamination of pasteurized milk,[441,448] estimation of lactic acid bacteria in fermented milks, detection of psychrotrophic bacteria in raw milk, and determination of microbial levels in powdered dairy products.[437]

A conductance method for the quantitative detection of coliforms in cheese has been developed by Khayat et al.[442] Also, a special selective medium (selenite–cystine broth) containing trimethylamine oxide (TMAO) and dulcitol was developed for detection of salmonella by the conductance method.[449–451] Recently, Cousins and Marlatt[452] evaluated a conductance monitoring method for the enumeration of *Enterobacteriaceae* in milk. Detection of <10 to 500 cfu/ml of *Enterobacteriaceae* in raw milk in 6 to 12 h was reported.[452]

Radiometry and microcalorimetry have been used to estimate numbers of microorganisms in clinical specimens and a variety of foods. The radiometric method deals with monitoring the production of radioactive CO_2 by microorganisms growing in a medium containing radioactive glucose. The $^{14}CO_2$ generated, which is directly proportional to the metabolic activity of the microorganisms present in a sample, is measured by an instrument such as the Bactec. The microcaloric method involves measurement of minute changes in heat using sensitive instruments such as the Bio Activity Monitor. The applications of radiometry and calorimetry in food microbiology have been discussed by Lampi et al.,[453] Rowley et al.,[454] and Gram and Sogaard.[455]

The Limulus Amoebocyte Lystate (LAL) method is a rapid (1 h) and sensitive test for detection of low levels of Gram-negative bacteria in milk and dairy products. All Gram-negative bacteria contain endotoxin (lipopolysaccharides, LPs) that can be determined by the LAL test. In the classic LAL test, serial dilutions of the sample are mixed with the LAL reagent (amoebocyte lystate of horseshoe crab, *Limulus*) and incubated at 37°C for 1 h. A positive reaction is indicated by formation of firm gel and levels of endotoxin (ng) are calculated based on the highest dilution showing a firm gel. Other LAL test procedures involving turbidimetric and colorimetric measurements of the LAL reaction have been developed, some for use with a robotic system for automatic handling of large numbers of samples. A microfiltration method for application of the limulus test in dairy bacteriology has been developed as a commercial kit.[456]

The LAL is a simple, rapid, and sensitive test for low levels of Gram-negative bacteria in milk and dairy products.[457,458] It is also useful in determining the previous history of the milk in investigating quality and shelf-life problems of heat-treated products such as UHT milk and dry milk powders. Further details of the LAL test and its applications in food microbiology may be found in a recent review by Jay[459] and by Heeschen et al.[460]

The catalase test is another rapid method for estimating microbial populations in certain foods. Because many psychrotrophic spoilage organisms, particularly *Pseudomonas* spp., important in causing spoilage of milk and dairy products are strongly catalase positive, this test may be used as a rapid screening test for assessing milk quality. Other important organisms such as *Staphylococcus*, *Micrococcus*, *E. coli*, and others are also catalase positive and may be detected by this test.

Recently, an instrument called the Catalasemeter was developed for rapid detection of catalase activity. This instrument is based on the simple and rapid estimation of catalase activity present in milk or culture filtrates. The principle is based on the flotation time of a paper filter disc containing catalase in a tube containing stabilized H_2O_2. On reaction, the evolved gases cause the disc to float. The time required for the disc to float (disc flotation time) is inversely proportional to the catalase activity. Because mastitic milk characteristically contains elevated levels of somatic cells and high catalase activity, the catalasemeter has been used for rapid screening of abnormal and poor quality milk[396,461,462] and for predicting milk quality and shelf life.[426]

Rapid screening methods for dairy microbiology also include the Direct Epifluorescent FIlter Technique (DEFT) test,[395,408,463] which involves filtering of a sample or dilution through a polycarbonate filter (0.6 μm size, 25 mm diameter) to concentrate bacteria on the filter followed by staining the filter using acridine orange dye. The filters are then examined with epifluorescent microscopy. The applications of the DEFT include rapid estimation of viable cells in milk[464] detection of postpasteurization contamination in cream,[395] and assessment of keeping quality of milk samples. However, it requires special equipment and skilled labor. Also, poor correlations between DEFT count and colony counts in products such as milk powder, pasteurized whey, and ripened cream butter limit the applications of the DEFT for microbiological analysis of milk and milk products. The principle equipment and applications of the DEFT test have been reviewed by Pettipher[408,463] and Pettipher et al.[397,464,465]

A reflectance colorimeter instrument has been developed for measurement of microbial and enzyme activities in milk and dairy products,[466] The instrument, Omnispec, consists of a reflective colorimeter, computer, and a robotic laboratory automations system. The instrument measures color changes in a microtest well containing sample at frequent intervals. The color change measurements are then related to biochemical changes caused by the activity of microorganisms or enzymes and converted to estimates of microbial numbers by a computer. The Omnispec may be used for traditional quality control tests in dairy industry including rapid estimation of microbial numbers, detection of antibiotics, screening abnormal milk, culture activities test, coliforms, staphylococcal and yeast and mold counts, and keeping quality tests.[466]

5.9.2.4 Rapid Methods for Detection and Identification of Pathogens and Toxins

Routine microbiological analysis of milk and dairy products seldom involve isolation and identification of microorganisms or detection of toxins. However, detection and characterization of pathogenic organisms and toxins is often necessary to ensure regulatory compliance and safety of milk and dairy products. Many diagnostic kits, for example, API, Micro ID, Enterotube, etc., developed during the 1970s for clinical applications are now being used to identify microbial isolates in the dairy industry.[467–471] More sophisticated tests such as the DNA probes and immunological assays (enzyme-linked immunosorbant assay, ELISA or EIA) and latex agglutination tests are available for rapid detection of pathogenic bacteria such as *Salmonella, Listeria, E. coli* 0517:H7, *S. aureus*, *Clostridium perfringens*, and toxins including aflatoxin, *B. cereus* toxin, and staphylococcal toxins.[410,468,472–474] Monitoring milk supply for aflatoxin and animal drug residues such as antibiotics and sulfamethazine has been facilitated tremendously by the ELISA-based and other rapid tests.[475]

Automated systems for rapid identification and characterization of microbial isolates include the Vitek System, the AMBIS system, and the HP Microbial Identification System. The Vitek Automicrobial System and the Vitek Jr. are computer-driven systems involving the use of specially designed test cards containing micro-wells lined with lyophilized media for specific biochemical tests. The test card is aseptically inoculated with a suspension of pure isolate, and loaded into the incubator equipped with a photometric reader/detector to detect turbidity or color difference indicating a positive/negative test result. The biochemical reactions of the test microorganisms are compared with data for known standard microorganisms and an identification is made. The Vitek System can allow characterization and identification of as many as 120 different isolates.

Table 5.24 SELECTION CRITERIA FOR AN IDEAL AUTOMATED MICROBIOLOGY ASSAY SYSTEM

1. Accuracy for the intended purpose sensitivity: minimal detectable limits specificity of test system versatility: potential applications comparison to referenced methods.
2. Speed-productivity in obtaining results number of samples processed per run; per day.
3. Cost initial, per test, reagents,others.
4. Acceptability by scientific community by regulatory agencies.
5. Simplicity of operation samples preparation operation of test equipment computer versatility.
6. Training on site; how long quality of training personnel.
7. Regents reagent preparation–stability–availability–consistency.
8. Company reputation
9. Technical service speed and availability cost scope of technical background
10. Utility and space requirements

The AMBIS microbiology system is based on a computerized comparison of peptide banding pattern or microbial "finger printing" of polypeptide patterns for known standard microorganisms. The pure colony is incubated in a medium containing L-[^{35}S]methionine, followed by sodium dodecyl sulfate-polyacrylamide gel electrophoresis of the cell free extract and automated comparison of the polypeptide banding patterns of the unknown against that of the known standard microorganism.

The HP microbial identification system is based on the determination of cellular fatty acid composition of unknown isolates by a computerized gas-chromatographic method. The HP microbial identification system is reportedly capable of differentiating between two otherwise indistinguishable pathovars of *Pseudomonas syringae*[410]

Rapid and automated methods are increasingly being adopted by the dairy industry. However, the main limitations appear to be the regulatory status (FDA or AOAC approval), familiarity with various systems available, and initial cost of equipment and supplies. Several important criteria of selection and adoption of rapid and automated methods in dairy laboratories are listed in Table 5.24. New methods may be justified based on reducing labor and expense and computerized handling, interpreting, and retrieving of microbiological data. Given the current industry trends for consolidation, reduction in work force, and implementation of new programs such as HACCP, use of rapid and automated methods for microbiological analysis of milk and dairy products will continue in the foreseeable future.

5.9.3 Microbiological Tests for Assessing Sanitation and Air Quality in Dairy Plant

Microbiological quality of milk and milk products often depends on the status of cleaning and sanitation practices, conditions of storage, and handling of raw and processed products as well as airborne contamination. Quality control programs include routine testing of plant and equipment surfaces, packaging material, and air for the presence of microorganisms.

Surface sampling methods, for example, swab, surface rinse, and adhesive tape methods are widely used in the dairy industry.[402] These methods involve transferring residual contamination on the designated area of the surface to be tested to sterile dilution blanks using cotton swabs, followed by the plate count method. Following specified incubation (e.g., 30°C/48 h), the colonies are counted to determine the level of contamination. Another method used for assessing microbiological contamination on dairy plant surfaces is the RODAC plate method which involves pressing of small plastic petri dish ($\pm$ 25 cm^2) containing solidified agar medium to the surface followed by incubation and counting of colonies. The RODAC plates are not suitable for wet or heavily contaminated surfaces.

Recently, rapid dip-stick type methods for determining total or coliform counts on dairy plant surfaces have been introduced. These methods may be used in conjunction with the swab or rinse method. They are preferred by some laboratories as they eliminate the need for using petri dishes.

5.9.4 Shelf-Life Tests

Traditionally, shelf life of pasteurized milk and milk products has been determined using the Mosley test,[58,400] which involves the comparison of the plate count of the sample on day zero and after 5 or 7 days of storage at 7°C. The Mosley count yields high correlation with the shelf life and is widely used in the dairy industry for categorizing milks as "poor," "marginal," or "good." However, it is impractical due to the time (up to 9 days) required to obtain results.

As the shelf life of milk and dairy products depends on the extent of postpasteurization contamination, particularly psychrotrophic bacteria, attempts have been made to devise a modified psychrotrophic bacterial counts. Methods based on preincubation of the sample, increasing the incubation temperature, use of selective enrichment designed to enumerate Gram-negative bacteria, or a combination of these have been developed for assessing shelf life of milk and milk products.[422,423,427] Also, several rapid and automated methods, for example DEFT, the catalasemeter, impedimetric evaluation, and the LAL test have been used for determining shelf life of milk and milk products.[398,422,426,427]

Recently, mathematical models have been used for monitoring product quality[107] and shelf-life prediction.[476–478] In this procedure regression equations are generated to predict the growth and relative growth rate of spoilage microorganisms at various product temperatures. One such model is the square root model of Ratkowsky et al.[479] This model has been used for predicting shelf life of pasteurized milk.[476,477,480]

5.10 Microbiology of Milk and Dairy Products

The microbiological spoilage of milk and dairy products will depend on the quality of the raw milk used to make the products, the contamination during processing, the processing that has been done to the products, the final pH and water activity of the products, the packaging and storage conditions, and the intended shelf life of the products. Zikakis[481] has reviewed these factors that affect the keeping quality of dairy products. Cooling and refrigeration have been extensively used to slow the growth of psychrotrophic microorganisms and stop the growth of mesophilic and thermophilic microorganisms. After milk reaches the processing plant, it can be pasteurized or sterilized to reduce some or all spoilage microorganisms, respectively. In addition milk can be fermented to make several different types of dairy products that have decreased pH and, in some cases, water activity when compared to fluid milk. A two-volume book on dairy microbiology has recently been revised and edited by Robinson.[482,483] In the first volume the microbiology of raw, heat-treated, dried, and concentrated milks is reviewed. The second volume focuses on the microbiology of cream, ice cream and frozen desserts, butter, cheese, and fermented milks. More details on the spoilage of these dairy products can be obtained from these books. This section will be a brief review of the microbiology and potential spoilage of dairy products.

5.10.1 Pasteurized Milk and Cream

The microbiological quality of the raw milk before processing will have an effect on the final milk quality after pasteurization. Cousin[6] has reviewed the growth and activity of psychrotrophs in milk. Generally, Gram-negative bacteria, such as species of *Pseudomonas*, *Moraxella*, *Flavobacterium*, *Acinetobacter*, and *Alcaligenes* predominate over Gram-positive bacteria in causing spoilage of pasteurized milks. These bacteria are part of the microflora of raw milk that can become resident in the dairy plant and contaminate the milk after it has been pasteurized because these Gram-negative bacteria are sensitive to heat and would be killed by normal pasteurization. Many Gram-negative bacteria produce proteinases and lipases that result in decreased product quality. *Acinetobacter* species can also produce slime in milk.[484] *Enterobacteriaceae*, such as, *Citrobacter freundii*, *Serratia liquefaciens*, *E. coli*, *Enterobacter agglomerans*, *Enterobacter cloacae*, and *Klebsiella ozaenae* have been isolated from milk.[485] In pure culture studies, these *Enterobacteriaceae* decreased the pH to 6, reduced the redox potential, and produced protoeolytic and lipolytic degradation in milk. Yeasts can be isolated from both raw and pasteurized milks.[99,486] Species of *Rhodotorula*, *Candida*, *Cryptococcus*, and *Kluyveryomyces* can be found in milk but they are readily out-competed by the psychrotrophic bacteria.

Gebre-Egziabher et al.[487] reported that raw milk could be held for 3 days at the farm if proper sanitation and storage conditions were followed. This milk would still be acceptable for processing into fluid milk and other dairy products. Milks with high psychrotrophic counts before pasteurization generally result in milk that spoils faster at refrigeration temperatures.[488] Off-flavors, particularly bitterness, are reported for these milks and are probably due to the proteinases produced by the psychrotrophs. Muir and Phillips[489] set the rejection level for raw milk at 5×10^6 cfu/ml after studying storage time and initial count to calculate the generation time.

Several studies have been done on the keeping quality of the milk once it has been pasteurized and stored. Schröder[490] studied the postpasteurization contamination of milk and reported that Gram-negative bacteria were not detectable immediately after pasteurization, but could be detected in the packaged milk samples. Psychrotrophic bacteria were recovered from storage tanks and filling equipment, suggesting that these were areas where postpasteurization contamination was occurring. Flavor defects were noted when psychrotrophic levels reached 10^7 cfu/ml. The temperature of pasteurized milk storage also plays a role in the overall spoilage of the product. Much research has been done to predict the keeping quality or shelf life of pasteurized milk. Griffiths et al.[57] found that psychrotrophic counts rather than total aerobic counts were better indicators for the shelf life of milk stored at 6°C; however, the prediction of shelf life was correct only 51 to 72% of the time. Hence, preincubation tests that took 24 to 50 h to complete were used to improve the predictability to 83 to 87% for correct identification of pasteurized milk and cream that would have an estimated shelf life. Bishop and White[423] suggested that the ideal test for estimating the shelf life of milk should be simple to do, indicate the exact number of microorganisms in the milk, produce results in a very short time, and be economical. Some of the new methods discussed included detection of metabolites (proteases, lipases, and endotoxins) and use of automated estimation of total numbers (impedance). Chen and Zall[491] suggested using a bar-coded polymer that shows temperature changes to determine potential spoilage of milk. The polymer reflectance changes correlated to the taste of the milk for some of the samples, but not for all samples. Therefore, more research needs to be done on the use of these temperature indicators. Mathematical models have been used to study bacterial growth.[107] Chandler and McMeekin[477,480] studied a temperature function integration model based on the square root to predict the spoilage of milk and found that at temperatures <15°C, the curve had a T_0 (conceptual temperature where the square root of growth is zero) of 264 K if the spoilage limit was set at $10^{7.5}$ cfu/ml for pseudomonads. This model takes into account temperature variations during storage and can be used to monitor a product continually. Obviously, more research needs to be done on the prediction of the keeping quality of pasteurized milk and cream.

The microflora of the pasteurized milk will also be a result of the pasteurization treatment that is given.[105,492] Cromie et al.[105] studied 15 temperatures between 72 and 88°C for 1 to 45 s followed by aseptic packaging. In these milks coryneform bacteria (*Microbacterium* and *Corynebacterium*) made up 83.8% of the population followed by 12.8% Gram-positive cocci (species of *Micrococcus*, *Aerococcus*, and *Streptococcus*) and by 3.4% *Bacillus* species. *B. circulans* was the predominant microorganisms isolated when these milk(s), regardless of pasteurization temperature, were spoiled. This suggests that aseptic packaging prevents the entrance into pasteurized milk of the normal Gram-negative postpasteurization spoilage microflora and only the thermoduric bacteria will be of concern. Psychrotrophic *Bacillus* species were found more frequently in the summer–autumn months than in the winter–spring months.[493] Psychrotrophic species most frequently identified were *B. cereus*, *B. circulans*, and *B. mycoides*. Therefore, *Bacillus* species become important when they are selected by temperature and aseptic conditions of packaging.

5.10.2 Dried Milk Powder

Dried milk powder does not support microbial growth, but microorganisms and their enzymes can be present and cause problems on use once rehydrated. Although there is not agreement on the quality of the dried milk powder, some specifications suggested for a quality milk powder are a total count of <50,000 cfu/g, a coliform count of <10 cfu/g, a spore count of <1000 spores/g, a yeast and mold count of <10 cfu/g, and the absence of *Salmonella* species.[494] The total count and thermoduric bacterial count generally decreased throughout the drying of skim milk with the greatest decreases at the higher temperatures of processing.[494,495] As the total solids content increased so did the thermal resistance of bacteria during spray drying.[496] The spray drying process did not result in straight lines for the plot of the natural log (initial number/number after time *t*) versus the reciprocal of the absolute outlet temperatures.[497] Death of microbes during spray drying is, therefore, complex. Stadhouders et al.[498] reported that *Bacillus* species, *Clostridium perfringens, Microbacterium lacticum, Streptococcus thermophilus,* and *Enterococcus* species (*E., faecium* and *E. faecalis*), *S. aureus*, and other thermoduric bacteria were isolated from spray dried milk. Chopra and Mathur[499] also isolated thermophilic *Bacillus* species from spray- and roller-dried skim milk powder. Chopin[500] reported that some *Lactococcus* bacteriophages were resistant to spray drying and were not reduced for 9 months during storage of milk powder. These bacteriophage could potentially cause problems in cultured products that are made with milks where bacteriophage have survived processing. Therefore, the survival of bacteria in dried milk could result in problems for subsequent products that are made from the rehydrated powder.

Psychrotrophs that have previously grown in milk can change the properties of dried milk. Burlingame-Frey and Marth[501] reported that freeze-dried milk made from milk with previous psychrotrophic growth and either increased or decreased dispersibility and foam production, depending on the type of psychrotroph and increased insolubility. These changes were attributed to the degradation of milk proteins. Previous microbial growth can, therefore, affect the functional properties of the final powder.

5.10.3 Evaporated Milk

Evaporated milk is heat processed in the can; therefore, the keeping quality will depend on the successful commercial sterilization. One problem that has been noted is flat sour due to *Bacillus* spores. Kalogridou-Vassiliadou et al.[502] isolated bacilli from 82.2% of spoiled evaporated milk samples. Most isolates were *B. coagulans, B. licheniformis*, and *B. stearothermophilus*. These bacilli reduced the pH to 4.7 to 5.3 and produced acid and cheesy odors/flavors and dark colored milk. The sources of these contaminants were studied during the processing of evaporated milk.[503] *B. coagulans* and *B. licheniformis* were isolated from all raw milk samples used in the processing and from some canned evaporated milks. *B. stearothermophilus* was isolated from some raw milk samples. *Enterococcus faecium* and *Bacillus subtilis* were isolated from an acid-coagulated evaporated milk.[504] Both acid and gas production that resulted in ruptured cans were noted in spoiled evaporated milk. Proteolysis by *B. subtilis* stimulated the *E. faecium*. The heat resistance of these isolates was not studied in the evaporated milk. Usually evaporated milk will spoil because the processing has not been adequate to inactivate the spore-forming *Bacillus* species in the milk. The evaporation will concentrate both the milk and spores and make the thermal processing more complicated.

5.10.4 Cottage Cheese

Cottage cheese is a nonripened cheese with a high water activity and pH around 5.0; therefore, it is susceptible to both bacterial and fungal spoilage. *Pseudomonas* species (*P. putida* and *P. fluorescens*) and *Enterobacter agglomerans* have been isolated from spoiled cottage cheese.[505,506] Most of these isolates grew at pH 4.9 at either 7 or 20°C. *E. agglomerans* grew at pH 4.6 and some strains grew at pH 3.8 at 7°C. Brocklehurst and Lund[507] studied the microbial changes in commercially produced creamed cottage cheese. Bacteria in the genera *Pseudomonas, Micrococcus, Bacillus,* and *Enterobacter* were detected in spoiled cottage cheese that was stored at 7°C. Yeasts species of *Trichosporon, Candida, Cryptococcus,* and *Sporobolomyces* were found at the end of the storage life of the creamed cottage cheese at 7°C. Fleet and Mian[486] reported that cottage cheese samples had 10^1 and 10^7 yeasts/g, mainly species of *Candida, Cryptococcus,* and *Rhodotorula*. Bigalke[508,509] suggested that raw milk standards for cottage cheese be <1000 psychrotrophs/ml and $<50,000$ total count/ml for good quality cottage cheese that has final product counts of $<1/50$g for psychrotrophs and yeasts/molds. The keeping quality of cottage cheese will depend on the contamination after processing by psychrotrophic bacteria and yeasts, and in some cases molds.

5.10.5 Mold-Ripened Cheeses

Most mold-ripened cheeses are soft to semisoft in texture. These include blue-veined cheeses, such as Roquefort, Blue, Gorgonzola, and Stilton, and Camembert and Brie. These cheeses can spoil due to bacteria, yeasts, and molds. In a survey of blue-veined cheeses in The Netherlands, de Boer and Kuik[510] isolated *Enterobacteriaceae, B. cereus,* and *Staphylococcus aureus* from 40%, 10%, and 5% of the cheeses, respectively. *Yersinia enterocolitica* and *L. monocytogenes* were isolated only from 1 and 2 cheese samples, respectively, out of 256 cheeses analyzed. The yeasts most frequently isolated were *Debaryomyces hansenii* followed by *Kluyveromyces marxianus, Sacchromyces cerevisiae, Yarrowia lipolytica,* and *Candida* species. Some of these yeasts may help to produce flavor in the cheeses and may not be spoilage microbes. Fleet[99] reported that species of *Torulopsis, Hansenula,* and *Pichia* can also be isolated from blue-veined cheeses. The lactose fermenting strains help to open the texture of the cheese for better penetration of *Penicillium roqueforti*. Yeasts produce alcohols for ester generation. Since many of these yeasts use lactic acid, the pH of the cheese increases and bacteria can then grow. Molds, such as *Geotrichum candidum, Penicillium camemberti, P. verrucosum,* and *Cladosporium macrocarpum,* also were isolated from the blue-veined cheeses.

Enterobacteriaceae were isolated from 85 and 88% of Brie and Camembert cheeses, respectively.[511] The highest number was in the rind as opposed to the core of the cheese. *Escherichia coli* was isolated in $>10^2$ cfu/g in 23 to 32% of the Camembert and Brie cheeses, respectively. Bacterial pathogens were found in relatively few samples. *S. aureus* was isolated from one Brie and three Camembert cheeses; *Y. enterocolitica* was isolated from only three cheeses; *B. cereus* was found in one Brie and four Camembert cheeses. Yeasts (mainly *Yarrowia lipolytica, Debaryomyces hansenii, Kluyveromyces marxianus,* and *Candida* spp.) were isolated from over 80% of the Brie and camembert cheeses. In addition to *P. camemberti,* other molds, such as *G. candidum, Cladosporium macrocarpum, Stachybotrys chartarum, Mucor plumbeus, Aspergillus niger,* and *Fusarium* were isolated from some of these cheeses. Mold contamination was minimal.

5.10.6 Hard Cheese

Molds are the most common contaminant of hard cheeses. Tsai et al.[206] examined surplus commodity cheese in United States warehouses and found that all isolates were *Penicillium* species. The major species isolated from these cheeses have been identified as *P. roqueforti, P. cyclopium, P. viridicatum,* and *P. crustosum.* Although molds are the major spoilage microorganisms for hard cheeses, there can be special problems with bacteria and yeasts. Previous growth of psychrotrophic bacteria in cheese milk can cause flavor and odor problems in the finished cheeses. Lipolytic changes have been implicated in the low flavor scores for cheese.[97,512,513] Proteolysis has also been attributed to the presence of bitter and unclean flavors in cheese.[6,97] In Edam, Emmenthal, Gouda, and similar cheeses, *Clostridium tyrobutyricum* can cause late swelling defects.[514] Rind rot of Swiss cheese was caused by *Pseudomonas putida* and *Klebsiella pneumoniae.*[515] This defect with soft white spots on the surface is seen during ripening of the cheese at 22 to 24°C for 4 to 6 weeks. In a study of processed cheese, Warburton et al.[516] found that only 24% of the samples had >500 aerobic sporeformers/g and 15% had over 500 anaerobic sporeformers/g, suggesting that good manufacturing practices were probably used in their manufacture.

Yeasts are infrequently reported as causing defects in hard cheeses.[99] In a study done by Fleet and Mian,[486] *Candida famata, Kluyveromyces marxianus, Candida diffluens, Cryptococcus flavus,* and *Saccharomyces cerevisiae* were isolated from 38, 19, 14, 8, and 8% of the Cheddar cheese samples, respectively. The determination of whether the presence of these yeasts constitutes spoilage depends on whether they can grow in the cheese. Horwood et al.[517] examined a 6-month-old commercial Cheddar cheese that had a "fermented yeasty" defect and noted that high levels of ethanol, ethylacetate, and ethyl butyrate were identified by gas chromatography. About 10^5 yeasts/g of cheese were enumerated and the yeasts were identified tentatively as *Candida* species. As this cheese had a high moisture content and low starter activity and salt content, *Candida* spp. or other yeasts could easily grow and produce off-flavors. The final spoilage of hard cheeses will depend on the pH, water activity, packaging, and storage conditions.

5.10.7 Yogurt and Cultured Milks

The pH of yogurt and fermented milks will normally limit the bacterial spoilage potential and select for mold and yeast growth. Generally yeasts limit the shelf life of yogurt because they can cause sufficient gas production at 10^5 to 10^6 yeasts/g to produce a swollen package.[99] Yeasts can contaminate the yogurt because of poor sanitation or due to contaminated ingredients, such as fruits, nuts, and sweeteners. Tamime et al.[518] surveyed yogurts in Ayrshire and found that 80% of the samples had <10 yeasts/g when examined after manufacture. However, after storage at 5°C, the counts increased to up to 10^4 yeasts/g depending on the season, source, and flavor. Some of the fruit-flavored yogurts in this study also had preservatives added, but that did not prevent the count from increasing to the high levels. A survey of liquid yogurt in Saudi Arabia revealed that very low levels of molds and yeasts (<100/g) were found in yogurts stored at 7°C; however, if stored at 10 to 15°C the counts increased to 10^4 to 10^6 yeasts and molds/g.[519] In several surveys of yogurts, yeasts belonging to the genera *Candida, Kloeckera, Kluyveryomyces, Pichia, Rhodotorula, Saccharomyces,* and *Torulopsis* have been isolated.[486,520–522] KcKay[523] isolated *Yarrowia lipolytica* from yogurt. In all of these studies few of the isolated yeasts were able to ferment lactose. Only *Kluyveryomyces* species[520–522] and *Torulopsis versatilis*[522] were able to ferment lactose. Many could use lactic acid and several fermented galactose and sucrose and most fermented glucose and fructose. Fleet and Mian[486] and Suriyarachchi and Fleet[522] reported that most isolates were not inhibited by sorbate or benzoate and could grow in yogurt with these preservatives. Langeveld and Bolle[524] isolated non-lactose-fermenting yeasts from yogurt and reported that the availability of oxygen was the limiting factor for potential growth. Banks and Board[124] isolated species of *Candida cryptococcus, Debaryomyces,* and *Rhodotorula* from dairy products, such as yogurt, cheese, butter, and quark.

Molds have also been isolated from yogurt. García and Fernández[525] found that the microflora of yogurt in Spain consisted of species of *Penicillium, Monilia, Cladosporium, Micelia sterilia, Alternaria, Rhizopus,* and *Aspergillus.* Both *Penicillium* and *Monilia* species were most frequently isolated. Only one toxigenic species, *Penicillium frequentans,* was isolated from these yogurt samples.[525]

Other cultured dairy products can be contaminated by microorganisms. Buttermilk can be contaminated by bacteria. Psychrotrophic bacteria can reduce diacetyl yielding flavor loss, off-odors and -flavors, and discoloration in buttermilk.[526] Generally, *Pseudomonas* spp. can grow if the pH is above 5.0 and cause these problems. When Hankin et al.[527] studied sour cream and sour dressings, they found that microbial contamination depended on the samples. Only 2 of 21 samples had high aerobic counts (mainly Gram-positive bacteria), 7 of the 21 samples had yeast levels >50/g, and over half the samples had >10 coliforms/g, suggesting poor processing and packaging techniques. Very few Gram-negative bacteria, which are able to degrade protein and fat, were isolated from the products.

5.10.8 Butter

Butter is a water-in-oil emulsion that contains over 80% fat. Generally, well made butter from pasteurized cream has few microbiological problems unless there is postprocessing contamination and storage at temperatures above refrigeration. Hankin and Hanna[528] did a survey on 32 butter samples and found that five had counts $>10^5$ aerobic bacteria/g, four of the samples had psychrotrophic counts above 1000 cfu/g, only four samples had more than 200 yeasts and molds/g, and only five samples had high lipolytic or proteolytic bacterial counts ($>5 \times 10^3$ cfu/g). Although there are no definite microbial standards for butter, most of these samples would be considered microbiologically acceptable and would be expected to have long shelf lives. The incidence of yeasts in butter is very low.[99,486] Jensen et al.[529] found that the storage temperature and salt had an inhibitory effect on yeasts in butter. Also, both coliform and other bacteria were reduced in number over time in salted butter. Mold growth in butter was effectively inhibited by 0.1% potassium sorbate with or without an added 2% sodium chloride.[530] The potential for microbial spoilage of butter will depend on the microorganisms in the water phase, the temperature of storage, and the amount of salt present.

5.10.9 Ice Cream and Frozen Dairy Desserts

Microorganisms cannot grow in ice cream and frozen dairy desserts as long as the temperatures remain below -10°C; however, the presence of microorganisms in these products can give information about the raw ingredient quality and the sanitary nature of processing and packaging. Bigalke[531] reported that ice cream can become contaminated by ingredients that are added postpasteurization and by improper sanitation of equipment and the environment. Hence, <10 coliforms/ml of ice cream have been set in the United States to show that both good quality ingredients and proper sanitation have been used in ice cream manufacture.[532] There have been some surveys in the last decade of the bacteriological quality of ice cream and related products. Ryan and Gough[533] surveyed soft-serve frozen dairy products over a 2-year period in Louisiana and found that 38.5% of the ready-to-serve samples had $>50{,}000$ bacteria/g and 51.2% of these products had >10 coliforms/g. These results suggested that there were sanitation problems associated with soft-serve frozen products. A study of the bacteriological quality of ice cream over three summers in the Netherlands revealed that 11% of the samples had $>10^5$ bacterial/ml which is the legal limit and 33% of the samples exceeded the Dutch law for coliforms.[534] *Staphylococcus aureus* was found in 7 of 89 samples, with the highest count at 2.2×10^4 cfu/ml and *Bacillus cereus* was isolated from 30 of 100 samples with the highest count at 2.8×10^2 cfu/ml. Massa et al.[535] surveyed Italian ice cream over 15 months and found that all ice cream samples had counts $<10^5$ cfu/g (Italian Standard) with most being $<10^2$ to 10^3 cfu/g. Only 6% of the samples had fecal coliforms exceeding the 100 cfu/g limit and only 3.2% of the samples exceeded the limit of 12 cfu/g for *S. aureus*; none of the isolates could produce enterotoxins A to D. Yeasts are reported only in low levels in ice cream, generally $<10^3$ cfu/g.[99,486]

5.11 Microbiological Considerations of New Processing Technologies

Processing technologies for dairy products are continually changing and being updated to meet the needs of consumers for new and improved foods that have acceptable sensory attributes and extended shelf life. If these technologies are to be effectively used, then their effects on microorganisms must be thoroughly evaluated and understood. During the past decade, some technologies that have been used to a limited extent began to gain more interest and commercial use in the dairy industry. Three of these technologies are ultrafiltration, reverse osmosis, and ultra high temperature (UHT) processing. Three new processing technologies that are not used commercially to any great extent are irradiation, microwave, and supercritical CO_2 processing. The microbiology of all these processing technologies will be briefly discussed.

5.11.1 Ulatrafiltration and Reverse Osmosis

Although they are not new technologies, ultrafiltration and reverse osmosis are still evolving and more information on the microbiological aspects has been generated over the past decade. Ultrafiltration (UF) is a fractionation and concentration process that is pressure driven and uses a semipermeable membrane with specific pore sizes that act as a molecular sieve. Molecules with molecular weights larger than the molecular weight cutoff of the membrane are retained (retentate) and molecules that are smaller pass through the membranes (permeate). UF has mainly been used to concentrate milk for production of soft cheeses (Camembert, Brie, Feta, Quarg, Ricotta, and cream cheese) and some hard cheeses (Mozzarella, Blue, Cheddar, Brick, and others). A recent review by Lelievre and Lawrence[536] gives more details on cheese manufacture with UF. Much research has been done on cheese, yogurt, and other dairy product manufacture, but not much research has been done on the microbiology of UF milk.

Bacterial cells, spores, and bacteriophages are retained and concentrated with the milk proteins.[537] The bacteria can grow during UF if the temperature is in the right range. Viellet-Poncet et al.[537] reported that both mesophilic and psychrotrophic bacteria increased eightfold when milk was concentrated to 4:1 at 35°C. Increases from 2.8- to 10-fold in the mesophilic and psychrotrophic counts have been reported during ultrafiltration.[538,539] When milk was ultrafiltered at 50°C, the bacterial count concentrated proportionally to that of the milk.[229] A spore-forming contaminant, tentatively identified as *Bacillus cereus*, also concentrated and caused problems later in the use of the milk. Eckner and Zottola[540] reported that UF of reconstituted skim milk at >50°C reduced or eliminated the levels of *Pseudomonas fragi* in retentates.

Barbano et al.[240] reported that levels of psychrotrophs from <100 to 14 × 10^6/ml did not change the flux during UF. Bacteriophage concentrated, mainly with the casein, at about the same rate as protein when concentrated twofold, but only 2.4:1 when concentrated fourfold.[541] These phages were destroyed at 85°C for 30 min. Zottola et al.[542] reported that bacteriophages did not pass through the UF membrane, but were trapped in the polysulfone membrane or remained in the retentate. The temperature of the ultrafiltration process can, therefore, affect the kinds and numbers of surviving microorganisms.

Limited research has been done on the growth of pathogens in retentates. Haggerty and Potter[543] noted that *Staphylococcus aureus*, *Streptococcus faecalis*, and *Escherichia coli* grew as well in a twofold retentate as in skim milk at 13°C. Enteropathogenic *E. coli* survived and grew better in UF retentates than in skim milk due to the high buffering capacity.[544] Growth of enteropathogenic *E. coli* could be prevented in Camembert cheese if milk was preacidified to pH 5.9 and an active starter was used[545] or if partial fermentation followed by diafiltration to reduce the buffering capacity was used.[546] *Salmonella typhimurium* var. *Hillfarm* grew in retentate concentrated twofold at 7 and 10°C, but *S. aureus* grew only at 10°C.[540] When grown in the presence of *Pseudomonas fragi*, *S. aureus* grew better probably due to the proteolysis by *P. fragi*. In high moisture Monterey Jack cheese, *S. aureus* levels remained stable and *Salmonella* spp. decreased during 6 months at 4.5°C.[547] The thermal resistance of *S. aureus* did not change from whole or skim milk to fourfold concentrated milk.[548] Similar results were reported by Haggerty and Potter[543] for twofold concentrated milk.

The growth of various lactic acid starter cultures has been studied in ultrafiltered milk[549] to determine whether they produce the same amount of acid and lower the pH to the same level as in nonfiltered milk. Hickey et al.[550] reported that strains of *Lactococcus lactis* subsp. *lactis* and *L. lactis* subsp. *cremoris* produced more lactic acid in UF retentates of 5:1 and 2.5:1 compared to whole milk. Although more acid was produced, the corresponding pH values did not decrease accordingly. Other researchers reported similar results.[229,549,551,552] The increased buffering capacity in the UF retentates prevented the pH from being decreased to its normal level. Mistry and Kosikowski[551] noted that increasing the inoculum level did not change the ability of the culture to lower the pH and retentates of four- to five-fold concentration could not have the pH reduced to 4.6 even after 11 h of fermentation. Srilaorkul et al.[549] reported that the maximum buffering capacity was pH 5 to 5.4 due to the protein and minerals, especially phosphate, calcium, and magnesium. The high buffering capacity could be overcome if a high inoculum level (up to 10%) of a very proteolytic starter was used. However, use of highly proteolytic strains can result in bitter flavor development. Other ways that have been used to overcome the buffering capacity are acidification of milk before UF or diafiltration of UF milk to reduce the mineral content. Ultrafiltration to five-fold decreases the B-vitamins, thiamin, riboflavin, niacin, pantothenate, and biotin by 85, 71, 87, 82, and 84%, respectively.[553] Free amino acids also decreased by 50 to 98% in five-fold retentates. Mistry et al.[554] found that neither mineral nor vitamin B addition to 2- to 2.4-fold retentates produced significant increases in lactic production by *L. lactis* subsp. *cremoris* or *L. lactis* subsp.

lactis. Qvist et al.[555] made Havarti cheese from UF five-fold retentates and found that the degradation of β-casein was retarded in UF cheeses and slower flavor development by diacetyl producers was noted as a result of slow protein breakdown. There results plus those given above for growth of lactic acid starters in UF retentates suggest that special concerns for pH, decreased moisture, and proper flavor development are needed when cheese is made from UF retentates.

Because UF retentates have high buffering capacity, they could be used as media for propagation of lactic starter cultures for dairy manufacture. Christopherson and Zottola[556] found that strains of *L. lactis* subsp. *lactis* and *L. lactis* subsp. *cremoris* generally grew to higher cell numbers in UF retentates with 12 to 13% total solids compared to nonfat dry milk reconstituted to 8.3 and 15% solids; therefore, retentates could serve as natural buffered media for starter culture propagation. Whey permeate could also serve as a medium to propagate lactic starter cultures because the decreased lactose and increased solids content compared to skim milk kept the pH higher.[557,558] Addition of 1% yeast extract to the permeate stimulate growth of the *Lactococcus* spp. Cheddar cheese whey permeate was used successfully to propagate strains of *L. lactis* subsp. *lactis* and *L. lactis* subsp. *cremoris* over several transfers for Colby cheese manufacture.[557] The pH and bacterial count from Colby cheese made with a two-fold retentate were comparable to cheese made from unconcentrated milks; however, the moisture content was higher.

Whey permeate has a high biochemical oxygen demand (BOD) that can result in high sewage treatment costs. Reinbold and Takemoto[559] showed that *Bacillus megaterium*, *Rhodopseudomonas sphaerroides*, and *Kluyveromyces fragilis* could reduce the BOD of permeate from 15,500 mg/L to 1580 mg/L. Further research is needed on the reduction of BOD in permeate by bacteria and yeasts.

Another concern of using UF technology is the ability to properly clean and sanitize the membranes after use.[560] Several reports have been published on the inability of commercial cleaners and sanitizers to effectively remove microorganisms from the membranes.[561–565] Bisulfite was not an effective sanitizer of unclean membranes because it needed a pH of 3.5 which resulted in corrosion and pitting of stainless steel fittings and rubber gaskets.[562] Even if the membranes were clean, none of the sanitizers (50 ppm available chlorine, 0.2% hydrogen peroxide, acid anionic surfactant at pH 2.5) were completely effective because of circulation problems.[563] A new sanitizer that releases chlorine dioxide and chlorous acid from a sodium chlorite solution at pH 2.7 effectively sanitized a polysulfone UF membrane; however, electron micrographs showed that the membranes were still plugged with particulates, such as protein and possibly nonviable bacteria.[561] More research needs to be done to improve the UF membranes, produce better cleaners than are now available, and manufacture acceptable sanitizers.

Reverse osmosis (RO) has not had as much acceptance as UF because the cellulose acetate membranes could withstand temperatures to only 35°C.[566] RO is a concentration method that allows most water to pass through the membane under pressure and retains most other components. Normal RO operating temperatures of 20 to 35°C allowed psychrotrophic and mesophilic microorganisms to grow. Now new composite membranes can be operated up to 50°C; however, little research has been done with them. Previous research with the cellulose acetate membranes had shown that RO could be used to manufacture yogurt that compared to conventionally manufactured products for culture growth, acid production, viscosity, and flavor.[567] RO has been used for experimental production of butter, reduced water content in fluid milks, yogurt, and skim milk powder.[568] Drew and Manners[569] showed that processing at 50 to 55°C reduced the bacterial population in RO concentrates; however, psychrotrophs grew at about the same rate in RO and raw milk at 5°C. Cromie et al.[566] reported that preheating milk to 50°C before RO of 2:1 reduced the psychrotrophic, proteolytic, lipolytic, and coliform bacteria, and yeasts and molds by 16 to 50%. In RO concentrates it took 3.5 days longer for the count to reach the same level as in the raw milk. As newer, more temperature-stable membranes are developed, there will be a need for more research on RO.

Microfiltration is a separation process that uses filters with pore sizes of 0.1 to 10 μm to remove microorganisms from liquid that results in a permeate (filtrate) and retentate (concentrate).[570] A small pressure differential is used across the membrane.[571] Microfiltration of milk reduced the *B. cereus* spore count by 99.95 to 99.98% and the total count by 99.99%.[571] Microfiltration units that can filter viscous liquids are pleated tangential crossflow cartridges.[572] This type of filtration can be used to separate bacteria from milk in addition to fat from milk and casein from milk protein. Microfiltration can remove bacteria and clostridial spores from milk better than by bactofugation. Trouvé et al.[573] showed that a 1.4 μm membrane retained 99.93 to 99.99% of the bacteria when milk was microfiltered. Microfiltration may find greater uses in the future for removing bacteria from milk for both fluid consumption and manufacturing uses.

5.11.2 Ultrahigh Temperature Sterilization of Milk and Dairy Products

Ultrahigh temperature (UHT) sterilization is not a new technology; however, it is plagued with some microbiological problems. Burton[574] reviewed 35 years of research and development in UHT processing of milk and dairy products. The bacteriology of UHT processing, especially resistance of spores to high temperatures, has been reviewed by Brown and Ayres[575] and Burton.[576] Two major concerns of UHT processing of milk and dairy products are the heat resistance of bacterial spores and bacterial or native enzymes, particularly proteases and lipases.

Cerf[577] reviewed the techniques for measuring heat resistance of bacterial spores for optimizing UHT processing. The best microbes to use for the thermal process calculations are natural thermophilic sporeformers from milk. The best process is to use the actual UHT equipment to determine the heat resistance of the spores. Duquet et al.[578] studied the thermal resistance of mesophilic and thermophilic spores during UHT processing of milk and found that the $D_{121°C}$ was 0.6 and 58s, respectively. Z values were 10 and 9.6 K, respectively. D values of *Bacillus stearothermophilus* spores in sterilized milk were 22.4 , 3.5, and 0.37 min at 115.5°C, 121.1°C, and 126.6°C, respectively.[579] The spores for these experiments were produced at 55°C in trypticase soy broth (pH 7.1) with 25 ppm of calcium, 31 ppm of iron, 30 ppm of manganese, and 11 ppm of magnesium. One concern with spores at the temperatures above 121°C is whether they have a *Z* value of 10°C. Brown and Gaze[580] studied the thermal resistance of *Clostridium botulinum* from 120 to 140°C and found that the *Z* value was 11°C; therefore, the traditional botulinal process can be safely extrapolated to UHT-processed foods. Lembke and Wartenberg[234] suggested using a bactofuge to remove bacteria from milk before it was UHT processed. Some *Bacillus* species that were isolated from spoiled UGT-processed milk were identified as mesophilic *Bacillus* species, *B. subtilis*, and *B. cereus*.[581] As these strains had *Z* values of 5.6 to 8.8°C, they should have been inactivated by the UHT process. This could suggest postprocess contamination of the UHT milk. The studies that have been done suggest that spores, even thermophilic ones, should be inactivated by the UHT processing.

Proteinases and lipases have not been inactivated by UHT processing and can cause problems in the milk during extended storage. Adams and Brawley[582] reported that lipase from a *Pseudomonas* sp. had *D* values of 1620 to 63 s at 100 to 150°C. The *Z* value was 38.4°C. Kroll[2] and Fox et al.[1] have reviewed the heat resistance of proteinases and lipases, especially those produced by *Pseudomonas* species. A modified UHT treatment of 140°C for 5 s followed by 60°C for 5 min reduced the proteolytic and lipolytic activity in milk.[583,584] The presence of proteinases and lipases in the milk used for UHT processing can therefore create problems with the final product. Gillis et al.[585] and Mottar et al.[586] reported that milk with high proteolytic and psychrotrophic counts, especially Gram-negative bacteria, showed more proteolysis in the final UHT milks. Mottar et al.[586] used in HPLC method to determine the proteolytic quality of milk for UHT processing. Two components, identified only as 2 and 3, were highly correlated to protein breakdown by bacterial proteinases. Gillis et al.[585] found that both the Hull and the trinitrobenzenesulfonic acid (TNBS) tests correlated proteolysis to milk samples with microbial populations between 10^5 and 10^6 cfu/ml. One of the problems is the ability to measure the proteolytic activity. Rollema et al.[587] did a collaborative study to compare several methods of detecting bacterial proteinases in milk. The 2-fluorescamine, azocoll, and TNBS assays were equally sensitive and gave comparable results. The results of the proteolytic assays needs to be compared to the keeping quality or shelf life of the UHT milk.

Several of the effects of proteinases and lipases have been reviewed by Cousin[100] and Mottar.[97] The keeping quality of the milk is related to the presence and activity of heat-resistant enzymes, especially proteinases. Bitter flavors and gelation are common factors in UHT milk spoilage. Keogh and Pettingill[588] found a highly significant correlation between proteolytic enzyme activity and age gelation of UHT milk. The increase in free amino groups during 4 weeks of storage at 21°C indicated that proteolytic enzymes were active in UHT milk.[589] Although *Pseudomonas* spp. are most frequently identified as the protease producers, Keogh and Pettingill[590] identified coryneform bacteria, such as *Arthrobacter* spp., as being involved in age gelation of UHT milk. Aseptically packaged UHT cream became bitter due to proteolytic enzymes that were optimally active at 30 to 37°C.[591] Heat-resistant lipases have caused rancid flavors in UHT milks,[592] but they are usually of lesser importance than proteinases.[97] The increase in lipolytic activity is normally followed by the acid degree value or increase in free fatty acids, especially those with C_4 to C_{12} chain lengths.[97,589,592] Both heat-resistant proteinases and lipases can affect the quality of UHT milk. Hence, good quality assurance programs are needed to ensure that UHT milk is of acceptable quality. Various aspects of quality assurance and final UHT product quality are reviewed by Cordier,[5] Dunkley and Stevenson,[593] Farahnik,[594] and Reinheimer et al.[595] Additional research is needed to determine new methods for the detection of heat-resistant proteinases and lipases in milk that is used for UHT processing. Also, new methods are needed for the final assessment of sterility of UHT-processed milk and dairy products.

5.11.3 Low-Dose Irradiation of Milk

Low-dose irradiation has been suggested for improving safety of foods by reducing populations of food pathogens, such as *Salmonella* spp., *Campylobacter* spp., *Listeria monocytogenes*, *Yersinia enterocolitica*, and others, and for increasing the shelf life of perishable foods.[596,597] Low-dose gamma-irradiation has been suggested for milk, cheese, yogurt, and other dairy products. Raj and Roy[598] reported that 10, 50, and 100 Krad increased the storage life of raw milk at 8 to 10°C by 33, 120, and 120 h, respectively. There was no change in either flavor or color after irradiation at these doses. Sadoun et al.[599] reported that irradiation of pasteurized milk above 0.5 KGy at 4°C resulted in objectionable off-flavors. At this level, the total population was reduced only by about 2 logrithmic cycles; however, if the milk was irradiated at room temperature and stored at 4°C, then the shelf life doubled. *Pseudomonas fluorescens* and other species were easily killed by irradiation in this study. Searle and McAthey[600] found that it took about 200 Gy in air to decrease *P. aeruginosa* by 5 logarithmic cycles and 600 Gy in nitrogen and that more than 1600 Gy were needed to sterilize the UHT milk with *P. aeruginosa* added. The absence of air during irradiation helps to lower the lipid peroxidation, but increases the time needed to kill bacteria. Because the irradiated milk spoiled within 21 days, these authors suggested that the bacteria were not being killed exponentially at higher irradiation levels. More research needs to be done to determine what is happening with the irradiation of milk because many of these experiments were done with different radiation doses, different temperatures, and either raw or pasteurized milks.

Gamma irradiation has been used experimentally to decrease microbial populations in several dairy products. A dose of 400 Krad decreased the total microbial count by 4 log cycles in fluid milk, but in whole milk powder this decreased the microbial count only by 2 log cycles.[601] The color and flavor were adversely affected by irradiation in the milk powder. When Gouda cheese was irradiated at 60 Krad for 1 h at 27, 40, and 48°C, coliforms, yeasts and molds, and psychrotrophs decreased by 2 to 3 logarithmic cycles as the temperature increased from 27 to 48°C. Additional results of research have shown that 0.75 Gy decreased microbial populations by 96 to 99% in Camembert and cottage cheeses.[602] Yüccer and Gündüz[603] suggested that irradiation could be used as a supplement to other preservation methods because levels of irradiation over 0.15 Mrad caused off-flavors and colors in Kashar cheese and yogurt. At doses of 0.02 to 0.04 Mrad for 8 min at a rate of 0.0025 to 0.005

Mrad/min, irradiation increased the shelf life of cheese by four- to five-fold and yogurt by three-fold. Irradiation at −78°C and 40 KGy sterilized ice cream and frozen yogurt, but not Mozzarella or Cheddar cheese.[604] The 12 *D* values for *B. cereus* spores in ice cream, frozen yogurt, and Mozzarella cheese were 49, 47.9, and 43.1 KGy, respectively. *Listeria monocytogenes* was inactivated by irradiation at −78°C using low doses.[604] The 12 *D* in Mozzarella cheese was 16.8 KGy. In ice cream, the 12 *D* was 24.4 KGy. The results of this research indicate that low-dose gamma irradiation can be used to lower the level of microorganisms and some pathogens in dairy products. More research is needed in this area to correlate levels that reduce microorganisms versus those levels that result in organoleptic changes.

5.11.4 Microwave Processing of Milk and Dairy Products

The use of microwaves for pasteurization and sterilization of foods has been researched for years. Although the use of microwaves has been suggested as a way to process milk and dairy products, its use is mainly in the laboratory phase.[605–609,625] Pasteurized milk had to be heated to 55 to 60°C by microwaves before significant inactivation of psychrotrophs was noted.[610,611] Jaynes[612] showed that a microwave treatment at 2450 MHz resulting in a temperature of 72°C for 15 s hold could be used as a continuous system for HTST pasteurization of milk. The use of microwaves to simulate low temperature–long time (LTLT) processing of 65°C for 30 min was as effective as conventional pasteurization.[84] Knutson et al.[613] found that a simulated HTST process by microwaves that achieved a temperature of 71.7°C for 15 s did not inactivate all cells of *Salmonella typhimurium*, *Pseudomonas fluorescens*, and *E. coli*. Similarly, the simulated microwave LTLT did not inactivate *Streptococcus faecalis* to the same level as seen in conventionally treated milk at 62.8°C for 30 min. It was suggested that nonuniform heating in microwave ovens caused these results. Tochman et al.[614] showed that microwave treatment of cottage cheese in the package could extend the shelf life by 1 month over that of the nontreated control. The microwave treatment resulted in a temperature of 48.8°C that reduced the spoilage microorganisms and did not affect the organoleptic quality of the cottage cheese. Additional suggested uses of microwave processing are for tempering and thawing of frozen milk or butter and drying or evaporation of dairy products.[609,615] Although there are several advantages for using microwaves in food processing, several disadvantages, especially the cost for equipment and operation, low efficiency of conversion of electrical energy to microwave energy, uneven product heating, and organoleptic changes in products, have prevented widespread adoption of this new technology.

5.11.5 Use of Carbon Dioxide and Supercritical Carbon Dioxide for Reduction of Microbial Populations

CO_2 in various atmospheres (modified atmospheric packaging) can affect populations of aerobic microorganisms. The use of CO_2 in packages with high barrier properties has been used to extend the shelf life of refrigerated foods. Chen and

Hotchkiss[616] reported that a headspace of 35 to 45% CO_2 resulted in cottage cheese in glass jars that had no increase in psychrotrophic microbial counts for 30 days at 7°C or 80 days at 4°C. Yeasts and molds were not detected in any cottage cheese with added CO_2. Research needs to be done with the plastic containers that are currently used for cottage cheese packaging. Modified atmospheric packaging of dairy products needs to be further investigated.

Supercritical extraction of milk fat using CO_2 has been evaluated because this technique offers advantages over current separation processes. In supercritical CO_2 extraction a range of both temperature and pressure of the gas is used at levels higher than the critical values.[617] This results in the separation based on molecular size. Little is known about the microbiological effects of using supercritical carbon dioxide extraction methods. Kamihira et al.[618] found the supercritical carbondioxide at 200 atm and 35°C could drastically reduce populations of wet cells of yeasts, *E. coli*, *S. aureus*, and *Aspergillus niger*, but no effect was seen with dry cells or spores of *Bacillus* species. The death of microorganisms by supercritical CO_2 still needs a lot more research before it can be suggested for use in pasteurizing or sterilizing milk and dairy products.

5.12 Assuring Microbiological Quality and Safety of Milk and Milk Products: HACCP Approach

Traditionally, quality and safety of milk and dairy products is evaluated in terms of the presence (and levels) or absence of certain microorganisms in raw or finished products. The traditional quality control programs emphasized inspection and end-product testing to determine compliance with standards, specifications, and regulations pertaining to milk and dairy products. The major goal of these programs was to reduce manufacturing defects in dairy foods through the use of Good Manufacturing Practices (GMPs) in processing, random inspections, and laboratory analysis of finished, packaged products, to ensure compliance with specifications and regulations. Recent incidences of pathogenic contaminations and recalls have clearly demonstrated limitations of traditional quality control programs and emphasized the need for a proactive, systematic approach to prevent defects from occurring in the first place by monitoring the manufacturing process and raw material rather than testing end products for defects or presence of contamination. The Hazard Analysis and Critical Control Points (HACCP) is an integral part of the total quality system (TQS) or total quality management (TQM) approach currently in vogue worldwide.

The HACCP system was pioneered in the 1960s by the Pillsbury Company, the U. S. Army Natick Research and Development Laboratories, and the National Aeronautics and Space Administration for designing pathogen-free foods for the space program.[619] Since the 1970s, the HACCP has been used for assuring safety of the low-acid canned foods.[620] The HACCP approach was adopted by major food companies and endorsed by national and international organizations and regulatory agencies[621–623] in the 1980s. The principle and basic elements of the HACCP system are briefly reviewed.

5.12.1 HACCP Principle

The HACCP involves two main aspects:

1. Hazard Analysis: A critical examination of entire food manufacturing process to determine every step, or point, where a possibility of physical, chemical, or microbiological contamination may enter the food and render it unsafe or unacceptable for human consumption.
2. Critical Control Points: A point in a food process where there is a high probability that the lack of control may cause, allow, or contribute to a hazard or to filth in the final food, or to decomposition of the final food.

Originally the HACCP included three principles: (1) hazard analysis and risk assessment, (2) determination of CCPs, and (3) monitoring of the CCPs. However, the U.S. National Advisory Committee on Microbiological Criteria of Food (NACMCF) expanded the original principles of the HACCP to seven principles: (1) conduct hazard analysis and risk assessment, (2) determine CCPs (including CCP_1 and CCP_2 where complete or partial control of a potential hazard is affected), (3) establish specifications for each CCP, (4) monitor each CCP, (5) establish corrective action to be taken if a deviation occurs at a CCP, (6) establish a record-keeping system, and (7) establish verification procedures.

5.12.2 Elements of the HACCP System

Some of the major elements of the HACCP system are as follows:

1. Develop an up-to-date plant flow diagram indicating clearly various streams—raw materials, processed products, CIP-lines, etc. The process flow diagram may consist of several subsystems with an overall flow diagram showing integrated systems. The product/process flow diagram must be accurate and match with plant engineering blue prints.
2. Monitor quality or raw products and ingredients to ensure compliance vendor agreements and specifications. This is particularly important for minimizing the potential hazard of microbial contamination, metal fragments, filth, and other impurities. Raw material quality control is the first line of defense against quality problems in finished products.
3. Determine process compliance by frequent, if possible, on-line monitoring of critical parameters such as temperature, pH, salt content, etc. It can be claimed that if quality control of raw material and ingredients is perfect and the manufacturing process is in compliance with set specifications for that process, the final product will be a quality product requiring very little end-product inspection and testing.
4. In addition to cleaning and sanitation of processing equipment, control of plant environment is critical to product safety and quality. Many organisms can be transmitted through airborne contamination. Therefore, monitoring heating, ventilating and air conditioning system, drains, screen traps, etc. is essential for a successful HACCP. Results of dairy plant surveillance by the industry and the

FDA had indicated that organisms such as *Listeria* may indeed be isolated from the plant environment. Isolating critical areas from main traffic flow and minimizing employee movement from the raw to the finished areas is critical in reducing the risk of pathogenic contamination.

5. Keep accurate records of critical control point monitoring and other process variable. Designate a specific location for these records and person(s) responsible for maintaining records of the critical control point monitoring.
6. Finally, plan a good product recall (retrieval program that is adequately tested). Designate a ''response team'' and a plan of action to be followed in the event of product contamination.

The HACCP approach provides a systematic way to minimize hazards associated with the raw or processed foods, including potential consumer abuse. Development and implementation of the HACCP by major dairy food processors worldwide indicate the desire of the industry to provide high-quality, safe dairy products to the consumer.

5.13 Conclusion

The significance of dairy microbiology vis-a-vis processing, manufacturing, safety, quality, and shelf life of milk and dairy products, particularly new technologies such as membrane processing, microwave, and UHT technology cannot be overemphasized. Much research has been done to understand the behavior of spoilage and pathogenic organisms in milk and dairy foods. Yet, much information is needed to devise practical ways of managing microbiological problems in the dairy industry. Genetic manipulations of starter bacteria offer much promise for development of strains with desirable properties for fermentation of conventional and novel dairy foods and ingredients. Understanding of the crucial role of dairy microbiology has led to regulations regarding the maximum levels of microbial contamination permitted. The increasing need for microbiological testing of milk and dairy products has also prompted developments of rapid and automated methods in dairy microbiology.

This chapter has only touched on the main areas of dairy microbiology as it is impossible to discuss in great detail the myriad of microorganisms that may be associated with milk and dairy products. Further information on any of the areas mentioned may be found in several recently published monographs, reviews, and reference books, some of which are listed in this chapter. It is our hope that those in the dairy industry interested in acquainting themselves with a basic knowledge of dairy microbiology as well as those seeking review of research dealing with microbiological aspects of milk and dairy products processing, quality, and safety will find the information presented here useful.

5.14 References

1. Fox, P. F., P. Power, and T. M. Cogan. 1989. Isolation and molecular characteristics. *In* R. C. McKellar (ed.), *Enzymes of Psychrotrophs in Raw Food*, pp. 57–120. CRC Press, Boca Raton, FL.

2. Kroll, S. 1989. Thermal stability. *In* R. C. McKellar (ed.), *Enzymes of Psychrotrophs in Raw Food*, pp. 121–152. CRC Press, Boca Raton, FL.

3. Adams, D. M., J. T. Barach, and M. L. Speck. 1975. Heat resistant proteases produced in milk by psychrotrophic bacteria of dairy origin. *J. Dairy Sci.* **58:**828–834.

4. Patel, T. R., F. M. Bartlett, and J. Hamid. 1983. Extracellular heat-resistant proteases of psychrotrophic pseudomonads. *J. Food Prot.* **46:**90–94.

5. Cordier, J. L. 1990. Quality assurance and quality monitoring of UHT processed foods. *J. Soc. Dairy Technol.* **43:**42–45.

6. Cousin, M. A. 1982. Presence and activity of psychrotrophic microorganisms in milk and dairy products: a review. *J. Food Prot.* **45:**172–207.

7. Dairy Research Foundation. 1986. The significance of psychrotrophic bacteria in milk and dairy products. *Dairy Res. Rev.* **2:**1–3.

8. Eddy, B. P. 1960. The use and meaning of the term ''psychrophilic.'' *J. Appl. Bacteriol.* **23:**189–190.

9. Herbert, R. A. 1981. A comparative study of the physiology of psychrotrophic and psychrophilic bacteria. *In* T. A. Roberts, G. Hobbs, J. H. B. Christian, and N. Skovgaard (eds.), pp. 3–16. Academic Press, New York.

10. Meer, R. R., J. Baker, F. W. Bodyfelt, and M. W. Griffiths. 1991. Psychrotrophic *Bacillus* spp. in fluid milk products: A review. *J. Food Prot.* **54:**969–979.

11. Witter, L. D. 1961. Psychrophilic bacteria. A review. *J. Dairy Sci.* **44:**983–1015.

12. McKay, L. 1986. Application of genetic engineering techniques for dairy starter culture improvement. *In* S. K. Harlander and T. P. Labuza (eds.), *Biotechnology in Food Processing,* pp. 145–155. Noyes Publications, Parkridge, NJ.

13. Bryan, F. L. 1983. Epidemiology of milk-borne diseases. *J. Food Prot.* **46:**637–649.

14. Vasvada, P. C. 1988a. Pathogenic bacteria in milk—a review. *J. Dairy Sci.* **71:**2809–2816.

15. Holt, J. G. 1986. *Bergey's Manual of Systematic Bacteriology*, Vol. 1. Williams & Wilkins, Baltimore.

16. Holt, J. G. 1986. *Bergey's Manual of Systematic Bacteriology*, Vol. 2. Williams & Wilkins, Baltimore.

17. Banwart, G. J. 1989. *Basic Food Microbiology*, 2nd edit. Van Nostrand Reinhold, New York.

18. Gilmour, A., and M. T. Rowe. 1990. Microorganisms associated with milk. *In* R. K. Robinson (ed.), *Dairy Microbiology*, Vol. 2, 2nd Edit., pp. 37–75. Elsevier Applied Science, London.

19. Jay, J. M. 1986. *Modern Food Microbiol*, 3rd edit. Van Nostrand Reinhold, New York.

20. Palumbo, S. A. 1986. *Campylobacter jejuni* in foods: its occurrence, isolation from foods and injury. *J. Food Prot.* **49:**161–166.

21. Stern, N. J. 1982. Methods of recovery of *Campylobacter jejuni* from foods. *J. Food Prot.* **45:** 1332–1337.

22. Stern, N. J., and S. U. Kazmi. 1989. *Campylobacter jejuni In* M. P. Doyle (ed.), pp. 71–110. *Foodborne Bacterial Pathogens*, Marcel Dekker, New York.

23. Anonymous. 1983. Campylobacteriosis associated with raw milk consumption—Pennsylvania. *Morbid. Mortal. Wkly. Rep.* **32:**337–338, 344.

24. Anonymous. 1984. Campylobacteriosis outbreak associated with certified raw milk products—California. *Morbid. Mortal. Wkly. Rep.* **33:**562.

25. Briesman, M. A. 1984. Raw milk consumption as a probable cause of two outbreaks of campylobacter infection. *N. Z. Med. J.* **97:**411–413.

26. Potter, M. E., M. J. Blaser, R. K. Sikes, A. F. Kaufmann, and J. G. Wells. 1983. Human campylobacteriosis infection associated with certified raw milk. *Am. J. Epidemiol.* **117:**475–483.

27. Lander, E. F., and K. P. Gill. 1980. Experimental infection of the bovine udder with *Campylobacter coli/jejuni. J. Hyg. Camb.* **84:**421–428.

28. Logan, E. F., S. D. Neill, and D. P. Macki. 1982. Mastitis in dairy cows associated with an aerotolerant campylobacter. *Vet. Rec.* **110:**229–230.

29. Bean, N. H., and P. M. Griffin. 1990a. Foodborne disease outbreaks in the United States, 1973–1987: pathogens, vehicles, and trends. *J. Food Prot.* **53:**804–817.

30. Bean, N. H., P. M. Griffin, J. S. Groulding, and C. B. Ivey. 1990b. Foodborne disease outbreaks, 5-year summary, 1983–1987. *J. Food Prot.* **53:**711–728.

31. Corbel, M. J., C. D. Bracewell, E. L. Thomas, and K. P. Grill. 1979. Techniques in the identification and classification of *Brucella. In* F. A. Skinner and D. W. Lovelock (ed.), *Identification Methods for Microbiologists*, 2nd edit., pp. 71–122. Academic Press, London.

32. Doyle, M. P., and V. V. Padhye. 1989. *Escherichia coli. In* M. P. Doyle (ed.), *Foodborne Bacterial Pathogens*, pp. 235–281. Marcel Dekker, New York.

33. Doyle, M. P., and J. L. Schoeni. 1987. Isolation of *Escherichia coli* 0157:H7 from retail fresh meats and poultry. *Appl. Environ. Microbiol.* **53:**2394–2396.

34. Stiles, M. E. 1989. Presumptive foodborne pathogenic bacteria. *In* M. P. Doyle (ed.), *Foodborne Bacterial Pathogens*, pp. 673–733. Marcel Dekker, New York.

35. Montogmerie, J. Z., D. B. Doak, D. E. M. Taylor, J. D. K. North, and W. J. Martin. 1970. Klebsiella in renal-transplant patients. *Lancet* **ii:**787–792.

36. Blood, D. C., and M. Radostitis 1989. *Veterinary Medicine*. 7th edit., pp. 501–514. W. B. Saunders, London.

37. Anonymous. 1987. *Current Concepts of Bovine Mastitis*, 3rd edit. The National Mastitis Council, Arlington, VA.

38. Anonymous. 1981b. Salmonellosis associated with raw milk—Montana. *Morbid. Mortal. Wkly. Rep.* **30:**211.

39. Anonymous. 1985b. Milk-borne salmonellosis—Illinois. *Morbid. Mortal. Wkly. Rep.* **34:**2000.

40. D'Aoust, J. Y. 1989. Manfacture of dairy products from unpasteurized milk: a safety assessment. *J. Food Prot.* **52:**906–914.

41. Bootone, E. J. 1981. *Yersinia enterocolitica.* CRC Press. Boca Raton, FL.

42. Schiemann, D. A. 1987. *Yersinia enterocolitica* in milk and dairy products. *J. Dairy Sci.* **70:**383–391.

43. Schiemann, D. A. 1989. *Yersinia enterocolitica* and *Yersinia pseudotuberculosis. In* M. P. Doyle (ed.), *Foodborne Bacterial Pathogens*, pp. 601–672. Marcel Dekker, New York.

44. Swaminathan, B., M. C. Harmon, and I. J. Mehlman. 1982. *Yersinia enterocolitica*: a review. *J. Appl. Bacteriol.* **52:**151–183.

45. Anonymous. 1982. Multistate outbreak of yersiniosis. *Morbid. Mortal. Wkly. Rep.* **31:**505.

46. Palumbo, S. A., D. R. Morgan, and R. L. Buchanan. 1985. The influence of temperature, NaCl and pH on the growth of *Aeromonas hydrophila. J. Food Sci.* **50:**1417–1421.

47. Palumbo, S. A. 1986. Is refrigeration enough to restrain foodborne pathogens? *J. Food Prot.* **49:**1003–1009.

48. Rouf, M. A., and M. M. Rigney. 1971. Growth temperature and temperature characteristics of *Aeromonas. Appl. Microbiol.* **22:**503–506.

49. Stelma, G. N. 1989. *Aeromonas hydrophila In* M. P. Doyle (ed.), *Foodborne Bacterial Pathogens*, pp. 1–19. Marcel Dekker, New York.

50. Hoover, D. G., S. R. Tatini, and J. B. Maltais 1983. Characterization of staphylococci. *Appl. Environ. Microbiol.* **46:**649–660.

51. Edwards, A. T., M. Roulson, M. J. Ironside, J. F. Barraclough, and G. Morgan. 1984. An outbreak of serious infection due to *Streptococcus zooepidemicus* (Lancefield Group C). *Commun. Dis. Rep.* **41:**3–6.

52. Gasson, M. 1983. Genetic transfer systems in lactic acid bacteria. *Antonie van Leeuwenhoek* **49:**275–282.

53. Gasson, M. J. 1984. Transfer of sucrose fermenting ability, nisin resistance and nisin production in *Streptococcus lactis* 712. *FEMS Microbiol. Lett.* **21:**7–10.

54. Gasson, M. J., and F. L. Davies. 1984. The genetics of dairy lactic acid bacteria. *In* F. L. Davis and B. A. Law (ed.), *Advances in the Microbiology and Biochemistry of Cheese and Fermented Milk*, pp. 99–126. Elsevier Applied Science, New York.

55. McKay, L. L. 1983. Functional properties of plasmids in lactic streptococci. *Antonie Van Leeuwenhoek* **49:**259–274.

56. McKay, L. L. 1985. Role of plasmids in starter cultures. *In* S. E. Gilliland (ed.), *Bacterial Starter Cultures for Foods*, pp. 159–174. CRC Press, Boca Raton, FL.

57. Griffiths, M. W., J. D. Phillips, and D. D. Muir. 1985. The quality of pasteurized milk and cream at the point of sale. *Dairy Indust. Int.* **50:**25, 27–28, 31.

58. Richardson, G. H. 1985. (ed.). *Standard Methods for the Examination of Dairy Products*, 15th edit. American Public Health Association. Washington, D.C.

59. Doyle, M. P. 1988. *Bacillus cereus. Food Technol.* **42:**199.

60. Johnson, K. M. 1984. *Bacillus cereus* foodborne illness: An update. *J. Food Prot.* **47:**145–153.

61. Kramer, J. M., and R. J. Gilbert. 1989. *Bacillus cereus* and other *Bacillus* species. *In* M. P. Doyle (ed.), *Foodborne Bacterial Pathogens*, pp. 21–70. Marcel Dekker, New York.

62. Hauschild, A. H. W. 1989. *Clostridium botulinum. In* M. P. Doyle (ed.), *Foodborne Bacterial Pathogens*, pp. 191–234. Marcel Dekker, New York.

63. Campbell, J. R., and R. T. Marshall. 1975. *The Science of Providing Milk for Man*, pp. 329–350. McGraw-Hill, New York.

64. Tamime, A. Y. 1990. Microbiology of starter cultures. *In* R. K. Robinson (ed.), *Dairy Microbiology*, Vol. 2, 2nd edit, pp. 131–201. Elsevier Applied Science, New York.

65. Donnelly, C. W. 1986. Listeriosis and dairy products—why now and why milk. *Hoard's Dairyman* **131:**663—687.

66. Pearson, L. J., and E. H. Marth. 1990. *Listeria monocytogens*—Threat to a safe food supply: a review. *J. Dairy Sci.* **73:**912–928.

67. Lovett, J., D. W. Francis, and J. M. Hunt. 1987. *Listeria monocytogenes* in raw milk: detection, incidence and pathogenicity. *J. Food. Prot.* **50:**188–192.

68. Seeliger, H. R. P. 1961. *Listeriosis.* Hafner, New York.

69. Gitter, M., R. Bradley, and P. H. Blampied. 1980. *Listeria monocytogens* infection in bovine mastitis. *Vet. Rec.* **107:**390–393.

70. Doyle, M. P., K. A. Glass, J. T. Berry, G. A. Garcia, D. J. Pollard, and R. D. Schultz 1987. Survival of *Listeria monocytogenes* in raw bovine milk during high-temperature short-time pasteurization. *Appl. Environ. Microbiol.* **53:**1433–1438.

71. Bunning, V. K., R. G. Crawford, J. G. Brodshaw, J. T. Peeler, J. T. Tierney, and R. M. Twedt. 1985. Thermal resistance of intracellular *Listeria monocytogenes* cells suspended in raw bovine milk. *Appl. Environ. Microbiol.* **52:**1398–1402.

72. Lovett, J. 1989. *Listeria monocytogens. In* M. P. Doyle (ed.), *Foodborne Bacterial Pathogens*, pp. 283–310. Marcel Dekker, New York.

73. Jenkins, P. A., L. R. Doodridge, C. H. Collins, and M. D. Yates. 1985. *Mycobacteria. In* C. H. Collins and J. M. Grange (eds.), *Isolations and Identification of Microorganisms of Medical and Veterinary Importance.* Academic Press, London.

74. Hosty, F. S., and C. I. McDurmont 1975. Isolation of acid fast organisms from milk and oysters. *Health Lab. Sci.* **12:**16–19.

75. Cogan, T. M., and J. P. Accolas. 1990. Starter cultures: types, metabolism and bacteriophage. *In* R. K. Robinson (ed.), *Dairy Microbiology*, Vol. 1, 2nd edit., pp. 77–114. Elsevier Applied Science, London.

76. Davies, F. L., and M. J. Gasson. 1984. Bacteriophage of dairy lactic acid bacteria. *In* F. L. Davies, and B. A. Law (eds.), *Advances in the Microbiology and Biochemistry of Cheese and Fermented Milk*, pp. 127–151. Elsevier Applied Science, London.

77. Bradley, D. E. 1967. Ultrastructure of bacteriophages and bacteriocins. *Bacteriol. Rev.* **31:**230–314.

78. Klaenhammer, T. R. 1984. Interactions of bacteriophages with lactic streptococci. *Adv. Appl. Microbiol.* **30:**1–29.

79. Klaenhammer, T. R. 1987. Plasmid-directed mechanisms for bacteriophage defense in lactic streptococci. *FEMS Microbiol. Rev.* **46:**313–325.

80. Sanders, M. E. 1988. Phage resistance in lactic acid bacteria. *Biochimie* **70:**411–422.

81. Sechaud, L., P. J. Cluzel, M. Rosseau, A. Baumgartner, and J. P. Accolas. 1988. Bacteriophages of lactobacilli. *Biochimie* **70:**401–410.

82. Ingraham, J. L., and J. L. Stokes. 1959. Psychrophilic bacteria. *Bacteriol. Rev.* **23:**97–108.

83. Mossel, D. A. A., and H. Zwart. 1960. The rapid tentative recognition of psychrotrophic types among *Enterobacteriaceae* isolated from foods. *J. Appl. Bacteriol.* **23:**185–188.

84. Morita, R. Y. 1975. Psychrophilic bacteria. *Bacteriol. Rev.* **39:**144–167.

85. Gounot, A. M. 1986. Psychrophilic and psychrotrophic microorganisms. *Experientia* **42:**1192–1197.

86. Herbert, R. A. 1986. The ecology and physiology of psychrophilic microorganisms. *In* R. A. Herbert and G. A. Codd (eds.), *Microbes in Extreme Environments*, pp. 1–23. Academic Press, New York.

87. Phillips, J. D., and M. W. Griffiths. 1987. The relation between temperature and growth of bacteria in dairy products. *Food Microbiol.* **4:**173–185.

88. Reichardt, W., and R. Y. Morita. 1982. Temperature characteristics of psychrotrophic and psychrophilic bacteria. *J. Gen. Microbiol.* **128:**565-568.

89. Stannard, C. J., A. P. Williams, and P. A. Gibbs. 1985. Temperature/growth relationships for psychrotrophic food-spoilage bacteria. *Food Microbiol.* **2:**115–122.

90. Suhren, G. 1989. Producer microorganisms. *In* R. C. McKellar (ed.), *Enzymes of Psychrotrophs in Raw Milk*, pp. 3–34. CRC Press, Boca Raton, FL.

91. Griffiths, M. W., J. D. Phillips, and D. D. Muir. 1984a. Detection of post heat-treatment contaminants using ATP photometry. *In Rapid Detection of Post-Pasteurization Contamination*, pp. 37–45. Hannah Research Bulletin No. 10.

91a. Spohr, M., and Schütz. 1990. Metabolism of glucose, pyruvate, lactate, and malate in refrigerated milk by *Pseudomonas* species before and during exponential growth. *Milchwissenschaft* **45:**145–148.

92. Griffiths, M. W., and J. D. Phillips. 1990c. Incidence, source and some properties of psychrotrophic *Bacillus* spp. found in raw and pasteurized milk. *J. Soc. Dairy Technol.* **43:**62–66.

93. Bloquel, R., and L. Veillet-Poncet. 1980. Évolution et détermination de la flore bactérienne d'un lait cru réfrigéré paucimicrobien en fonction du temps. *Le Lait* **60:**474–486.

94. Shelley, A. W., H. C. Deeth, and I. C. MacRae. 1987. A numerical taxonomic study of psychrotrophic bacteria associated with lipolytic spoilage of raw milk. *J. Appl. Bacteriol.* **62:**197–207.

95. Kroll, S., B. Kellerer, M. Busse, and H. Klostermeyer. 1984. Die proteolytisch aktiven psychrotrophen Bakterien in Rohmilch. 1. Differenzierung aufgrund stoffwechselphysiologischor Merkmale. *Milchwissenschaft* **39:**538–540.

96. Cogan, T. M. 1980. Heat resistant lipases and proteinases and the quality of dairy products. *IDF Bull.* **118:**26–32.

97. Mottar, J. F. 1989. Effect on the quality of dairy products. *In* R. C. McKellar (ed.), *Enzymes of Psychrotrophs in Raw Food*, pp. 227–243. CRC Press, Boca Raton, FL.

98. Walker, S. J. 1988. Major spoilage microorganisms in milk and dairy products. *J. Soc. Dairy Technol.* **41:**91–92.

99. Fleet, G. H. 1990. Yeasts in dairy products. *J. Appl. Bacteriol.* **68:**199–211.

100. Cousin, M. A. 1989. Physical and biochemical effects of microbial enzymes on milk components. *In* R. C. Mckellar (ed.), *Enzymes of Psychrotrophs in Raw Food*, pp. 205–225. CRC Press, Boca Raton, FL.

101. Coghill, D., and H. S. Juffs. 1979. Incidence of psychrotrophic spore-forming bacteria in pasteurized milk and cream products and effect of temperature on their growth. *Aust. J. Dairy Technol.* **34:**150–153.

102. McKinnon, C. H., and G. L. Pettipher. 1983. A survey of sources of heat-resistant bacteria in milk with particular reference to psychrotrophic spore-forming bacteria. *J. Dairy Res.* **50:**163–170.

103. Collins, E. B. 1981. Heat resistant psychrotrophic microorganisms. *J. Dairy Sci.* **64:**157–160.

104. Cromie, S. J., D. Schmidt, and T. W. Dommett. 1989b. Effect of pasteurization and storage conditions on the microbiological, chemical and physical quality of aseptically packaged milk. *Aust. J. Dairy Technol.* **44:**25–30.

105. Cromie, S. J., T. W. Dommett, and D. Schmidt. 1989a. Changes in the microflora of milk with different pasteurization and storage conditions and aseptic packaging. *Aust. J. Dairy Technol.* **44:**74–77.

106. Engle, G., and M. Teuber. 1991. Heat resistance of ascospores of *Byssochlamys nivea* in milk and cream. *Int. J. Food Microbiol.* **12:**225–234.

107. Griffiths, M. W., and J. D. Phillips. 1988a. Modelling the relation between bacterial growth and storage temperature in pasteurized milks of varying hygienic quality. *J. Soc. Dairy Technol.* **41:**96–102.

108. McKellar, R. C. 1989b. Regulation and control of synthesis. *In* R. C. McKellar (ed.), *Enzymes of Psychrotrophs in Raw Food*, pp. 153–172. CRC Press, Boca Raton, FL.

109. Griffiths, M. W. 1989. Effect of temperature and milk fat on extracellular enzymes synthesis by psychrotrophic bacteria during growth in milk. *Milchwissenschaft* **44:**539–543.

110. McKellar, R. C. 1982. Factors influencing the production of extracellular proteinase by *Pseudomonas fluorescens*. *J. Appl. Bacteriol.* **53:**305-316.

111. Stead, D. 1987. Production of extracellular lipases and proteinases during prolonged growth of strains of psychrotrophic bacteria in whole milk. *J. Dairy Res.* **54:**535–543.

112. Andersson, R. E. 1980. Microbial lipolysis at low temperatures. *Appl. Environ. Microbiol.* **39:**36–40.

113. Bucky, A. R., P. R. Hayes, and D. S. Robinson. 1988a. Lipase production by a strain of *Pseudomonas fluorescens* in whole milk and skimmed milk. *Food Microbiol.* **3:**37–44.

114. Jooste, P. J., and T. J. Britz. 1986. The significance of flavobacteria as proteolytic psychrotrophs. *Milchwissenschaft* **41:**618–621.

115. Griffiths, M. W., and J. D. Phillips. 1984. Effect of aeration on extracellular enzyme synthesis by psychrotrophs growing in milk during refrigeration. *J. Food Prot.* **47:**697–702.

116. Murray, S. K., K. K. H. Kwan, B. J. Skura, and R. C. McKellar. 1983. Effect of nitrogen flushing on the production of proteinase by psychrotrophic bacteria in raw milk. *J. Food Sci.* **48:**1166–1169.

117. McKellar, R. C. (ed.). 1989a. *Enzymes of Psychrotrophs in Raw Foods*. CRC Press, Boca Raton, FL.

118. Fairbairn, D. J., and B. A. Law. 1986. Proteinases of psychrotrophic bacteria: their production, properties, effects, and control. *J. Dairy Sci.* **53:**139–177.

119. Stead, D. 1986. Microbial lipases: their characteristics, role in food spoilage and industrial use. *J. Dairy Res.* **53:**481–505.

120. Janzen, J. J., J. R. Bishop, and A. B. Bodine. 1982. Relationship of protease activity to shelf-life of skim and whole milk. *J. Dairy Sci.* **65:**2237–2240.

121. Juven, B. J., S. Gordin, and A. Laufer. 1979. Significance of psychrotrophic strains of *Serratia liquefaciens* in milk. *J. Food Prot.* **42:**938–941.

122. Mottar, J. 1984. Thermorésistance des bactériés psychrotrophes du lait cru et de leurs proteinases. *Le Lait* **64:**356–367.

123. Aylward, E. B., J. O'Leary, and B. E. Langlois. 1980. Effect of milk storage on cottage cheese yield. *J. Dairy Sci.* **63:**1819–1825.

124. Banks, J. G., and R. G. Board. 1987. Some factors influencing the recovery of yeasts and moulds from chilled foods. *Int. J. Food Microbiol.* **4:**197–206.

125. Ellis, B. R., and E. H. Marth. 1984. Growth of *Pseudomonas* or *Flavobacterium* in milk reduced yield of Cheddar cheese. *J. Food Prot.* **47:**713–716.

126. Hicks, C. L., C. Onuorah, J. O'Leary, and B. E. Langlois. 1986. Effects of milk quality and low temperature storage on cheese yield—a summation. *J. Dairy Sci.* **69:**649–657.

127. Hicks, C. L., M. Allanddin, B. E. Langlois, and J. O'Leary. 1982. Psychrotrophic bacteria reduce cheese yield. *J. Food Prot.* **45:**331–334.

128. Weber, G. H., and W. A. Broich. 1986. Shelf-life extension of cultured dairy foods. *Cult. Dairy Prod. J.* **21:**19–21, 23.

129. Yan, L., B. E. Langolois, J. O'Leary, and C. Hicks. 1983. Effect of storage conditions of grade A raw milk on proteolysis and cheese yield. *Milchwissenschaft* **38:**715–717.

130. Burlingame-Frey, J. P., and E. H. Marth. 1984. Changes in size of casein micelles caused by growth of psychrotrophic bacteria in raw skim milk. *J. Food Prot.* **47:**16–19.

131. Linden, G. 1986. Biochemical aspects. *IDF Bull.* **200:**17–21.

132. Diermayr, P., S. Kroll, and H. Klostermeyer. 1987. Mechanism of heat inactivation of a proteinase from *Pseudomonas fluorescens* biotype I. *J. Dairy Res.* **54:**51–60.

133. Leinmüller, R., and J. Christophersen. 1982. Beobachtungen zur Hitzeinaktivierung und Charakterisierung thermoresistenter Proteasen aus Pseudomonas fluorescens. *Milchwissenschaft* **37:**472–476.

134. Kumura, H., K. Mikawa, and Z. Saito. 1991. Influence of concomitant protease on the thermostability of lipase of psychrotrophic bacteria. *Milchwissenschaft* **46:**144–149.

135. Ekstrand, B. 1989. Antimicrobial factors in milk—a review. *Food Biotechnol.* **3:**105–126.

136. Reiter, B. 1985a. The biological significance and exploitation of the non-immunoglobulin protective proteins in milk: lysozyme, lactoferrin, lactoperoxidase, xanthine oxidase. *Int. Dairy Fed. Bull.* **191:**2–35.

137. Reiter, B. 1985b. The biological significance of the nonimmunoglobulin protective proteins in milk: lysozyme, lactoferrin, lactoperoxidase. *In* P. F. Fox (ed.), *Developments in Dairy Chemistry-3*, pp. 281–336. Elsevier Applied Science, New York.

138. Reiter, B., and B. G. Härnulv. 1982. The preservation of refrigerated and uncooled milk by its natural lactoperoxidase system. *Dairy Indust. Int.* **47:**13, 15, 17, 19.

139. Reiter, B., and B. G. Härnulv. 1984. Lactoperoxidase antibacterial system: natural occurrence, biological functions and practical application. *J. Food Prot.* **47:**724–732.

140. Reiter, B. 1985c. The lactoperoxidase system of bovine milk. *In* K. M. Pruitt and J. D. Tenovuo (ed.), *The Lactoperoxidase System. Chemistry and Biological Significance*, pp. 123–141. Marcel Dekker, New York.

141. Hernandez, M. C. M., B. W. van Markwijk, and H. J. Vreeman. 1990. Isolation and properties of lactoperoxidase from bovine milk. *Netherlands Milk Dairy J. 44:213–231.*

142. Law, B. A., and P. John. 1981. Effect of the lactoperoxidase bactericidal system on the formation of the electrochemical proton gradient in *E. coli. FEMS Microbiol. Lett.* **10:**67–70.

143. Pruitt, K. M., and B. Reiter. 1985. Biochemistry of perioxidase system: antimicrobial effects. *In* K. M. Pruitt and J. O. Tenovuo (eds.), *The Lactoperoxidase System: Chemistry and Biological Significance*, pp. 143–178. Marcel Dekker, New York.

144. Reiter, B. 1981. The impact of the lactoperoxidase system on the psychrotrophic microflora in milk. *In* T. A. Roberts, G. Hobbs, J. H. B. Christian, and N. Skovgaard (eds.), *Psychrotrophic Microorganisms in Spoilage and Pathogenicity*, pp. 73–85. Academic Press, New York.

145. Zajac, M., J. Gladys, M. Skarzynska, G. Härnulv, and L. Björck. 1983a. Changes in the bacteriological quality of raw milk stabilized by activation of its lactoperoxidase system and stored at different temperatures. *J. Food Prot.* **46:**1065–1068.

146. Zajac, M., J. Gladys, M. Skarzynska. G. Härnulv, and K. Eilersten. 1983b. Milk quality preservation by heat treatment or activation of the lactoperoxidase system in combination with refrigerated storage. *Milchwissenschaft* **38:**645–648.

147. Martinez, C. E., P. G. Mendoza, F. J. Alacron, and H. S. Garcia. 1988. Reactivation of the lactoperoxidase system during raw milk storage and its effect on the characteristics of pasteurized milk. *J. Food Prot.* **51:**558–561.

148. Kamau, D. N., S. Doores, and K. M. Pruitt. 1991. Activation of the lactoperoxidase system prior to pasteurization for shelf-life extension of milk. *Milchwissenschaft* **46:**213–214.

149. Ekstrand, B., W. M. A. Mullan, and A. Waterhouse. 1985. Inhibition of the antibacterial lactoperoxidase-thiocyanate–hydrogen peroxide system by heat-treated milk. *J. Food Prot.* **48:**494–498.

150. Ahrné, L., and L. Björck. 1985. Effect of lactoperoxidase system on lipoprotein lipase activity and lipolysis in milk. *J. Dairy Res.* **52:**513–520.

151. Lara, R., A. Mendoza, I. De La Cruz, and H. S. Garcia. 1987. Effect of the lactoperoxidase system on yield and characteristics of fresh type cheese. *Milchwissenschaft* **42:**773–775.

152. Zall, R. R., J. H. Chen, and D. J. Dzurec. 1983a. Effect of thiocyanate and hydrogen peroxide in cultured products. *Milchwissenschaft* **38:**264–266.

153. Zall, R. R., J. H. Chen, and D. J. Dzurec. 1983b. Effect of thiocyanate-lactoperoxidase-hydrogen peroxide system and farm heat treatment on the manufacturing of cottage cheese and Cheddar cheese. *Milchwissenschaft* **38:**203–206.

154. Kamau, D. N., and M. Kroger. 1984. Preservation of raw milk by treatment with hydrogen peroxide and by activation of the lactoperoxidase (LP) system. *Milchwissenschaft* **39:**658–661.

155. Earnshaw, R. G., J. G. Banks, D. Defrise, and C. Francotte. 1989. The preservation of cottage cheese by an activated lactoperoxidase system. *Food Microbiol.* **6:**285–288.

156. Zajac, M., L. Björck, and O. Claesson. 1981. Antibacterial effect of the lactoperoxidase system against *Bacillus cereus. Milchwissenschaft* **36:**417–418.

157. Beumer, R. R., A. Noomen, J. A. Marijs, and E. H. Kampelmacher. 1985. Antibacterial action of the lactoperoxidase system on *Campylobacter jejuni* in cow's milk. *Netherlands Milk Dairy J.* **39:**107–114.

158. Bibi, W., and M. R. Bachmann. 1990. Antibacterial effect of the lactoperoxidase–thiocyanate–hydrogen peroxide system on the growth of *Listeria* spp. in skim milk. *Milchwissenschaft* **45:**26–28.

159. El-Shenawy, M. A., H. S. Garcia, and E. H. Marth. 1990. Inhibition and inactivation of *Listeria monocytogens* by the lactoperoxidase system in raw milk, buffer or a semisynthetic medium. *Milchwissenschaft* **45:**638–641.

160. Kamau, D. N., S. Doores, and K. M. Pruitt. 1990. Enhanced thermal destruction of *Listeria monocytogens* and *Staphylococcus aureus* by the lactoperoxidase system. *Appl. Environ. Microbiol.* **56:**2711–2716.

161. Björck, L., O. Claesson, and W. Schulthess. 1979. The lactoperoxidase/thiocyanate/hydrogen peroxidase system as a temporary preservative for raw milk in developing countries. *Milchwissenschaft* **34:**726–729.

162. Ridley, S. C., and P. L. Shalo. 1990. Farm application of lactoperoxidase treatment and evaporative cooling for the immediate preservation of unprocessed milk in Kenya. *J. Food Prot.* **53:**592–597.

163. Härnulv, B. G., and C. Kandasamy. 1982. Increasing the keeping quality of milk by activation of its lactoperoxidase system. Results from Sri Lanka. *Milchwissenschaft* **37:**454–457.

164. Ellison, R. T., T. J. Giehl, and F. M. Laforce. 1988. Damage of the outer membrane of enteric Gram-negative bacteria by lactoferrin and transferrin. *Infect. Immun.* **56:**2774–2781.

165. Wasserfall, F., and M. Teuber. 1979. Action of egg white lysozyme on Clostridium tyrobutyricum. *Appl. Environ. Microbiol.* **38:**197–199.

166. Crawford, R. J. M. 1987. The use of lysozyme in the prevention of late blowing in cheese. *Int. Dairy Fed. Bull.* **216:**1–16.

167. Lodi, R. 1990. The use of lysozyme to control butyric acid fermentation. *Int. Dairy Fed. Bull.* **251:**51–54.

168. Bester, B. H., and S. H. Lombard. 1990. Influence of lysozyme on selected bacteria assocaited with Gouda cheese. *J. Food Prot.*, **53:**306–311.

169. El-Gendy, S. M., T. Nassib, H. Abed-El-Gellel, and N. E. H. Hanafy. 1980. Survival and growth of *Clostridium* species in the presence of hydrogen peroxide. *J. Food Prot.* **43:**431–432.

170. Griffiths, M. W., and J. D. Phillips. 1990a. Strategies to control the outgrowth of spores of psychrotrophic *Bacillus* spp. in dairy products. I. Use of naturally-occurring materials. *Milchwissenschaft* **45:**621–626.

171. Roginski, H., M. C. Broome, D. Hungerford, and M. W. Hickey. 1984b. Non-phage inhibition of group N streptocococci in milk 2. The effects of some inhibitory compounds. *Aust. J. Dairy Technol.* **39:**28–32.

172. Daeschel, M. A. 1989. Antimicrobial substances from lactic acid bacteria for use as food preservative. *Food Technol.* **43:**164–167.

173. Gibbs, P. A. 1987. Novel uses for lactic acid fermentation in food preservation. *J. Appl. Bacteriol. Sym. Suppl.* 51s–58s.

174. Martin, D. R., and S. E. Gilliland. 1980. Inhibition of psychrotrophic bacteria in refrigerated milk by lactobacilli isolated from yogurt. *J. Food Prot.* **43:**675–678.

175. Champagne, C. P., F. Girard, and N. Morin. 1990. Inhibition of psychrotrophic bacteria of raw milk by addition of lactic acid bacteria. *J. Food Prot.* **53:**400–403.

176. Kivanc, M. 1990. Antagonistic action of lactic cultures toward spoilage and pathogenic microorganisms in food. *Nahrung* **3:**273–277.

177. Batish, V. K., R. Lal, and S. Grover. 1991. Interaction of *S. lactis* subsp. *diacetylactis* DRC-1 with *Aspergillus parasiticus* and *A. fumigatus* in milk. *Cult. Dairy Prod. J. **26**:13–14.*

178. Klaenhammer, T. R. 1988. Bacteriocin of lactic acid bacteria. *Biochimie* **70:**337–349.

179. Spelhaug, S. R., and S. K. Harlander. 1989. Inhibition of foodborne bacterial pathogens by bacteriocins from *Lactococcus lactis* and *Pediococcus pentosaceous. J. Food Prot.* **52:**856–862.

180. Davidson, P. M., and V. K. Juneja. 1990. Antimicrobial agents. *In* A. L., Branen, P. M. Davidson, and S. Salminen (eds.), *Food Additives*, pp. 83–137. Marcel Dekker, New York.

181. Delves-Broughton, J. 1990a. Nisin and its application as a food preservative. *J. Soc. Dairy Technol.* **43:**73–76.

182. Delves-Broughton, J. 1990b. Nisin and its uses as a food preservative. *Food Technol.* **44:**100, 102, 104, 106, 108, 111–112, 117.

183. Henning, S., R. Metz, and W. P. Hammes. 1986. Studies on the mode of action of nisin. *Int. J. Food Microbiol.* **3:**121–134.

184. Hurst, A. 1981. *Nisin. Adv. Appl. Microbiol.* **27:**85–123.

185. Hurst, A. 1983. Nisin and other inhibitory substances from lactic acid bacteria. *In* A. L. Branen and P. M. Davidson (ed.), *Antimicrobials in Foods*, pp. 327–351. Marcel Dekker, New York.

186. Magdoub, M. N. I., A. E. Shehata, Y. A. El-Samragy, and A. A. Hassan. 1984. Interaction of heat shock, magnesium, L-alanine, β-alanine and nisin in spore germination of some psychrotrophic *Bacillus* strains in milk. *Milchwissenschaft* **39:**159–162.

187. Somers, E. B., and S. L. Taylor. 1987. Antibotulinal effectiveness of nisin in pasteurized process cheese spreads. *J. Food Prot.* **50:**842–848.

188. Anonymous. 1985. Nisin preservation of chilled desserts. *Dairy Indust. Int.* **50:**41, 43.

189. Phillips, J. D., M. W. Griffiths, and D. D. Muir. 1983. Effect of nisin on the shelf-life of pasteurized double cream. *J. Soc. Dairy Technol.* **36:**17–21.

190. Stadhouders, J. 1990b. Alternate methods of controlling butyric acid fermentation in cheese. *Int. Dairy Fed. Bull.* **251:**55–58.

191. Bhunia, M., C. Johnson, B. Ray, and N. Kalchayanand. 1991. Mode of action of pediocin AcH from *Pediococcus acidilactici* H. on sensitive bacterial strains. *J. Appl. Bacteriol.* **70:**25–33.

192. Pucci, M. J., E. V. Vedamuthu, B. S. Kunka, and P. A. Vandenbergh. 1988. Inhibition of *Listeria monocytogens* by using bacteriocin PA-1 produced by *Pediococcus acidilactici* PAC 1.0. *Appl. Environ. Microbiol.* **54:**2349–2353.

193. Barefoot, S. F., and T. R. Klaenhammer. 1983. Detection and activity of lactacin B, a bacteriocin produced by *Lactobacillus acidophilus. Appl. Environ. Microbiol.* **45:**1808–1815.

194. Pulusani, S. R., D. R. Rao, and G. R. Sunki. 1979. Antimicrobial activity of lactic cultures: partial purification and characterization of antimicrobial compounds(s) produced by *Streptococcus thermophilus. J. Food Sci.* **44:**575–578.

195. Anand, S. K., R. A. Srinivasan, and L. K. Rao. 1984. Antibacterial activity associated with *Bifidobacterium bifidum. Cult. Dairy Prod. J.* **19:**6–8.

196. Salih, M. A., W. E. Sandine, and J. W. Ayres. 1990. Inhibitory effects of Microgard® on yogurt and cottage cheese spoilage organisms. *J. Dairy Sci.* **73:**887–893.

197. Al-Zoreky, N., J. W. Ayres, and W. E. Sandine. 1991. Antimicrobial activity of Microgard® against food spoilage and pathogenic microorganisms. *J. Dairy Sci.* **74:**758–763.

198. Morris, H. A., and H. B. Castberg. 1980. Control of surface growth on blue cheese using pimaricin. *Cult. Dairy Prod. J.* **15:**21–23.

199. de Ruig, W. G., and G. van den Berg. 1985. Influence of the fungicides sorbate and natamycin in cheese coatings on the quality of the cheese. *Netherlands Milk Dairy J.* **39:**165–172.

200. Stadhouders, J. 1990a. Prevention of butyric acid fermentation by the use of nitrate. *Int. Dairy Fed. Bull.* **251:**40–46.

201. Bullerman, L. B. 1983. Effects of potassium sorbate on growth and aflatoxin production by *Aspergillus parasiticus* and *Aspergillus flavus. J. Food Prot.* **46:**940–942.

202. Bullerman, L. B. 1984. Effects of potassium sorbate on growth and patulin production by *Penicillium patulum* and *Penicillium roqueforti. J. Food Prot.* **47:**312–316.

203. Bullerman, L. B. 1985. Effects of potassium sorbate on growth and ochratoxin production by *Aspergillus ochraceus* and *Penicillium* species. *J. Food Prot.* **48:**162–165.

204. Liewen, M. B., and E. H. Marth. 1984. Inhibition of penicillia and aspergilli by potassium sorbate. *J. Food Prot.* **47:**554–556.

205. Yousef, A. E., and E. H. Marth. 1981. Growth and synthesis of aflatoxin by *Aspergillus parasiticus* in the presence of sorbic acid. *J. Food Prot.* **44:**736–741.

206. Tsai, W. Y. J., M. B. Liewen, and L. B. Bullerman. 1988. Toxicity and sorbate sensitivity of molds isolated from surplus commodity cheeses. *J. Food Prot.* **51:**457–462.

207. Liewen, M. B., and E. H. Marth. 1985a. Growth and inhibition of microorganisms in the presence of sorbic acid: a review. *J. Food Prot.* **48:**364–375.

208. Ahmad, S., and A. L. Branen. 1981. Inhibition of mold growth by butylated hyroxyanisole. *J. Food Sci.* **46:**1059–1063.

209. Liewen, M. B., and E. H. Marth 1985b. Production of mycotoxins by sorbate-resistant molds. *J. Food Prot.* **48:**156–157.

210. Gilliland, S. E., and H. R. Ewell. 1983. Influence of combinations of *Lactobacillus lactis* and potassium sorbate on growth of psychrotrophs in raw milk. *J. Dairy Sci.* **66:**974–980.

211. Mistry, V. V., and F. V. Kosikowski. 1985a. Influence of potassium sorbate and hydrogen peroxide on psychrotrophic bacteria in milk. *J. Dairy Sci.* **68:**605–608.

212. King, J. S., and L. A. Mabbitt. 1982. Preservation of raw milk by the addition of carbon dioxide. *J. Dairy Res.* **49:**439–447.

213. King, J. S., and L. A. Mabbitt. 1987. The use of carbon dioxide for the preservation of milk. *In* R. G. Board, M. C. Allwood, and J. G. Banks (eds.), *Preservatives in the Food Pharmaceutical and Environmental Industries*, pp. 35–43. Blackwell Scientific Publications, Boston.

214. Law, B. A., and L. A. Mabbitt. 1983. New methods for controlling the spoilage of milk and milk products. *In* T. A. Roberts and F. A. Skinner (eds.), *Food Microbiology: Advances and Prospects*, pp. 131–150. Academic Press, New York.

215. Guirguis, A. H., M. W. Griffiths, and D. D. Muir. 1984. Spore forming bacteria in milk. II. Effect of carbon dioxide addition on heat activation of spores of *Bacillus* species. *Milchwissenschaft* **39:**144–146.

216. Coghill, D. M., I. D. Mutzelburg, and S. J. Birch. 1982. Effect of thermization on the bacteriological and chemical quality of milk. *Aust. J. Dairy Technol.* **37:**48–50.

217. Gilmour, A., R. S. Machelhinney, D. E. Johnston, and R. J. Murphy. 1981. Thermisation of milk. Some microbiological aspects. *Milchwissenschaft* **36:**457–461.

218. Humbert, E. S., J. N. Campbell, G. Blankengagel, and A. Gebre-Egziabher. 1985. Extended storage of raw milk. II. The role of thermization. *Can. Inst. Food Sci. Technol. J.* **18:**302–305.

219. Muir, D. D. 1986. The yield and quality of Cheddar cheese produced from thermised milk. *Dairy Indust. Int.* **51:**31, 32, 34–35.

220. Johnston, D. E., R. J. Murphy, A. Gilmore, J. T. M. McGuiggan, M. T. Rowe, and W. M. A. Mullan. 1987. Manufacture of Cheddar cheese from thermized cold-stored milk. *Milchwissenschaft* **42:**226–231.

221. Johnston, D. E., R. J. Murphy, A. Gilmore, J. T. M. McGuiggan, and W. M. A. Mullan. 1988. Maturation and quality of Cheddar cheese from thermized, cold stored milk. *Milchwissenschaft* **43:**211–215.

222. West, I. G., M. W. Griffiths, J. D. Phillips, A. W. M. Sweetsur, and D. D. Muir. 1986. Production of dried skim milk from thermised milk. *Dairy Indust. Int.* **51:**33–34.

223. Gilmour, A., and J. T. M. McGuiggan. 1989. Thermization of milk. Safety aspects with respect to *Yersinia enterocolitica. Milchwissenschaft* **44:**418–422.

224. Stephaniak, L., E. Zakrzewski, and T. Sorhaug. 1991. Inactivation of heat-stable proteinase from *Pseudomonas fluorescens* Pl at pH 4.5 and 55°C. *Milchwissenschaft* **46:**139–142.

225. Guamis, B., T. Huerta, and E. Garay. 1987. Heat-inactivation of bacterial proteases in milk before UHT-treatment. *Milchwissenschaft* **42:**651–653.

226. Senyk, G. F., R. R. Zall, and W. F. Shipe. 1982. Pasteurization heat treatment to inactivate lipase and control bacterial growth in raw milk. *J. Food Prot.* **45:**513–515.

227. Fitz-Gerald, C. H., H. C. Deeth, and D. M. Coghill. 1982. Low temperature inactivation of lipases from psychrotrophic bacteria. *Aust. J. Dairy Technol.* **37:**51–54.

228. Griffiths, M. W., and J. D. Phillips. 1990b. Strategies to control the outgrowth of spores of psychrotrophic *Bacillus* spp. in dairy products. II. Use of heat treatments. *Milchwissenschaft* **45:**719–721.

229. Premaratne, R. J., and M. A. Cousin. 1991b. Microbiological analysis and starter culture growth in retentates. *J. Dairy Sci.* **74:**3284–3292.

230. Sillén, G. 1987. Modern bactofuges in the dairy industry. *Dairy Indust. Int.* **52:**27–29.

231. Waes, G., and A. Van Heddeghem. 1990. Prevention of butyric acid fermentation by bacterial centrifugation of the cheese milk. *Int. Dairy Fed. Bull.* **251:**47–50.

232. Lembke, A., and E. Wartenberg. 1982. Verbesserung von UHT-Milch durch Einsatz einer Baktofuge. *Milchwissenschaft* **37:**737–741.

233. Schalm, O. W., E. J. Carrol, and N. C. Jain. 1971. *Bovine Mastitis.* Lea & Febiger, Philadelphia.

234. Kitchen, B. J. 1981. Review of the progress of dairy science: Bovine Mastitis: Milk compositional changes and related diagnostic tests. *J. Dairy Res.* **48:**167–188.

235. Schultz, L. H. 1977. Somatic cells in milk—physiological aspects and relationship to amount and composition of milk. *J. Food Prot.* **40:**125–131.

236. Jurczak, M. E., and A. Sciubisz. 1981. Studies on the lipolytic changes in milk from cows with mastitis. *Milchwissenschaft* **36:**217–219.

237. Luhtala, A., and M. Antila. 1968. Lipases and lipolysis of milk. *Fette Seifen. Anstrichmittel.* **70:**280–288.

238. Weihrauch, J. C. 1988. Lipids in milk: deterioration. *In* N. P. Wong, R. Jenness, M. Keeney, and E. H. Marth (eds.), *Fundamentals of Dairy Chemistry*, 2nd edit., pp. 219–220. Van Nostrand Reinhold, New York.

239. Salih, A. M. A., and M. Anderson. 1979. Observations on the influence of high cell count on lipolysis in bovine milk. *J. Dairy Res.* **46:**453–462.

240. Barbano, D. M., Rudan, M. A., and R. R. Rasmessen. 1989. Influence of milk composition and somatic cell and psychrotrophic bacteria counts on ultrafiltration flux. *J. Dairy Sci.* **72:**1118–1123.

241. International Dairy Federation. 1984. Recommended methods for somatic cell counts in milk. Document 168, Ed. IDF, Bruxelles.

242. Anonymous. 1981a. Outbreak of enteritis associated with raw milk—Kansas. *Morbid. Mortal. Wkly. Rep.* **30:**218.

243. Barza, M. 1985. Listeriosis in milk. *N. Engl. J. Med.* **312:**438–440.

244. Vasavada, P. C. 1988b. A lesson in listeriosis. *Dairy Herd Management.* **29:**40.

245. Black, R. E., J. Jackson, T. Tasi, M. Medvesky, M. Shayegoni, J. C. Feeley, K. I. F. MacLeod, and A. M. Wakalee. 1978. Epidemic *Yersinia enterocolitica* infection due to contaminated chocolate milk. *N. Engl. J. Med.* **298:**76–79.

246. Vasavada, P. C., and W. C. Mahanna. 1984. The campylobacter connection. *Dairy Herd Management* **21:**22–24.

247. Lecos, C. 1986b. A closer look at dairy safety. *Dairy Food Sanit.* **6:**240–242.

248. Hayes, P. S., J. L. Feeley, L. M. Graves, G. W. Ajello, and D. W. Fleming. 1986. Isolation of *Listeria monocytogenes* from raw milk. *Appl. Environ. Microbiol.* **51:**438.

249. Gray, M. L., and A. H. Killinger. 1966. *Listeria monocytogens* and listeric infections. *Bacteriol. Rev.* **30:**309–382.

250. Aulisio, C. C. G., I. J. Mehlman, and A. C. Sanders. 1980. Alkali methods for rapid recovery of *Yersinia enterocolitica* and *Yersinia pseudotuberculosis* from foods. *Appl. Environ. Microbiol.* **39:**135–140.

251. Larkin, L. L., P. C. Vasavada, and E. H. Marth. 1991. Incidence of *Yersinia enterocolitica* in raw milk as related to its quality. *Milchwissenschaft* **46:**500–502.

252. Moustafa, M. K., A. A. H. Ahmed, and E. H. Marth. 1983. Occurrence of *Yersinia enterocolitica* in raw and pasteurized milk. *J. Food Prot.* **46:**276–278.

253. Schiemann, D. A. 1982. Development of a two-step enrichment procedure for recovery of *Yersinia enterocolitica* from food. *Appl. Environ. Microbiol.* **43:**14–27.

254. Stern, N. J., M. D. Pierson, and A. M. Kotula. 1980. Growth and competitive nature of *Yersinia enterocolitica* in whole milk. *J. Food Sci.* **45:**972–974.

255. Vasavada, P. C., L. Larkin, and E. H. Marth. 1985. Incidence of *Yersinia enterocolitica* in relation to quality of milk. *J. Dairy Sci.* **68**(*Suppl. 1*):82.

256. Vidon, D. J. M., and C. L. Delmas. 1981. Incidence of *Yersinia enterocolitica* in raw milk in eastern France. *Appl. Environ. Microbiol.* **41:**355–359.

257. Walker, S. J., and A. Gilmour. 1990. Production of enterotoxin by *Yersinia* species isolated from milk. *J. Food Prot.*, **53:**751.

258. Francis, D. W., P. L. Spaulding, and J. Lovett. 1980. Enterotoxin production and thermal resistance of *Yersinia enterocolitica* in milk. *Appl. Environ. Microbiol.* **40:**174–176.

259. Larkin, L. L. P. C. Vasavada, and E. H. Marth. 1991. Incidence of *campylobacter jejuni* in raw milk as related to its quality. *Milchwissenschaft* **46:**428–430.

260. Oosterom, J., G. B. Engel, R. Peters, and R. Pot. 1982. *Campylobacter jejuni* in cattle and raw milk in the Netherlands. *J. Food Prot.* **45:**1212–1213.

261. Vasavada, P. C., L. Larkin, and E. H. Marth. 1986. Incidence of *Campylobacter jejuni* in relation to quality of raw milk. *J. Dairy Sci.* **69**(*Suppl. 1*):67.

262. Wyatt, C. J., and E. M. Timm. 1982. Occurrence and survival of Campylobacter jejuni in milk and turkey. *J. Food Prot.* **45:**1218–1220.

263. Doyle, M. P., and D. J. Roman. 1982. Prevalence and survival of *Campylobacter jejuni* in unpasteurized milk. *Appl. Environ. Microbiol.* **44:**1154.

264. Kornacki, J. L., and E. H. Marth. 1982. Foodborne illness caused by *Escherichia coli*: a review. *J. Food Prot.* **45:**1051.

265. Anonymous. 1985a. Hemorrhagic colitis in a nursing home. Ontario, Canada. *Dis. Wkly. Rep.* **11:**169.

266. Maier, R., J. G. Wells, R. C. Swenson, and I. J. Mehlman. 1973. An outbreak of enteropathogenic *Escherichia coli* foodborne disease traced to imported French cheese. *Lancet* **ii:**1376.

267. Foster, E. M. 1988. *Escherichia coli. Food Technol.* **40:**20.

268. Karmali, M. A., B. T. Steele, M. Petrie, and C. Lin. 1983. Sporadic cases of hemolytic-uremic syndrome associated with fecal cytotoxin and cytotoxin producing *Escherichia coli* in stools. *Lancet* **1:**619.

269. Martin, M. L., L. D. Shipman, J. G. Wells, M. E. Potter, K. Hedlberg, I. K. Wachsmuth, R. V. Tauxe, J. P. Davis, J. Arnoldi, and J. Tilleli. 1986. Isolation of an *Escherichia coli* 0157:H7 from dairy cattle associated with two cases of hemolytic uremic syndrome. *Lancet* **ii:**1043.

270. Harmon, S. M. 1984. *Bacillus cereus In Bacteriological Analytical Manual*, 6th edit., pp. 16.01–16.08. Association of Official Analytical Chemists, Arlington, VA.

271. Turnbull, P. C. B. 1981. *Pharmacol. Ther.* **13:**453–505.

272. Archer, D. L., and J. E. Kvenberg. 1985. Incidence and cost of foodborne diarrheal diseases in the United States. *J. Food Prot.* **48:**887–894.

273. Lecos, C. 1986a. Of microbes and milk: probing America's worst *Salmonella* outbreak. *Dairy Food Sanit.* **6:**136–140.

274. James, S. M., L. Fannin, B. A. Agree, B. Mall, E. Parker, J. Vogt, G. Run, J. Williams, L. Gieh, C. Salminen, T. Pendernot, S. B. Werner, and J. Chin. 1985. Listeriosis outbreak associated with Mexican-style cheese—California. *Morbid. Mortal. Wkly. Rep.* **34:**357–359.

275. Todd, E. C. 1985. Economic loss from foodborne disease and non-illness related recalls because of mishandling by food processors. *J. Food Prot.* **48:**621–633.

276. Morris, H. A., and S. R. Tatini. 1987. Progress in cheese technology—safety aspects with microbial emphasis. *Milk—the Vital Force*, D. Reidel, pp. 187–194.

277. Stoloff, L. 1980b. Aflatoxins in perspective. *J. Food Prot.* **43:**226–230.

278. Chapman, H. R., and M. E. Sharpe. 1990. Microbiology of Cheese. *In* R. K. Robinson (ed.), *Dairy Microbiology*, 2nd edit., pp. 213–215. Elsevier Applied Science, London.

279. Applebaum, R. A., R. E. Brackett, D. W. Wiseman, and E. H. Marth. 1982. Aflatoxin: toxicity to dairy cattle and occurrence in milk and milk products—a review. *J. Food Prot.* **45:**752–777.

280. Bullerman, L. B. 1981. Public health significance of molds and mycotoxins in fermented dairy products. *J. Dairy Sci.* **64:**2439–2452.

281. Scott, P. M. 1978. Mycotoxins in feeds and ingredients and their origin. *J. Food Prot.* **41:**385–398.

282. Scott, P. M. 1984. Effect of food processing on mycotoxins. *J. Food Prot.* **47:**488–499.

283. Scott, P. M. 1989. Mycotoxigenic fungal contaminants of cheese and other dairy products. *In* H. P. van Egmond (ed.), *Mycotoxins in Dairy Products*, pp. 193–259. Elsevier Applied Science, New York.

284. Edwards, S. T., and W. E. Sandine. 1981. Public health significance of amines in cheese. *J. Dairy Sci.* **64:**2431–2438.

285. Elsworth, J. D., J. Glover, G. P. Reynolds, M. Sandler, A. J. Lees, P. Phaupradit, K. M. Shaw, G. M. Stern, and P. Kumar. 1978. Deprenyl administration in man: a selective monoamine oxidase B inhibitor without the "cheese effect." *Psychropharmacology* **57:**33.

286. Marley, E., and B. Blackwell. 1970 Interactions of monoamine oxidase inhibitors, amines and food stuffs. *Adv. Pharmacol. Chemother.* **8:**185.

287. Douglas, M. H. G., J. Huisman, and J. P. Nater. 1967. Histamine intoxications after cheese. *Lancet* **ii:**1361.

288. Voight, M. N., and R. R. Eitenmiller. 1977. Production of tyrosine and histidine decarboxylase in dairy related bacteria. *J. Food Prot.* **40:**241–245.

289. Vadillo, S., M. J. Paya, M. T. Cutuli, and G. Suárez. 1987. Mycoflora of milk after several types of pasteurization. *Le Lait* **67:**265–273.

290. Goto, T. 1990. Mycotoxins: current situation. *Food Rev. Int.* **6:**265–290.

291. van Egmond, H. P. 1983. Mycotoxins in dairy products. *Food Chem.* **11:**289–307.

292. van Egmond, H. P., and W. E. Paulsch. 1986. Mycotoxins in milk and milk products. *Netherlands Milk Dairy J.* **40:**175–188.

293. Hsieh, D. P. H. 1989. Carcinogenic potential of mycotoxins in foods. *In* S. L. Taylor and R. A. Scanlan (eds.), *Food Toxicology. A Perspective on the Relative Risks*, pp. 11–30. Marcel Dekker, New York.

294. Scott, P. M. 1981. Toxins of *Penicillium* species used in cheese manufacture. *J. Food Prot.* **44:**702–710.

295. van Egmond, H. P. (ed.). 1989b. *Mycotoxins in Dairy Products.* Elsevier Applied Science, New York.

296. Applebaum, R. A., and E. H. Marth. 1982b. Inactivation of aflatoxin M_1 in milk using hydrogen peroxide and hydrogen peroxide plus riboflavin or lactoperoxidase. *J. Food Prot.* **45:**557–560.

297. Applebaum, R. A., and E. H. Marth. 1982a. Fate of aflatoxin M_1 in cottage cheese. *J. Food Prot.* **45:**903–904.

298. Schreeve, B. J., D. S. P. Patterson, and B. A. Roberts. 1979. The 'carry-over' of aflatoxin, ochratoxin and zearaleonone from naturally contaminated feed to tissues, urine and milk of dairy cows. *Food Cosmet. Toxicol.* **17:**151–152.

299. Patterson, D. S. P., E. M. Glancy, and B. A. Roberts. 1980. The 'carry over' of aflatoxin M_1 into the milk of cows fed rations containing a low concentration of aflatoxin B_1. *Food Cosmet. Toxicol.* **18:**35–37.

300. Munksgaard, L., J. Larsen, H. Werner, P. E. Andersen, and B. T. Viuf. 1987. Carry over of aflatoxin from cows' feed to milk and milk products. *Milchwissenschaft* **42:**165–167.

301. Price, R. L., J. H. Paulson, O. G. Lough, C. Gingg, and A. G. Kurtz. 1985. Aflatoxin conversion by dairy cattle consuming naturally-contaminated whole cottonseed. *J. Food Prot.* **48:**11–15.

302. Frobish, R. A., B. O. Bradley, D. D. Agner, P. E. Long-Bradley, and H. Hairston. 1986. Aflatoxin residues in milk of dairy cows after ingestion of naturally contaminated grain. *J. Food Prot.* **49:**781–785.

303. Corbett, W. T., C. F. Brownie, S. B. Hagler, and W. H. Hagler. 1988. An epidemiological investigation associating aflatoxin M_1 with milk production in dairy cattle. *Vet. Hum. Toxicol.* **30:**5–8.

304. Fremy, J. M., J. P. Gautier, M. P. Herry, C. Terrier, and C. Calet. 1987. Effect of ammoniation on the 'carry-over' of aflatoxins into bovine milk. *Food Add. Contamin.* **5:**39–44.

305. Park, K. Y., and L. B. Bullerman. 1983a. Effect of cycling temperature on aflatoxin production by *Aspergillus parasiticus* and *Aspergillus flavus* in rice and Cheddar cheese. *J. Food Sci.* **48:**889–896.

306. Park, K. Y., and L. B. Bullerman. 1983b. Effects of substrate and temperature on aflaxtoxin production by *Aspergillus parasiticus* and *Aspergillus flavus*, *J. Food Prot.* **46:**178–184.

307. El-Gendy, S. M., and E. H. Marth. 1981. Growth and aflatoxin production by *Aspergillus parasiticus* in the presence of *Lactobacillus casei. J. Food Prot.* **44:**211–212.

308. Mohran, M. A., S. E. Megalla, and M. R. Said. 1984. Effect of aflatoxin B_1 on the proteolytic activity of some lactic-acid bacteria. *Mycopathologia* **86:**99–101.

309. Veringa, H. A., G. van den Berg, and C. B. G. Daamen. 1989. Factors affecting the growth of Aspergillus versicolor and the production of sterigmatocystin on cheese. *Netherlands Milk Dairy J.* **43:**311–326.

310. Northholt, M. D., H. P. van Egmond, P. Soentoro, and E. Deijll. 1980. Fungal growth and the presence of sterigmatocystin in hard cheese. *J. Assoc. Off. Anal. Chem.* **63:**115–119.

311. Northolt, M. D., H. P. van Egmond, and W. E. Paulsch. 1979b. Penicillic acid production by some fungal species in relation to water activity and temperature. *J. Food Prot.* **42:**476–484.

312. Northolt, M. D., H. P. van Egmond, and W. E. Paulsch. 1979a. Ochratoxin A production by some fungal species in relation to water activity and temperature. *J. Food Prot.* **42:**485–490.

313. Engle, G., and M. Teuber. 1989. Toxic metabolites from fungal cheese starter cultures (*Penicillium camemberti* and *Penicillium roqueforti*). *In* H. P. van Egmond (ed.), pp. 163–192. *Mycotoxins in Dairy Products*, Elsevier Applied Science, New York.

314. Engel, G., K. E. von Milczewski, D. Prokopek, and M. Teuber. 1982. Strain-specific synthesis of mycophenolic acid by *Penicillium roqueforti* in blue-veined cheese. *Appl. Environ. Microbiol.* **43:**1034–1040.

315. LeBars, J. 1979. Cyclopiazonic acid production by *Penicillium camemberti* Thom and natural occurrence of this mycotoxin in cheese. *Appl. Environ. Microbiol.* **38:**1052–1055.

316. Geisen, R., E. Gleen, and L. Leistner. 1990. Two *Penicillium camemberti* mutants affected in the production of cyclopiazonic acid. *Appl. Environ. Microbiol.* **56:**3587-3590.

317. Wiseman, D. W., R. S. Applebaum, R. E. Brackett, and E. H. Marth. 1983. Distribution and resistance to pasteurization of aflatoxin M_1 in naturally contaminated whole milk, cream and skim milk. *J. Food Prot.* **46:**530–532.

318. Wiseman, D. W., and E. H. Marth. 1983c. Heat and acid stability of aflatoxin M_1 in naturally and artificially contaminated milk. *Milchwissenschaft* **38:**464–466.

319. Brackett, R. E., and E. H. Marth. 1982a. Fate of aflatoxin M_1 in Cheddar cheese and in process cheese spread. *J. Food Prot.* **45:**549–552.

320. Brackett, R. E., R. A. Applebaum, D. W. Wiseman, and E. H. Marth. 1982. Fate of aflatoxin M_1 in Brick and Limburger-like cheese. *J. Food Prot.* **45:**553–556.

321. Brackett, R. E., and E. H. Marth. 1982b. Fate of aflatoxin M_1 in Parmesan and Mozzarella cheese. *J. Food Prot.* **45:**597–600.

322. Wiseman, D. W., and E. H. Marth. 1983a. Behavior of aflatoxin M_1 during manufacture and storage of Queso Blanco and bakers' cheese. *J. Food Prot.* **46:**910–913.

323. Blanco, J. L., L. Domínquez, E. Gómez-Lucía, J. F. F. Garayzabal, J. Goyache, and G. Suárez. 1988b. Behavior of aflatoxin during the manufacture, ripening and storage of manchego-type cheese. *J. Food Sci.* **53:**1373–1376.

324. Wiseman, D. W., and E. H. Marth. 1983b. Behavior of aflatoxin M_1 in yogurt, buttermilk and kefir. *J. Food Prot.* **46:**115–118.

325. Wiseman, D. W., and E. H. Marth. 1983d. Stability of aflatoxin M_1 during manufacture and storage of a butter-like spread, non-fat dried milk and dried buttermilk. *J. Food Prot.* **46:**633–636.

326. Megalla, S. E., and M. A. Mohran. 1984. Fate of aflatoxin B_1 in fermented dairy products. *Mycopathologia* **88:**27–29.

327. Yousef, A. E., and E. H. Marth. 1989. Stability and degradation of aflatoxin M_1. *In* H. P. van Egmond (ed.), *Mycotoxins in Dairy Products*, pp. 127–161. Elsevier Applied Science, New York.

328. Doyle, M. P., R. S. Applebaum, R. E. Brackett, and E. H. Marth. 1982. Physical, chemical, and biological degradation of mycotoxins in foods and agricultural commodities. *J. Food Prot.* **45:**964–971.

329. Yousef, A. E., and E. H. Marth. 1986. Use of ultraviolet energy to degrade aflatoxin M_1 in raw or heated milk with and without added peroxide. *J. Dairy Sci.* **69:**2243–2247.

330. Yousef, A. E., and E. H. Marth. 1985. Degradation of aflatoxin M_1 in milk by ultraviolet energy. *J. Food Prot.* **48:**697–698.

331. Samarajeewa, U., A. C. Sen, M. D. Cohen, and C. I. Wei. 1990. Detoxification of aflatoxins in foods and feeds by physical and chemical methods. *J. Food Prot.* **53:**489–501.

332. Price, R. L., O. G. Lough, and W. H. Brown. 1982. Ammoniation of whole cottonseed at atmospheric pressure and ambient temperature to reduce aflatoxin M_1 in milk. *J. Food Prot.* **45:**341–344.

333. Ray, L. L., and L. B. Bullerman. 1982. Preventing growth of potentially toxic molds using antifungal agents. *J. Food Prot.* **45:**953–963.

334. Labuza, T. P. 1983. Regulation of mycotoxins in food. *J. Food Prot.* **46:**260–265.

335. Stoloff, L. 1980a. Aflatoxin control: Past and present. *J. Assoc. Off. Anal. Chem.* **63:**1067–1073.

336. van Egmond, H. P. 1989a. Current situations on regulations for mycotoxins. Overview of tolerances and status of standard methods of sampling and analysis. *Food Add. Contam.* **6:**139–188.

337. Bullerman, L. B. 1980. Incidence of mycotoxic molds in domestic and imported cheeses. *J. Food Safety* **2:**47–58.

338. Turcksess, M. W., and S. W. Page. 1986. Examination of imported cheeses for aflatoxin M_1. *J. Food Prot.* **49:**632–633.

339. Kivanc, M. 1990. Mold growth and presence of aflatoxin in some Turkish cheeses. *J. Food Safety.* **10:**287–294.

340. Blanco, J. L., L. Domínguez, E. Gómez-Lucía, J. F. F. Garayzabal, J. A. Garciía, and G. Suárez. 1988a. Presence of aflatoxin M_1 in commercial ultra-high temperature treated milk. *Appl. Environ. Microbiol.* **54:**1622–1623.

341. Wood, G. E. 1989. Aflatoxins in domestic and imported foods and feeds. *J. Assoc. Off. Anal. Chem.* **72:**543–548.

342. Cogan, T. M. 1983. Some aspects of the metabolism of dairy starter cultures. *Irish J. Food Sci. Technol.* **7:**1–13.

343. Cogan, T. M., and C. Daly. 1987. Cheese starter cultures. *In* P. F. Fox (ed.), *Cheese—Chemistry, Physics, and Microbiology*, Vol. 1, 2nd edit., pp. 179–249. Elsevier Applied Science, London.

344. Gilliland, S. E. 1985. *Bacterial starter cultures for foods.* CRC Press, Boca Raton, FL.

345. Huggins, A. R. 1984. Progress in dairy starter culture technology. *Food Technol.* **38:**41–50.

346. Shahani, K. M., and B. A. Friend. 1983. Properties of and prospects for cultured dairy foods. *In* T. A. Roberts and E. A. Skinner (eds.), *Food Microbiol.: Advances and Prospects*, pp. 257–269. Academic Press, New York.

347. Lawrence, R. C., and H. A. Heap, 1986. *In* Special addresses given at IDF Annual Sessions—Auckland, New Zealand, 1985, Doc. No. 199, International Dairy Federation, Brussels, Belgium, pp. 4–20.

348. Thunnell, R. K., and W. E. Sandine. 1985. Types of starter cultures. *In* S. E. Gilliland (ed.), *Bacterial Starter Cultures for Foods*, pp. 127–144. CRC, FL.

349. Stadhouders, J. 1986. The control of cheese starter activity. *Netherlands Milk Dairy J.* **40:**155–173.

350. Park, Y. H., and L. L. McKay. 1982. Distinct galactose phosphoenol-pyruvate dependent phosphotransferase system in *Streptococcus lactis*. *J. Bacteriol.* **149:**420–425.

351. Crow, V. L., G. P. Davey, L. E. Pearce, and T. D. Thomas. 1983. Plasmid linkage of the D-tagatose-6-phosphate pathway in *Streptococcus lactis*. Effects of lactose and galactose metabolism. *J. Bacteriol.* **153:**76–83.

352. Frank, J. F., and E. H. Marth. 1988. Fermentations. *In*: N. P. Wong, R. Jenness, M. Keeney and E. H. Marth. (eds.), *Fundamentals of Dairy Chemistry*, 2nd edit., pp. 665–738. Van Nostrand Reinhold, New York.

353. Steffen, C., E. Fluedciger, J. D. Bosset, and M. Ruegg. 1987. Swiss-type varieties. *In* P. F. Fox (ed.), *Cheese—Chemistry, Physics and Microbiology*, pp. 93–120. Elsevier Applied Science, London.

354. Ibrahim, G. F. 1978. Inhibition of *Staphylococcus aureus* under simulated cheese making conditions. *Aust J. Dairy Technol.* **33:**102–108.

355. Norholt, M. D. 1984. Growth and inactivation of pathogenic microorganisms during manufacture and storage of fermented milk products. *Netherlands Milk Dairy J.* **38:**135–150.

356. Lewis, J. E. 1987. *Cheese Starters—Development and Application of the Lewis System.* Elsevier Applied Science, London.

357. Tamime, A. Y., and R. K. Robinson, 1985. *Yogurt—Science and Technology.* Pergamon Press, Oxford.

358. Richardson, G. H., C. T. Cheng, and R. Young. 1977. Lactic bulk culture system utilizing a whey based bacteriophage inhibitory medium and pH control I. Applicability to American style cheese. *J. Dairy Sci.* **60:**378–386.

359. Wright, S. L., and G. H. Richardson. 1982. Optimization of whey-based or non fat dry-milk based media for production of pH controlled bulk lactic cultures. *J. Dairy Sci.* **65:**1882–1889.

360. Mermelstein, N. H. 1982. Advanced bulk starter medium improves fermentation process. *Food Technol.* **36:**69–76.

361. Sandine, W. E., and J. W. Ayers. 1983. Method and starter compositions for the growth of acid producing bacteria and bacterial compositions produced thereby. U. S. Patent 4,382,965. May 10.

362. Sandine, W. E., and J. W. Ayers. 1981. Method and starter compositions for the growth of acid producing bacteria and bacterial compositions produced thereby. U. S. Patent 4,282,255. August 4.

363. Willrett, D. L., W. E. Sandine, and J. W. Ayers. 1982. Evaluation of pH controlled starter media including a new product for Italian and Swiss type cheese. *Cult. Dairy Prod. J.* **17:**5–9.

364. Hargrove, R. E. 1962. Control of bacteriophage. U. S. Patent 3, 04, 248, June 26.

365. Christensen, V. W. 1967. Dry starter composition. U. S. Patent 3, 354, 649, November 21.

366. Christensen, V. W. 1969. Cheese starter culture. U. S. Patent 3, 483, 087, Dec. 9.

367. Christensen, V. W. 1971. Production of cell culture concentrates. U. S. Patent 3, 592,740, July 13.

368. Richardson, G. H. 1982. Need for standards for culture tanks and culture control equipment. *In Proceeding of the 5th Biennial Cheese Industry Conference*, Utah State University, Loyan, UT, Sept. 1–2.

369. Pearce, L. E. 1978. The effect of host-controlled modification of the replications rate of a lactic streptococcal bacteriophage. *N. Z. J. Dairy Sci. Technol.* **13:**166–171.

370. Cogan, T. M. 1972. Susceptibility of cheese and yogurt starters to antibiotics. *J. Appl. Microbiol.* **23:**960.

371. Hsu, H. Y., F. F. Jewett, and S. E. Charm. 1987. What is killing the bugs in your starter culture? *Cult. Dairy Products J.* **22:**18.

372. Tamime, A. Y., and H. C. Deeth. 1980. Yogurt: Technology and Biochemistry. *J. Food Prot.* **43:**939–977.

373. Dunsmore, D. G., D. Makin, and R. Arkins. 1985. Effect of residues of five disinfectants in milk on acid production by strains of lactic starters used for Cheddar cheese making and on organoleptic properties of the cheese. *J. Dairy Res.* **52:**287.

374. Pearce, L. E. 1969. Activity test for cheese starter cultures. *N. Z. J. Dairy Sci. Technol.* **4:**246.

375. Guirguis, N., and M. W. Hickey. 1987. Factors affecting the performance of thermophilic starters. 2. Sensitivity to the lactoperoxidase system. *Aust. J. Dairy Technol.* **42:**14–16, 26.

376. Roginski, H., M. C. Broome, D. Hungerford, and M. W. Hickey 1984a. Non-phage inhibition of group N streptococci in milk 1. The incidence of inhibition in bulk milk. *Aust. J. Dairy Technol.* **39:**23–27.

377. Le Blanc, D. J., V. L. Crow, and L. N. Lee. 1980. Plasmid mediated carbohydrate catabolic enzymes among stains of *Streptococcus lactis, In* C. Stultand and K. R. Rozee (eds.), *Plasmids and Transposons: Environmental Effects and Maintenance Mechanisms*, pp. 34–41.

378. Gonzalez, C. F., and B. S. Kunka, 1985. Transfer of sucrose-fermenting ability and nisin production phenotype among lactic streptococci. *Appl. Environ. Microbiol.* **49:**627–633.

379. Kok, J., J. Maarten van Dijl, J. M. B. M. van der Vossen, and G. Venema. 1985. Cloning and expression of *Streptococcus cremoris* proteinase in *Bacillus subtilis* and *Streptococcus lactis. Appl. Environ. Microbiol.* **50:**94–101.

380. Kempler, G. M., and L. L. McKay. 1981. Biochemistry and genetics of citrate utilization in *Streptococcus lactis* subsp. *diacetylactis. J. Dairy Sci.* **64:**1527–1539.

381. Scherwitz, K. M., K. A. Baldwin, and L. L. McKay. 1983. Plasmid linkage of a bacteriocin-like substance in *Streptococcus lactis* subsp. *diacetylactis WM4* and its transferability to *Streptococcus lactis. Appl. Environ. Microbiol.* **45:**1506-1512.

382. Scherwitz, K., and L. L. McKay. 1987. Restriction enzyme analysis of lactose and bacteriocin plasmids from *Streptococcus lactis* subsp. *diacetylactis* WM4 and cloning of BC-I fragments coding for bacteriocin production. 1987. *Appl. Environ. Microbiol.* **53:**1171–1174.

383. McKay, L. L., and K. A. Baldwin. 1984. Conjugative 40-megadalton plasmid in *Streptococcus lactis* subsp. *diacetylactis DRC3* is associated with resistance to nisin and bacteriophage. *Appl. Environ. Microbiol.* **47:**68–74.

384. Sanders, M. E., and T. R. Klaenhammer. 1984. Phage resistance in a phage insensitive strain of *Streptococcus lactis*: Temperature dependent phage development and host-controlled phage replication. *Appl. Environ. Microbiol* **47:**979–985.

385. Chopin, A. M., C. Chopin, A. Moillo-Batt, and P. Langella. 1984. Two plasmid determined restriction and modification systems in *Streptococcus lactis. Plasmid* **11:**260–263.

386. deVos, W. M., H. M. Underwood, and F. L. Davies. 1984. Plasmid encoded bacteriophage resistance in *Streptococcus cremoris* SK11. *FEMS Microbiol Lett.* **23:**175–178.

387. Sanders, N. G., and T. R. Klaenhammer. 1981. Evidence of plasmid linkage of restriction and modification in *Streptococcus cremoris* KH. *Appl. Environ. Microbiol.* **42:**944–950.

388. Wesley, P. 1982. At Miles. Genetic strategies for inhibition of bacteriophages in cheese fermentation. *Genet. Engin. News,* May–June, pp. 16.

389. Bishop, J. R., C. R. White, and R. Firstenberg-Eden. 1984. A rapid impedimetric method for determining the potential shelf-life of pasteurized whole milk. *J. Food Prot.* **47:**471–475.

390. Bishop, J. R., and C. H. White. 1985. Estimation of potential shelf-life of pasteurized fluid milk utilizing bacterial numbers and metabolites. *J. Food Prot.* **48:**663–667.

391. Bossuyt, R. G. 1981. Determination of bacteriological quality of raw milk by an ATP assay technique. *Milchwissenschaft* **36:**257–260.

392. Eden, R., and G. Eden. 1984. *Impedance Microbiology.* Research Studies Press, Letchworth, England.

393. Entis, P. 1983. Enumeration of coliforms in nonfat dry milk and canned custard by hydrophobic grid membrane filter method. *J. Assoc. Off. Anal. Chemists* **66:**897–904.

394. Fung, D. Y. C. 1991. Rapid methods and automation for food microbiology. *In* D. Y. C. Fung and R. F. Matthews (eds.), *Instrumental Methods for Quality Assurance in Foods*, pp. 1–38. Marcel Dekker, New York.

395. Griffiths, M. W., J. D. Phillips, and D. D. Muir. 1984b. Detection of post heat-treatment contaminants using DEFT. *In Rapid Detection of Post-Pasteurization Contamination*, pp. 28–36. Hannah Research Bulletin No. 10.

396. Johnson, K. K., and P. C. Vasavada. 1988. Evaluation of raw milk quality by the Wisconsin Mastitis Test and catalase activity in milk. *J. Food Prot.* **51:**825.

397. Pettipher, G. L., and V. M. Rodrigues. 1982. Rapid enumeration of microorganisms in foods by the direct epifluorescent filter technique. *Appl. Environ. Microbiol.* **44:**809–813.

398. Phillips, J. D., and M. W. Griffiths. 1985. Bioluminescence and impedimetric methods for assessing shelf-life of pasteurized milk and cream. *Food Microbiol.* **2:**39–51.

399. Sharpe, A. N., and D. S. Clark (eds.). 1978. *Mechanizing Microbiology.* Charles C. Thomas, Springfield, IL.

400. Elliker, R. R., E. L. Singh, L. J. Christensen, and W. E. Sandine. 1964. Psychrophilic bacteria and keeping quality of pasteurized dairy products. *J. Milk Food Technol.* **27:**69–75.

401. Edmonson, J. E., R. Golden, and D. B. Wedle. 1985. Reduction methods. *In* G. H. Richardson (ed.), *Standard Methods for Examination of Dairy Products*, 15th edit., pp. 259–264. American Public Health Association, Washington, D.C.

402. Luck, H., and H. Gavron. 1990. Quality control in the dairy industry. *In* R. K. Robinson (ed.), *Dairy Microbiology* Vol. 2, 2nd edit., pp. 345–392. Elsevier Applied Science, London.

403. Fung D. Y. C., and R. F. Matthews. 1991. *Instrumental Methods for Quality Assurance in Foods.* Marcel Dekker, New York.

404. Tilton, R. C. (ed.). 1982. *Rapid Methods and Automation in Microbiology.* American Society for Microbiology, Washington, D.C.

405. Habermehl, K. O. (ed.). 1985. *Rapid Methods and Automation in Microbiology and Immunology.* John Wiley & Sons, New York.

406. Pierson, M. D., and N. J. Stern. 1986. *Foodborne Microorganisms and Their Toxins: Developing Methodology.* Marcel Dekker, New York.

407. Adams, M. R., and C. F. A. Hope. 1989. *Rapid Methods in Food Microbiology.* Elsevier, Amsterdam.

408. Pettipher, G. L. 1983. *The Direct Epifluorescent Filter Technique for the Rapid Enumeration of Microrrganisms.* Research Studies Press, Letchworth, England.

409. International Dairy Federation. 1981. Liquid milk—enumeration of microorganisms colony count technique at 30°C. *FIL-IDF* **10:**1981.

410. Sharpe, A. N., and D. C. Kilsby. 1971. A rapid, inexpensive bacterial count technique using agar droplets. *J. Appl. Bacteriol.* **34:**435–440.

411. Sharpe, A. N., E. J. Dyelt, A. K. Jackson, and D. C. Kilsby. 1972. Techniques and apparatus for rapid and inexpensive enumeration of bacteria. *Appl. Microbiol.* **24:**4–7.

412. Wright, E. O., G. W. Reinbold, L. Burmeister, and J. Mellon. 1970. Prediction of standard plate count of manufacturing-grade raw milk from the plate loop count. *J. Milk Food Technol.* **33:** 168–170.

413. Entis, P. 1986. Membrane filtration systems. *In* M. D. Pierson and N. J. Stern (eds.), *Foodborne Microorganisms and Their Toxins: Developing Methodology*, pp. 91–106. Marcel Dekker, New York.

414. Sharpe, A. N., and P. I. Peterkin. 1988. *Membrane Filter Food Microbiology*. Research Studies Press, Letchworth, England.

415. Dziezak, J. D. 1987. Rapid methods for microbiological analysis of food. *Food Technol.* **41:**56–73.

416. Ginn, R. E., V. S. Packard, and T. L. Fox. 1986. Enumeration of total bacteria and coliforms in milk by dry rehydratable film method: collaborative study. *J. Assoc. Off. Anal. Chem.* **69:**527–531.

417. Ginn, R. E., V. S. Packard, and T. L. Fox. 1984. Evaluation of the 3M dry medium culture plate (Petrifilm SM) method for determining number of bacteria in raw milk. *J. Food Prot.* **47:**753–755.

418. McAllister, J. S., M. S. Ramos, and T. L. Fox. 1987. Evaluation of dry milk film (Petrifilm SM) method for enumerating bacteria in processed fluid milk samples. *Dairy Food Sanit.* **7:**632–635.

419. Smith, L. B., T. L. Fox, and F. F. Busta. 1986. Comparison of a dry medium culture plate (Petrifilm SM plates) method to the aerobic plate count method for enumeration of mesophilic aerobic colony-forming units in fresh ground beef. *J. Food Prot.* **48:**1044–1045.

420. Gilchrist, J. E., J. E. Campbell, B. C. Donnelly, J. J. Peeler, and J. M. Delaney. 1973. Spiral plate method for bacterial determination. *Appl. Microbiol.* **25:**244–252.

421. Jarvis, B., V. H. Lach, and J. M. Wood. 1977. Evaluation of the spiral plate maker for the enumeration of microorganisms in foods. *J. Appl. Bacteriol.* **43:**149–157.

422. Bishop, J. R., and J. Y. Juan. 1988. Improved methods for quality assessment of raw milk. *J. Food Prot.* **51:**955–957.

423. Bishop, J. R., and C. H. White. 1986. Assessment of dairy product quality and potential shelf-life—a review. *J. Food Prot.* **49:**739–753.

424. Phillips, J. D., M. W. Griffiths, and D. D. Muir. 1984. Preincubation test to rapidly identify post-pasteurization contamination of milk and single cream. *J. Food Prot.* **47:**391–393.

425. Bishop, J. R., and C. H. White. 1985. Estimation of potential shelf-life to cottage cheese utilizing bacterial numbers and metabolites. *J. Food Prot.* **48:**663–667.

426. Byrne, R. D., J. R. Bishop, and J. W. Boling. 1989. Estimation of potential shelf-life of pasteurized fluid milk using a selective preliminary incubation. *J. Food Prot.* **52:**805–807.

427. Griffiths, M. W., J. D. Phillips, and D. D. Muir. 1984c. Improvement for assay of post heat-treatment of contaminants using plate-counting procedures. *In Rapid Detection of Post-Pasteurization Contamination*, pp. 14–27. Hannah Research Bulletin No. 10.

428. Chain, V. S., and D. Y. C. Fung. 1991. Comparison of Redigel, Petrifilm, Spiral Plate System, Isogrid, and Aerobic Plate Count for determining numbers of aerobic bacteria in selected foods. *J. Food Prot.* **54:**208–211.

429. Chain, V. S. 1988. *Comparison of Redigel, Petrifilm, Spiral Plate System, Isogrid and Standard Plate Count for the Aerobic Plate Count in Selected Foods*. M.S. Thesis, Manhattan, KS, Kansas State University.

430. Roth, J. N. 1988. Temperature independent pectin gel method for aerobic plate count in dairy and nondairy food products: collaborative studies. *J. Assoc. Off. Anal. Chem.* **71:**343–349.

431. Roth, J. N., and G. L. Bontrager. 1989. Temperature-independent pectin gel method for coliform determination in dairy products: Collaborative study. *J. Assoc. Off. Anal. Chem.* **72:**298–302.

432. Bossuyt, R. G. 1982. A 5-minute ATP platform test for judging the bacteriological quality of raw milk. *Netherlands Milk Dairy J.* **36:**355–364.

433. Waes, G. M., and R. G. Bossuyt. 1981. A rapid method to detect post pasteurization contamination in pasteurized milk. *Milchwissenschaft* **36:**548–522.

434. Waes, G. M., and R. G. Bossuyt. 1982. Usefulness of the benzalkon crystal-violet ATP method for predicting the keeping quality of pasteurized milk. *J. Food Prot.* **45:**928–931.

435. LaRocco, K. A., K. J. Littel, and M. D. Pierson. 1986. The bioluminescent ATP assay for determining the microbial quality of foods. *In* M. D. Person and N. J. Stern (eds.), *Foodborne Microorganisms and Their Toxins: Developing Methodology*, pp. 145–174. Marcel Dekker, New York.

436. Stannard, C. J. 1989. ATP estimation. *In* M. R. Adams and C. F. A. Hope (eds.), *Rapid Methods in Food Microbiology*, pp. 1–18. Elsevier, Amsterdam.

437. Easter, M. C., and D. M. Gibson. 1989. Detection of microorganisms by electrical measurements. *In* M. R. Adams and C. F. A. Hope (eds.), *Rapid Methods in Food Microbiology*, pp. 57–100. Elsevier, Amsterdam.

438. Khayat, F. A., and G. H. Richardson. 1986. Detection of abnormal milk with impedance microbiology instrumentation. *J. Food Prot.* **49:**519–522.

439. Bossuyt, R. G., and G. M. Waes. 1983. Impedance measurements to detect post-pasteurization contamination of pasteurized milk. *J. Food Prot.* **46:**622–624.

440. Gnan, S., and L. O. Luedecke. 1982. Impedance measurements in raw milk as an alternative to the standard plate count. *J. Food Prot.* **45:**4–7.

441. Visser, I. J. R., and J. deGroote. 1984a. Prospects for the use of conductivity as an aid in the bacteriological monitoring of pasteurized milk. *Antonie van Leewenhoek* **50:**202–206.

442. Khayat, F. A., J. C. Bruhn, and G. H. Richardson. 1988. A survey of coliforms and *Staphylococcus aureus* in cheese using impedimetric and plate count methods. *J. Food Prot.* **51:**53–55.

443. Okibo, L. M., and G. H. Richardson. 1985. Detection of penicillin and streptomycin in milk by impedance microbiology. *J. Food Prot.* **68:**971–981.

444. Okibo, L. M., C. J. Oberg, and G. H. Richardson. 1985. Lactic culture activity tests using pH and impedance instrumentation. *J. Dairy Sci.* **68:**2521–2526.

445. Tsai, K. P., and L. O. Luedecke. 1989. Impedance measurement of change in activity of lactic cheese starter cultures after storage at 4°C *J. Dairy Sci.* **72:**2239–2241.

446. D'Ombrain, L., S. Toyne, P. C. Vasavada, and R. R. Hull. 1990. Impedance measurement to determine activity and bacteriophage content of lactic starter culture. *In Proceedings of the Xth International Dairy Congress*, Montreal, Canada, October. Short Communication No. 945.

447. Waes, G. M., and R. G. Bossuyt. 1984. Impedance measurement to detect bacteriophage in cheddar cheesemaking. *J. Food Prot.* **47:**349–351.

448. Visser, I. J. R., and J. deGroote. 1984b. The Malthus microbiological growth analyzer as an aid in the detection of post-pasteurization contamination of pasteurized milk. *Netherlands Milk Dairy J.* **38:**151–156.

449. Easter, M. C., and D. M. Gibson. 1985. Rapid and automated detection of *Salmonella* by electrical measurements. *J. Hyg.* **94:**245–262.

450. Gibson, D. M. 1987. Some modification to the media for rapid automated detection of salmonellas by conductance measurement. *J. Appl. Bacteriol.* **63:**299–304.

451. Ogden, I. D. 1988. A conductance medium to distinguish between *Salmonella* and *Citrobacter* spp. *Int. J. Food Microbiol.* **7:**287–297.

452. Cousins, D. L., and F. Marlatt. 1990. An evaluation of a conductance method for the enumeration of *Enterobacteriaceae* in milk. *J. Food Prot.* **53:**568–570.

453. Lampi, R. A., D. A. Mikelson, D. B. Rowley, J. J. Previte, and R. E. Wells. 1974. Radiometry and microcalorimetry techniques for the rapid detection of foodborne microorganisms. *Food Technol.* **28:**52–58.

454. Rowley, D. R., P. Vandemark, D. Johnson, and E. Shatluck. 1979. Resuscitation of stressed fecal coliforms and their subsequent detection of radiometric and impedance techniques. *J. Food Prot.* **42:**335–341.

455. Gram, L., and H. Sogaard. 1986. Microcalorimetry as a rapid method for estimation of bacterial levels in ground meat. *J. Food Prot.* **48:**341–345.

456. Sudi, J., G. Suhren, W. Heeschen, and A. Tolle. 1981. Entwicklung eines minaturisierten Limulus-Tests im Mikrotiter-System zum quantitativen Nachweis gram-negativer Bakterien in Milch and Milchprodukten. *Milchwissenschaft* **36:**193–198.

457. Hansen, K., T. Mikkelsen, and A. Moller-Madsen. 1982. Use of the Limulus test to determine the hygienic status of milk products as characterized by levels of gram-negative LPS present. *J. Dairy Res.* **49:**323–328.

458. Mikolajczik, E. M., and R. B. Brucker. 1983. LAL assay. A rapid test for the assessment of raw and pasteurized milk quality. *Dairy Food Sanit.* **3:**129–131.

459. Jay, J. M. 1989. The limulus amoebocyte lysate (LAL) test. *In* M. R. Adams and C. F. A. Hope (eds.), *Rapid Methods in Food Microbiology*, pp. 101–120. Elsevier Science Publishers, Amsterdam.

460. Heeschen, W., J. Sudi, and G. Suhren. 1985. Application of the Limulus test for detection of Gram-negative microorganisms in milk and dairy products. *In* K. O. Habermehl (ed.), *Rapid Methods and Automation in Microbiology and Immunology*, pp. 638–648. John Wiley & Sons, Inc., New York.

461. Fischer, J. E., and P. C. Vasavada. 1988. Rapid method for detection of abnormal milk by the Catalase test. *J. Dairy Sci.* **70** (*Suppl. 1*):75.

462. Vasavada, P. C., T. A. Bon, and L. Bauman. 1988. The use of the Catalasemeter in assessing abnormality in raw milk. *J. Dairy Sci.* **71**(*Suppl. 1*):113.

463. Pettipher, G. L. 1986. Review: the direct epifluorescent filter technique. *J. Food Technol.* **21:**535–546.

464. Pettipher, G. L., R. Mansell, C. H. McKinnon, and C. M. Cousins. 1980. Rapid membrane filter epifluorescent microscopy technique for the direct enumeration of bacteria in raw milk. *Appl. Environ. Microbiol.* **39:**423–429.

465. Pettipher, G. L., and V. M. Rodrigues. 1980. Rapid enumeration of bacteria in heat treated milk and milk products using a membrane filtration-epifluorescent microscopy technique. *J. Appl. Bacteriol.* **50:**157–166.

466. Richardson, G. H., R. Grappin, and T. C. Yuan. 1988. A reflectance colorimeter instrument for measurement of microbial and enzymatic activities in milk and dairy products. *J. Food Prot.* **51:**778–785.

467. Bailey, J. S., N. A. Cox, J. E. Thomson, and D. C. Fung. 1985. Identification of *Enterobacteriaceae* in foods with the Automicrobic System. *J. Food Prot.* **48:**147–149.

468. Cox, N. A., D. Y. C. Fung, J. S. Bailey, P. A. Hartman, and P. C. Vasavada. 1988. Miniaturized kits, immunoassays and DNA hybridization for recognition and identification of food-borne bacteria. *Dairy Food Sanit.* **7:**628–631.

469. Cox, N. A., J. McHan, and D. Y. C. Fung. 1987. Commercially available minikits for the identification of *Enterobacteriaceae*: a review. *J. Food Prot.* **40:**866–872.

470. Fung, D. Y. C., M. C. Goldschmidt, and N. A. Cox. 1984. Evaluation of bacterial diagnostic kits and systems at an instructional workshop. *J. Food Prot.* **47:**68–73.

471. Fung, D. Y. C., and N. A. Cox, 1981. Rapid identification systems in the food industry: present and future. *J. Food Prot.* **44:**877–880.

472. Fitts, R. 1985. Development of a DNA-DNA hybridization test for the presence of *Salmonella* in foods. *Food Technol.* **39:**95–102.

473. Flowers, R. S. 1985. Comparison of rapid *Salmonella* screening methods and the conventional culture method. *Food Technol.* **39:**103–108.

474. Fung, D. Y. C., N. A. Cox, and J. S. Bailey. 1988. Rapid methods and automation in microbiology. *Dairy Food Sanit.* **8:**292–296.

475. Dilley, C. L., and D. Dixon-Holland. 1990. Rapid residue test for aflatoxin M_1 and sulfamethazine in dairy products. *Food Technol.* **44:**32.

476. Chandler, R. E., and T. A. McMeekin. 1989. Temperature function integration as the basis of an accelerated method to predict the shelf-life of pasteurized, homogenized milk. *Food Microbiol.* **6:**105–111.

477. Chandler, R. E., and T. A. McMeekin. 1985b. Temperature function integration and the prediction of shelf-life of milk. *Aust. J. Dairy Technol.* **40:**10–13.

478. Griffiths, M. W., and J. D. Phillips. 1988b. Prediction of the shelf-life of pasteurized milk at different storage temperatures *J. Appl. Bacteriol.* **65:**269–278.

479. Ratkowsky, D. A., J. Olley, T. A. McMeekin, and A. Ball. 1982. Relationship between temperature and growth rate of bacterial cultures. *J. Bacteriol.* **149:**1–5.

480. Chandler, R. E., and T. A. McMeekin. 1985a. Temperature function integration and its relationship to the spoilage of pasteurized, homogenized milk. *Aust. J. Dairy Technol.* **40:**37–40.

481. Zikakis, J. P. 1986. Factors affecting the shelf life of dairy products. *In* G. Charalambous (ed.), *The Shelf Life of Foods and Beverages*, pp. 313–334. Elsevier Applied Science, New York.

482. Robinson, R. K. (ed.). 1990a. *Dairy Microbiology*, Vol. 1: *The Microbiology of Milk*, 2nd edit. Elsevier Applied Science, New York.

483. Robinson, R. K. (ed.). 1990b. *Dairy Microbiology*, Vol. 2: *The Microbiology of Milk Products*, 2nd edit. Elsevier Applied Science, New York.

484. Morton, D. J., and E. L. Barrett. 1982. Gram-negative respiratory bacteria which cause ropy milk constitute a distance cluster within the genus *Acinetobacter*. *Curr. Microbiol.* **7:**107–112.

485. Juven, B. J., S. Gordin, I. Rosenthal, and A. Laufer. 1981. Changes in refrigerated milk caused by *Enterobacteriaceae*. *J. Dairy Sci.* **64:**1781–1784.

486. Fleet, G. H., and M. A. Mian. 1987. The occurrence and growth of yeasts in dairy products. *Int. J. Food Microbiol.* **4:**145–155.

487. Gebre-Egziabher, A., A. K. Cheong, G. Blankenagel, and E. S. Humbert. 1985. Extended storage of raw milk. I. Effect on microbiological quality. *Can. Inst. Food Sci. Technol. J.* **18:**247–250.

488. Baker, S. K. 1983. The keeping quality of refrigerated pasteurized milk. *Aust. J. Dairy Technol.* **38:**124–127.

489. Muir, D. D., and J. D. Phillips. 1984. Prediction of shelf-life of raw milk during refrigerated storage. *Milchwissenschaft* **39:**7–11.

490. Schröder, M. J. A. 1984. Origins and levels of post pasteurization contamination of milk in the dairy and their effects on keeping quality. *J. Dairy Res.* **51:**59–67.

491. Chen, J. H., and R. R. Zall. 1987. Packaged milk, cream and cottage cheese can be monitored for freshness using polymer indicator labels. *Dairy Food Sanit.* **7:**402–404.

492. Griffiths, M. W., J. D. Phillips, and D. D. Muir. 1986. The effect of sub-pasteurization heat treatments on the shelf-life of milk. *Dairy Indust. Int.* **51:**31, 33–35.

493. Phillips, J. D., and M. W. Griffiths. 1986. Factors contributing to the seasonal variation of *Bacillus* spp. in pasteurized dairy products. *J. Appl. Bacteriol.* **61:**275–285.

494. Kwee, W. S., T. W. Dommett, J. E. Giles, R. Roberts, and R. A. D. Smith. 1986a. Microbiological parameters during powdered milk manufacture. 1. Variation between processes and stages. *Aust. J. Dairy Technol.* **41:**3–6.

495. Kwee, W. S., T. W. Dommett, J. E. Giles, R. A. D. Smith, and R. Roberts. 1986b. Microbiological parameters during powdered milk manufacture. 2. Relationships and predictability among counts. *Aust. J. Dairy Technol.* **41:**6–12.

496. Daemen, A. L. H. 1981. The destruction of enzymes and bacteria during the spray-drying of milk and whey. I. The thermoresistance of some enzymes and bacteria in milk and whey with various total solids contents. *Netherlands Milk Dairy J.* **35:**133–144.

497. Daemen, A. L. H., and H. J. van der Stege. 1982. The destruction of enzymes and bacteria during the spray-drying of milk and whey. 2. The effect of the drying conditions. *Netherlands Milk Dairy J.* **36:**211–229.

498. Stadhouders, J., G. Hup, and F. Hassing. 1982. The conceptions index and indicator organisms discussed on the basis of the bacteriology of spray-dried milk powder. *Netherlands Milk Dairy J.* **36:**231–260.

499. Chopra, A. K., and D. K. Mathur. 1984. Isolation, screening and characterization of thermophilic *Bacillus* species isolated from dairy products. *J. Appl. Bacteriol.* **57:**263–271.

500. Chopin, M. C. 1980. Resistance of 17 mesophilic lactic *Streptococcus* bacteriophages to pasteurization and spray-drying. *J. Dairy Res.* **47:**131–139.

501. Burlingame-Frey, J. P., and E. H. Marth. 1984. Changes in some functional properties of freeze-dried milk made from skim milk that supports growth of psychrotrophic bacteria. *J. Food Prot.* **47:**288–292.

502. Kalogridou-Vassiliadou, D., N., Tzanetakis, and K. Manolkidis. 1989. *Bacillus* species isolated from flat sour evaportated milk. *Lebensm. Wiss. U.-Technol.* **22:**287–291.

503. Kalogridou-Vassiliadou, D. 1990. Sources of evaporated milk contamination by 'flat sour' bacilli. *Lebensm.-Wiss. U.-Technol.* **23:**285–288.

504. McKellar, R. C., and D. Nichols-Nelson. 1984. Acid-coagulation of evaporated milk by a coculture of *Enterococcus faecium* and *Bacillus subtilis. J. Food Prot.* **47:**853–855.

505. Brocklehurst, T. F., and B. M. Lund. 1988. The effect of pH on the initiation of growth of cottage cheese spoilage bacteria. *Int. J. Food Microbiol.* **6:**43–49.

506. Marshall, R. T. 1979. Psychrotrophic bacteria—their relationship to raw milk quality and keeping quality of cottage cheese. *Proc. First Biennial Marshall Int. Cheese Conf.* 423–434.

507. Brocklehurst, T. F., and B. M. Lund. 1985. Microbiological changes in cottage cheese varieties during storage at +7°C. *Food Microbiol.* **2:**207–233.

508. Bigalke, D. 1984a. Cottage cheese quality—the importance of ingredient quality. *Dairy Food Sanit.* **4:**482–483.

509. Bigalke, D. 1985. Cottage cheese quality process control and process standards. *Dairy Food Sanit.* **5:**23–24.

510. de Boer, E., and D. Kuik. 1987. A survey of the microbiological quality of blue-veined cheeses. *Netherlands Milk Dairy J.* **41:**227–237.

511. Nooitgedagt, A. J., and B. J. Hartog. 1988. A survey of the microbiological quality of Brie and Camembert cheese. *Netherlands Milk Dairy J.* **42:**57–72.

512. Banks, J. M., M. W. Griffiths, J. D. Phillips, and D. D. Muir. 1986. The yield and quality of Cheddar cheese produced from thermised milk. *Dairy Indust. Int.* **51:**31–32, 34–35.

513. Weatherup, W. W., W. Michael, A. Mullan, and J. Kormos. 1988. Effect of storing milk at 3° and 7°C on the quality and yield of Cheddar cheese. *Dairy Indust. Int.* **53:**16–17, 25.

514. Bourgeois, C. M., O. LeParc, B. Abgrall, and J. J. Cleret. 1984. Membrane filtration of milk for counting spores of *Clostridium tryobutyricum. J. Dairy Sci.* **67:**2493–2499.

515. Soehnlen, J. S., J. B. Lindamood, and E. M. Mikolajcik. 1989. Characterizations of organisms involved with Swiss cheese rind rot defect. *Cult. Dairy Prod. J.* **24:**24–26, 33.

516. Warburton, D. W., P. I. Peterkin, and K. F. Weiss. 1986. A survey of the microbiological quality of processed cheese products. *J. Food Prot.* **49:**229–230.

517. Horwood, J. F., W. Stark, and R. R. Hull. 1987. A ''fermented, yeasty'' flavour defect in Cheddar cheese. *Aust. J. Dairy Technol.* **42:**25–26.

518. Tamime, A. Y., G. Davies, and M. P. Hamilton. 1987. The quality of yogurt on retail sale in Ayrshire. *Dairy Indust. Int.* **52:**19–21.

519. Salji, J. P., S. R. Saadi, and A. Mashhadi. 1987. Shelf life of plain liquid yougurt manufactured in Saudi Arabia. *J. Food Prot.* **50:**123–126.

520. Green, M. D., and S. N. Ibe. 1987. Yeasts as primary contaminants in yogurts produced commercially in Lagos, Nigeria. *J. Food Prot.* **50:**193–198.

521. Spillmann, H., and O. Geiges. 1983. Identification von Hefen und Schimmelpilzen aus bombierten Joghurt-Puckungen. *Milchwissenschaft* **38:**129–132.

522. Suriyarachchi, V. R., and G. H. Fleet. 1981. Occurrence and growth of yeasts in yogurts. *Appl. Environ. Microbiol.* **42:**574–579.

523. McKay, A. M. 1991. Strain differentiation of yeasts associated with dairy products by agarose gel electrophoresis of nucleic acids. *Milchwissenschaft* **46:**79–81.

524. Langeveld, L. P. M., and A. C. Bolle. 1989. Oxygen availability, carbon dioxide concentration and growth of yeasts in fermented milk products; implications for growth during cold storage and for rapid enrichment. *Netherlands Milk Dairy J.* **43:**407–422.

525. Garcia, A. M., and G. S. Fernández. 1984. Contaminating mycoflora in yogurt: general aspects and special reference to the genus *Penicillium. J. Food Prot.* **47:**629–636.

526. Vedamuthu, E. R. 1985. What is wrong with cultured buttermilk today? *Dairy Food Sanit.* **5:**8–13.

527. Hankin, L., D. Shields, and J. G. Hanna. 1982. Quality of sour cream and non-butterfat sour dressing. *Dairy Food Sanit.* **2:**232–234.

528. Hankin, L., and J. G. Hanna. 1983. Quality of butter and blends of butter with oleomargarine. *Dairy Food Sanit.* **3:**458–460.

529. Jensen, H., H. Danmark, and G. Mogensen. 1983. Effect of storage temperature on microbiological changes in different types of butter. *Milchwissenschaft* **38:**482–484.

530. Kaul, A., J. Singh, and R. K. Kuila. 1979. Effect of potassium sorbate on the microbiological quality of butter. *J. Food Prot.* **42:**656–657.

531. Bigalke, D. 1984b. Ice cream microbiological quality. Part II. Recommended laboratory procedures for monitoring and controlling microbiological quality of ice cream. *Dairy Food Sanit.* **4:**398–399.

532. Bigalke, D., and A. Chappel. 1984. Ice cream microbiological quality. Part I. Controlling coliform and other microbial contamination in ice cream. *Dairy Food Sanit.* **4:**318–319.

533. Ryan, J. J., and R. H. Gough. 1982. Bacteriological quality of soft-serve mixes and frozen products. *J. Food Prot.* **45:**279–280.

534. Tamminga, S. K., R. R. Beumer, and E. H. Kampelmacher. 1980. Bacteriological examination of ice-cream in The Netherlands: Comparative Studies on methods. *J. Appl. Bacteriol.* **49:**239–253.

535. Massa, S., G. Poda, D. Cesaroni, and L. D. Trovatelli. 1989. A bacteriological survey of retail ice cream. *Food Microbiol.* **6:**129–134.

536. Lelievre, J., and R. C. Lawrence. 1988. Manufacture of cheese from milk concentrated by ultrafiltration. *J. Dairy Res.* **55:**465–478.

537. Veillet-Poncet, L., A. Tayfour, and J. B. Milliere. 1980. Etude bacteriologique de l'ultrafiltration du lait et due stockage au froid du retentate. *Le Lait* **60:**351–374.

538. Benard, S., J. L. Maubois, and A. Tareck. 1981. Ultrafiltration—thermisation du lait a la production: aspects bacteriologiques. *Le Lait* **61:**435–457.

539. Tayfour, A., J. B. Milliere, and L. Veillet-Poncet. 1982. Growth and proteolytic activity of *Pseudomonas fluorescens* 28 P 12 in milk retantates stored at low temperatures. *Milchwissenschaft* **37:**720–723.

540. Eckner, K. F., and E. A. Zottola. 1989. Behavioral response of spoilage and pathogenic microorganisms inoculated into reconstituted skim milk concentrated by ultrafiltration. *Milchwissenschaft.* **44:**208–212.

541. Mistry, V. V., and F. V. Kosikowski. 1986. Influence of milk ultrafiltration on bacteriophages of lactic acid bacteria. *J. Dairy Sci.* **69:**2577–2582.

542. Zottola, E. A., T. M. Cogan, and J. Kelley. 1987. Partition of lactic streptococcal bacteriophage during the ultrafiltration concentration of milk and whey. *J. Dairy Sci.* **70:**2013–2021.

543. Haggerty, P., and N. N. Potter. 1986. Growth and death of selected microorganisms in ultrafiltered milk. *J. Food Prot.* **49:**233–235.

544. Rash, K. E., and F. V. Kosikowski. 1982b. Influence of diafiltration, lactose hydrolysis and carbon dioxide on enteropathogenic *Escherichia coli* in Camembert cheese made from ultrafiltered milk. *J. Food Sci.* **47:**733–736.

545. Rash, K. E., and F. V. Kosikowski. 1982a. Behavior of enteropathogenic *Escherichia coli* in Camembert cheese made from ultrafiltered milk. *J. Food Sci.* **47:**728–732, 736.

546. Rash, K. E., and F. V. Kosikowski. 1982c. Influence of lactic acid starter bacteria on enteropathogenic *Escherichia coli* in ultrafiltration prepared Camembert cheese. *J. Dairy Sci.* **65:**537–543.

547. Eckner, K. F., and E. A. Zottola. 1991. The behavior of selected microorganisms during the manufacture of high moisture jack cheeses from ultrafiltered milk. *J. Dairy Sci.* **74:**2820–2830.

548. Kornacki, J. L., and E. H. Marth. 1989. Thermal inactivation of *Staphylococcus aureus* in retentates from ultrafiltered milk. *J. Food Prot.* **52:**631–637.

549. Srilaorkul, S., L. Ozimek, and M. E. Stiles. 1989. Growth and activity of *Lactococcus lactis* ssp. *cremoris* in ultrafiltered skim milk. *J. Dairy Sci.* **72:**2435–2443.

550. Hickey, M. W., H. Roginski, and M. C. Broom. 1983. Growth and acid production of group N streptococci in ultrafiltered milk. *Aust. J. Dairy Technol.* **38:**138–143.

551. Mistry, V. V., and F. V. Kosikowski. 1985b. Fermentation of ultrafiltered skim milk retentates with mesophilic lactic cheese starters. *J. Dairy Sci.* **68:**1613–1617.

552. Mistry, V. V., and F. V. Kosikowski. 1985c. Growth of lactic acid bacteria in highly concentrated ultrafiltered skim milk retentates. *J. Dairy Sci.* **68:**2536–2543.

553. Premaratne, R. J., and M. A. Cousin. 1991a. Changes in the chemical composition during ultrafiltration of skim milk. *J. Dairy Sci.* **74:**788–795.

554. Mistry, V. V., F. V. Kosikowski, and W. D. Bellamy. 1987. Improvement of lactic acid production in ultrafiltered milk by the addition of nutrients. *J. Dairy Sci.* **70:**2220–2225.

555. Qvist, K. B., D. Thomsen, and E. Hoier. 1987. Effect of ultrafiltered milk and use of different starters on the manufacture, fermentation and ripening of Havarti cheese. *J. Dairy Res.* **54:**437–446.

556. Christopherson, A. T., and E. A. Zottola. 1989a. Growth and activity of mesophilic lactic acid streptococci in ultrafiltered skim milk and in reconstituted nonfat dry milk of differing total solids contents. *J. Dairy Sci.* **72:**2856–2861.

557. Christopherson, A. T., and E. A. Zottola. 1989b. The use of whey permeate as starter media in cheese production. *J. Dairy Sci.* **72:**2862–2868.

558. Christopherson, A. T., and E. A. Zottola. 1989c. Whey permeate as a medium for mesophilic lactic acid streptococci. *J. Dairy Sci.* **72:**1701–1706.

559. Reinbold, R. S., and J. Takemoto. 1988. Use of Swiss cheese whey permeate by *Kluyveromyces fragilis* and mixed culture of *Rhodopseudomonas spheroids* and *Bacillus megaterium*. *J. Dairy Sci.* **71:**1799–1802.

560. Beaton, N. C. 1979. Ultrafiltration and reverse osmosis in the dairy industry—an introduction to sanitary considerations. *J. Food Prot.* **42:**584–590.

561. Bohner, H. F., and R. L. Bradley. 1990. Effective control of microbial populations in polysulfone ultrafiltration membrane systems. *J. Dairy Sci.* **73:**2309–2317.

562. Smith, K. E., and R. L. Bradley, Jr. 1986. Ineffective cleaning of polysulfone ultrafiltration membrane systems and corrosion by bisulfite used as a sanitizer. *J. Dairy Sci.* **69:**1232–1240.

563. Smith, K. E., and R. L. Bradley, Jr. 1987a. Efficiency of sanitizers using unsoiled spiral-wound polysulfone ultrafiltration membrances. *J. Food Prot.* **50:**567–572.

564. Smith, K. E., and R. L. Bradley, Jr. 1987b. Evaluation of efficiency of four commercial enzyme-based cleaners of ultrafiltration systems. *J. Dairy Sci.* **70:**1168–1177.

565. Smith, K. E., and R. L. Bradley, Jr. 1988. Evaluation of three different cleaners recommended for ultrafiltration systems by direct observations of commercial-scale spiral-wound ultrafiltration membranes. *J. Food Prot.* **51:**89–104.

566. Cromie, S. J., D. Schmidt, and J. E. Giles. 1986. The effect of reverse osmosis concentration and subsequent storage on the microflora of raw milk. *N. Z. J. Dairy Sci. Technol.* **21:**1–7.

567. Davies, F. L., P. A. Shankar, and H. M. Underwood. 1977. The use of milk concentrated by reverse osmosis for the manufacture of yogurt. *J. Soc. Dairy Technol.* **30:**23–28.

568. Dixon, D. B. 1985. Dairy products prepared from reverse osmosis concentrate—market milk products, butter, skim milk powder and yoghurt. *Aust. J. Dairy Technol.* **40:**91–95.

569. Drew, P. G., and J. G. Manners. 1985. Microbiological aspects of reverse osmosis concentration of milk. *Aust. J. Dairy Technol.* **40:**108–112.

570. Kosikowski, F. V., and V. V. Mistry. 1990. Microfiltration, ultrafiltration, and centrifugation separation and sterilization processes for improving milk and cheese quality. *J. Dairy Sci.* **73:**1411–1419.

571. Olesen, N., and F. Jensen. 1989. Microfiltration. The influence of operation parameters on the process. *Milchwissenschaft* **44:**476–479.

572. Merin, U., and G. Daufin. 1990. Crossflow microfiltration in the dairy industry: state-of-the-art. *Le Lait* **70:**281–291.

573. Trouvé, E., J. L. Maubois, M. Piot, M. N. Madec, J. Fauquant, A. Ronault, J. Tabard, and G. Brinkman. 1991. Rétention de différentes espèces microbiennes lors de l'épuration du lait par microfiltration en flux tangentiel. *Le Lait* **71:**1–13.

574. Burton, H. 1985. Thirty-five years on—a story of UHT research and development. *Chem. Indust. Aug.* **19:**546–553.

575. Brown, K. L., and C. A. Ayres. 1982. Thermobacteriology of UHT processed foods. *In* R. Davies (ed.), *Developments in Food Microbiology.* Vol. 1, pp. 119–152. Elsevier Applied Science, Essex, England.

576. Burton, H. 1988. *Ultra-High-Temperature Processing of Milk and Milk Products.* Elsevier Applied Science, New York.

577. Cerf, O. 1987. Revue bibliographique: caractérisation de la thermorésistance des spores bactériennes pour l'optimisation des traitements UHT. *Le Lait* **67:**97–109.

578. Duquet, J. P., A. Trouvat, A. Mouniqua, G. Odet, and O. Cerf. 1987. Les spores thermorésistantes du lait utilisé pour la fabrication de laits de longue conservation. *Le Lait* **67:**393–402.

579. Yildiz, F., and D. C. Westhoff. 1989. Sporulation and thermal resistance of *Bacillus* stearothermophilus spores in milk. *Food Microbiol.* **6:**245–250.

580. Brown, K., and J. Gaze. 1988. High temperature resistance of bacterial spores. *Dairy Indust. Int.* **53:**37, 39.

581. Westhoff, D. C., and S. L. Dougherty. 1981. Characterization of *Bacillus* species isolated from spoiled ultrahigh temperature processed milk. *J. Dairy Sci.* **64:**572–580.

582. Adams, D. M., and T. G. Brawley. 1981. Heat resistant bacterial lipases and ultra-high temperature sterilization of dairy products. *J. Dairy Sci.* **64:** 1951–1957.

583. Bucky, A. R., P. R. Hayes, and D. S. Robinson. 1987. A modified ultrahigh temperature treatment for reducing microbial lipolysis in stored milk. *J. Dairy Res.* **54:**275–282.

584. Bucky, A. R., P. R. Hayes, and D. S. Robinson. 1988b. Enhanced inactivation of bacterial lipases and proteinases in whole milk by a modified ultra-high temperature treatment. *J. Dairy Res.* **55:**373–380.

585. Gillis, W. T., M. F. Cartledge, I. R. Rodriguez, and E. J. Suarez. 1985. Effect of raw milk quality on ultra-high temperature processed milk. *J. Dairy Sci.* **68:**2875–2879.

586. Mottar, J., R. Van Renterghem, and J. DeVilder. 1985. Evaluation of the raw material for UHT milk by determining the degree of protein breakdown through HPLC. *Milchwissenschaft* **40:**717–721.

587. Rollema, H. S., R. C. McKellar, T. Sorhaug, G. Suhren, J. G. Zadow, B. A. Law, J. K. Poll, L. Stepaniak, and G. Vagias. 1989. Comparison of different methods for the detection of bacterial proteolytic enzymes in milk. *Milchwissenschaft* **44:**491–496.

588. Keogh, B. P., and G. Pettingill. 1984. Influence of enzyme activity of bacteria in leucocytes in raw milk on age gelation after UHT processing. *J. Food Prot.* **47:**105–107.

589. Christen, G. L., W. C. Wang, and T. J. Ren. 1986. Comparison of the heat resistance of bacterial lipases and proteases and the effects on ultra-high temperature milk quality. *J. Dairy Sci.* **69:** 2769–2778.

590. Keogh, B. P., and G. Pettingill. 1982. Possible role of coryneform bacteria in age gelation of ultrahigh-temperature-processed milk. *Appl. Environ. Microbiol.* **43:**1495–1497.

591. Richter, R. L., R. H. Schmidt, K. L. Smith, L. E. Mull, and S. L. Henry. 1979. Proteolytic activity in ultra-pasteurized, aseptically packaged whipping cream. *J. Food Prot.* **42:**43–45.

592. Andersson, R. E., G. Danielsson, C. B. Hedlund, and S. G. Svensson. 1981. Effect of a heat-resistant microbial lipase on flavor of ultra-high temperature sterilized milk. *J. Dairy Sci.* **64:**375–379.

593. Dunkley, W. L., and K. E. Stevenson. 1987. Ultra-high temperature processing and aseptic packaging of dairy products. *J. Dairy Sci.* **70:**2192–2202.

594. Farahnik, S. 1982. A quality control program recommendation for UHT processing and aseptic packaging of milk and milk products. *Dairy Food Sanit.* **2:**454–457.

595. Reinheimer, J. A., and M. R. Denkow. 1990. Comparison of rapid tests for assessing UHT milk sterility. *J. Dairy Res.* **57:**239–243.

596. Farkas, J. 1989. Microbiological safety of irradiated foods. *Int. J. Food Microbiol.* **9:**1–15.

597. Patterson, M. F. 1990. The potential for food irradiation. *Lett. Appl. Microbiol.* **11:**55–61.

598. Raj, D., and M. K. Roy. 1987. Preservation of milk by gamma-irradiation. *J. Nucl. Agric. Biol.* **16:**227–229.

599. Sadoun, D., C. Couvercelle, A. Strasser, A. Egler, and C. Hasselmann. 1991. Low dose irradiation of liquid milk. *Milchwissenschaft* **46:**295–299.

600. Searl, A. J. F., and P. McAthey. 1989. Treatment of milk by gamma irradiation—effect of anoxia on lipid peroxidation and the survival of *Pseudomonas aeruginosa*. *J. Sci. Food Agric.* **48:**361–367.

601. Rosenthal, I., M. Martinot, P. Lindner, and B. J. Juven. 1983. A study of ionizing irradiation of dairy products. *Milchwissenschaft* **38:**467–470.

602. Jones, T. H., and P. Jelen. 1988. Low dose τ-irradiation of Camembert, cottage cheese and cottage cheese whey. *Milchwissenschaft* **43:**233–235.

603. Yüccer, S., and G. Gündüz. 1980. Preservation of cheese and plain yogurt by low-dose irradiation. *J. Food Prot.* **43:**114–118.

604. Hashisaka, A. E., J. R. Matches, Y. Batters, F. P. Hungate, and F. M. Dong. 1990. Effects of gamma irradiation at −78°C on microbial populations in dairy products. *J. Food Sci.* **55:**1284–1289.

605. Decareau, R. V. 1985. *Microwaves in the Food Processing Industry*. Academic Press, New York.

606. Knutson, K. M., E. H. Marth, and M. K. Wagner. 1987. Microwave heating of food. *Lebensm.-Wiss. U.-Technol.* **20:**101–110.

607. Sims, L. A., P. C. Vasavada, R. R. Hull, R. A. Chandler, and E. H. Marth. 1991. Impedimetric analysis of quality and shelf-life of milk pasteurized by a continuous microwave treatment. *J. Dairy Sci.* **74**(*Suppl. 1*):139.

608. Stearns, G., and P. C. Vasavada. 1986. Effect of microwave processing on quality of milk. *J. Food Prot.* **49:**853.

609. Vasavada, P. C. 1990. Microwave processing for the dairy industry. *Food Aust.* **42:**562–564.

610. Chiu, C. P., K. Tateishi, F. V. Kosikowski, and G. Armbruster. 1984. Microwave treatment of pasteurized milk. *J. Microwave Power* **19:**269–272.

611. Chiu, P., K. Tateishi, F. Kosikowski, and G. Armbruster. 1982. Microwave treatment of pasteurized milk. *J. Microwave Power* **17:**316–317.

612. Jaynes, H. O. 1975. Microwave pasteurization of milk. *J. Milk Food Technol.* **38:**386–387.

613. Knutson, K. M., E. H. Marth, and M. K. Wagner. 1988. Use of microwave ovens to pasteurize milk. *J. Food Prot.* **51:**715–719.

614. Tochman, L. M., C. M. Stine, and B. R. Harte. 1985. Thermal treatment of cottage cheese "in-package" by microwave heating. *J. Food Prot.* **48:**932–938.

615. Young, G. S., and P. G. Jolly. 1990. Microwaves: the potential for use in dairy processing. *Aust. J. Dairy Technol.* **45:**34–37.

616. Chen, J. H., and J. H. Hotchkiss. 1991. Effect of dissolved carbon dioxide on the growth of psychrotrophic organisms in cottage cheese. *J. Dairy Sci.* **74:**2941–2945.

617. Kankare, V., V. Antila, T. Harvala, and V. Komppa. 1989. Extraction of milk fat with supercritical carbon dioxide. *Milchwissenschaft* **44:**407–411.

618. Kamihira, M., M. Taniguchi, and T. Kobayashi. 1987. Sterilization of microorganisms with supercritical carbon dioxide. *Agric. Biol. Chem.* **51:**407–412.

619. Sperber, W. H. 1991. The model HACCP System. *Food Technol.* **45:**116–118, 120.

620. Baumann, H. E. 1974. The HACCP concept and microbiological hazard categories. *Food Technol.* **28:**28,30,32,34, 79.

621. Bryan, F. L. 1988. Hazard analysis and critical control point: what the system is and what it is not. *J. Environ. Health.* **50:**400–401.

622. ICMSF. 1988. *Microorganisms in Foods, Vol. 4, Application of the Hazard Analysis and Critical Control Point System to ensure microbiological safety and quality.* International Commission on Microbiological Standards for Foods. Blackwell Scientific Publications, Oxford.

623. NACMCF. 1989. Hazard Analysis and Critical Control Point System. National Advisory Commission on Microbiological Criteria for Foods. Food safety and inspection service. U. S. Department of Agriculture, Washington, D.C.

624. Merin, U., and I. Rosenthal. 1984. Pasteurization of milk by microwave irradiation. *Milchwissenschaft* **39:**643–644.

APPENDIX

Food and Drug Administration Part 135—Frozen Desserts April 1, 1992*

AUTHORITY: Secs. 201, 401, 403, 409, 701, 706 of the Federal Food, Drug, and Cosmetic Act (21 U.S.C. 321, 341, 343, 348, 371, 376).

Subpart A—General Provisions

§ 135.3 Definitions.

For the purposes of this part, a pasteurized mix is one in which every particle of the mix has been heated in properly operated equipment to one of the temperatures specified in the table in this section and held continuously at or above that temperature for the specified time (or other time/temperature relationship which has been demonstrated to be equivalent thereto in microbial destruction):

Temperature	Time
155°F	30 min.
175°F	25 sec.

[42 FR 19132, Apr. 12, 1977]

Subpart B—Requirements for Specific Standardized Frozen Desserts

§ 135.110 Ice cream and frozen custard.

(a) *Description.* (1) Ice cream is a food produced by freezing, while stirring, a pasteurized mix consisting of one or more of the

*In order to provide the most recent standards for frozen desserts, this legal document is reproduced as an appendix to this volume instead of to Chapter 2.

optional dairy ingredients specified in paragraph (b) of this section, and may contain one or more of the optional caseinates specified in paragraph (c) of this section subject to the conditions hereinafter set forth, and other safe and suitable nonmilk-derived ingredients; and excluding other food fats, except such as are natural components of flavoring ingredients used or are added in incidental amounts to accomplish specific functions. Ice cream is sweetened with nutritive carbohydrate sweeteners and may or may not be characterized by the addition of flavoring ingredients.

(2) Ice cream contains not less than 1.6 pounds of total solids to the gallon, and weighs not less than 4.5 pounds to the gallon. Ice cream contains not less than 10 percent milkfat, nor less than 10 percent nonfat milk solids, except that when it contains milkfat at 1 percent increments above the 10 percent minimum, it may contain the following milkfat-to-nonfat milk solids levels:

Percent milkfat	Minimum percent nonfat milk solids
10	10
11	9
12	8
13	7
14	6

Except that when one or more bulky flavors are used, the weights of milkfat and total milk solids are not less than 10 percent and 20 percent, respectively, of the remainder obtained by subtracting the weight of the bulky flavors from the weight of the finished food; but in no case is the weight of milkfat or total milk solids less than 8 percent and 16 percent, respectively, of the weight of the finished food. Except in the case of frozen custard, ice cream contains less than 1.4 percent of egg yolk solids by weight of the food, exclusive of the weight of any bulky flavoring ingredients used. Frozen custard shall contain 1.4 percent egg yolk solids by weight of the finished food: *Provided, however*, That when bulky flavors are added the egg yolk solids content of frozen custard may be reduced in proportion to the amount by weight of the bulky flavors added, but in no case is the content of egg yolk solids in the finished food less than 1.12 percent. A product containing egg yolk solids in excess of 1.4 percent, the maximum set forth in this paragraph for ice cream, may be marketed if labeled as specified by paragraph (e)(1) of this section.

(3) When calculating the minimum amount of milkfat and nonfat milk solids required in the finished food, the solids of chocolate or cocoa used shall be considered a bulky flavoring ingredient. In order to make allowance for additional sweetening ingredients needed when certain bulky ingredients are used, the weight of chocolate or cocoa solids used may be multiplied by 2.5; the weight of fruit or nuts used may be multiplied by 1.4; and the weight of partially or wholly dried fruits or fruit juices may be multiplied by appropriate factors to obtain the original weights before drying and this weight may be multiplied by 1.4.

(b) *Optional dairy ingredients.* The optional dairy ingredients referred to in paragraph (a) of this section are: Cream, dried cream, plastic cream (sometimes known as concentrated milkfat), butter, butter oil, milk, concentrated milk, evaporated milk, sweetened condensed milk, superheated condensed milk, dried milk, skim milk, concentrated skim milk, evaporated skim milk, condensed skim milk, superheated condensed skim milk, sweetened condensed skim milk, sweetened condensed part-skim milk, nonfat dry milk, sweet cream buttermilk, condensed sweet cream buttermilk, dried sweet cream buttermilk, skim milk that has been concentrated and from which part of the lactose has been removed by crystallization, skim milk in concentrated or dried form that has been modified by treating the concentrated skim milk with calcium hydroxide and disodium phosphate, and whey and those modified whey products (e.g., reduced lactose whey, reduced minerals whey, and whey protein concentrate) that have been determined by FDA to be generally recognized as safe (GRAS) for use in this type of food. Water may be added, or water may be evaporated from the mix. The sweet cream buttermilk and the concentrated sweet cream buttermilk or dried sweet cream buttermilk, when adjusted with water to a total solids content of 8.5 percent, has a titratable acidity of not more than 0.17 percent, calculated as lactic acid. The term "milk" as used in this section means cow's milk. Any whey and modified whey products used contribute, singly or in combination, not more than 25 percent by weight of the total nonfat milk solids content of the finished food. The modified skim milk, when adjusted with water to a total solids content of 9 percent, is substantially free of lactic acid as determined by titration with 0.1 *N* NaOH, and it has a pH value in the range of 8.0 to 8.3.

(c) *Optional caseinates.* The optional caseinates referred to in paragraph (a) of this section that may be added to ice cream mix containing not less than 20 percent total milk solids are: Casein prepared by precipitation with gums, ammonium caseinate, calcium caseinate, potassium caseinate, and sodium caseinate. Caseinate may be added in liquid or dry form, but must be free of excess alkali.

(d) *Methods of analysis.* The fat content shall be determined by the method prescribed in "Official Methods of Analysis of the Association of Official Analytical Chemists," 13th Ed. (1980), sections 16.287 and 16.059, under "Fat, Roese-Gottlieb Method—Official Final Action," which is incorporated by reference. Copies may be obtained from the Association of Official Analytical Chemists, 2200 Wilson Blvd., Suite 400, Arlington, VA 22201-3301, or may be examined at the Office of the Federal Register, 1100 L St. NW., Washington, DC 20408.

(e) *Nomenclature.* (1) The name of the food is "ice cream"; except that when the egg yolk solids content of the food is in excess of that specified for ice cream by paragraph (a) of this section, the name of the food is "frozen custard" or "french ice cream" or "french custard ice cream".

(2) (i) If the food contains no artificial flavor, the name on the principal display panel or panels of the label shall be accompanied by the common or usual name of the characterizing flavor, e.g., "vanilla", in letters not less than one-half the height of the letters used in the words "ice cream".

(ii) If the food contains both a natural characterizing flavor and an artificial flavor simulating it, and if the natural flavor predominates, the name on the principal display panel or panels of the label shall be accompanied by the common name of the characterizing flavor, in letters not less than one-half the height of the letters used in the words "ice cream", followed by the word "flavored", in letters not less than one-half the height of the letters in the name of the characterizing flavor, e.g., "Vanilla flavored", or "Peach flavored", or "Vanilla flavored and Strawberry flavored".

(iii) If the food contains both a natural characterizing flavor and an artificial flavor simulating it, and if the artificial flavor predominates, or if artificial flavor is used alone the name on the principal display panel or panels of the label shall be accompanied by the common name of the characterizing flavor in letters not less than one-half the height of the letters used in the words "ice cream", preceded by "artificial" or "artificially flavored", in letters not less than one-half the height of the letters in the name of the characterizing flavor, e.g., "artificial Vanilla", or "artificially flavored Strawberry" or "artificially flavored Vanilla and artificially flavored Strawberry".

(3)(i) If the food is subject to the requirements of paragraph (e)(2)(ii) of this section or if it contains any artificial flavor not simulating the characterizing flavor, the label shall also bear the words "artificial flavor added" or "artificial ——— flavor added", the blank being filled with the common name of the flavor simulated by the artificial flavor in letters of the same size and prominence as the words that precede and follow it.

(ii) Wherever the name of the characterizing flavor appears on the label so conspicuously as to be easily seen under customary conditions of purchase, the words prescribed by this paragraph shall immediately and conspicuously precede or follow such name, in a size reasonably related to the prominence of the name of the characterizing flavor and in any event the size of the type is not less than 6-point on packages containing less than 1 pint, not less than 8-point on packages containing at least 1 pint but less than one-half gallon, not less than 10-point on packages containing at least one-half gallon but less than 1 gallon, and not less than 12-point on packages containing 1 gallon or over: *Provided, however,* That where the characterizing flavor and a trademark or brand are presented together, other written, printed, or graphic matter that is a part of or is associated with the trademark or brand, may intervene if the required words are in such relationship with the trademark or brand as to be clearly related to the characterizing flavor: *And provided further,* That if the finished product contains more than one flavor of ice cream subject to the requirements of this paragraph, the statements required by this paragraph need appear only once in each statement of characterizing flavors present in such ice cream, e.g., ''Vanilla flavored, Chocolate, and Strawberry flavored, artificial flavors added''.

(4) If the food contains both a natural characterizing flavor and an artificial flavor simulating the characterizing flavor, any reference to the natural characterizing flavor shall, except as otherwise authorized by this paragraph, be accompanied by a reference to the artificial flavor, displayed with substantially equal prominence, e.g., ''strawberry and artificial strawberry flavor''.

(5) An artificial flavor simulating the characterizing flavor shall be deemed to predominate:

(i) In the case of vanilla beans or vanilla extract used in combination with vanillin if the amount of vanillin used is greater than 1 ounce per unit of vanilla constituent, as that term is defined in § 169.3(c) of this chapter.

(ii) In the case of fruit or fruit juice used in combination with artificial fruit flavor, if the quantity of the fruit or fruit juice used is such that, in relation to the weight of the finished ice cream, the weight of the fruit or fruit juice, as the case may be (including water necessary to reconstitute partially or wholly dried fruits or fruit juices to their original moisture content) is less than 2 percent in the case of citrus ice cream, 6 percent in the case of berry or cherry ice cream, and 10 percent in the case of ice cream prepared with other fruits.

(iii) In the case of nut meats used in combination with artificial nut flavor, if the quantity of nut meats used is such that, in relation to the finished ice cream the weight of the nut meats is less than 2 percent.

(iv) In the case of two or more fruits or fruit juices, or nut meats, or both, used in combination with artificial flavors simulating the natural flavors and dispersed throughout the food, if the quantity of any fruit or fruit juice or nut meat is less than one-half the applicable percentage specified in paragraph (e)(5) (ii) or (iii) of this section. For example, if a combination ice cream contains less than 5 percent of bananas and less than 1 percent of almonds it would be ''artificially flavored banana-almond ice cream''. However, if it contains more than 5 percent of bananas and more than 1 percent of almonds it would be ''banana-almond flavored ice cream''.

(6) If two or more flavors of ice cream are distinctively combined in one package, e.g., ''Neopolitan'' ice cream, the applicable provisions of this paragraph shall govern each flavor of ice cream comprising the combination.

(f) *Label declaration.* Each of the optional ingredients used shall be declared on the label as required by the applicable sections of Part 101 of this chapter, except that sources of milkfat or milk solids not fat may be declared in descending order of predominance either by the use of all the terms ''milkfat and nonfat milk'' when one or any combination of two or more of the ingredients listed in § 101.4(b) (3), (4), (8), and (9) of this chapter are used or alternatively as permitted in § 101.4 of this chapter. Under section 403(k) of the Federal Food, Drug, and Cosmetic Act, artificial color need not be declared in ice cream, except as required by § 101.22(c) of this chapter. Voluntary declaration of all colors used in ice cream and frozen custard is recommended.

[43 FR 4598, Feb. 3, 1978, as amended at 45 FR 63838, Sept. 26, 1980; 46 FR 44433, Sept. 4, 1981; 47 FR 11826, Mar. 19, 1982; 49 FR 10096, Mar. 19, 1984; 54 FR 24894, June 12, 1989]

§ 135.115 Goat's milk ice cream.

(a) *Description.* Goat's milk ice cream is the food prepared in the same manner prescribed in § 135.110 for ice cream, and complies with all the provisions of § 135.110, except that the only optional dairy ingredients that may be used are those in paragraph (b) of this section; caseinates may not be used; and paragraphs (e)(1) and (f) of § 135.110 shall not apply.

(b) *Optional dairy ingredients.* The optional dairy ingredients referred to in paragraph (a) of this section are goat's skim milk, goat's milk, and goat's cream. These optional dairy ingredients may be used in liquid, concentrated, and/or dry form.

(c) *Nomenclature.* The name of the food is ''goat's milk ice cream'' or, alternatively, ''ice cream made with goat's milk'', except that when the egg yolk solids content of the food is in excess of that specified for ice cream in paragraph (a) of § 135.110, the name of the food is ''goat's milk frozen custard'' or, alternatively, ''frozen custard made with goat's milk'', or ''goat's milk french ice cream'', or, alternatively, ''french ice cream made with goat's milk'', or ''goat's milk french custard ice cream'', or, alternatively, ''french custard ice cream made with goat's milk''.

(d) *Label declaration.* Each of the optional ingredients used shall be declared on the label as required by the applicable sections of Part 101 of this chapter.

[47 FR 41526, Sept. 21, 1982]

§ 135.120 Ice milk.

(a) *Description.* Ice milk is the food prepared from the same ingredients and in the same manner prescribed in § 135.110 for ice cream and complies with all the provisions of § 135.110 (including the requirements for label statement of optional ingredients), except that:

(1) Its content of milkfat is more than 2 percent but not more than 7 percent.

(2) Its content of total milk solids is not less than 11 percent.

(3) Caseinates may be added when the content of total milk solids is not less than 11 percent.

(4) The provision for reduction in milkfat and nonfat milk solids content from the addition of bulky flavors in § 135.110(a) applies, except that in no case will the milkfat content be less than 2 percent, nor the nonfat milk solids content be less than 4 percent. When the milkfat content increases in increments of 1 percent above the 2 percent minimum, it may contain the following milkfat-to-nonfat milk solids levels:

Percent milkfat	Minimum percent nonfat milk solids
2	9
3	8
4	7
5	6
6	5
7	4

(5) The quantity of food solids per gallon is not less than 1.3 pounds.

(6) When any artificial coloring is used in ice milk, directly or as a component of any other ingredient, the label shall bear the statement ''artificially colored'', ''artificial coloring added'', ''with added artificial color'', or ''————, an artificial color added''; the blank being filled in with the common or usual name of the artificial color; or in lieu thereof, in case the artificial color is a component of another ingredient, ''———— artificially colored''.

(7) If both artificial color and artificial flavoring are used, the label statements may be combined.

(b) *Nomenclature.* The name of the food is ''ice milk''.

[43 FR 4599, Feb. 3, 1978, as amended at 48 FR 13024, Mar. 29, 1983]

§ 135.125 Goat's milk ice milk.

(a) *Description.* Goat's milk ice milk is the food prepared in the same manner prescribed in § 135.115 for goat's milk ice cream, except that paragraph (c) shall not apply, and which complies with all the requirements of § 135.120(a) (1), (2), (4), (5), (6), and (7) for ice milk.

(b) *Nomenclature.* The name of the food is ''goat's milk ice milk'' or, alternatively, ''ice milk made with goat's milk''.

[47 FR 41526, Sept. 21, 1982]

§ 135.130 Mellorine.

(a) *Description.* (1) Mellorine is a food produced by freezing, while stirring, a pasteurized mix consisting of safe and suitable ingredients including, but not limited to, milk-derived nonfat solids and animal or vegetable fat, or both, only part of which may be milkfat. Mellorine is sweetened with nutritive carbohydrate sweetener and is characterized by the addition of flavoring ingredients.

(2) Mellorine contains not less than 1.6 pounds of total solids to the gallon, and weighs not less than 4.5 pounds to the gallon. Mellorine contains not less than 6 percent fat and 2.7 percent protein having a protein efficiency ratio (PER) not less than that of whole milk protein (108 percent of casein) by weight of the food, exclusive of the weight of any bulky flavoring ingredients used. In no case shall the fat content of the finished food be less than 4.8 percent or the protein content be less than 2.2 percent. The protein to meet the minimum protein requirements shall be provided by milk solids, not fat and/or other milk-derived ingredients.

(3) When calculating the minimum amount of milkfat and protein required in the finished food, the solids of chocolate or cocoa used shall be considered a bulky flavoring ingredient. In order to make allowance for additional sweetening ingredients needed when certain bulky ingredients are used, the weight of chocolate or cocoa solids used may be multiplied by 2.5; the weight of fruit or nuts used may be multiplied by 1.4; and the weight of partially or wholly dried fruits or fruit juices may be multiplied by appropriate factors to obtain the original weights before drying and this weight may be multiplied by 1.4.

(b) *Fortification.* Vitamin A is present in a quantity which will ensure that 40 international units (IU) are available for each gram of fat in mellorine, within limits of good manufacturing practice.

(c) *Methods of analysis.* Fat and protein content, and the PER shall be determined by following the methods contained in ''Official Methods of Analysis of the Association of Official Analytical Chemists,'' 13th Ed. (1980), which is incorporated by reference. Copies may be obtained from the Association of Official Analytical Chemists, 2200 Wilson Blvd., Suite 400, Arlington, VA 22201-3301, or may be examined at the Office of the Federal Register, 1100 L St. N.W, Washington, DC 20408.

(1) Fat content shall be determined by the method: ''Fat, Roese-Gottlieb Method—Official Final Action,'' section 16.287.

(2) Protein content shall be determined by one of the following methods: ''Nitrogen—Official Final Action,'' Kjeldahl Method,

section 16.285, or Dye Binding Method, section 16.286.

(3) PER shall be determined by the method: "Biological Evaluation of Protein Quality—Official Final Action," sections 43.212-43.216.

(d) *Nomenclature*. The name of the food is "mellorine". The name of the food on the label shall be accompanied by a declaration indicating the presence of characterizing flavoring in the same manner as is specified in § 135.110(c).

(e) *Label declaration*. The common or usual name of each of the ingredients used shall be declared on the label as required by the applicable sections of Part 101 of this chapter, except that sources of milkfat or milk solids not fat may be declared, in descending order of predominance, either by the use of the terms "milkfat, and nonfat milk" when one or any combination of two or more ingredients listed in § 101.4(b) (3), (4), (8), and (9) of this chapter are used, or alternatively as permitted in § 101.4 of this chapter.

[42 FR 19137, Apr. 12, 1977, as amended at 47 FR 11826, Mar. 19, 1982; 49 FR 10096, Mar. 19, 1984; 54 FR 24894, June 12, 1989]

§ 135.140 Sherbet.

(a) *Description*. (1) Sherbet is a food produced by freezing, while stirring, a pasteurized mix consisting of one or more of the optional dairy ingredients specified in paragraph (b) of this section, and may contain one or more of the optional caseinates specified in paragraph (c) of this section subject to the conditions hereinafter set forth, and other safe and suitable nonmilk-derived ingredients; and excluding other food fats, except such as are added in small amounts to accomplish specific functions or are natural components of flavoring ingredients used. Sherbet is sweetened with nutritive carbohydrate sweeteners and is characterized by the addition of one or more of the characterizing fruit ingredients specified in paragraph (d) of this section or one or more of the nonfruit-characterizing ingredients specified in paragraph (e) of this section.

(2) Sherbet weighs not less than 6 pounds to the gallon. The milkfat content is not less than 1 percent nor more than 2 percent, the nonfat milk-derived solids content not less than 1 percent, and the total milk or milk-derived solids content is not less than 2 percent nor more than 5 percent by weight of the finished food. Sherbet that is characterized by a fruit ingredient shall have a titratable acidity, calculated as lactic acid, of not less than 0.35 percent.

(b) *Optional dairy ingredients*. The optional dairy ingredients referred to in paragraph (a) of this section are: Cream, dried cream, plastic cream (sometimes known as concentrated milkfat), butter, butter oil, milk, concentrated milk, evaporated milk, superheated condensed milk, sweetened condensed milk, dried milk, skim milk, concentrated skim milk, evaporated skim milk, condensed skim milk, sweetened condensed skim milk, sweetened condensed part-skim milk, nonfat dry milk, sweet cream buttermilk, condensed sweet cream buttermilk, dried sweet cream buttermilk, skim milk that has been concentrated and from which part of the lactose has been removed by crystallization, and whey and those modified whey products (e.g., reduced lactose whey, reduced minerals whey, and whey protein concentrate) that have been determined by FDA to be generally recognized as safe (GRAS) for use in this type of food. Water may be added, or water may be evaporated from the mix. The sweet cream buttermilk and the concentrated sweet cream buttermilk or dried sweet cream buttermilk, when adjusted with water to a total solids content of 8.5 percent, has a titratable acidity of not more than 0.17 percent calculated as lactic acid. The term "milk" as used in this section means cow's milk.

(c) *Optional caseinates*. The optional caseinates referred to in paragraph (a) of this section which may be added to sherbet mix are: Casein prepared by precipitation with gums, ammonium caseinate, calcium caseinate, potassium caseinate, and sodium caseinate. Caseinates may be added in liquid or dry form, but must be free of excess alkali, such caseinates are not considered to be milk solids.

(d) *Optional fruit-characterizing ingredients.* The optional fruit-characterizing ingredients referred to in paragraph (a) of this section are any mature fruit or the juice of any mature fruit. The fruit or fruit juice used may be fresh, frozen, canned, concentrated, or partially or wholly dried. The fruit may be thickened with pectin or other optional ingredients. The fruit is prepared by the removal of pits, seeds, skins, and cores, where such removal is usual in preparing that kind of fruit for consumption as fresh fruit. The fruit may be screened, crushed, or otherwise comminuted. It may be acidulated. In the case of concentrated fruit or fruit juices, from which part of the water is removed, substances contributing flavor volatilized during water removal may be condensed and reincorporated in the concentrated fruit or fruit juice. In the case of citrus fruits, the whole fruit, including the peel but excluding the seeds, may be used, and in the case of citrus juice or concentrated citrus juices, cold-pressed citrus oil may be added thereto in an amount not exceeding that which would have been obtained if the whole fruit had been used. The quantity of fruit ingredients used is such that, in relation to the weight of the finished sherbet, the weight of fruit or fruit juice, as the case may be (including water necessary to reconstitute partially or wholly dried fruits or fruit juices to their original moisture content), is not less than 2 percent in the case of citrus sherbets, 6 percent in the case of berry sherbets, and 10 percent in the case of sherbets prepared with other fruits. For the purpose of this section, tomatoes and rhubarb are considered as kinds of fruit.

(e) *Optional nonfruit characterizing ingredients.* The optimal nonfruit characterizing ingredients referred to in paragraph (a) of this section include but are not limited to the following:

(1) Ground spice or infusion of coffee or tea.

(2) Chocolate or cocoa, including sirup.

(3) Confectionery.

(4) Distilled alcoholic beverage, including liqueurs or wine, in an amount not to exceed that required for flavoring the sherbet.

(5) Any natural or artificial food flavoring (except any having a characteristic fruit or fruit-like flavor).

(f) *Nomenclature.* (1) The name of each sherbet is as follows:

(i) The name of each fruit sherbet is "——— sherbet", the blank being filled in with the common name of the fruit or fruits from which the fruit ingredients used are obtained. When the names of two or more fruits are included, such names shall be arranged in order of predominance, if any, by weight of the respective fruit ingredients used.

(ii) The name of each nonfruit sherbet is "——— sherbet", the blank being filled in with the common or usual name or names of the characterizing flavor or flavors; for example, "peppermint", except that if the characterizing flavor used is vanilla, the name of the food is "——— sherbet", the blank being filled in as specified by § 135.110(e)(2) and (5)(i).

(2) When the optional ingredients, artificial flavoring, or artificial coloring are used in sherbet, they shall be named on the label as follows:

(i) If the flavoring ingredient or ingredients consist exclusively of artificial flavoring, the label designation shall be "artificially flavored".

(ii) If the flavoring ingredients are a combination of natural and artificial flavors, the label designation shall be "artificial and natural flavoring added".

(iii) The label shall designate artificial coloring by the statement "artificially colored", "artificial coloring added", "with added artificial coloring", or "———, an artificial color added", the blank being filled in with the name of the artificial coloring used.

(g) *Characterizing flavor(s).* Wherever there appears on the label any representation as to the characterizing flavor or flavors of the food and such flavor or flavors consist in whole or in part of artificial flavoring, the statement required by paragraph (f)(2) (i) and (ii) of this section, as appropriate, shall immediately and conspicuously precede or follow such representation, without intervening written, printed, or graphic matter (except that the word "sherbet" may intervene) in a size reasonably related to the prominence of the name of the characterizing flavor and in any event the size of the type is not less than 6-point on packages containing less than 1 pint, not less than 8-point on packages containing at least 1 pint but less than one-half gallon, not less than 10-point on packages containing at least one-half gallon but less than 1 gallon, and not less than 12-point on packages containing 1 gallon or over.

(h) *Display of statements required by paragraph* (f)(2). Except as specified in paragraph (g) of this section, the statements required by paragraph (f)(2) of this section shall be set forth on the principal display panel or panels of the label with such prominence and conspicuousness as to render them likely to be read and understood by the ordinary individual under customary conditions of purchase and use.

(i) *Label declaration.* Each of the optional ingredients used shall be declared on the label as required by the applicable sections of Part 101 of this chapter.

[43 FR 4599, Feb. 3, 1978, as amended at 46 FR 44434, Sept. 4, 1981]

§ 135.160 Water ices.

(a) *Description.* Water ices are the foods each of which is prepared from the same ingredients and in the same manner prescribed in § 135.140 for sherbets, except that the mix need not be pasteurized, and complies with all the provisions of § 135.140 (including the requirements for label statement of optional ingredients) except that no milk or milk-derived ingredient and no egg ingredient, other than egg white, is used.

(b) *Nomenclature.* The name of the food is "——— ice", the blank being filled in, in the same manner as specified in § 135.140(f)(1) (i) and (ii), as appropriate.

[42 FR 19132, Apr. 12, 1977]